Scientific Computation

Springer
Berlin
Heidelberg
New York
Barcelona
Budapest
Hong Kong
London
Milan
Paris
Santa Clara
Singapore
Tokyo

Scientific Computation

A Computational Method in Plasma Physics
F. Bauer, O. Betancourt, P. Garabedian

Implementation of Finite Element Methods for Navier-Stokes Equations
F. Thomasset

Finite-Difference Techniques for Vectorized Fluid Dynamics Calculations
Edited by D. Book

Unsteady Viscous Flows D. P. Telionis

Computational Methods for Fluid Flow R. Peyret, T. D. Taylor

Computational Methods in Bifurcation Theory and Dissipative Structures
M. Kubicek, M. Marek

Optimal Shape Design for Elliptic Systems O. Pironneau

The Method of Differential Approximation Yu. I. Shokin

Computational Galerkin Methods C. A. J. Fletcher

Numerical Methods for Nonlinear Variational Problems
R. Glowinski

Numerical Methods in Fluid Dynamics Second Edition
M. Holt

Computer Studies of Phase Transitions and Critical Phenomena
O. G. Mouritsen

Finite Element Methods in Linear Ideal Magnetohydrodynamics
R. Gruber, J. Rappaz

Numerical Simulation of Plasmas Y. N. Dnestrovskii, D. P. Kostomarov

Computational Methods for Kinetic Models of Magnetically Confined Plasmas
J. Killeen, G. D. Kerbel, M. C. McCoy, A. A. Mirin

Spectral Methods in Fluid Dynamics Second Edition
C. Canuto, M. Y. Hussaini, A. Quarteroni, T. A. Zang

Computational Techniques for Fluid Dynamics 1 Second Edition
Fundamental and General Techniques C. A. J. Fletcher

Computational Techniques for Fluid Dynamics 2 Second Edition
Specific Techniques for Different Flow Categories C. A. J. Fletcher

Methods for the Localization of Singularities in Numerical Solutions of Gas Dynamics Problems E. V. Vorozhtsov, N. N. Yanenko

Classical Orthogonal Polynomials of a Discrete Variable
A. F. Nikiforov, S. K. Suslov, V. B. Uvarov

Flux Coordinates and Magnetic Field Structure: A Guide to a Fundamental Tool of Plasma Theory
W. D. D'haeseleer, W. N. G. Hitchon, J. D. Callen, J. L. Shohet

Monte Carlo Methods in Boundary Value Problems
K. K. Sabelfeld

Computer Simulation of Dynamic Phenomena
M. L. Wilkins

The Least-Squares Finite Element Method
Theory and Applications in Computational Fluid Dynamics and Electromagnetics
Bo-nan Jiang

Bo-nan Jiang

The Least-Squares Finite Element Method

Theory and Applications in Computational Fluid Dynamics and Electromagnetics

With 130 Figures and 11 Tables

Springer

Dr. Bo-nan Jiang

Institute for Computational Methods in Propulsion
NASA Lewis Research Center
Cleveland, OH 44135, USA

ISSN 0172-5726
ISBN 978-3-642-08367-9

Library of Congress Cataloging-in-Publication Data. Jiang, Bo-nan, 1940– The least-squares finite element method: theory and applications in computational fluid dynamics and electromagnetics / Bo-nan Jiang. p. cm. – (Scientific computation, ISSN 0172-5726) Includes bibliographical references and index. 1. Fluid mechanics–Mathematics. 2. Electromegnetics–Mathematics. 3. Finite element method. 4. Least squares. 5. Differential equaions, Partial–Numerical solutions. I. Title. II. Series. QC151.J53 1998 532'.051'01515353–dc21 97-51980

Printed in Germany

Cover design: *design & production* GmbH, Heidelberg

Preface

The Rayleigh–Ritz finite element method, which is based on the principle of minimum potential energy, has proved to be extremely successful and become a dominating computational technique in solid mechanics, heat transfer and static electromagnetics. Because the Rayleigh–Ritz method is applicable only for equations with self-adjoint operators, the Galerkin method, which is based on the principle of virtual work or the weighted residual form, has been much emphasized since the early 1970s and considered a universal approach to construction of finite element schemes. However, attempts to apply the Galerkin method to non-self-adjoint equations in fluid dynamics and other transport problems encounter serious difficulties, including oscillations and instabilities of the solution and poor approximation of its derivatives. The least-squares finite element method (LSFEM) based on simply minimizing the L_2 norm of the residuals of a first-order system of differential equations promises to eliminate these drawbacks, and is receiving increasing attention.

The basic idea of the least-squares method for numerical solution of differential equations is well known. However, the true power of the LSFEM has not been exploited until quite recently. The objective of this book is to provide a comprehensive introduction to the LSFEM including its theory and applications.

Although the idea of the LSFEM is simple, its theory is not so simple, and is still evolving. The mathematical theory of the LSFEM for elliptic first-order systems developed by Wendland (1979) relies on the modern theory of complex functions, and thus is applicable only to two-dimensional problems. A more general analysis developed by Aziz et al. (1985) is based on the Agmon–Douglis–Nirenberg (1964) theory of elliptic partial differential equations. Their theory invokes high-level mathematics, and thus is suitable mainly for mathematicians.

It is the author's opinion that the theoretical basis of the general least-squares method is the bounded inverse theorem of linear operators. This explains why LSFEM can provide numerical solutions for all types of partial differential equations within one mathematical/computational framework without any special treatment. The analysis of the least-squares method for most partial differential equations (not only limited to elliptic problems) in engineering and physics can be based on the bounded inverse theorem and the

Friedrichs inequalities related to grad, div and curl operators. Of course, the analysis can also be based on the famous Lax–Milgram theorem. However, the proof of the Lax–Milgram theorem needs knowledge of functional analysis and its application is limited to elliptic problems. Therefore, in this book we emphasize the approach based on the bounded inverse theorem. This approach gives a clear picture of the least-squares method without specialized mathematical knowledge beyond calculus and elementary differential equations. Following this principle, this book establishes an almost self-sufficient and reasonably rigorous mathematical framework for the least-squares method, while the mathematics has been kept as simple as possible.

This book is written mainly for both engineers and physicists as well as researchers. Most parts of the book cover diverse applications of the LSFEM.

It is presumed that the reader has basic knowledge about the finite element method for second-order elliptic partial differential equations. The introduction of the finite element method is reduced to a minimum in order to keep the book short. For those readers who are not familiar with finite elements, many excellent introductory books are available. For example, the readers may consult Zienkiewicz and Morgan (1983) or Becker et al. (1983).

The book contains fifteen chapters and four appendices.

Part I (Chaps. 1–3) gives some basic ideas about the least-squares method.

Chapter 1 provides an overview of the LSFEM.

Through simple one-dimensional examples, Chap. 2 explains why the standard Galerkin method or the central difference method fails and why the LSFEM doesn't need upwinding and is perfectly suitable for convective transport problems described by first-order derivatives.

Chapter 3 compares the LSFEM with the mixed Galerkin method for first-order elliptic systems and shows why the LSFEM can accommodate equal-order elements.

Part II (Chaps. 4–6) introduces some theoretical aspects of the LSFEM.

Chapter 4 provides the mathematical foundation and general formulation of the LSFEM based on a first-order system of partial differential equations.

Chapter 5 deals with the div–curl system which is fundamental for studying the incompressible Navier–Stokes equations in fluid dynamics and the Maxwell equations in electromagnetics. This chapter shows that the three-dimensional div–curl system is not overdetermined and the LSFEM is the best one for seeking its solution. This chapter introduces the div–curl method and the least-squares method for deriving equivalent second-order equations and their boundary conditions.

Chapter 6 deals with diffusion problems which are usually described by a second-order scalar elliptic equation. A second-order equation can be decomposed into a grad–div system or a div–curl–grad system. Both the mixed Galerkin method and the conventional least-squares method are based on the grad–div system. This chapter shows that the LSFEM based on the div–curl–grad system has significant advantages: accommodation of equal-order elements, symmetric positive-definite matrices, and optimal rates of convergence.

Part III (Chaps. 7–13) covers the application of LSFEM to a broad range of problems in fluid dynamics.

Chapter 7 treats inviscid irrotational flows for both incompressible and subsonic compressible cases. The LSFEM directly determines the velocity components instead of the potential to gain better accuracy and efficiency. This chapter shows that due to the continuity of the velocity, a branch cut is not needed for lifting airfoil problems in the LSFEM.

Chapter 8 deals with steady and transient incompressible viscous flows including flow-heat coupling such as surface-tension-driven convection. Usually theoretical analysis and numerical solution of the incompressible Navier–Stokes equations are based on the velocity–pressure formulation and conducted via the mixed Galerkin method which leads to difficult saddle-point problems. This chapter shows that the principle part of the incompressible Navier–Stokes equations in the velocity–pressure–vorticity formulation consists of two coupled div–curl systems. From this completely new point of view, this chapter systematicly and rigorously derives permissible non-standard boundary conditions and equivalent second-order Navier–Stokes equations. Through various examples, this chapter demonstrates that the corresponding LSFEM has advantages over other methods: special treatments, such as non-equal-order interpolation, upwinding, artificial compressibility, and operator splitting are not needed.

In Chap. 9 a variety of methods for convective transport problems are compared. It demonstrates that the LSFEM inherently contains a streamline upwind mechanism to stablize the solution. Finding accurate approximation of discontinuous solution of hyperbolic equations has been a persistently difficult task. This chapter shows that the L_1 method and the reweighted LSFEM can resolve contact discontinuities in one element without error even for coarse meshes.

Chapter 10 deals with rotational inviscid flows. The Euler equations governing incompressible rotational flows look simple, but in fact remain one of the most difficult problems in computational fluid dynamics, since they are neither elliptic nor hyperbolic. This chapter demonstrates that the loss of kinetic energy that is common in other approaches does not occur in the LSFEM.

Chapter 11 is devoted to the simulation of low-speed compressible viscous flows. It is essential for the design of combustion chambers and chemical vapor deposition reactors where heat addition induces significant temperature and density variations. The pressure-based finite difference or finite volume methods need the use of staggered grids, while the density-based method requires preconditioning. This chapter shows that the LSFEM based on the velocity–vorticity–pressure–compressibility–temperature–heat flux formulation does not require any special treatment.

Chapter 12 presents the simulation of two-fluid flows. To capture interfaces, conventional methods resort to iteration between two distinct methodologies, namely, a flow equation solver usually based on projection methods and a pure convection equation solver to identify the materials by Lagrangian or upwinding methods. This chapter demonstrates that the LSFEM solves the convective color function in conjunction with the velocity and pressure in a unified and fully implicit manner without any special treatment.

Chapter 13 deals with high-speed compressible gas flows governed by the Euler equations. This chapter illustrates that backward time-differencing together with the LSFEM can capture shocks and yield high-resolution shocks when combined with adaptive remeshing techniques. In contrast to existing methods, the LEFEM does not use upwinding, directional splitting and Riemann solvers, and its implementation is thus considerably simple.

In Part IV (Chap. 14) the LSFEM is applied to electromagnetics problems. It is commonly believed that the divergence equations in the Maxwell equations are "redundant" for transient and time-harmonic problems, therefore most numerical methods in computational electromagnetics solve only two first-order curl equations, or the second-order curl–curl equations. This chapter shows that this misconception is the true origin of spurious modes, inaccurate solutions, and failure of iterative solvers. This chapter clarifies that the first-order full Maxwell equations are not "overdetermined" and the divergence equations must always be included to maintain the ellipticity of the system in the space domain, to guarantee the uniqueness of the solution and the accuracy of the numerical methods and to eliminate the infinitely degenerate eigenvalue. This chapter also shows that the common derivation and usage of the second-order curl–curl equations are incorrect and that the solution of the Helmholtz equations needs the divergence condition to be enforced on the boundary. This chapter explains that the div–curl method and the least-squares method can provide a rigorous derivation of the equivalent second-order Maxwell equations and their boundary conditions, as well as their corresponding variational principles. This chapter demonstrates that the node-based LSFEM can solve the first-order full Maxwell equations directly and without spurious solutions.

Application of the LSFEM always leads to a symmetric positive-definite system of linear algebraic equations. To compute the solution efficiently, in the last chapter we introduce the matrix-free element-by-element preconditioned conjugate gradient method. It is worth mentioning that the conjugate gradient method is a kind of least-squares method.

In this book we carry on the least-squares idea from the beginning to the end, i.e., from constructing the formulation to solving the resulted discretized equations.

The book reflects recent developments of the LSFEM. Most of the results in this book have been obtained by the author and his collaborators, and many of them are not published elsewhere.

I would like to express my sincere gratitude to Dr. Louis A. Povinelli, the director of Institute for Computational Mechanics in Propulsion, NASA Lewis Research Center, and Prof. Theo G. Keith, the vice president of Ohio Aerospace Institute, who have constantly supported research on the LSFEM. My appreciation also goes to Prof. J. Tinsley Oden who introduced me to the finite element theory and is always available when I need support and advice. I thank my colleagues and friends who have contributed to the development of this subject. I am particularly grateful to Mr. Jia-zhen Chai, with whose help I initiated this research, and Dr. Tsung-Liang Lin, who wrote most of our early version of the LSFEM code for large-scale computations. I thank Drs. Jie Wu, Sheng-Tao Yu, Vijay Sonnad, Srinivas Chari, and Ching Yuen Loh, whose significant contributions have been included in this book. Special thanks are also due to Prof. Yuesheng Xu, Prof. Yanzhao Cao, and Dr. Yunhe Zhao for reading a part of the text and making helpful suggestions. I am particularly indebted to Dr. Vijay Sonnad for his enthusiasm about the LSFEM. In 1990, Dr. Sonnad invited me to work with him in IBM on the p-version LSFEM. He has read the entire manuscript and made many valuable comments which helped to improve the book.

It is a pleasure to acknowledge the tremendous effect provided by Prof. W. Beiglböck, Mr. F. Holzwarth, Ms. B. Reichel-Mayer, and Mrs. P. Treiber at Springer-Verlag in the production of this book. Finally, I wish to thank my wife Dr. Mei-yu Yu for her encouragement.

Cleveland, January 1998

Bo-nan Jiang
bnjiang@yahoo.com

Contents

Part II. Fundamentals of LSFEM

Part III. LSFEM in Fluid Dynamics

Part I

Basic Concepts of LSFEM

1. Introduction

In this chapter we briefly overview the basic ideas and features of the least-squares finite element method. In the following chapters we will elaborate on the view points discussed in this chapter.

1.1 Why Finite Elements?

The finite element method is one of the most general techniques for the numerical solution of differential equations. It has been astonishingly successful. "Perhaps no other family of approximation methods has had a greater impact on the theory and application of numerical methods during the twentieth century" (Oden 1991). The finite element method has now been used throughout the fields of engineering and applied science.

The finite element method does not operate directly on the differential equations; instead, the continuous boundary and initial value problems are put into equivalent variational forms. The solution appears in the integral of a quantity over a domain. The integral of a function over an arbitrary domain can be broken up into the sum of integrals over an arbitrary collection of subdomains called finite elements. As long as the subdomains are sufficiently small, polynomial functions can adequately represent the local behavior of the solution. According to remarks made by Zienkiewicz (1975) and Oden (1991), the most important features of finite elements can be summarized as:

(1) *Arbitrary geometries.* The finite element method is essentially independent of geometry. It can be applied to domains of complex shape and with quite arbitrary boundary conditions.

(2) *Unstructured meshes.* In finite element analyses a global coordinate transformation is not needed. Finite elements can be placed anywhere in physical domains. In engineering practice one often modifies the original design to satisfy different requirements. Finite element analysts can add or delete elements without changing the global data structure. If iterative solvers are employed, the element and nodal numbering can be arbitrary without sacrificing efficiency.

(3) *Flexible and general purpose format of program.* The clear structure and versatility of the finite element method makes it possible to construct general purpose software for application.

(4) *Mathematical foundation.* Because of the extensive work on the mathematical theory during the past two decades, the finite element method now enjoys a rich and solid mathematical basis. This gives added reliability and in many cases makes it possible to mathematically analyze and estimate the accuracy of finite element solutions.

1.2 Why Least-Squares?

According to the underlining variational principle, the finite element method can be classified into three major groups: the Rayleigh–Ritz method, the Galerkin method and the least-squares method.

The Rayleigh–Ritz method seeks to minimize the total potential energy, and hence the numerical solution of the Rayleigh–Ritz finite element method possesses the best approximation property. That is, the difference between the finite element solution and the exact solution is minimized with respect to a certain energy norm. Moreover, the Rayleigh–Ritz finite element method leads to symmetric and positive-definite systems of linear algebraic equations. This formulation has proven eminently successful in application to problems in solid mechanics and in other situations, such as heat conduction, governed by self-adjoint, second- or fourth-order elliptic diffusion-type equations. In fact, today, most widely used commercial finite element codes in engineering analysis are based on the Rayleigh–Ritz method.

The Galerkin method is based on the weighted residual form. To illustrate the basic idea of the Galerkin method, we consider a mathematical problem defined by a set of partial differential equations in the form

$$\boldsymbol{A}\boldsymbol{u} = \boldsymbol{f} \qquad \text{in } \Omega, \tag{1.1a}$$

$$\boldsymbol{B}\boldsymbol{u} = \boldsymbol{0} \qquad \text{on } \Gamma, \tag{1.1b}$$

where $\boldsymbol{A}$ is the linear differential operator, $\boldsymbol{B}$ is the boundary operator, $\boldsymbol{u}$ is the dependent unknown vector, $\boldsymbol{f}$ is the force vector, Ω is the domain, and Γ is the boundary of Ω.

The approximation process of the weighed residual method can be written as follows. First, the function is approximated by a set of unknown parameters $\boldsymbol{u}_j$ and trial (basis) functions $\boldsymbol{\Phi}_j(\boldsymbol{x})$, where the vector $\boldsymbol{x}$ stands for the independent variables,

$$\boldsymbol{u} \simeq \boldsymbol{\Phi}_j \boldsymbol{u}_j \qquad j = 1, ..., n. \tag{1.2}$$

Second, the algebraic equation permitting a numerical solution is formed as a "weighted residual",

$$\int_\Omega \boldsymbol{v}_i^T (\boldsymbol{A}\boldsymbol{u} - \boldsymbol{f}) d\Omega + \int_\Gamma \overline{\boldsymbol{v}}_i^T \boldsymbol{B}\boldsymbol{u} d\Gamma = 0, \tag{1.3}$$

where $\boldsymbol{v}_i$ and $\overline{\boldsymbol{v}}_i$ are "suitably chosen" test functions, and T denotes the transpose. In the conventional Galerkin method, the choice is

$$v_i = \overline{v}_i = \boldsymbol{\Phi}_i. \tag{1.4}$$

For equations with self-adjoint and positive definite operators, the Galerkin formulation results in the same system of equations as in the Rayleigh–Ritz formulation. However, the Galerkin method appears more general, because it is applicable for non-self-adjoint equations such as arise in fluid mechanics. This generality of the Galerkin finite element method provided, in the early 1970s, a strong impetus for the utilization of the method in fluid dynamics. It was thought that the significant advantages gained in solid mechanics and diffusion-type problems by using the Rayleigh–Ritz method would again be open to exploitation in the area of fluid flow by using the Galerkin method. In reality, this proved to be too optimistic, especially for modeling convection dominated flow problems. Convection operators are of first order, and thus non-self-adjoint; as a result, the Galerkin method does not exhibit the best approximation property.

In practice, solutions to convection dominated transport problems by the Galerkin method are often corrupted by spurious oscillations or "wiggles". These can only be removed by severe mesh refinements which clearly undermine the practical utility of the method.

The classic Galerkin method also behaves poorly for high-speed compressible flow and shallow water wave problems. In these cases, the flows are governed by nonlinear first-order hyperbolic equations; their solutions may be discontinuous, that is, shocks often develop even if the initial flow field is smooth.

Notorious difficulties arise even for the solution of elliptic problems by the Galerkin mixed method. One example is incompressible irrotational flow problems governed by Laplace or Poisson equations of the potential. In common practice, the primal variable, i.e., the potential is solved by the Rayleigh–Ritz method, then a *posteriori* numerical differentiation is required to obtain the dual variables, i.e. the velocity components which are more important. The computed velocity is not continuous across element boundaries, and in general its accuracy is one order lower than that of the potential. The Galerkin mixed method based on the first-order equations was devised in the hope of obtaining better accuracy for both primal and dual variables. Here the term 'mixed' means that both primal variables and dual variables are approximated as fundamental unknowns. Unfortunately, the Galerkin mixed method brings perhaps more troubles than benefits, at least for second-order elliptic problems. First, the original simple minimization problem is turned into a difficult saddle-point problem, and one must use different elements to interpolate primal variables and dual variables, and often lower-order elements must be employed for dual variables to satisfy the celebrated Ladyzhenskaya–Babuška–Brezzi (LBB) condition (see, e.g., Babuška 1971, Brezzi 1974, and Oden and Carey 1983). As a consequence, better accuracy for dual variables may not be achieved. Second, the Rayleigh–Ritz method leads to symmetric positive-definite matrices, while the Galerkin mixed method produces non-positive-definite matrices which have been hard to solve for large-scale problems.

Another well-known example is viscous flow problems governed by the incompressible Navier–Stokes equations. Here one encounters the same instabilities when the Galerkin mixed method is used for the solution of velocity and pressure.

In computational electromagnetics (CEM), the most popular finite element methods for static and eddy currents problems are based on the use of potentials that involve difficulties related to the appropriate gauging method and the loss of accuracy and inter-element continuity of the calculated field intensity by numerical differentiation. It is commonly believed that the divergence equations in the Maxwell equations are redundant. Therefore, conventional time-domain methods are based only on the two curl equations, and frequency methods are based only on the curl–curl equations. Due to the appearance of spurious modes which violate the divergence-free equations, the direct solution of the field intensity vectors in the first-order Maxwell equations by using the node-based Galerkin-type procedure has not been successful.

It is not an exaggeration to say that during the passed three decades much of finite element research in fluid mechanics was devoted to modification and improvement of the Galerkin method. The achievement of mathematical analysis of the Galerkin method is remarkable; however, the application of the Galerkin method in computational fluid dynamics (CFD) and computational electromagnetics (CEM) has not produced entirely satisfactory results. This is a major reason why the finite difference and finite volume methods are much in vogue in CFD and time-domain CEM.

It is the author's opinion that for non-self-adjoint systems, such as arise in fluid mechanics and electromagnetics, the application of the least-squares finite element method (LSFEM) is the right direction to go.

The LSFEM is based on the minimization of the residuals in a least-squares sense. The method seeks the minimizer of the following functional

$$I(\boldsymbol{u}) = \int_{\Omega} (\boldsymbol{A}\boldsymbol{u} - \boldsymbol{f})^2 d\Omega \tag{1.5}$$

within the constraint of a given boundary condition (1.1b). That is, the least-squares solution is calculated from the following variational statement:

$$\int_{\Omega} (\boldsymbol{A}\boldsymbol{\Phi}_i)^T (\boldsymbol{A}\boldsymbol{u} - \boldsymbol{f}) d\Omega = \boldsymbol{0}. \tag{1.6}$$

In contrast to the Galerkin formulation (1.3) which is only conceptual without much practical meaning and thus in most cases needs further problem-dependent mathematical manipulation in order to obtain a realistic computational formulation for a particular problem, the least-squares formulation, (1.6), is a final mathematical as well as computational formulation for any

problem. Although the LSFEM formulation is simple, it offers significant merits. The following features of LSFEM will be elaborated in the text:
(1) *Universality.* It has become standard practice to employ different numerical schemes for different types of differential equations. In finite difference literature, one can find central difference schemes used for elliptic (diffusion) problems, upwind difference schemes for hyperbolic (convection) problems, and many special schemes developed for particular problems. Often the latter are named after their inventors, such as the Beam–Warming, Lax–Wendroff, MacCormack schemes, etc. In finite element literature, there exist many different Galerkin methods: the classic Galerkin, Galerkin mixed, Petrov–Galerkin, Taylor–Galerkin, Galerkin/least-squares, etc. All are devised for different problems and have very different principles and structures. In contrast, the LSFEM has a unified formulation for the numerical solution of all types of partial differential equations. It does not matter whether the equations are elliptic, parabolic, hyperbolic or mixed. As long as the equations have a unique solution, the LSFEM always can determine a good approximate solution. For this reason, the LSFEM is able to simulate fluid dynamics problems in all flight regimes, from incompressible/subsonic through transonic, supersonic and hypersonic. And it does this within one mathematical/computational framework.
(2) *Efficiency.* The LSFEM is naturally suited for first-order differential operators. In many areas of engineering and applied science, the original governing partial differential equations obtained from physical laws are of first order or can be turned into a set of first-order equations. However, it is difficult to deal with first-order differential operators by conventional methods, because such equation systems generally lead to nonsymmetric matrices for all other numerical methods. On the other hand, the LSFEM, for linear partial differential equations, always leads to symmetric positive-definite matrices which can be efficiently solved by matrix-free iterative methods, such as the preconditioned conjugate gradient method. Consequently, large-scale three-dimensional problems can be solved within reasonable computer memory and in relatively small amount of time. Moreover, the parallelization is trivial.
(3) *Robustness.* When the LSFEM is employed, special treatments, such as upwinding, artificial dissipation, staggered grid or non-equal-order elements, artificial compressibility, operator-splitting, and operator-preconditioning, etc. are unnecessary; moreover, user-tunable parameters that are prevalent in traditional schemes are avoided. In all areas where the Galerkin mixed method is invoked, the LSFEM can be applied without the limitation of the LBB condition, that is, equal-order elements, which make the programming much easier, can be used. For the solution of problems in pure convection and high-speed compressible flows, the LSFEM inherently contains a mechanism to automatically capture discontinuities or shocks. In addition, for pure convection problems, the reweighted LSFEM can capture contact discontinuities in one element with machine-accuracy.

(4) *Optimality.* In many cases it can be rigorously proven that the LSFEM solution is the best approximation, that is, the error of the LSFEM solution has the same order as the interpolation error. In many areas of engineering and applied science, it is as important to evaluate the accuracy of an approximate solution as it is to obtain that solution. The least-squares method meets the need for *posterior* error analysis by supplying an error indicator in the form of the residuals that are minimized by the procedure. This very reliable error indicator can be used for adaptive refinement to achieve optimal solutions. No other approximate method can supply this information without additional calculation.
(5) *Concurrent simulation of multiple physics.* Since the LSFEM is a unified method for the approximate solution of differential equations governing diverse physical phenomena, a one-algorithm, one-code and one-pass simulation tool for concurrent analysis of different disciplines, such as conjugate heat transfer and solid–fluid interaction, can be developed.
(6) *General-purpose coding.* As mentioned above, the LSFEM is formulated in a very general setting. Therefore, the LSFEM can be programmed systematically, such that for a new application, one need only add simple subroutines to supply the coefficients, the load vector, and the boundary conditions for the first-order system. This can drastically reduce the time, cost and programming errors in code development.

The least-squares method also has a great potential for the theoretical study of the mathematical properties of partial differential equations including uniqness and stability of solutions and the permissibility of boundary conditions, as well as the derivation of equivalent high-order versions of equations.

The mathematical study of the incompressible Navier–Stokes equations has been mainly confined to the standard case in which velocity components are prescribed on the boundary, and traditionally conducted via the Galerkin method which leads to a difficult saddle-point problem. On the other hand, computational researchers often choose non-standard boundary conditions based on physical intuition. Although the Navier–Stokes equations were established more than one hundred years ago, there has been a lack of systematic study about their permissible boundary conditions. Recently, by considering the div–curl structure of the incompressible Stokes and Navier–Stokes equations in the velocity–pressure–vorticity formulation and by using the least-squares method, for the first time, mathematically permissible boundary conditions have been systematically and rigorously derived and analyzed.

The Maxwell equations in electromagnetics were established by James Clerk Maxwell in 1873. The original first-order full Maxwell equations, which consist of two curl equations (the Faraday and Ampere laws) and two divergence equations (the Gauss laws), reflect general and independent laws of physics. These laws can not be deduced from each other. They govern all electromagnetic phenomena, no matter whether the problem is static, time-harmonic or transient. Surprisingly, in almost all enginering literature, the full first-order Maxwell equations are considered as "overspecified", and the two divergence equations are thus regarded as "dependent" and often neglected in numerical computation. However, by considering the div–curl structure and using the least-squares method it can be shown that the Maxwell equations are not "overspecified" and that the divergence equations must always be included in analytical and numerical formulations to guarantee the uniqueness of the solution and the accuracy of the numerical methods, and to eliminate spurious solutions.

Satisfaction of the divergence equations in the Maxwell equations can be easily accomplished by the use of node-based LSFEM. The introduction of complicated and inefficient edge elements and the use of potentials (except for multi-layered materials) are unnecessary.

In fluid dynamics, as well as in electromagnetics, there are some good reasons why other higher-order versions of the Navier–Stokes equations and Maxwell equations are often useful. These are all derived from the "primitive" equations by differentiation and often offer additional insights regarding fluid flow and electromagnetic fields. They also serve as a starting point for devising alternative numerical schemes. For example, the pressure Poisson equation is obtained by applying the divergence operator to the momentum equation in the velocity–pressure formulation; the vorticity transport equation is obtained by applying the curl operator to the momentum equation; the second-order curl–curl equations in electromagnetics are derived from the first-order Maxwell curl equations by applying the curl operator. A key question is whether a derived equation obtained by simple differentiation admits more solutions than do its progenitors, in other words, whether spurious solutions or spurious modes in electromagnetic waveguides are generated. Only with careful selection of additional equations and boundary conditions, which are usually based on heuristics and numerical experience, will the solution of the derived equations also solve the original equations. However, the least-squares method and the related div–curl method have settled the question definitely and provide systematic and rigorous ways to derive higher-order versions of differential equations and corresponding boundary conditions without generating spurious solutions. That is, the Euler–Lagrange equations associated with the least-squares variational formulations are the most appropriate derived equations.

In summary, the least-squares method provides not only a powerful technique for numerical solution but also a useful tool for theoretical analysis of partial differential equations. The LSFEM is a simple and universal method. In the LSFEM there is nothing beyond the finite element interpolation and the least-squares principle. The basic features of finite element methodology and the attractive advantages of the least-squares method represent the most desirable properties of any numerical scheme designed to handle real-world problems. For these reasons, we believe that the LSFEM will play an important role in computational engineering and science in the twenty first century.

2. First-Order Scalar Equation in One Dimension

In fluid dynamics, convective transport described by first-order derivatives in the governing equations plays an important role. The Maxwell equations in electromagnetics are of first order. First-order differential operators which seem very simple, are in fact difficult to deal with numerically. For the solution of first-order differential equations, the classic Galerkin method or the central difference method fails miserably. In this chapter we use a very simple one-dimensional model problem to explain why the classic Galerkin method generates oscillatory solutions, and why the least-squares method does not need upwinding and is perfectly suitable for the solution of first-order differential equations.

2.1 A Model Problem

Let us consider a first-order scalar ordinary differential equation in the interval $[0,1]$

$$u'(x) = f(x), \tag{2.1a}$$

$$u(0) = 0, \tag{2.1b}$$

where $u' = du/dx$, and f is a given continuous function.

For convenience we introduce the function space $C^m(\Omega)$. In this chapter Ω is a bounded one-dimensional domain. If all $u(x), u'(x), ..., d^m u/dx^m$ are continuous in the closed interval $\overline{\Omega}$, then u is said to belong to $C^m(\Omega)$, that is denoted by $u \in C^m(\Omega)$.

The classic solution of the initial or the one-point boundary value problem (2.1) is a function that satisfies the equation (2.1a) and the boundary condition (2.1b). By integrating $u' = f$ once, it is easy to see that this problem has a unique classic solution u. The smoothness or the regularity of the classic solution $u(x)$ depends on the smoothness of the right-hand side function f. If $f \in C^m(\Omega), m \geq 0$, then the classic solution $u \in C^{m+1}(\Omega)$.

2.2 Function Spaces $H^m(\Omega)$

We shall now convert the problem (2.1) into a variational problem. For this we must turn to new function spaces $H^m(\Omega)$ that are larger (i.e., contain somewhat more functions) than $C^m(\Omega)$.

First we define the space of "square integrable" functions on Ω:

$$H^0(\Omega) = L_2(\Omega) = \{\, u : u \text{ is defined on } \Omega \text{ and } \int_\Omega u^2 dx < \infty\}. \tag{2.2}$$

The norm of u is denoted by

$$\|u\|_0 = \left\{\int_\Omega |u(x)|^2 dx\right\}^{1/2}. \tag{2.2}$$

Continuous functions, piecewise continuous functions (particularly, piecewise constants) are typical functions belonging to $H^0(\Omega)$. Of course, $H^0(\Omega)$ contains more than those functions. Note that "square integrable" is a stronger requirement than "integrable". For example, $u = x^{-1/2}$ is integrable on $[0,1]$, but its square is not integrable.

Assume that $u, v \in H^0(\Omega)$, then their inner product is defined as

$$(u,v) = \int_\Omega u(x)v(x)dx. \tag{2.3}$$

Now let us prove the Schwarz inequality:

$$|\,(u,v)\,| \le \|u\|_0 \|v\|_0, \tag{2.4}$$

and the triangle inequality:

$$\|u+v\|_0 \le \|u\|_0 + \|v\|_0. \tag{2.5}$$

Consider a quadratic polynomial of any real number λ:

$$\int_\Omega (\lambda u + v)^2 dx = \lambda^2 \int_\Omega u^2 dx + 2\lambda \int_\Omega uv dx + \int_\Omega v^2 dx \ge 0,$$

its discriminant must be nonnegative, i.e.

$$\left(\int_\Omega uv dx\right)^2 - \int_\Omega u^2 dx \cdot \int_\Omega v^2 dx \le 0,$$

which leads to (2.4). Then from (2.4),

$$\begin{aligned}\|u+v\|_0^2 &= \|u\|_0^2 + 2(u,v) + \|v\|_0^2 \\ &\le \|u\|_0^2 + 2\|u\|_0\|v\|_0 + \|v\|_0^2 = (\|u\|_0 + \|v\|_0)^2.\end{aligned}$$

The square roots of both sides of the above inequality leads to (2.5).

If u is continuous on $\overline{\Omega}$, the derivative $u'(x) \in H^0(\Omega)$ and $u(x) - u(x_o) = \int_{x0}^{x} u'(t)dt$, then u is said to belong to the function space $H^1(\Omega)$. The corresponding norm is

$$\|u\|_1 = \left\{\int_\Omega (|u(x)|^2 + |u'(x)|^2)dx\right\}^{1/2}. \tag{2.6}$$

For the function space $H^1(\Omega)$ we also define the semi-norm:

$$|u|_1 = \left\{\int_\Omega |u'(x)|^2 dx\right\}^{1/2}.$$

Continuous functions $C^1(\Omega)$ and piece-wise differentiable continuous functions belong to $H^1(\Omega)$. For example,

$$f(x) = \begin{cases} 2x/l, & \text{if } 0 \le x \le l/2; \\ 1, & \text{if } l/2 \le x \le l \end{cases} \tag{2.7}$$

is continuous in $[0, l]$, and its derivative is continuous on $[0, l/2]$ and $[l/2, l]$, so $f \in H^1(\Omega)$. Its norm is

$$\|f\|_1 = (2/l + 2l/3)^{1/2}.$$

Generally we define the function space $H^m(\Omega)$, $m > 0$. If the function $u \in C^{m-1}(\Omega)$ and $u^{(m)} \in H^0(\Omega)$, then it is said that $u \in H^m(\Omega)$.

In the following we introduce two important inequalities.

(1) If $u \in H^1(0, l)$ and $u(0) = 0$, then the Friedrichs inequality holds

$$\left(\int_0^l |u(x)|^2 dx\right)^{1/2} \le l\left(\int_0^l |u'(x)|^2 dx\right)^{1/2}. \tag{2.8}$$

Proof: Since

$$u(x) = u(x) - u(0) = \int_0^x u'(t)dt,$$

upon using the Schwarz inequality (2.4) we have

$$|u(x)|^2 = \left|\int_0^x 1 \cdot u'(t)dt\right|^2 \le \int_0^x 1dt \cdot \int_0^x |u'(t)|^2 dt \le l \cdot \int_0^l |u'(x)|^2 dx.$$

Integrating both sides of the above inequality leads to (2.8). □

(2) If $u \in H^1(0, l)$, then the Sobolev inequality holds

$$|u(x)| \le C\|u(x)\|_1, \qquad C = (2/l + 2l/3)^{1/2}. \tag{2.9}$$

Proof: Assume that $x \in [l/2, l]$. Consider $v = uf$ in which f is defined by (2.7). Obviously, $v(0) = 0$, and

$$u(x) = v(x) - v(0) = \int_0^x v'(t)dt = \int_0^x (u'f + uf')dt.$$

By using the Schwarz inequality (2.4) and the Cauchy inequality:

$$a_1b_1 + a_2b_2 \le (a_1^2 + a_2^2)^{1/2}(b_1^2 + b_2^2)^{1/2},$$

we have

$$|u(x)| \leq \int_0^x |u'f + uf'|dt \leq \|f\|_1 \|u\|_1 \leq C\|u\|_1.$$

By changing $f(x)$ we can prove (2.9) if $x \in [0, l/2]$. □

The Basic Lemma of Variational Principles. Assume that $f(x) \in C^0(\Omega)$ and

$$\int_\Omega f(x)\phi(x)dx = 0 \qquad \forall \phi \in C^0(\Omega),$$

then $f(x) = 0$, where the notation $\forall$ denotes "for all".

Proof: Assume that $f(x) \neq 0$, then there exists a point x_0 in Ω such that $f(x_0) \neq 0$ (assuming that $f(x_0) > 0$ without loss of generality). From continuity we know that $f(x) > 0$ in a neighborhood $K_\rho = \{|x - x_0| < \rho\}$ for some $\rho > 0$. Let

$$\phi(x) = \begin{cases} \rho^2 - (x - x_0)^2, & \text{if } x \in K_\rho; \\ 0, & \text{otherwise.} \end{cases}$$

Obviously $\phi \in C^0(\Omega)$ and $\phi > 0$ in K_ρ, therefore

$$\int_\Omega f(x)\phi(x)dx = \int_{K_\rho} f(x)\phi(x)dx > 0.$$

This implies a contradiction, and thus we must have $f(x) = 0$. □

This lemma holds also for two and three-dimensional cases in which the neighborhood K_ρ in the proof should be taken as a circle or a sphere.

2.3 The Classic Galerkin Method – Global Approximation

At first, let us try to use the classic Galerkin method for finding the approximate solution of the problem (2.1). The Galerkin method is a member of the class of weighted residual methods. The starting point for a weighted residual method is to assume an approximate solution written as

$$u(x) = a_1\phi_1(x) + a_2\phi_2(x) + \dots + a_n\phi_n(x) \qquad x \in \Omega = [0, 1], \tag{2.10}$$

where the basis functions $\phi_1, \dots, \phi_n$ are known. That is, the approximate (trial) solution is expressed as a linear combination of the basis functions. Of course, $u(x)$ should satisfy the boundary condition $u(0) = 0$. In one spatial dimension, the simple basis functions might be polynomials or trigonometric functions, e.g.,

$$\phi_j(x) = x^j \qquad \text{or} \qquad \phi_j(x) = \sin(j\pi x) \qquad j = 1, 2, \dots.$$

The coefficients a_j are unknowns to be determined. If the approximate solution (2.10) is substituted into (2.1), it will not generally be identically zero. Thus we can write

$$R(x) = u'(x) - f(x), \tag{2.11}$$

where $R(x)$ is referred to as the equation residual.

In the method of weighted residual, the coefficients a_j are determined by requiring that the integral of the weighted residual over the computational domain is zero, i.e.,

$$\int_{\Omega} v(x)R(x)dx = 0. \tag{2.12}$$

If the integral in (2.12) is zero for any weight (test) function $v(x) \in C^0(\Omega)$, due to the basic lemma of variational principles, the residual R will be identical to zero, i.e., the differential equation (2.1) is satisfied by the solution $u(x)$. However, in practice, it is impossible or unnecessary to test the residual R by all C^0 functions in $[0, 1]$, since the approximate solution $u(x)$ itself has only n degrees of freedom, and n test functions would be enough. Now the question is how to choose the test functions. Different choices for the test functions $v(x)$ in (2.12) give rise to different methods. In the Galerkin method, the test functions are chosen from the same family as the approximating (trial) functions. If the basis functions form a complete set (on $[0, 1]$ a complete set of trigonometric functions would be $1, \sin(\pi x), \cos(\pi x), \sin(2\pi x), \cos(2\pi x), \dots, \sin(n\pi x), \cos(n\pi x)$), (2.12) indicates that the residual is orthogonal to every member of a complete set. Consequently, one hopes that as n tends to infinity the approximate solution will converge to the exact solution.

Unfortunately, in solving first-order differential equations, the classic Galerkin method does not behave as expected. Let us choose the following simple problem as an example:

$$u' = 2\pi\cos(\pi x) \qquad \text{in } [0, 1], \tag{2.13a}$$

$$u = 0 \qquad \text{at } x = 0. \tag{2.13b}$$

Obviously,

$$u_{\mathbf{exact}} = 2\sin(\pi x).$$

The one-point boundary value problem (2.13) can now be given in terms of the following Galerkin formulation: Find $u \in V = \{u \in H^1(0, 1) : u = 0 \text{ at } x = 0\}$ such that

$$\int_0^1 \{u' - 2\pi\cos(\pi x)\}v dx = 0 \qquad \forall v \in V. \tag{2.14}$$

We choose the trial function:

$$u = a\sin(\pi x), \tag{2.15}$$

thus in the Galerkin method the test function should be

$$v = \sin(\pi x). \tag{2.16}$$

To determine the unknown coefficient a in (2.15), we substitute (2.15) and (2.16) into the Galerkin formulation (2.14), and obtain

$$a\pi \int_0^1 \cos(\pi x)\sin(\pi x)dx = 2\pi \int_0^1 \cos(\pi x)\sin(\pi x)dx. \tag{2.17}$$

Since $\cos(\pi x)$ and $\sin(\pi x)$ are orthogonal on $[0,1]$, (2.17) becomes:

$$a \times 0 = 0.$$

Surprisingly, the coefficient a cannot be determined by the Galerkin method! This trouble comes from the fact that the Galerkin method takes the test functions only from the same family of trial functions.

In order to overcome this difficulty, one may take the test functions from a space which is different from the space of trial functions where the approximate solution is sought. Such a method is called the Petrov–Galerkin method. For the problem (2.13) one may take

$$v = \cos(\pi x) \tag{2.18}$$

as a test function. Substituting (2.15) and (2.18) into (2.14), we obtain

$$a\pi \int_0^1 \cos(\pi x)\cos(\pi x)dx = 2\pi \int_0^1 \cos(\pi x)\cos(\pi x)dx, \tag{2.19}$$

and thus

$$a = 2,$$

which is the correct solution. However, the Petrov–Galerkin method itself cannot answer the question: How to choose an appropriate space of test functions for general problems? Therefore, the Petrov–Galerkin method is not always a satisfactory method.

2.4 The Least-Squares Method – Global Approximation

An alternative choice for the approximate solution of the model problem (2.1) is the least-squares method. The basic idea in the least-squares method is to determine the coefficients a_j in (2.10) by minimizing the integral of the square of the residual (2.11) over the computational domain. Therefore, we construct a quadratic functional

$$I(u) = \|R(u)\|_0^2 = \int_0^1 \{u'(x) - f(x)\}^2 dx, \tag{2.20}$$

over all $u \in V = \{u \in H^1(0,1) : u = 0 \text{ at } x = 0\}$. We refer to I in (2.20) as a functional to emphasize that its domain is the space V of admissible functions, and its range (i.e., its set of values) is a set of real numbers, i.e., a subset of $\mathbb{R}$. Here we generally use the symbol $\mathbb{R}$ to refer the real number system. We express these facts symbolically by writing

$$I : V \to \mathbb{R}$$

which is read "I maps the space of admissible functions V into $\mathbb{R}$".

A necessary condition that $u \in V$ be a minimizer of the functional I in (2.20) is that its first variation vanishes at u for all admissible v. That is,

$$\lim_{t\to 0} \frac{d}{dt} I(u + tv) \equiv 2 \int_0^1 \{u' - f(x)\} v' dx = 0 \qquad \forall v \in V,$$

or

$$(u', v') = (f, v') \qquad \forall v \in V. \tag{2.21}$$

Obviously, the least-squares method expressed by (2.21) can be interpreted as a Petrov–Galerkin method in which v' is chosen as a test function in stead of v.

We shall now use the least-squares method to solve the first-order differential equation (2.13). As in Sect. 2.3, we select the trial function

$$u = a\sin(\pi x). \tag{2.22}$$

Thus the test function is

$$v' = \pi\cos(\pi x). \tag{2.23}$$

To determine a we substitute (2.22) and (2.23) into the least-squares formulation (2.21) and obtain

$$a\pi^2 \int_0^1 \cos(\pi x)\cos(\pi x) dx = 2\pi^2 \int_0^1 \cos(\pi x)\cos(\pi x) dx,$$

or

$$a \times \frac{\pi^2}{2} = 2 \times \frac{\pi^2}{2}.$$

That is

$$a = 2.$$

In this case the least-squares method gives the exact solution.

2.5 One-Dimensional Finite Elements

In the methods discussed in Sects. 2.3 and 2.4, the global smooth functions are taken as trial functions, while in the finite element method, continuous piecewise polynomials are chosen as trial functions.

We shall now construct a finite-dimensional subspace V_h of the space V defined above consisting of piecewise linear functions. We use the nodes

$$0 = x_0 < x_1 < x_2 < ... < x_{n-1} < x_n = 1$$

to divide the interval $[0, 1]$ into n subintervals or elements $e_j = (x_{j-1}, x_j), j = 1, 2, ..., n$, of length $h_j = x_j - x_{j-1}$, and set $h = \max h_j$. The quantity h is then a measure of how fine the partition is.

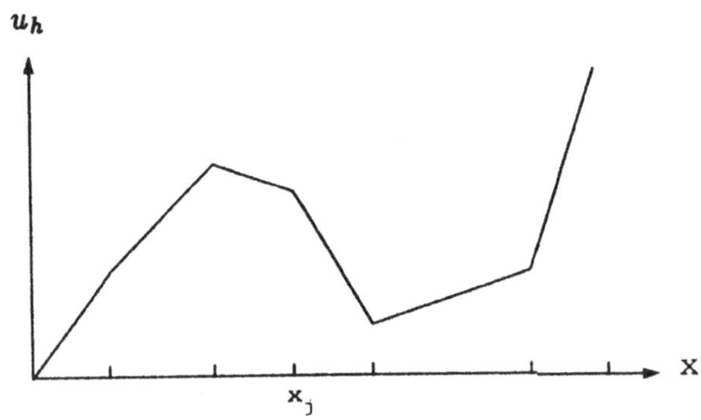

Fig. 2.1. Example of a function $u_h \in V_h$

We now construct a trial function u_h such that u_h is linear on each element e_j, and u_h is continuous on $[0, 1]$ and satisfies the boundary condition $u_h(0) = 0$, as illustrated in Fig. 2.1. These functions constitute the subspace V_h, and $V_h \subset V$. As parameters to describe a function $u_h \in V_h$, we may choose the values $u_0, u_1, u_2, ..., u_n$ at the nodes x_j. On each element $e_j = (x_{j-1}, x_j)$, $u_h(x)$ can be expressed as

$$u_h(x) = \psi_1^{(j)}(x)u_{j-1} + \psi_2^{(j)}(x)u_j \qquad x \in e_j, \tag{2.24}$$

in which the shape functions

$$\psi^{(j)}(x) = \begin{pmatrix} \psi_1^{(j)} \\ \psi_2^{(j)} \end{pmatrix} = \begin{pmatrix} (x_j - x)/h_j \\ (x - x_{j-1})/h_j \end{pmatrix}. \tag{2.25}$$

Therefore, on the entire domain [0,1], $u_h(x)$ can be written as

$$u_h(x) = \phi_0(x)u_0 + \phi_1(x)u_1 + ... + \phi_n(x)u_n, \tag{2.26}$$

in which

$$\phi_j(x) = \begin{cases} \psi_2^{(j)}(x), & x \in e_j; \\ \psi_1^{(j+1)}(x), & x \in e_{j+1};\ 1 \le j \le n-1 \\ 0, & \text{otherwise}; \end{cases}$$

$$\phi_0(x) = \begin{cases} \psi_1^{(1)}(x), & x \in e_1; \\ 0, & \text{otherwise}; \end{cases} \tag{2.27}$$

$$\phi_n(x) = \begin{cases} \psi_2^{(n)}(x), & x \in e_n; \\ 0, & \text{otherwise}; \end{cases}$$

are the so-called basis functions. $\phi_j(x)$ is a piecewise continuous linear function with a value unity at node x_j and null at other nodes, as illustrated in Fig. 2.2. Therefore, $u_h(x_j) = u_j$. Especially, if we take $u_0 = 0$, then (2.26) satisfies the boundary condition (2.1b), and other parameters (nodal values) $u_1, u_2, ..., u_n$ can be arbitrary. All functions $u_h(x)$ constitute the space of trial functions V_h.

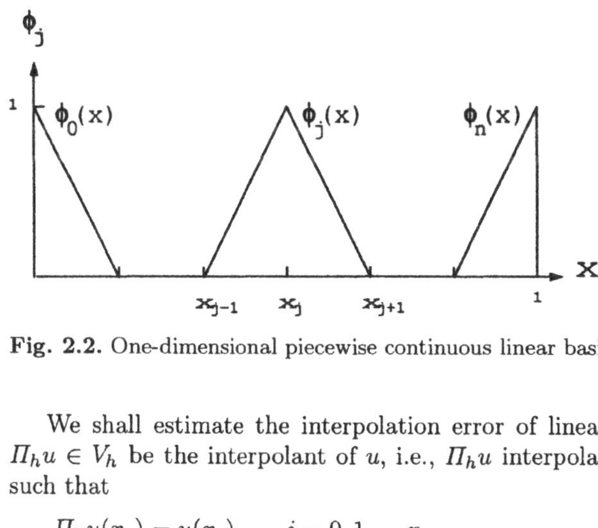

Fig. 2.2. One-dimensional piecewise continuous linear basis functions

We shall estimate the interpolation error of linear finite elements. Let $\Pi_h u \in V_h$ be the interpolant of u, i.e., $\Pi_h u$ interpolates u at the nodes x_j such that

$$\Pi_h u(x_j) = u(x_j) \qquad j = 0, 1, ..., n.$$

Let $E = u - \Pi_h u$ be the interpolation error function and consider an arbitrary element e_j with points $x_{j-1} \le x \le x_j$ in the mesh. We assume that $u \in H^2(0,1)$, and let

$$|u|_2 = \left\{ \int_0^1 u''(x)^2 dx \right\}^{1/2}.$$

This is the semi-norm of $H^2(0,1)$ introduced in Sect. 2.2. Since $\Pi_h u$ is the interpolant of u, the error E vanishes at both endpoints x_{j-1} and x_j. By virtue of the Rolle theorem, there exists at least one point $\bar{x}$ between x_{j-1} and x_j at which $E'(\bar{x}) = 0$. Then for any x,

$$E'(x) = \int_{\bar{x}}^{x} E''(t)dt.$$

Since $\Pi_h u$ is linear, $E = u - \Pi_h u$ implies $E'' = u'' - (\Pi_h u)'' = u''$ within e_j. Applying the Schwarz inequality (2.4), we find

$$\begin{aligned} |E'(x)| &\le \left\{\int_{\bar{x}}^{x} 1^2 dt\right\}^{1/2} \left\{\int_{\bar{x}}^{x} u''(t)^2 dt\right\}^{1/2} \\ &\le h_j^{1/2}\left\{\int_{x_{j-1}}^{x_j} u''(t)^2 dt\right\}^{1/2} \quad x \in e_j. \end{aligned} \tag{2.28}$$

To derive a bound on $|E(x)|$, we write

$$E(x) = \int_{x_{j-1}}^{x} E'(t)dt$$

and use (2.28) to obtain

$$|E(x)| \le \int_{x_{j-1}}^{x} |E'(t)|dt \le h_j^{3/2}\left\{\int_{x_{j-1}}^{x_j} E''(t)^2 dt\right\}^{1/2} \qquad x \in e_j. \tag{2.29}$$

Squaring (2.28) and (2.29), integrating, summing over all elements, and taking square roots, we arrive at the estimates of the error:

$$\|(u - \Pi_h u)'\|_0 \le h|u|_2, \tag{2.30a}$$

$$\|u - \Pi_h u\|_0 \le h^2|u|_2. \tag{2.30b}$$

Finally since

$$\|u - \Pi_h u\|_1^2 = \|u - \Pi_h u\|_0^2 + \|(u - \Pi_h u)'\|_0^2,$$

we find easily that

$$\|u - \Pi_h u\|_1 \le \sqrt{2}h|u|_2. \tag{2.31}$$

Note that the bounds (2.30) and (2.31) are satisfactory in the sense that the powers of h are the best possible. To obtain an improved result, by expanding the error $E(x)$ in a sine series, one can verify that (Strang and Fix 1973, p.45)

$$\|(u - \Pi_h u)'\|_0 \le \pi^{-1}h|u|_2, \tag{2.32}$$

$$\|u - \Pi_h u\|_0 \le \pi^{-2}h^2|u|_2. \tag{2.33}$$

2.6 The Classic Galerkin Finite Element Method

The Galerkin finite element method for the first-order equation (2.1) can now be formulated as follows: Find $u_h \in V_h$ such that

$$\int_0^1 \{u_h' - f(x)\}v_h dx = 0 \qquad \forall v_h \in V_h. \tag{2.34}$$

Since

$$u_h(x) = \sum_{j=1}^{n} \phi_j(x) u_j, \tag{2.35}$$

and (2.34) must be valid by taking $v_h = \phi_i(x)$, $i = 1, 2, ..., n$ as the test functions, we can write (2.34) as

$$\sum_{j=1}^{n} (\phi_i, \phi_j') u_j = (\phi_i, f) \qquad i = 1, 2, ..., n, \tag{2.36}$$

which is a system of linear algebraic equations with n equations in n unknown $u_1, u_2, ..., u_n$. In the matrix form the linear system (2.36) can be written as

$$\boldsymbol{KU} = \boldsymbol{F}, \tag{2.37}$$

where the global matrix $\boldsymbol{K} = (K_{ij})$ is $n \times n$ matrix with entries $K_{ij} = (\phi_i, \phi_j')$, the global unknown vector $\boldsymbol{U}^T = (u_1, u_2, ..., u_n)$ and the global load vector $\boldsymbol{F}^T = (F_1, F_2, ..., F_n)$ with $F_i = (\phi_i, f)$. Here the basis functions $\phi_j(x)$ are employed to derive $\boldsymbol{K}$ and $\boldsymbol{F}$. In general-purpose finite element programs the global matrix $\boldsymbol{K}$ and the global force vector $\boldsymbol{F}$ are assembled from the element matrices and the element force vectors, respectively, by using the element shape functions ψ_j, see any text book on finite elements, e.g., Becker et al. (1983) or Johnson (1987).

The entries $K_{ij} = (\phi_i, \phi_j')$ in the global matrix $\boldsymbol{K}$ can easily be computed. We first observe that $(\phi_i, \phi_j') = 0$ if $|i - j| > 1$, since in this case for all $x \in [0, 1]$ either $\phi_i(x)$ or $\phi_j(x)$ is equal to zero. Thus, the matrix $\boldsymbol{K}$ is tri-diagonal, i.e., only the entries in the main diagonal and the two adjacent diagonals may be different from zero. We have for $i = 1, 2, ..., n-1$

$$\begin{aligned} K_{ii} = (\phi_i, \phi_i') &= \int_{x_{i-1}}^{x_i} \frac{(x - x_{i-1})}{h_i} \cdot \frac{1}{h_i} dx + \int_{x_i}^{x_{i+1}} \frac{(x_{i+1} - x)}{h_{i+1}} \cdot \frac{-1}{h_{i+1}} dx \\ &= \frac{1}{2} - \frac{1}{2} = 0, \end{aligned}$$

and for $i = 2, ..., n-1$

$$K_{i,i-1} = (\phi_i, \phi_{i-1}') = \int_{x_{i-1}}^{x_i} \frac{(x - x_{i-1})}{h_i} \cdot \frac{-1}{h_i} dx = -\frac{1}{2},$$

$$K_{i,i+1} = (\phi_i, \phi_{i+1}') = \int_{x_i}^{x_{i+1}} \frac{(x_{i+1} - x)}{h_{i+1}} \cdot \frac{1}{h_{i+1}} dx = \frac{1}{2},$$

and

$$K_{nn} = (\phi_n, \phi_n') = \int_{x_{n-1}}^{x_n} \frac{(x - x_{n-1})}{h_n} \cdot \frac{1}{h_n} dx = \frac{1}{2}.$$

We are not able to accurately calculate F_j unless the analytical expression of the function $f(x)$ is given. For the purpose of illustration, we use the one-point Simpson rule to perform the quadrature. We obtain for $i = 1, 2, ..., n-1$

$$
\begin{aligned}
F_i = (\phi_i, f) &= \int_{x_{i-1}}^{x_i} \frac{(x - x_{i-1})}{h_i} \cdot f dx + \int_{x_i}^{x_{i+1}} \frac{(x_{i+1} - x)}{h_{i+1}} \cdot f dx \\
&\simeq \frac{1}{2} f_i h_i + \frac{1}{2} f_i h_{i+1},
\end{aligned}
$$

and

$$
F_n = (\phi_n, f) = \int_{x_{n-1}}^{x_n} f \cdot \frac{(x - x_{n-1})}{h_n} dx \simeq \frac{1}{2} f_n h_n.
$$

In the special case of a uniform partition with $h_i = h = 1/n$, the system (2.36) takes the form

$$
\frac{1}{2}\begin{pmatrix}
0 & 1 & 0 & 0 & . & . & . & 0 \\
-1 & 0 & 1 & 0 & . & . & . & 0 \\
0 & -1 & 0 & 1 & . & . & . & 0 \\
. & . & . & . & . & . & . & . \\
. & . & . & . & . & . & . & . \\
0 & 0 & 0 & 0 & -1 & 0 & 1 \\
0 & 0 & 0 & 0 & 0 & -1 & 1
\end{pmatrix}
\begin{pmatrix} u_1 \\ u_2 \\ u_3 \\ . \\ . \\ . \\ u_n \end{pmatrix}
= h \begin{pmatrix} f_1 \\ f_2 \\ f_3 \\ . \\ . \\ . \\ 0.5 f_n \end{pmatrix}. \tag{2.38}
$$

After division by h the first three equations in (2.38) may be written as

$$
\frac{u_2 - 0}{2h} = f_1,
$$

$$
\frac{u_3 - u_1}{2h} = f_2,
$$

$$
\frac{u_4 - u_2}{2h} = f_3.
$$

The above is, incidentally, identical to the standard central difference approximation obtained by putting

$$
u' \simeq \frac{u_{i+1} - u_{i-1}}{2h}.
$$

The matrix is obviously non-symmetric with zeros on the diagonal, and has odd-even decoupling which leads to oscillatory solutions. This trouble persists even after the mesh is refined. In order to overcome this bad approximation problem, finite difference practitioners use one-sided or upwind differencing

$$
u' \simeq \frac{u_i - u_{i-1}}{h} = \frac{u_{i+1} - u_{i-1}}{2h} - \frac{h}{2} \frac{u_{i+1} - 2u_i + u_{i-1}}{h^2},
$$

which is equivalent to adding some numerical dissipation into the central difference scheme. That is, they solve, instead of (2.1), the following second-order equation by the central difference method

$$
u' - \frac{h}{2} u'' = f.
$$

2.7 The Least-Squares Finite Element Method

2.7.1 The Least-Squares Formulation

For the first-order differential equation (2.1) the least-squares finite element method is formulated according to (2.21) as follows: Find $u_h \in V_h$ such that

$$\int_0^1 \Big\{u_h' - f(x)\Big\} v_h' dx = 0 \qquad \forall v_h \in V_h. \tag{2.39}$$

Since

$$u_h(x) = \phi_1(x)u_1 + ... + \phi_n(x)u_n, \tag{2.40}$$

(2.39) can be written as

$$\sum_{j=1}^{n} (\phi_i', \phi_j')u_j = (\phi_i', f) \qquad i = 1, 2, ..., n, \tag{2.41}$$

or in matrix form as

$$\boldsymbol{KU} = \boldsymbol{F}.$$

By using one-point Simpson quadrature we obtain for $i = 1, 2, ..., n-1$

$$\begin{aligned} F_i = (\phi_i', f) &= \int_{x_{i-1}}^{x_i} \frac{1}{h_i} f dx + \int_{x_i}^{x_{i+1}} \frac{-1}{h_{i+1}} f dx \\ &\simeq \frac{(f_{i-1} + f_i)}{2} \frac{1}{h_i} h_i - \frac{(f_i + f_{i+1})}{2} \frac{1}{h_{i+1}} h_{i+1} \\ &= -\frac{(f_{i+1} - f_{i-1})}{2}, \end{aligned}$$

and

$$F_n = (\phi_n', f) = \int_{x_{n-1}}^{x_n} \frac{1}{h_n} f dx \simeq \frac{(f_{n-1} + f_n)}{2}.$$

The global matrix $\boldsymbol{K}$ is tri-diagonal, so we have

$$\begin{aligned} K_{ii} = (\phi_i', \phi_i') &= \int_{x_{i-1}}^{x_i} \frac{1}{h_i^2} dx + \int_{x_i}^{x_{i+1}} \frac{1}{h_{i+1}^2} dx \\ &= \frac{1}{h_i} + \frac{1}{h_{i+1}} \qquad i = 1, 2, ..., n-1, \end{aligned}$$

$$K_{i,i-1} = (\phi_i', \phi_{i-1}') = -\int_{x_{i-1}}^{x_i} \frac{1}{h_i^2} dx = -\frac{1}{h_i} \qquad i = 2, ..., n-1,$$

$$K_{i,i+1} = (\phi_i', \phi_{i+1}') = -\int_{x_i}^{x_{i+1}} \frac{1}{h_{i+1}^2} dx = -\frac{1}{h_{i+1}} \qquad i = 2, ..., n-1,$$

and

$$K_{nn} = (\phi'_n, \phi'_n) = \int_{x_{n-1}}^{x_n} \frac{1}{h_n^2} dx = \frac{1}{h_n}.$$

Note that the matrix $\boldsymbol{K}$ is symmetric and positive-definite, since

$$(\phi'_i, \phi'_j) = (\phi'_j, \phi'_i),$$

and because

$$u_h(x) = \sum_{i=1}^{n} \phi_i(x) u_i$$

we have

$$\boldsymbol{U}^T \boldsymbol{K} \boldsymbol{U} = \Big(\sum_{i=1}^{n} u_i \phi'_i, \sum_{j=1}^{n} u_j \phi'_j \Big) = (u'_h, u'_h) \geq 0$$

with equality only if $u' \equiv 0$. Since a positive-definite matrix is always invertible, it follows that the linear system (2.41) has a unique solution.

In the special case of a uniform partition with $h_i = h = 1/n$, the system (2.41) takes the form

$$-\frac{1}{h^2} \begin{pmatrix} 2 & -1 & 0 & 0 & . & . & . & 0 \\ -1 & 2 & -1 & 0 & . & . & . & 0 \\ 0 & -1 & 2 & -1 & . & . & . & 0 \\ . & . & . & . & . & . & . & . \\ . & . & . & . & . & . & . & . \\ 0 & 0 & 0 & 0 & 0 & -1 & 2 & -1 \\ 0 & 0 & 0 & 0 & 0 & 0 & -1 & 1 \end{pmatrix} \begin{pmatrix} u_1 \\ u_2 \\ u_3 \\ . \\ . \\ . \\ u_n \end{pmatrix}$$

$$= \frac{1}{2h} \begin{pmatrix} f_2 - f_0 \\ f_3 - f_1 \\ f_4 - f_2 \\ . \\ . \\ . \\ -f_{n-1} - f_n \end{pmatrix}. \tag{2.42}$$

Obviously, the left-hand side of (2.42) can be interpreted as a standard central difference for u'' and the right-hand side as a central difference for f'. We clearly see again that the least-squares finite element method for the first-order differential equation leads to a symmetric and positive-definite system of linear algebraic equations.

2.7.2 The Euler–Lagrange Equation

In order to understand how the least-squares method performs, we shall derive the Euler–Lagrange equation associated with the least-squares formulation: Find $u \in V$ such that

$$\int_0^1 (u' - f)v'dx = 0 \qquad \forall v \in V. \tag{2.43}$$

Suppose that u and v are sufficiently smooth. Upon integration by parts, (2.43) yields

$$(u' - f)v|_{x=1} - \int_0^1 (u'' - f')vdx = 0 \qquad \forall v \in V. \tag{2.44}$$

By virtue of the basic lemma of variational principle (Sect. 2.2), we obtain the Euler–Lagrange equation

$$u'' - f' = 0 \qquad \text{in } \Omega, \tag{2.45a}$$

$$u' - f = 0 \qquad \text{at } x = 1, \tag{2.45b}$$

$$u = 0 \qquad \text{at } x = 0. \tag{2.45c}$$

Here the second-order equation (2.45a) is the derivative of the original first-order equation (2.1a); (2.45b) requires that the original equation (2.1a) be satisfied at the other end of the domain as the natural boundary condition; the original boundary condition (2.45c) serves as the essential boundary condition.

Obviously, (2.44) is the Galerkin formulation for the second-order problem (2.45). We now see that the first-order differential equation (2.1), the least-squares formulation (2.43), the second-order differential equation (2.45), and the Galerkin formulation (2.44) are equivalent to each other.

The least-squares method for the first-order differential equation (2.1) corresponds to using the Galerkin method to solve the second-order differential equation (2.45). It is well known that the Galerkin method (2.44) is a perfect method for the second-order differential equation (2.45) (see Chap. 3). Therefore, the least-squares method (2.43) is a perfect method for the first-order differential equation (2.1). Here the word "perfect" refers to the fact that the method yields a symmetric and positive-definite system of algebraic equations and has an optimal rate of convergence (see the next section for a direct proof). In other words, the least-squares method converts the difficult first-order differential equation into an easy second-order equation that everyone would rather solve.

It is correct that the least squares problem is formally equivalent to a higher-order problem. But that it "introduces the possibility of spurious solutions if incorrect boundary conditions are used" (Zienkiewicz and Morgan 1983, p.260; Zienkiewicz and Taylor 1988, p.254) is a common misconception about the least-squares method. Our derivation clearly shows that the least-squares method does not require any additional boundary conditions, since it seeks a solution in $V = \{u \in H^1(0,1) : u = 0 \text{ at } x = 0\}$. Only if the associated Euler–Lagrange equation (2.45a) is to be solved by using, e.g. the finite difference method, should be the additional natural boundary condition (2.45b) needed. In this finite difference method the original first-order differential equation serves as the natural boundary condition at $x = 1$.

In fact, the least-squares method provides a rigorous mathematical tool to derive an equivalent high-order system and its boundary condition with a guarantee of freedom of spurious solutions. We will elaborate on this point in Chap. 5 for the div-curl system, in Chap. 8 for the incompressible Navier–Stokes equations, and in Chap. 14 for the Maxwell equations.

2.7.3 Error Estimates

We shall now study the error of the least-squares finite element solution for the model problem (2.1). Let u be the solution of

$$\int_0^1 (u' - f)v'dx = 0 \qquad \forall v \in V, \tag{2.46}$$

and u_h be its finite element approximation, i.e. the solution of

$$\int_0^1 (u_h' - f)v_h'dx = 0 \qquad \forall v_h \in V_h. \tag{2.47}$$

The error in this approximation is defined as the function

$$e = u - u_h.$$

Setting $v = v_h$ in (2.46) (which is possible because V_h is a subspace of V) and subtracting (2.47) from (2.46), we find that e satisfies the orthogonal condition

$$(e', v_h') = 0 \qquad \forall v_h \in V_h. \tag{2.48}$$

We choose V_h to consist of piecewise linear functions on $[0, 1]$ introduced in Sect. 2.5. Let $\Pi_h u \in V_h$ be the interpolant of u satisfying (2.30) and (2.31), and write $e = u - \Pi_h u + \Pi_h u - u_h$. Then we have

$$\begin{aligned} \|e'\|_0^2 &= (e', e') = (e', u' - \Pi_h u') + (e', \Pi_h u' - u_h') = (e', u' - \Pi_h u') \\ &\leq \|e'\|_0 \|u' - \Pi_h u'\|_0. \end{aligned}$$

Here we have used the orthogonal condition (2.48) in the third step and the Schwarz inequality (2.4) in the fourth step. Thus

$$\|e'\|_0 \leq \|u' - \Pi_h u'\|_0. \tag{2.49}$$

This result establishes that the magnitude of the derivative of the error $u - u_h$, measured in the L_2-norm, is bounded above by the error of the finite element interpolant.

Combining (2.49) and (2.30a) we now obtain the following estimate for the derivative of the error:

$$\|e'\|_0 \leq h|u|_2, \tag{2.50}$$

if $u \in H^2(0, 1)$.

Since $e(0) = u(0) - u_h(0) = 0$, by virtue of the Friedrichs inequality (2.8), from (2.50) we establish the following estimate for the error itself:

$$\|e\|_0 \le h|u|_2. \tag{2.51}$$

It is sometimes desirable to measure the rate of convergence in other norms. Combining (2.50), (2.51) together with the Sobolev inequality (2.9) gives a pointwise error estimate:

$$|e| \le Ch|u|_2, \tag{2.52}$$

where C denotes a constant independent of the mesh parameter h, with possibly different values in each appearance.

We observe that these latter estimates (2.51) and (2.52), are less sharp than the estimate (2.30b) for the interpolation error where we have a factor h^2.

Next, we shall show that in fact the least-squares finite element method also gives a factor h^2 for the error $u - u_h$ in L_2 norm. To do this we adopt the Aubin–Nitsche trick, see, e.g., Oden and Carey (1983). Assume that w is the solution of the following second-order differential equation

$$-w'' = e \qquad \text{in } (0,1),$$

$$w = 0 \qquad \text{at } x = 0,$$

$$w' = 0 \qquad \text{at } x = 1,$$

where e is the error of the least-squares finite element solution of (2.47). From Sect. 2.5 we know that the finite element interpolant $\Pi_h w$ satisfies

$$\|(w - \Pi_h w)'\|_0 \le Ch\|w''\|_0 = Ch\|e\|_0. \tag{2.53}$$

Using integration by parts and the fact that $e = 0$ *at* $x = 0$ and $w' = 0$ *at* $x = 1$,

$$(e, e) = -(e, w'') = (e', w') = (e', w' - \Pi_h w'),$$

where the last equality follows from (2.48), since $\Pi_h w \in V_h$ so that $(e', \Pi_h w') = 0$. Applying the interpolation estimate (2.53) to w and using also (2.50), we find

$$\|e\|_0^2 \le \|e'\|_0 \|(w - \Pi_h w)'\|_0 \le Ch|u|_2 h\|e\|_0.$$

Dividing by $\|e\|_0$ we finally obtain

$$\|e\|_0 \le Ch^2|u|_2. \tag{2.54}$$

The result (2.54) indicates that the least-squares finite element method for first-order equations has an optimal rate of convergence in the sense that the error measured in L_2 norm is of the same order of h as the finite element interpolation.

2.7.4 Condition Number

The least-squares finite element method yields a symmetric and positive-definite system of linear algebraic equations that can be solved by iterative methods such as the conjugate gradient method. It is well known that the convergence of the conjugate gradient method depends on the condition number κ of the matrix $\boldsymbol{K}$. The larger the condition number κ is, the slower the conjugate gradient method converges. Here κ is defined to be the ratio of the largest eigenvalue of $\boldsymbol{K}$ to the smallest, that is

$$\kappa = \frac{\lambda_{\max}}{\lambda_{\min}},$$

where

$$\lambda_{\max} = \max_{U \neq 0} \frac{\boldsymbol{U}^T \boldsymbol{K} \boldsymbol{U}}{\boldsymbol{U}^T \boldsymbol{U}}, \qquad \lambda_{\min} = \min_{U \neq 0} \frac{\boldsymbol{U}^T \boldsymbol{K} \boldsymbol{U}}{\boldsymbol{U}^T \boldsymbol{U}},$$

in which $\boldsymbol{U}$ denotes the global vector of nodal values.

We can prove that there exist constants α, β and C, such that $\forall u_h = \sum_{i=1}^{n} \phi_i u_i \in V_h$ we have

$$\beta h \boldsymbol{U}^T \boldsymbol{U} \leq \|u_h\|_0^2 \leq \alpha h \boldsymbol{U}^T \boldsymbol{U}, \tag{2.55}$$

$$|u_h|_1 \leq C h^{-1} \|u_h\|_0. \tag{2.56}$$

The estimate (2.56) is the so called inverse estimate; here we estimate the L_2-norm of the gradient of u_h in terms of the L_2-norm of u_h itself. This is not possible for a general function $u \in H^1(\Omega)$, but it is possible for the function u_h in V_h at the price of the factor h^{-1}. It suffices to verify the validity of (2.55) and (2.56) for a single element. For any given element, for example, for the linear element $e_j = (x_{j-1}, x_j)$ in Fig. 2.1, this can be demonstrated by simple evaluation of $\|u_h\|_{0,e_j}$ and $|u_h|_{1,e_j}$ in which u_h is expressed by (2.24).

Combining (2.55) and (2.56) gives

$$(u_h', u_h') = |u_h|_1^2 \leq C h^{-2} \|u_h\|_0^2 \leq C_1 h^{-1} \boldsymbol{U}^T \boldsymbol{U}. \tag{2.57}$$

On the other hand, by using the Friedrichs inequality (2.8) and the inequality (2.56), we have

$$(u_h', u_h') = |u_h|_1^2 \geq C \|u_h\|_0^2 \geq C_2 h \boldsymbol{U}^T \boldsymbol{U}. \tag{2.58}$$

We recall that

$$(u_h', u_h') = \boldsymbol{U}^T \boldsymbol{K} \boldsymbol{U}, \tag{2.59}$$

so that from (2.57), (2.58) and (2.59) we have

$$C_2 h \leq \lambda_{\min}, \qquad \lambda_{\max} \leq C_1 h^{-1},$$

which lead to the desired result

$$\kappa(\boldsymbol{K}) = \frac{\lambda_{\max}}{\lambda_{\min}} \leq C h^{-2}. \tag{2.60}$$

We remark that if a second-order differential equation is approximated using the Galerkin finite element method with a uniform mesh h, the condition number of the resulting global matrix is $O(h^{-2})$. Therefore, the common speculation that the system of linear algebraic equations generated by the least-squares method has a bad condition number is also a misconception. The reason is simple: the least-squares method studied in this book is based on the first-order differential equations.

2.7.5 A Numerical Example

A further comparison of LSFEM with the Galerkin method and the finite difference method is made by solving the first-order ordinary differential equation:

$$u'(x) = \left[1 - \exp\left(-\frac{1}{\epsilon}\right)\right]^{-1} \frac{1}{\epsilon} \exp\left(-\frac{1-x}{\epsilon}\right) \qquad x \in [0,1], \tag{2.61a}$$

$$u(0) = 0, \tag{2.61b}$$

with $0 < \epsilon << 1$. The exact solution of this problem is given by

$$u(x) = 1 - \left[1 - \exp\left(-\frac{1}{\epsilon}\right)\right]^{-1} \left[1 - \exp\left(-\frac{1-x}{\epsilon}\right)\right].$$

For small ϵ, $u(x)$ is close to 0 except in a boundary layer at $x = 1$ of width $O(\epsilon)$ where u increases from 0 to 1 as shown in Fig. 2.3.

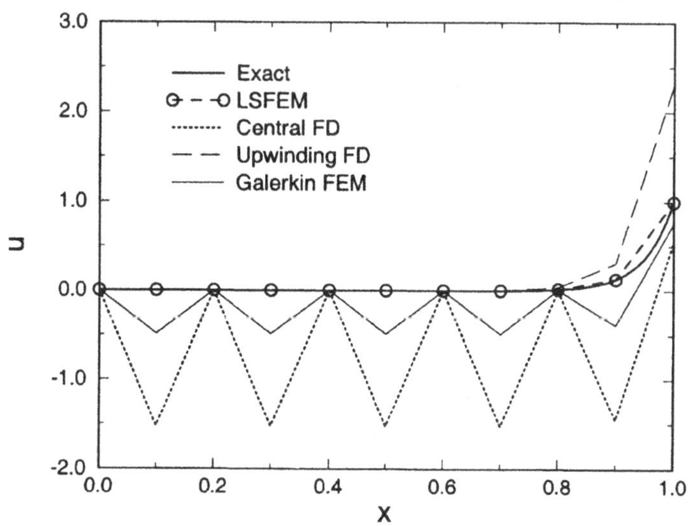

Fig. 2.3. Solution of (2.61) with $\epsilon = 0.05$

We apply the Galerkin method (2.36) by using piecewise linear elements on a uniform mesh with length $h = 0.1$. The global matrix is the same as given in (2.38). We divide each element into 10 segments and use the Simpson quadrature to calculate the right-hand side term (ϕ_i, f). We obtain the Galerkin solution that oscillates violently in the whole region and is not close to the exact solution as shown in Fig. 2.3.

Then we apply the least-squares method (2.41). We again use the same Simpson quadrature to calculate (ϕ_i', f). Although the problem is quite difficult for other methods, the least-squares solution with only 10 linear elements is very smooth and accurate as illustrated in Fig. 2.3.

For reference, in Fig. 2.3 we also show the results of the central finite difference method (which is identically equivalent to solving (2.38)) and the upwind finite difference method. Obviously, the upwinding finite difference method is too diffusive and not accurate.

2.8 Concluding Remarks

We have compared in this chapter the least-squares method with the Galerkin method for the solution of first-order scalar differential equations.

The Galerkin method is identical to the common central finite difference approximation. The algebraic equations generated by the Galerkin method are obviously non-symmetric, non-positive-definite and odd-even decoupled, and the solution is thus purely oscillatory and bears almost no relation to the underlying problem.

The least-squares method for first-order scalar differential equations is formally equivalent to the Galerkin method for corresponding second-order problems, and hence has an optimal rate of convergence. The least-squares approach yields a symmetric and positive-definite system that has significant computational advantages. That "the least-squares method needs extra boundary conditions and produces an ill-conditioned matrix" is a misconception. We have shown that the least-squares method does not need any additional boundary condition and that the condition number of the algebraic equations resulting from the least-squares method based on the first-order equations has an order of h^{-2} which is the same as in the classic Galerkin method for second-order equations.

For purely convective problems a common technique in finite difference methods is upwinding which introduces excessive dissipation and reduces the accuracy of the approximate solution. Upwinding and free parameters turn out to be unnecessary when the least-squares method is employed.

3. First-Order System in One Dimension

In fluid dynamics another important phenomenon is diffusion or conduction which is described by second-order derivatives in the governing equations. The Laplace or Poisson equation can be considered as the standard form of an equation describing isotropic diffusion in all space directions. For the Laplace or Poisson equation, the Rayleigh–Ritz method yields a symmetric and positive-definite system of linear algebraic equations and has an optimal rate of convergence. However, in practice, one is often interested in not only the primal variable (e.g., the temperature in heat conduction, the potential in irrotational incompressible flows, and the electric or magnetic potential in electromagnetics), but also the dual variable (e.g., the flux in heat conduction, the velocity in fluid flows, and the electric or magnetic field intensity in electromagnetics). The solution of dual variables computed by *a posteriori* numerical differentiation has low accuracy in general and is discontinuous across the element boundary.

The mixed Galerkin method was devised in the hope of obtaining better accuracy for dual variables. Here the term "mixed" refers to the fact that both the primal variable and the dual variables are approximated as fundamental unknowns. Obviously, in order to apply the mixed Galerkin method, the second-order scalar equation should be reduced to a first-order system.

In this chapter we use a very simple one-dimensional model problem to explain why the mixed Galerkin method is not satisfactory and the least-squares method is again perfectly suitable for diffusion problems.

3.1 A Model Problem

Let us consider a second-order ordinary differential equation

$$-u''(x) = f(x) \qquad x \in \Omega = [0,1], \tag{3.1a}$$

$$u(0) = u(1) = 0, \tag{3.1b}$$

where f is a given continuous function. The classic solution of the boundary value problem (3.1) is a function $u(x) \in C^2(\Omega)$ that satisfies the equation (3.1a) and the boundary conditions (3.1b). By integrating the equation, $-u'' = f$, twice, it is easy to see that this problem has a unique classic solution u.

3.2 The Rayleigh–Ritz Method

We have pointed out in Chap. 2 that the Galerkin method is based on the basic lemma of variational principles. From the viewpoint of mechanics the basic lemma of variational principles corresponds to the principle of virtual work (see, e.g., Washizu 1975). It is well-known that for diffusion and elasticity problems, the Galerkin method results in the same finite element scheme as in the Rayleigh–Ritz method which is based on the principle of minimum potential energy.

To formulate the Galerkin method for the problem (3.1) we first introduce the function space

$$H = \{u \in H^1(0,1) : u(0) = u(1) = 0\}.$$

Upon multiplying the equation (3.1a) by an arbitrary test function $v \in H$ and integrating over the interval $[0,1]$, we have

$$-(u'', v) = (f, v).$$

We now integrate the left-hand side by parts and consider the fact that $u(0) = u(1) = 0$ to obtain the following formulation: Find $u \in H$ such that

$$(u', v') = (f, v) \qquad \forall v \in H. \tag{3.2}$$

Further, one can show that the method (3.2) is the Rayleigh–Ritz method which seeks the minimizer of the functional

$$J(v) = \frac{1}{2}(v', v') - (f, v) \tag{3.3}$$

over all $v \in H$. The quantity $J(v)$ represents, for example, the total potential energy associated with the displacement $v \in H$ in an elastic bar. The term $(v', v')/2$ represents the internal elastic energy and (f, v) the load potential.

Following the same procedures used in Sects. 2.7 and 2.9, it is easy to show that the Rayleigh–Ritz finite element method leads to a linear system of equations with a symmetric and positive-definite global matrix, and the error is bounded above by the interpolation error, i.e., the finite element solution u_h has an optimal rate of convergence. If linear elements are used, the error estimates are

$$\|u'(x) - u_h'(x)\|_0 = O(h),$$

$$\|u - u_h\|_0 = O(h^2).$$

Once the nodal values $u_0, u_1, \ldots, u_n$ are obtained by the Rayleigh–Ritz finite element method, one can compute the derivatives in each element:

$$u_h'(x) = (u_i - u_{i-1})/h_i \qquad x \in e_i.$$

However, from the above error estimates we know that, in general, the accuracy of so-computed derivative $u_h'(x)$ is of only $O(h)$. Moreover, the derivative $u_h'(x)$ is not continuous at the nodes. This means that the Rayleigh–Ritz method is still not completely satisfactory.

3.3 The Mixed Galerkin Method

A complete discussion of the mixed Galerkin method is beyond the scope of this book even for problems in one dimension. In this section we offer a flavor of the mixed Galerkin method to illustrate its difficulties.

By introducing the flux $p = u'$ as an additional unknown variable, problem (3.1) can be decomposed into the following first-order system:

$$p - u' = 0 \qquad x \in [0,1], \tag{3.4a}$$

$$-p' = f(x) \qquad x \in [0,1], \tag{3.4b}$$

$$u(0) = u(1) = 0. \tag{3.4c}$$

Now multiplying the equation (3.4a) by an arbitrary test function $q \in S$ and integrating, and multiplying (3.4b) by $v \in H$ and integrating by parts, we are led to the mixed Galerkin formulation corresponding to (3.4): Find $u \in H = \{v \in H^1(0,1) : v(0) = 0, v(1) = 0\}$ and $p \in S = \{q \in L_2(0,1) : \int_0^1 q dx = 0\}$, such that

$$(p,q) - (u',q) = 0 \qquad \forall q \in S, \tag{3.5a}$$

$$-(p,v') = -(f,v) \qquad \forall v \in H. \tag{3.5b}$$

It is not difficult to verify that the solution of (3.5) provides a stationary value of the following functional:

$$L : H \times S \to \mathbb{R}$$

$$L(v,q) = \frac{1}{2}(q,q) - (f,v) + (v' - q, q). \tag{3.6}$$

In fact, it can be shown (see, e.g., Carey and Oden 1983a, p.106) that the solution pair $\{u,p\}$ is a saddle-point of the functional (3.6).

Let us now study first the stability of the problem (3.5). Here, stability means that the solution (u,p) depends continuously on the force term f, that is, if the force varies a little, so does the solution. A natural stability inequality for (3.5) would be the following: There is a constant C such that if $\{u,p\} \in H \times S$ satisfies (3.5), then

$$\|u\|_1 + \|p\|_0 \leq C\|f\|_{-1}, \tag{3.7}$$

where

$$\|f\|_{-1} = \sup_{0 \neq v \in H} \frac{(f,v)}{\|v\|_1} \tag{3.8}$$

in which "sup" denotes the least upper bound.

To be able to conclude (3.7), we need the Ladyzhenskaya–Babuška–Brezzi (LBB) condition (Babuška 1971, Brezzi 1974): Given any $q \in S$, there exists a constant $\gamma > 0$ such that

$$\sup_{0 \neq v \in H} \frac{(q, v')}{\|v\|_1} \geq \gamma \|q\|_0, \tag{3.9}$$

where γ may be chosen independent of the particular choice of $q \in S$.

This condition may be equivalently expressed in the form: Given any $q \in S$, there exists a nonzero $v \in H$ such that

$$(q, v') \geq \gamma \|q\|_0 \|v\|_1. \tag{3.10}$$

Of course, for each q a different v may be chosen in order to satisfy (3.10).

The estimate (3.7) is obtained as follows: Choosing such a v in (3.5b) that the LBB condition (3.10) holds, we have

$$\gamma \|p\|_0 \|v\|_1 \leq (p, v') = (f, v) \leq \|f\|_{-1} \|v\|_1.$$

Dividing both sides of the above inequality by $\|v\|_1$ leads to

$$\gamma \|p\|_0 \leq \|f\|_{-1}. \tag{3.11}$$

Taking $q = u'$ in (3.5a) and $v = u$ in (3.5b) we have

$$(p, u') - (u', u') = 0, \tag{3.12a}$$

$$(p, u') = (f, u). \tag{3.12b}$$

From (3.12) and the Friedrichs inequality (2.8) we obtain

$$\|u\|_1^2 = \|u\|_0^2 + |u|_1^2 \leq 2|u|_1^2 = 2(p, u') = 2(f, u) \leq 2\|f\|_{-1} \|u\|_1.$$

Dividing both sides of the above inequality by $\|u\|_1$ leads to

$$\|u\|_1 \leq 2\|f\|_{-1}. \tag{3.13}$$

Combining (3.11) and (3.13) yields the stability estimate (3.7). Now the importance of the LBB condition becomes clear: the LBB condition is sufficient to guarantee the stable solution of the mixed Galerkin method.

In the following let us show that the LBB condition (3.10) is valid. For a given $q \in S$, let

$$v' = q, \tag{3.14}$$

then

$$v = \int_0^x q(\xi) d\xi, \tag{3.15}$$

and $v(0) = v(1) = 0$ due to the fact that q satisfies the constraint $\int_0^1 q(\xi) d\xi = 0$, that is, v belongs to H. Obviously,

$$\|v\|_0^2 = \int_0^1 \left(\int_0^x q(\xi) d\xi \right)^2 dx \leq \|q\|_0^2. \tag{3.16}$$

We have

$$\|v\|_1^2 = \|v\|_0^2 + \|v'\|_0 = \|v\|_0^2 + \|q\|_0^2 \leq 2\|q\|_0^2.$$

Therefore,

$$(q, v') = \|q\|_0^2 \geq \frac{1}{\sqrt{2}} \|q\|_0 \|v\|_1. \tag{3.17}$$

That is, the LBB condition (3.10) indeed holds and $\gamma = 1/\sqrt{2}$.

To further understand the features of the mixed Galerkin method, we consider a discrete solution of (3.5). A natural idea to obtain a discrete analogue of (3.5) is now to replace H and S by finite element subspace H_h and S_h which satisfy the discrete LBB condition:

$$\sup_{0 \neq v_h \in H_h} \frac{(q_h, v_h')}{\|v_h\|_1} \geq \gamma \|q_h\|_0 \qquad \forall q_h \in S_h. \tag{3.18}$$

This gives the following mixed finite element method: Find $u_h \in H_h$ and $p_h \in S_h$ such that

$$(p_h, q_h) - (u_h', q_h) = 0 \qquad \forall q_h \in S_h, \tag{3.19a}$$

$$-(p_h, v_h') = -(f, v_h) \qquad \forall v_h \in H_h. \tag{3.19b}$$

Introducing the basis functions $\{\phi_1, \cdots, \phi_n\}$ and $\{\psi_1, \cdots, \psi_m\}$ for S_h and H_h, respectively, the discrete problem (3.19) can be written in matrix form:

$$\begin{pmatrix} \boldsymbol{A} & -\boldsymbol{B} \\ -\boldsymbol{B}^T & \boldsymbol{0} \end{pmatrix} \begin{pmatrix} \boldsymbol{P} \\ \boldsymbol{U} \end{pmatrix} = \begin{pmatrix} \boldsymbol{0} \\ \boldsymbol{F} \end{pmatrix}, \tag{3.20}$$

where

$$\boldsymbol{A} = \begin{pmatrix} a_{11} & \dots & a_{1n} \\ \vdots & \ddots & \vdots \\ a_{n1} & \dots & a_{nn} \end{pmatrix}, \qquad \boldsymbol{B} = \begin{pmatrix} b_{11} & \dots & b_{1m} \\ \vdots & \ddots & \vdots \\ b_{n1} & \dots & b_{nm} \end{pmatrix},$$

$$\boldsymbol{P} = \begin{pmatrix} p_1 \\ \vdots \\ p_n \end{pmatrix}, \boldsymbol{U} = \begin{pmatrix} u_1 \\ \vdots \\ u_m \end{pmatrix}, \boldsymbol{F} = \begin{pmatrix} f_1 \\ \vdots \\ f_m \end{pmatrix},$$

with $a_{ij} = (\phi_i, \phi_j)$, $b_{ij} = (\phi_i, \psi_j')$ and $f_i = -(\psi_i, f)$

In order for (3.20) to have a unique solution, the global matrix in (3.20) should have a full rank (see Sect. 4.9.1 for the definition of rank). Therefore, the row vectors of $(\boldsymbol{B}^T, \boldsymbol{0})$ must be linearly independent, that is,

$$\text{rank}\boldsymbol{B} = \text{rank}\boldsymbol{B}^T = m.$$

Since the matrix $\boldsymbol{B}$ has n rows, we must have

$$\text{rank}\boldsymbol{B} = m \leq n. \tag{3.21}$$

This means that in order for $\boldsymbol{P}$ and $\boldsymbol{U}$ to be uniquely determined from (3.20), we should have

$$\dim H_h \leq \dim S_h.$$

We shall show that if the rank condition (3.21) holds, then (3.20) has a unique solution. Here we need only to prove that, under the condition (3.21), the solution of the homogeneous version of (3.20) is null. To do this we assume that $\{\boldsymbol{P}, \boldsymbol{U}\}$ is the solution of the corresponding homogeneous equation of (3.20), then $\boldsymbol{B}^T\boldsymbol{P} = \boldsymbol{0}$. Multiplying the upper part in (3.20) by $\boldsymbol{P}^T$ yields

$$\boldsymbol{P}^T\boldsymbol{A}\boldsymbol{P} - \boldsymbol{P}^T\boldsymbol{B}\boldsymbol{U} = 0.$$

Therefore,

$$\boldsymbol{P}^T\boldsymbol{A}\boldsymbol{P} = 0.$$

Since $\boldsymbol{A}$ is positive-definite, we have $\boldsymbol{P} = \boldsymbol{0}$, and thus $\boldsymbol{B}\boldsymbol{U} = \boldsymbol{0}$. Due to condition (3.21), $\boldsymbol{U} = \boldsymbol{0}$.

In the above we have proved that the rank condition (3.21) is a sufficient and necessary condition for (3.20) to have a unique solution.

We now investigate the relationship between the rank condition (3.21) and the following condition:

$$\boldsymbol{P}^T\boldsymbol{B}\boldsymbol{U} > 0. \tag{3.22}$$

If condition (3.21) holds, then the only solution of $\boldsymbol{B}\boldsymbol{U} = \boldsymbol{0}$ is $\boldsymbol{U} = \boldsymbol{0}$. If $\boldsymbol{U} \neq \boldsymbol{0}$, then $\boldsymbol{B}\boldsymbol{U} \neq \boldsymbol{0}$. Therefore, there exists a $\boldsymbol{P} \in \mathbb{R}^n$, such that (3.22) is valid. Conversely, if (3.21) does not hold, i.e., $n < m$, then the equation $\boldsymbol{B}\boldsymbol{U} = \boldsymbol{0}$ has a non-zero solution $\boldsymbol{U}$. In such a case, for all $\boldsymbol{P} \in \mathbb{R}^n$, we have $\boldsymbol{P}^T\boldsymbol{B}\boldsymbol{U} = 0$, and thus (3.22) is not valid. Therefore (3.22) and (3.21) are equivalent.

It is clear that if condition (3.21) does not hold, then $\boldsymbol{P}^T\boldsymbol{B}\boldsymbol{U} = (q_h, v_h') = 0$, i.e., the discrete LBB condition cannot be realized. If the rank condition (3.21) holds, then $\boldsymbol{P}^T\boldsymbol{B}\boldsymbol{U} = (q_h, v_h') > 0$ which does not mean that the discrete LBB condition (3.18) is verified. In other words the rank condition (3.21) is only a necessary condition for the discrete LBB condition (3.18) to hold. The rank condition (3.21) guarantees only the existence of the discrete solution, but does not guarantee the stability of the solution. As a comment we note that the parameter γ appearing in the condition (3.18) in general is related to the mesh size h. Once the parameter γ in (3.18) is proved to be unrelated to h, the error in the finite element solution will depend only on the ability of the chosen finite element subspaces to approximate the exact solution. However, for particular choices of S_h and H_h, it is not an easy matter to verify whether the LBB condition holds or not. There are different ways in which arbitrarily chosen finite element spaces may fail to satisfy the LBB condition. In particular, mixed methods with equal-order finite elements are, in general, unstable in that the stability parameter γ tends to zero as $h \to 0$.

Now we investigate some particular finite element pairs of S_h and H_h. In all cases let the interval $[0, 1]$ be divided into N cells ($N > 3$). An example of instability is the following seemingly natural choice. For the v approximations, we choose H_h = set of piecewise continuous quadratic functions. For discrete

q, we choose S_h = set of piecewise linear functions. In this case, $\dim H_h = 2N-2 > \dim S_h = N+1$, so the rank condition (3.21) is violated.

We next consider equal-order finite elements. We now choose both H_h and S_h to consist of piecewise linear functions. One easily sees that $\dim H_h = N-1 < \dim S_h = N+1$. That is, the rank condition (3.21) is satisfied. However, for $q_h = \{1, -1, 1, -1, ...\}$(i.e., the nodal values are 1, -1, 1, -1,...), we have that $(q_h, v_h') = 0$ for all $v_h \in H_h$(since in each element v_h' is a constant, $\int_{\Omega_e} q_h v_h' dx = 0$), so that $\gamma = 0$ in (3.18). That is, the linear-linear pair fails to pass the LBB condition.

A good pair is the linear v and the constant q. Since $\dim H_h = N-1 < \dim S_h = N$, the rank condition is satisfied. The satisfaction of the discrete LBB condition can also be verified by the same argument as in the continuous case.

Provided that the LBB condition is satisfied, the following error estimate can be derived (see, e.g., Brezzi and Fortin 1991, Roberts and Thomas 1991):

$$\|u-u_h\|_1 + \|p-p_h\|_0 \le C(\|u-\Pi_h u\|_1 + \|p-\Pi_h p\|_0), \tag{3.23}$$

where $\Pi_h u \in H_h$ is the interpolant of u, and $\Pi_h p \in S_h$ is the interpolant of p.

If we choose H_h = set of piecewise linear continuous functions and S_h = set of piecewise constant functions, then the rate of convergence is (Roberts and Thomas 1991, p.578):

$$\|e_u\|_1 + \|e_p\|_0 \le Ch(|u|_2 + |p|_1). \tag{3.24}$$

From (3.23) we see that the objective of getting better accuracy for the dual variable p is not achieved by the mixed Galerkin method (3.5). One may say that the mixed Galerkin method brings more troubles than benefits, at least for the second-order diffusion problems. (For the fourth-order elliptic problems, such as, for beam, plate and shell problems, the use of the mixed Galerkin method is of some advantage, since it can avoid the need for C^1 elements. The definition of C^1 element will be given in Sect. 4.5.1.) Mixed Galerkin methods remain a delicate class of finite element methods in that it is not easy to choose the approximation spaces and verify the satisfaction of the LBB condition. The resulting matrix in (3.20) is not positive-definite, and thus it is difficult to solve (3.20) iteratively in an efficient way for large-scale problems.

3.4 The Least-Squares Finite Element Method

3.4.1 The Least-Squares Formulation

Before introducing the least-squares formulation let us recall a few simple concepts. Assume that V is a linear space. We say that L is a linear form on V, if $L : V \to \mathbb{R}$, i.e., $L(v) \in \mathbb{R}$ for $v \in V$, and L is linear, i.e., for all $v, w \in V$ and $\beta, \theta \in \mathbb{R}$

$$L(\beta v + \theta w) = \beta L(v) + \theta L(w).$$

Furthermore, we say that B(u,v) is a bilinear form on $V \times V$, if $B : V \times V \to \mathbb{R}$, i.e., $B(u, v) \in \mathbb{R}$ for $u, v \in V$, and B is linear in each argument, i.e., for all $u, v, w \in V$ and $\beta, \theta \in \mathbb{R}$ we have

$$B(u, \beta v + \theta w) = \beta B(u, v) + \theta B(u, w),$$

$$B(\beta u + \theta v, w) = \beta B(u, w) + \theta B(v, w).$$

The bilinear form $B(u, v)$ on $V \times V$ is said to be symmetric, if

$$B(u, v) = B(v, u) \qquad \forall u, v \in V.$$

Now let us consider the least-squares method for the solution of (3.4) which can be written as

$$\boldsymbol{Au} = \boldsymbol{f}, \tag{3.25}$$

in which

$$\boldsymbol{Au} = \begin{pmatrix} 0 & -1 \\ -1 & 0 \end{pmatrix} \frac{\partial \boldsymbol{u}}{\partial x} + \begin{pmatrix} 1 & 0 \\ 0 & 0 \end{pmatrix} \boldsymbol{u},$$

$$\boldsymbol{u} = \begin{pmatrix} p \\ u \end{pmatrix}, \qquad \boldsymbol{f} = \begin{pmatrix} 0 \\ f \end{pmatrix}.$$

We assume that $f \in L_2(0, 1)$.

We construct the following quadratic functional that is defined in terms of the L_2 norms of the equation residuals:

$$I : X \times H \to \mathbb{R},$$

$$\begin{aligned} I(p, u) &= \frac{1}{2}\|\boldsymbol{Au} - \boldsymbol{f}\|_0^2 \\ &= \frac{1}{2}(\boldsymbol{Au} - \boldsymbol{f}, \boldsymbol{Au} - \boldsymbol{f}) \\ &= \frac{1}{2}\int_0^1 \left[(p - u')^2 + (p' + f)^2\right] dx, \end{aligned}$$

where $X = \{q \in H^1(0, 1)\}$ and $H = \{v \in H^1(0, 1) : v(0) = 0, v(1) = 0\}$. A necessary condition that $\boldsymbol{u} = \{p, u\} \in X \times H$ be a minimizer of the functional I is that its first variation vanishes at $\boldsymbol{u}$ for all admissible $\boldsymbol{v} = \{q, v\} \in X \times H$. This leads to the least-squares variational formulation: Find $\boldsymbol{u} = \{p, u\} \in X \times H$ such that

$$\begin{aligned} \lim_{t \to 0} \frac{d}{dt} I(\boldsymbol{u} + t\boldsymbol{v}) &\equiv (\boldsymbol{Au} - \boldsymbol{f}, \boldsymbol{Av}) \\ &\equiv \int_0^1 [(p - u')(q - v') + (p' + f)q']\, dx = 0, \end{aligned}$$

$$\forall \boldsymbol{v} = \{q, v\} \in X \times H. \tag{3.26}$$

Let

$$B(\boldsymbol{u}, \boldsymbol{v}) = (\boldsymbol{A}\boldsymbol{u}, \boldsymbol{A}\boldsymbol{v}) = (p - u', q - v') + (p', q'), \tag{3.27}$$

$$L(\boldsymbol{v}) = (\boldsymbol{f}, \boldsymbol{A}\boldsymbol{v}) = -(f, q'), \tag{3.28}$$

then (3.26) can be written as

$$(\boldsymbol{A}\boldsymbol{u}, \boldsymbol{A}\boldsymbol{v}) = (\boldsymbol{f}, \boldsymbol{A}\boldsymbol{v}) \qquad \forall \boldsymbol{v} = \{q, v\} \in X \times H, \tag{3.29a}$$

or

$$B(\boldsymbol{u}, \boldsymbol{v}) = L(\boldsymbol{v}) \qquad \forall \boldsymbol{v} = \{q, v\} \in X \times H. \tag{3.29b}$$

3.4.2 Stability Estimate

We shall now analyze the least-squares method (3.29). The reader will fully understand these analyses after studying the bounded inverse theorem in Chap. 4. The analysis can also be based on the famous Lax–Milgram theorem (see Appendix D). Formulation (3.29a) is convenient for the analysis based on the bounded inverse theorem, while (3.29b) is popular for the application of the Lax–Milgram theorem. For both approaches the essential issue in the analyses is the same: the boundedness of the operator $\boldsymbol{A}$ or the coerciveness of the bilinear form $B(\boldsymbol{u}, \boldsymbol{v})$. In the following we use both notations.

Apparently, $B(\boldsymbol{u}, \boldsymbol{v})$ is symmetric. It is straightforward to verify that

$$|(\boldsymbol{A}\boldsymbol{u}, \boldsymbol{A}\boldsymbol{v})| = |B(\boldsymbol{u}, \boldsymbol{v})| \leq C\|\boldsymbol{u}\|_1 \cdot \|\boldsymbol{v}\|_1, \tag{3.30}$$

where $\|\boldsymbol{u}\|_1^2 = \|p\|_1^2 + \|u\|_1^2$. Therefore $\boldsymbol{A}$ or $B(\cdot, \cdot)$ is bounded on $X \times H$.

Now let us prove that $\boldsymbol{A}$ is bounded below or $B(\boldsymbol{v}, \boldsymbol{v})$ is coercive: There exists a constant $\alpha > 0$ such that

$$\|\boldsymbol{A}\boldsymbol{v}\|_0^2 = B(\boldsymbol{v}, \boldsymbol{v}) \geq \alpha\|\boldsymbol{v}\|_1^2 \qquad \forall \boldsymbol{v} \in X \times H. \tag{3.31}$$

We note that

$$B(\boldsymbol{v}, \boldsymbol{v}) = \|q - v'\|_0^2 + \|q'\|_0^2. \tag{3.32}$$

Consequently,

$$B(\boldsymbol{v}, \boldsymbol{v}) \geq \|q'\|_0^2, \tag{3.33}$$

$$B(\boldsymbol{v}, \boldsymbol{v}) \geq \|q - v'\|_0^2. \tag{3.34}$$

From (3.32) we have

$$\begin{aligned} B(\boldsymbol{v}, \boldsymbol{v}) &= \|q'\|_0^2 + \|q\|_0^2 + \|v'\|_0^2 - 2(q, v') \\ &\geq \|q'\|_0^2 + \|q\|_0^2 + \|v\|_0^2 + 2(q', v), \end{aligned}$$

where in the second step we have used integration-by-parts, the boundary conditions (3.4c), and the Friedrichs inequality (2.8), i.e., $\|v'\|_0^2 \geq \|v\|_0^2$, $\forall v \in H$. Thus

$$B(\boldsymbol{v},\boldsymbol{v}) \geq \|q' + v\|_0^2 + \|q\|_0^2. \tag{3.35}$$

From (3.35) we have

$$B(\boldsymbol{v},\boldsymbol{v}) \geq \|q' + v\|_0^2, \tag{3.36}$$

$$B(\boldsymbol{v},\boldsymbol{v}) \geq \|q\|_0^2. \tag{3.37}$$

The combination of (3.33) and (3.36) leads to

$$2(B(\boldsymbol{v},\boldsymbol{v}))^{1/2} \geq \|q'\|_0 + \|q' + v\|_0 = \|-q'\|_0 + \|q' + v\|_0 \geq \|v\|_0,$$

so that

$$4B(\boldsymbol{v},\boldsymbol{v}) \geq \|v\|_0^2. \tag{3.38}$$

Similarly, the combination of (3.34) and (3.37) gives

$$2\left(B(\boldsymbol{v},\boldsymbol{v})\right)^{1/2} \geq \|q - v'\|_0 + \|q\|_0 = \|v' - q\|_0 + \|q\|_0 \geq \|v'\|_0,$$

so that

$$4B(\boldsymbol{v},\boldsymbol{v}) \geq \|v'\|_0^2. \tag{3.39}$$

By combining (3.33), (3.37), (3.38) and (3.39), we finally obtain the boundedness (below), or the coerciveness

$$\|\boldsymbol{A}\boldsymbol{v}\|_0^2 = B(\boldsymbol{v},\boldsymbol{v}) \geq \frac{1}{10}\left(\|q\|_1^2 + \|v\|_1^2\right) = \frac{1}{10}\|\boldsymbol{v}\|_1^2. \tag{3.40}$$

Using (3.40) we can obtain the stability estimate of the solution $\boldsymbol{u}$ as follows. We have

$$\begin{aligned}\frac{1}{10}\|\boldsymbol{u}\|_1^2 \leq \|\boldsymbol{A}\boldsymbol{u}\|_0^2 &= B(\boldsymbol{u},\boldsymbol{u}) = L(\boldsymbol{u}) = (\boldsymbol{f},\boldsymbol{A}\boldsymbol{u}) \\ &= -(f,p') \leq \|f\|_0\|p'\|_0 \leq \|f\|_0\|\boldsymbol{u}\|_1.\end{aligned}$$

Dividing by $\|\boldsymbol{u}\|_1$ and squaring we obtain the desired stability estimate

$$(\|p\|_1^2 + \|u\|_1^2) \leq C\|f\|_0^2, \tag{3.41a}$$

or

$$(\|p\|_1 + \|u\|_1) \leq C\|f\|_0. \tag{3.41b}$$

Equation (3.41) shows that if $f \in L_2(0,1)$, then a unique and stable solution exists in $X \times H$. If f is a Dirac delta function and thus is not square-integrable, the least-squares method is not valid. This is one of the essential differences between the least-squares formulation and the mixed Galerkin formulation. Fortunately, in fluid dynamics the case in which the body force f is a Dirac delta function is seldom seen.

In the model problem (3.1) only the boundary conditions on u itself are given. When the problem (3.1) is reduced to the first-order system (3.4) by introducing the flux p, no boundary condition on p is required. This is an important observation and is contrary to a common misunderstanding that says "whenever a new variable is introduced to reduce a higher order differential equation to a lower order system, a boundary condition on this new variable is needed."

For the same problem with other boundary conditions such as:

(1) $u(0) = 0, \quad p(1) = 0$;

(2) $u(0) = 0, \quad p(0) = 0$.

We can prove the boundedness (below) of the $\boldsymbol{A}$ following the same procedure as above, see Jiang and Chang (1990).

The symmetry of the bilinear form $(\boldsymbol{Au}, \boldsymbol{Av})$ and the boundedness (below) of $\boldsymbol{A}$ guarantee that the matrix for the least-squares finite element method is symmetric and positive-definite. This is an important advantage of the least-squares method.

3.4.3 Error Analysis

The corresponding finite element problem is then to find $\boldsymbol{u}_h = \{p_h, u_h\} \in X_h \times H_h$ such that

$$(\boldsymbol{Au}_h, \boldsymbol{Av}_h) = (\boldsymbol{f}, \boldsymbol{Av}_h) \qquad \forall \boldsymbol{v}_h = \{q_h, v_h\} \in X_h \times H_h, \tag{3.42}$$

where

$$(\boldsymbol{Au}_h, \boldsymbol{Av}_h) = (p_h - u_h', q_h - v_h') + (p_h', q_h'), \tag{3.43}$$

$$(\boldsymbol{f}, \boldsymbol{Av}_h) = -(f, q_h'). \tag{3.44}$$

Since $X_h \times H_h \subset X \times H$ we have from (3.29) in particular

$$(\boldsymbol{Au}, \boldsymbol{Av}_h) = (\boldsymbol{f}, \boldsymbol{Av}_h) \qquad \forall \boldsymbol{v}_h = \{q_h, v_h\} \in X_h \times H_h, \tag{3.45}$$

so that after subtracting (3.42) we obtain the orthogonal condition

$$(\boldsymbol{A}(\boldsymbol{u} - \boldsymbol{u}_h), \boldsymbol{Av}_h) = 0 \qquad \forall \boldsymbol{v}_h = \{q_h, v_h\} \in X_h \times H_h. \tag{3.46}$$

Let $\Pi_h \boldsymbol{u} = \{\Pi_h p, \Pi_h u\}$ be the interpolant of $\boldsymbol{u}$, by using (3.40) and (3.46) we have

$$\begin{aligned}
\frac{1}{10}\|\boldsymbol{u} - \boldsymbol{u}_h\|_1^2 &\le (\boldsymbol{A}(\boldsymbol{u} - \boldsymbol{u}_h), \boldsymbol{A}(\boldsymbol{u} - \boldsymbol{u}_h)) \\
&= (\boldsymbol{A}(\boldsymbol{u} - \boldsymbol{u}_h), \boldsymbol{A}(\boldsymbol{u} - \Pi_h \boldsymbol{u})) \\
&\quad + (\boldsymbol{A}(\boldsymbol{u} - \boldsymbol{u}_h), \boldsymbol{A}(\Pi_h \boldsymbol{u} - \boldsymbol{u}_h)) \\
&= (\boldsymbol{A}(\boldsymbol{u} - \boldsymbol{u}_h), \boldsymbol{A}(\boldsymbol{u} - \Pi_h \boldsymbol{u})) \\
&\le C\|\boldsymbol{u} - \boldsymbol{u}_h\|_1 \|\boldsymbol{u} - \Pi_h \boldsymbol{u}\|_1,
\end{aligned}$$

where the last inequality follows from (3.30). Dividing the above inequality by $\|\boldsymbol{u} - \boldsymbol{u}_h\|_1$ we obtain

$$\|\boldsymbol{u} - \boldsymbol{u}_h\|_1 \leq C\|\boldsymbol{u} - \Pi_h \boldsymbol{u}\|_1. \tag{3.47}$$

From (3.47) we can obtain the error estimate

$$\|p - p_h\|_1 + \|u - u_h\|_1 \leq C(\|p - \Pi_h p\|_1 + \|u - \Pi_h u\|_1). \tag{3.48}$$

We have proved the following theorem about the rate of convergence of the least-squares finite element solution.

Theorem 3.1 Assume f is smooth enough and the finite element interpolation error estimates hold (see Sect. 4.6.3); that is,

$$\|p - \Pi_h p\|_1 \leq C_p h^{\ell} |p|_{\ell+1}, \qquad \|u - \Pi_h u\|_1 \leq C_u h^m |u|_{m+1}, \tag{3.49}$$

then

$$\|p - p_h\|_1 + \|u - u_h\|_1 \leq Ch^r \left(|p|_{\ell+1} + |u|_{m+1}\right), \tag{3.50}$$

where ℓ and m denote the orders of polynomials for p and u, respectively; $r = \min(\ell, m)$ and C_p, C_u and C are the constants which do not depend on the mesh size h.

We may utilize the Aubin–Nitsche trick (see Sect. 2.9) to obtain the optimal L_2-estimates of the error:

$$\|p - p_h\|_1 + \|u - u_h\|_1 \leq Ch^{r+1} \left(|p|_{\ell+1} + |u|_{m+1}\right). \tag{3.51}$$

We remark that for the least-squares method the choice of interpolations for p and u is not subject to any restriction as long as $\ell \geq 1$, $m \geq 1$. In particular, the equal-order finite elements are permissible. For example, if the linear finite element is chosen, the accuracy of the least-squares solution is of $O(h^2)$ for both p and u. However, inspection of the equation (3.4a) shows that in order to make the residual of this equation equal to zero throughout, one may choose the interpolation for u as one order higher than that for p, such as a pair of quadratic (u) and linear (p) elements. As seen in Sect. 3.3, this reasonable combination is not allowed in the mixed Galerkin method.

3.4.4 Numerical Results

To show the convergence of the least-squares finite element method, we performed numerical experiments for three simple, one-dimensional, boundary-value problems. We used a uniform mesh containing elements of length h. We are interested in the behavior of the error $e_p = p - p_h$ and $e_u = u - u_h$ in L_2 norm for various choices of polynomials of degree ℓ for p_h and m for u_h. Results of numerical experiments are shown in Fig. 3.1 and summarized in Table 3.1. We obtained the rate of convergence of the method by calculating the norm of the error for each h, plotting $\log \|\text{error}\|$ versus $\log h$, and calculating the slope of this line. All of the computed rates of convergence agree with the estimate (3.51).

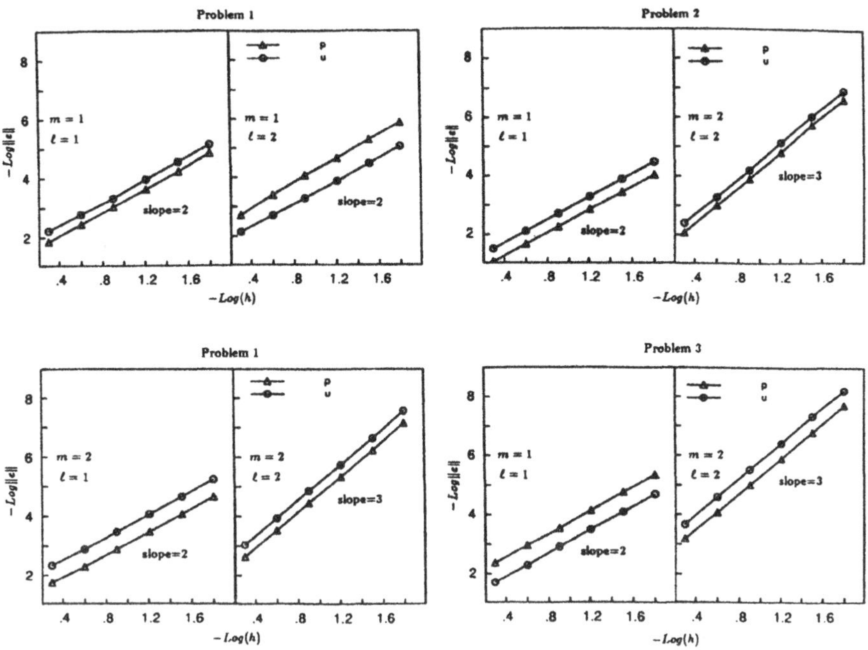

Fig. 3.1. Computed convergence rates for one-dimensional problems (From Jiang and Chang 1988)

Table 3.1. Computed convergence rates for model problems (From Jiang and Chang 1988)

Problem 1: $u' = p, \quad p' = -x^3, \quad u(0) = u(1) = 0$

order ℓ	order m	$\|e_p\|$	$\|e_u\|$
1	1	$0(h^2)$	$0(h^2)$
2	1	$0(h^2)$	$0(h^2)$
1	2	$0(h^2)$	$0(h^2)$
2	2	$0(h^3)$	$0(h^3)$

Problem 2: $u' + 3u = p,\ p' - 2u = -2x^2 + 6x - 2,\ u(0) = u(1) = 0$
Problem 3: $-(a(x)u)' = p, \quad p' = f, \quad u(0) = u(1) = 0$

$$a(x) = \tfrac{1}{\alpha} + \alpha(x - x_0)^2$$

$$f(x) = 2 + 2\alpha(x - x_0)\{\tan^{-1}[\alpha(x - x_0)] + \tan^{-1}(\alpha x_0)\}$$

$$\alpha = 0.5, \ x_0 = 0.5$$

order ℓ	order m	$\|e_p\|$	$\|e_u\|$
1	1	$0(h^2)$	$0(h^2)$
2	2	$0(h^3)$	$0(h^3)$

3.5 Concluding Remarks

For the solution of diffusion-type differential equations, both the mixed Galerkin method and the least-squares method are based on the first-order system generated by introducing the dual variable p (the flux) as an additional unknown. The mixed Galerkin method leads to a saddle-point problem, so that the choice of elements for p and u must satisfy the Ladyzhenskaya–Babuška–Brezzi condition, and in general equal-order elements are not allowed. Moreover, the system of linear algebraic equations resulting from the mixed Galerkin method is not positive-definite and thus it is not easy to solve iteratively in an efficient way.

The least-squares method leads to a minimization problem and thus the interpolations for p and u can be independently chosen without the limitation of the LBB condition. In particular, the equal-order interpolations which are simple in implementation are permissible and thus both p and u can have the same accuracy. The most important advantage of the least-squares method is that the resulting matrix is symmetric and positive-definite, and thus a simple iterative method such as the conjugate gradient method can be used to solve the problem efficiently.

The idea, analysis and numerical results of the least-squares finite element method presented in this section first appeared in Jiang and Chang (1988, 1990). A similar analysis for slightly more general first-order systems of ordinary differential equations including superconvergence estimates for the solution at the interelement nodes can be found in Pehlivanov et al. (1993).

Part II

Fundamentals of LSFEM

4. Basis of LSFEM

In this chapter an attempt is made to present a unified theory and formulation of the least-squares finite element method for multi-dimensional problems so that in principle, it is not necessary to repeat the same argument in different instances. In the following chapters, these theorems, inequalities and formulations will be extensively utilized. The mathematics is kept as simple as possible, since our purpose is to become familiar with certain modern concepts that will help readers understand the basic theory, formulation and properties of the least-squares method.

4.1 Function Spaces

In order to give a precise definition of the least-squares finite element method for two- and three-dimensional problems, we need to review some basic concepts of function spaces and associated norms. More detailed accounts concerning these spaces may be found in, e.g., Prenter (1975), Oden and Demkowicz (1996).

A linear space $\mathcal{U}$ equipped with a norm is called a normed space and is denoted by the pair $\{\mathcal{U},\ \|\cdot\|_u\}$. We use this notation to emphasize that the particular norm $\|\cdot\|_u$ is associated with the space $\mathcal{U}$. Whenever it is clear what norm belongs to $\mathcal{U}$, we write $\mathcal{U}$ instead of $\{\mathcal{U},\ \|\cdot\|_u\}$.

We recall that a Cauchy sequence in $\mathcal{U}$ is a sequence $u_1, u_2, u_3, ...,$ of elements in $\mathcal{U}$ which satisfies the following property: For any $\epsilon > 0$ there exists a natural number N such that the distance $\|u_i - u_j\|_u < \epsilon$ if $i, j > N$. Further, a sequence in $\mathcal{U}$ is said to converge to an element u if the distance $\|u_i - u\|_u \to 0$ as $i \to \infty$. A space $\mathcal{U}$ is said to be complete if any Cauchy sequence converges to a $u \in \mathcal{U}$. A complete normed space is called a Banach space.

Let $\Omega \subset \mathbb{R}^{n_d}$ be an open bounded domain with a piecewise smooth boundary Γ, where $n_d = 2$ or 3 represents the number of space dimensions, and $\boldsymbol{x} = (x_1, ..., x_{n_d})$, or (x, y, z) be a point in Ω. By a piecewise smooth boundary, we mean that: (1) Ω lies one side of Γ and Γ can be represented by a finite number of arcs, each of which is the graph of an infinitely differentiable function with respect to a suitable chosen local coordinate system; (2) the interior angles between the left and right tangents at the breakpoints of the arcs are greater than zero (see, e.g., Oden and Reddy 1976, p.59 or Křížek and Neittaanmäk 1990, p.4). A domain with a piecewise smooth boundary represents a sufficiently wide class of domains necessary for most practical purposes. Simple domains such as a ball, cube, torus, triangle, polygon, polyhedron,..., etc., certainly have a piecewise smooth boundary.

As usual, $L_p(\Omega)$, $1 \leq p \leq \infty$ denotes the space of of functions u defined on Ω whose absolute value have pth powers which are Lebesque-integrable on Ω. The norm on $L_p(\Omega)$ is given by

$$\|u\|_{L_p(\Omega)} = \left(\int_\Omega |u|^p d\Omega \right)^{1/p}.$$

The Lebesque space $L_p(\Omega)$ is a Banach spaces. For the purpose of this book the spaces $L_1(\Omega)$ and $L_2(\Omega)$ are of particular importance.

A Banach space $\mathcal{U}$ is called a Hilbert space, if there exists an inner product $(\cdot, \ \cdot)$ on $\mathcal{U}$ such that $\|u\|_u = (u,u)^{1/2}$ for all $u \in \mathcal{U}$. We note that $L_1(\Omega)$ is not a Hilbert space.

$L_2(\Omega)$ is a Hilbert space, since $L_2(\Omega)$ is the space of square-integrable functions defined over Ω and equipped with the inner product

$$(u,v) = \int_\Omega uv d\Omega \qquad u,v \in L_2(\Omega)$$

and the norm

$$\|u\|_0 = (u,u)^{1/2} \qquad u \in L_2(\Omega).$$

As in Sect. 2.2 we can prove the Schwarz inequality

$$|(u,v)| \leq \|u\|_0 \|v\|_0. \tag{4.1}$$

Next, for any non-negative integer k, we define the Sobolev space as:

$$H^k(\Omega) = \{u \in L_2(\Omega) : D^{\boldsymbol{\alpha}} u \in L_2(\Omega), \ \text{for } |\boldsymbol{\alpha}| \leq k\}$$

where we use the multi-index notation,

$$D^{\boldsymbol{\alpha}} u = \frac{\partial^{|\alpha|} u}{\partial x_1^{\alpha_1} \partial x_2^{\alpha_2} ... \partial x_{n_d}^{\alpha_{n_d}}},$$

$$|\boldsymbol{\alpha}| = |(\alpha_1, ..., \alpha_{n_d})| = \alpha_1 + \alpha_2 + ... \alpha_{n_d}.$$

Thus $H^k(\Omega)$ consists of functions whose derivatives up to order k are square-integrable. $H^k(\Omega)$ is equipped with the norm

$$\|u\|_k = (\|u\|_0^2 + \sum_{|\boldsymbol{\alpha}| \leq k} \|D^{|\boldsymbol{\alpha}|} u\|_0^2)^{1/2}. \tag{4.2}$$

We shall note the fact that $H^k(\Omega)$ is also a Hilbert space.

Clearly $H^0(\Omega) = L_2(\Omega)$. Of particular interest is the space $H^1(\Omega)$ consisting of functions with square-integrable first-order derivatives and its subspace

$$H_0^1(\Omega) = \{u \in H^1(\Omega) : u = 0 \text{ on } \Gamma\}$$

whose elements have square-integrable first-order derivatives over Ω and vanish on the boundary Γ. These spaces have the associated norm

$$\|u\|_1 = \left(\|u\|_0^2 + \sum_{i=1}^{n_d} \left\| \frac{\partial u}{\partial x_i} \right\|_0^2\right)^{1/2}. \tag{4.3}$$

For the function space $H^1(\Omega)$ we may also introduce the semi-norm

$$|u|_1 = \left(\sum_{i=1}^{n_d} \left\| \frac{\partial u}{\partial x_i} \right\|_0^2\right)^{1/2}. \tag{4.4}$$

Also we denote by $H^{-1}(\Omega)$ the dual space consisting of bounded linear functionals on $H_0^1(\Omega)$, i.e., $u \in H^{-1}(\Omega)$ implies that $(u, v) < \infty$ for all $v \in H_0^1(\Omega)$. A norm for $H^{-1}(\Omega)$ is given by

$$\|u\|_{-1} = \sup_{0 \neq v \in H_0^1(\Omega)} \frac{(u, v)}{|v|_1}, \tag{4.5}$$

where sup denotes the least upper bound.

We will also use the trace spaces, which consist of the restriction, to the boundary Γ, of functions belonging to $H^k(\Omega)$. For example, $H^{1/2}(\Gamma)$ consists of traces of functions belonging to $H^1(\Omega)$; a norm for functions belonging to $H^{1/2}(\Gamma)$ may be defined by

$$\|q\|_{1/2,\Gamma} = \inf_{\substack{u \in H^1(\Omega) \\ u=q \text{ on } \Gamma}} \|u\|_1, \tag{4.6}$$

where inf denotes the maximum lowest bound.

For the vector-valued function $\boldsymbol{u}$ with m components, we have the product spaces

$$\boldsymbol{L}_2(\Omega) = [L_2(\Omega)]^m = \{(u_1, u_2, ..., u_m) : u_j \in L_2(\Omega)\},$$

$$\boldsymbol{H}^1(\Omega) = [H^1(\Omega)]^m = \{(u_1, u_2, ..., u_m) : u_j \in H^1(\Omega)\},$$

and the corresponding norm

$$\|\boldsymbol{u}\|_0^2 = \sum_{j=1}^{m} \|u_j\|_0^2,$$

$$\|\boldsymbol{u}\|_1^2 = \sum_{j=1}^{m} \|u_j\|_1^2.$$

Also, the inner product for functions belonging to $\boldsymbol{L}_2(\Omega) = \boldsymbol{H}^0(\Omega) = [L_2(\Omega)]^m$ is given by

$$(\boldsymbol{u}, \boldsymbol{v}) = \int_\Omega \boldsymbol{u} \cdot \boldsymbol{v} d\Omega.$$

4.2 Linear Operators

Let us review some basic concepts and notations of linear operators. Let $\{\mathcal{U}, \|\cdot\|_u\}$ and $\{\mathcal{V}, \|\cdot\|_v\}$ denote two normed linear spaces. We recall that an operator A from $\mathcal{U}$ to $\mathcal{V}$ is linear if and only if

$$A(\alpha u_1 + \beta u_2) = \alpha A(u_1) + \beta A(u_2) \qquad \forall u_1, u_2 \in \mathcal{U}, \quad \forall \alpha, \beta \in \mathbb{R}. \tag{4.7}$$

If there exists a nonnegative real number M such that

$$\|Au\|_v \leq M\|u\|_u \qquad \forall u \in \mathcal{U}, \tag{4.8}$$

then the linear operator $A : \mathcal{U} \to \mathcal{V}$ is said to be bounded above. If for every sequence $\{u_n\}$ of elements in the domain $\mathcal{U}$ that converges to u relative to the norm $\|\cdot\|_u$, the sequence $\{Au_n\}$ converges to Au in $\mathcal{V}$ in the sense of the norm $\|\cdot\|_v$, then A is said to be continuous.

It is easy to show that continuity and boundedness (above) are equivalent. Moreover, the continuity of A is equivalent to the boundedness (above) of $\|Au\|_v$ on the unit sphere $\|u\|_u \leq 1$. In fact, if M is the upper bound of $\|Au\|_v$ on the unit sphere $\|u\|_u \leq 1$, then for any element u in the space $\mathcal{U}$, $u/\|u\|_u$ is on the surface of this unit sphere, therefore

$$\|A(u/\|u\|_u)\|_v \leq M\|(u/\|u\|_u)\|_u = M,$$

and thus

$$\|Au\|_v \leq M\|u\|_u.$$

If we take

$$M = \sup_{\|u\|_u \leq 1} \|Au\|_v,$$

i.e., M is the least upper bound of $\|Au\|_v$ on $\|u\|_u \leq 1$, then

$$\|Au\|_v \leq M\|u\|_u \quad \forall u \in \mathcal{U}.$$

This $\sup_{\|u\|_u \leq 1}\|Au\|_v$ is defined as the norm of the operator A and denoted by $\|A\|$.

Let us now choose a linear subspace Φ of the Hilbert space $L_2(\Omega)$ by imposing certain additional conditions which every function $u \in \Phi$ must satisfy. For example, we may require some specified smoothness conditions, boundary conditions on Γ, etc. These conditions, however, must be sufficient to guarantee that an operator A, if given, maps the subspace Φ into $L_2(\Omega)$. The subspace Φ is called the domain of the operator A and denoted by $\Phi(A)$.

Consider next the adjoint operator A^* defined by the identity

$$(Au, v) = (u, A^*v), \tag{4.9}$$

where $u \in \Phi(A), v \in \Phi(A^*)$.

The subspaces $\Phi(A)$ and $\Phi(A^*)$ of the Hilbert space $L_2(\Omega)$ do not coincide in general, despite the fact that the functions in these subspaces are defined on the same region Ω.

The operator A is called self-adjoint if $Au = A^*u$ for all $u \in \Phi(A)$ and $\Phi(A) = \Phi(A^*)$.

It is easy to verify that the first-order ordinary differential operators defined in (2.1) and (3.4) are not self-adjoint, but the second-order ordinary differential operator defined in (3.1) is self-adjoint. In fact, all first-order differential operators are not self-adjoint. This is the reason why the Galerkin method does not work well for first-order differential equations.

4.3 The Bounded Inverse Theorem

We shall now investigate why the problem of finding the solution of linear operator equations can be solved by minimizing a certain norm of the equation residuals.

We consider a linear operator equation

$$Au = f \qquad f \in \mathcal{V}. \tag{4.10}$$

Equation (4.10) may be a set of linear algebraic equations, linear ordinary differential equations, linear partial differential equations, or linear integral equations, etc. The solution of (4.10) is denoted by $u = A^{-1}f$; A^{-1} is called the inverse operator of A. If for any $f \in \mathcal{V}$ there exists one and only one $u \in \mathcal{U}$ such that $Au = f$, then $A : \mathcal{U} \to \mathcal{V}$ is called one-to-one.

In studying the abstract linear equation (4.10), we are often concerned with the existence, uniqueness and continuity of the solutions. We need to know whether (4.10) has a solution; if the solution exists, whether it is unique; if f is changed a little, whether the solution u is also changed a little. The following theorem answers these questions.

Theorem 4.1 (The Bounded Inverse Theorem) The sufficient and necessary condition for a linear operator $A : \mathcal{U} \to \mathcal{V}$ to have a continuous inverse operator is that A is bounded below, i.e., there exists a positive constant α such that

$$\alpha\|u\|_u \leq \|Au\|_v \qquad \forall u \in \mathcal{U}. \tag{4.11}$$

Proof: Necessity: Assume that A^{-1} exists and is continuous on $\mathcal{V}$. Then there is a constant $\alpha > 0$ such that $\|A^{-1}v\|_u \leq 1/\alpha\|v\|_v$. Setting $v = Au$ shows that A is bounded below.

Sufficiency: Suppose that A is bounded below. Then $u_1 \neq u_2$ means $Au_1 \neq Au_2$, therefore there is one and only one u in $\mathcal{U}$ for each Au in $\mathcal{V}$. Thus A^{-1} exists on $\mathcal{V}$. To show that A^{-1} is continuous, simply note that

$$\|A^{-1}v\|_u = \|u\|_u \leq \frac{1}{\alpha}\|Au\|_v = \frac{1}{\alpha}\|v\|_v.$$

Hence A^{-1} is bounded above, and thus continuous. □

The notion of bounded below operators provides for a simple interpretation of equivalence of norms. Normed spaces $\{\mathcal{U}, \|\cdot\|_u\}$ and $\{\mathcal{V}, \|\cdot\|_v\}$ are topologically equivalent if and only if there exists a linear operator $A: \mathcal{U} \to \mathcal{V}$, and positive constants α and M such that

$$\alpha\|u\|_u \leq \|Au\|_v \leq M\|u\|_u \qquad \forall u \in \mathcal{U}. \tag{4.12}$$

The interpretation of these inequalities is clear: $\alpha\|u\|_u \leq \|Au\|_v$ means that A is bounded below and, therefore, has a continuous inverse defined in $\mathcal{V}$; $\|Au\|_v \leq M\|u\|_u$ indicates that A is continuous.

The bounded inverse theorem is important in studying the existence of solutions of abstract linear equations on Banach spaces. The problem (4.10) is said to be well-posed, if there exists a unique solution u that depends continuously on the data f. If A is bounded below, then the problem (4.10) is well posed, because when A is bounded below, we have

$$\|u\|_u \leq \frac{1}{\alpha}\|Au\|_v = \frac{1}{\alpha}\|f\|_v,$$

i.e., the solution u clearly depends continuously on the data f.

The bounded inverse theorem is also useful in examining convergence and error estimate of approximate solutions. Suppose that an approximate solution u_n of (4.10) is obtained by a certain method. Since A is a linear operator, replacing u in (4.11) by $u_n - u$ results in

$$\|u_n - u\|_u \leq \frac{1}{\alpha}\|Au_n - Au\|_v.$$

If u is the exact solution of the equation, then $Au = f$ and the equation residual $R_n = Au_n - f$, therefore the distance between the approximate solution and the exact solution can be estimated by the equation residual:

$$\|u_n - u\|_u \leq \frac{1}{\alpha}\|Au_n - f\|_v. \tag{4.13}$$

This leads us to conclude that as the norm of the residual of the approximate solution approaches zero, i.e. $\|Au_n - f\|_v \to 0$, the approximate solution converges to the exact solution, i.e. $\|u_n - u\|_u \to 0$.

From the above investigation, we have in fact established a very important theorem which lays down the foundation of the general residual-minimization method for the solution of linear operator equations.

Theorem 4.2 (The Residual Minimization Theorem) Provided that a linear operator equation on a Banach space is well-posed, minimizing the equation residual measured in a proper norm always leads to a convergent solution.

In the light of Theorem 4.2 expressed mathematically by (4.13), the least-squares method discussed in this book, which minimizes the squared L_2 norm of the residual of first-order linear partial differential equations on Hilbert spaces, may be viewed as a special case of this very general method. The

L_1 method introduced in Sect. 9.2 for seeking discontinuous solution of first-order linear hyperbolic equations, and the H^{-1} method proposed by Bramble et al. (1997) for second-order elliptic problems also belong to this general least-squares method.

4.4 The Friedrichs Inequality

As in the one-dimensional case, the following general Friedrichs inequality and the Poincaré inequality play an essential role in showing that a particular linear first-order differential operator is bounded below or in proving the coerciveness of the bilinear forms generated by the least-squares method.

Theorem 4.3 (The Friedrichs Inequality) If $u \in H^1(\Omega)$ and satisfies $u = 0$ on Γ_1, where Γ_1 is a part of the boundary Γ, then there exists a constant C which is not related to u such that

$$\|u\|_0 \leq C|u|_1. \tag{4.14}$$

Proof: We follow the proof given by Chen (1982). Let us consider the two dimensional case. The proof for the three dimensional case is similar. Assume that $u = 0$ outside of $\bar{\Omega}$, where $\bar{\Omega}$ denotes the closure of Ω. We use a circle with diameter D to surround the domain Ω. Let the boundary Γ_1 be located as shown in Fig. 4.1 (this can always be done by rotating the coordinates), such that the set of y coordinates of Γ_1 contains $\alpha \leq y \leq \beta$.

We choose a point A on Γ_1. The ordinate of A is $\bar{y}$.

Through A we draw a straight line segment AB which is parallel to the x axis. We have

$$u(B) = u(B) - u(A) = \int_A^B \frac{\partial u(\xi, \bar{y})}{\partial \xi} d\xi.$$

By virtue of the Schwarz inequality we have

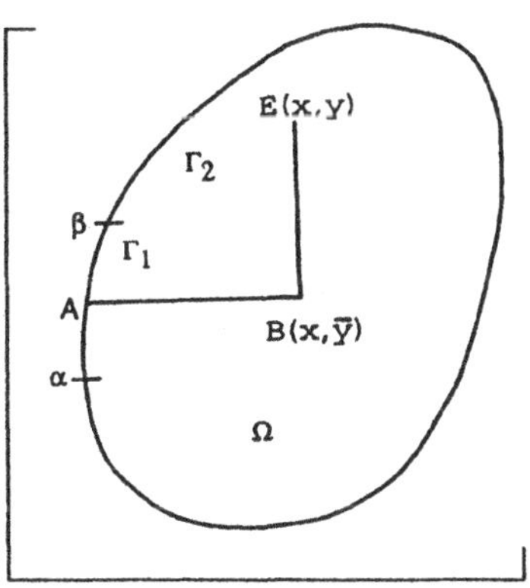

Fig. 4.1. A two-dimensional domain

$$|u(x,\bar{y})|^2 = \left(\int_A^B \frac{\partial u}{\partial \xi} d\xi\right)^2 \le D \int_A^B \left(\frac{\partial u}{\partial \xi}\right)^2 d\xi.$$

Integrating the above inequality with respect to $\bar{y}$ yields

$$\int_\alpha^\beta |u(x,\bar{y})|^2 d\bar{y} \le D \int_\Omega \left(\frac{\partial u}{\partial x}\right)^2 dxdy. \tag{4.15}$$

If the domain Ω is included in the strip $\alpha \le y \le \beta$, integrating (4.15) with respect to x leads to the inequality (4.14). Otherwise, through $B(x,\bar{y})$ we draw a straight line segment BE which is parallel to the y axis. We have

$$u(E) = u(B) + \int_{\bar{y}}^{y} \frac{\partial u(x,\eta)}{\partial \eta} d\eta.$$

From the above equation and by using the Cauchy inequality

$$(a+b)^2 \le 2(a^2+b^2) \tag{4.16}$$

and the Schwarz inequality we have

$$\begin{aligned} |u(x,y)|^2 &\le 2\left[|u(x,\bar{y})|^2 + \left(\int_{\bar{y}}^{y} \frac{\partial u(x,\eta)}{\partial \eta} d\eta\right)^2\right] \\ &\le 2|u(x,\bar{y})|^2 + 2D \int_{-\infty}^{\infty} \left(\frac{\partial u}{\partial \eta}\right)^2 d\eta. \end{aligned}$$

Integrating both sides of the above inequality with respect to $\bar{y}$ we have

$$|u(x,y)|^2(\beta-\alpha) \le 2\int_\alpha^\beta |u(x,\bar{y})|^2 d\bar{y} + 2D(\beta-\alpha)\int_{-\infty}^{\infty} \left(\frac{\partial u}{\partial y}\right)^2 dy.$$

By considering (4.15) the above inequality becomes

$$|u(x,y)|^2(\beta-\alpha) \le 2D\int_\Omega \left(\frac{\partial u}{\partial x}\right)^2 dxdy + 2D(\beta-\alpha)\int_{-\infty}^{\infty} \left(\frac{\partial u}{\partial y}\right)^2 dy.$$

Finally integrating the above inequality with respect to x and y yields the Friedrichs inequality (4.14). □

By adding $|u|_1$ to both sides of (4.14) we obtain

$$\|u\|_1 \le C|u|_1, \tag{4.17}$$

which implies that for functions belonging to $H_0^1(\Omega)$ or to $H^1(\Omega)$ and satisfying $u = 0$ on a part of the boundary, the semi-norm (4.4) defines a norm equivalent to (4.3), and for such functions, the semi-norm may be used instead of the norm.

4.5 The Poincaré Inequality

Theorem 4.4 (The Poincaré Inequality) If $u \in H^1(\Omega)$, then there exists a constant C which is not related to u such that

$$\|u\|_0^2 \leq C\Big\{|u|_1^2 + \Big(\int_\Omega u d\Omega\Big)^2\Big\}. \tag{4.18}$$

Proof: For simplicity let us prove its validity for two-dimensional rectangular domains, see Fig. 4.2.

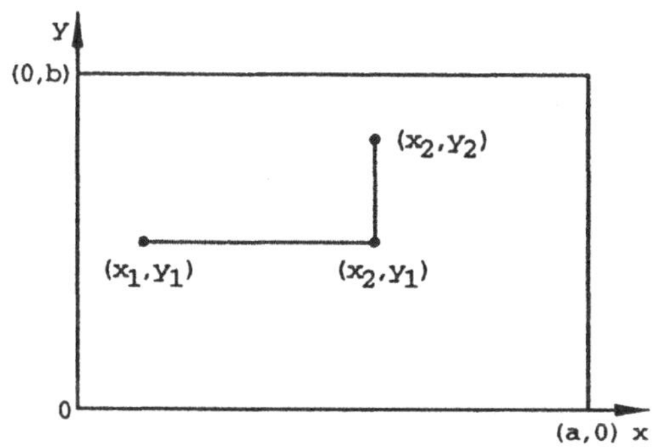

Fig. 4.2. A rectangular domain

Assume that Ω is a rectangle $0 \leq x \leq a,\ 0 \leq y \leq b$. Let (x_1, y_1) and (x_2, y_2) be two points in Ω, then

$$\begin{aligned} u(x_2,y_2) - u(x_1,y_1) &= u(x_2,y_2) - u(x_2,y_1) + u(x_2,y_1) - u(x_1,y_1) \\ &= \int_{x_1}^{x_2} \frac{\partial u(x,y_1)}{\partial x} dx + \int_{y_1}^{y_2} \frac{\partial u(x_2,y)}{\partial y} dy. \end{aligned}$$

Squaring both sides and using the Cauchy inequality (4.16) yield

$$\begin{aligned} \{u(x_2,y_2) - u(x_1,y_1)\}^2 &\leq 2\Big\{\Big(\int_{x_1}^{x_2} \frac{\partial u(x,y_1)}{\partial x} dx\Big)^2 \\ &+ \Big(\int_{y_1}^{y_2} \frac{\partial u(x_2,y)}{\partial y} dy\Big)^2\Big\}. \end{aligned}$$

For the right hand side by using the Schwarz inequality we have

$$\begin{aligned} u^2(x_2,y_2) &- 2u(x_2,y_2)u(x_1,y_1) + u^2(x_1,y_1) \\ &\leq 2a \int_0^a \Big(\frac{\partial u(x,y_1)}{\partial x}\Big)^2 dx + 2b \int_0^b \Big(\frac{\partial u(x_2,y)}{\partial y}\Big)^2 dy. \end{aligned}$$

Integrating both sides of the above equation with respect to x_1, y_1, x_2, y_2 yields

$$2ab\int_\Omega u^2 d\Omega - 2\Big(\int_\Omega u d\Omega\Big)^2 \le 2a\cdot a^2 b\int_\Omega\Big(\frac{\partial u}{\partial x}\Big)^2 d\Omega + 2b\cdot ab^2\int_\Omega\Big(\frac{\partial u}{\partial y}\Big)^2 d\Omega,$$

or

$$\int_\Omega u^2 d\Omega - \frac{1}{ab}\Big(\int_\Omega u d\Omega\Big)^2 \le a^2\int_\Omega\Big(\frac{\partial u}{\partial x}\Big)^2 d\Omega + b^2\int_\Omega\Big(\frac{\partial u}{\partial y}\Big)^2 d\Omega,$$

and thus

$$\int_\Omega u^2 d\Omega \le \max(a^2, b^2)\int_\Omega (u_x^2 + u_y^2) d\Omega + \frac{1}{ab}\Big(\int_\Omega u d\Omega\Big)^2.$$

This is the Poincaré inequality. □

4.6 Finite Element Spaces

A crucial step in the finite element analysis of a given problem is the choice of adequate finite element spaces V_h. These spaces consist of piecewise polynomial functions on subdivisions or "triangulations" T_h of a bounded domain $\Omega \subset \mathbb{R}^{n_d}$, $n_d = 1, 2, 3$ into elements K. For $n_d = 1$, the elements will be intervals. We have discussed one-dimensional linear elements in Sect. 2.5. For $n_d = 2$ the elements may be triangles or quadrilaterals, and for $n_d = 3$ tetrahedrons or hexahedrons.

Throughout this book, $P_r(K)$ denotes the space of polynomials of order less than or equal to r defined on K, and $Q_r(K)$ the functions that are polynomials of order less than or equal to r in each of the coordinate directions, e.g., $Q_1(K)$, for $K \subset \mathbb{R}^2$, denotes piecewise bilinear functions with respect to quadrilaterals K.

We assume that readers are familiar with finite elements, so we will not enter into details of various elements. Instead, we review some basic ideas and derive expressions for the interpolation errors only for linear triangular elements using an elementary method. For rigorous analysis of finite element interpolation errors, we refer to Ciarlet (1991).

4.6.1 Regularity Requirements

The least-squares finite element method will need either $V_h \subset H^1(\Omega)$ or $V_h \subset H^2(\Omega)$ depending on solving first-order or second-order boundary value problems. Since the space V_h consists of piecewise polynomials, we have

$$V_h \subset H^1(\Omega) \Leftrightarrow V_h \subset C^0(\bar\Omega), \tag{4.19}$$

$$V_h \subset H^2(\Omega) \Leftrightarrow V_h \subset C^1(\bar\Omega), \tag{4.20}$$

where $\bar\Omega = \Omega \cup \Gamma$. Thus, $V_h \subset H^1(\Omega)$ if and only if the functions $v \in V_h$ are continuous, and $V_h \subset H^2(\Omega)$ if and only if the functions $v \in V_h$ and their first-order derivatives are continuous. Equivalence (4.19) holds, since the functions $v \in V_h$ are piecewise polynomials on each element K, so that if v is continuous across the common boundary of adjoining elements, then the first-order derivatives $D^\alpha v(|\alpha| = 1)$ exist, and are piecewise continuous, therefore $v \in H^1(\Omega)$. On the other hand, if v is not continuous across a certain inter-element boundary, i.e. $v \notin C^0(\bar{\Omega})$, then the derivatives $D^\alpha v, |\alpha| = 1$ would be a δ function, and thus are not square-integrable on Ω and therefore $v \notin H^1(\Omega)$. Similarly we realize that (4.20) is valid.

Finite elements which satisfy the requirement (4.19) are called C^0 elements. Finite elements with continuous first-order derivatives across inter-element boundaries are called C^1 elements. If one applies the least-squares method directly to second-order differential equations (Bramble and Shatz 1970a,b), then the second-order derivatives should be integrable over Ω, and thus according to (4.20), C^1 elements should be employed. However, it is difficult and complicated to construct C^1 elements. Furthermore, the use of the C^1 least-squares method will leads to severely ill-conditioned matrices. For this reason, in this book we pursue application of the C^0 least-squares finite element method based on first-order systems.

In fact, first-order systems are more natural than higher-order systems. Conservative laws and constitutive laws in physics are in general governed by first-order systems. For historical reasons (convenience for hand calculation and analysis), equations in a first-order system are combined into a high-order partial differential equation (or equations) with one or fewer unknowns. For example, for incompressible and irrotational flows, by introducing the potential (or the stream function in 2D cases), the incompressibility and irrotationality are combined into a second-order Laplace or Poisson equation. Because the potential or the stream function is not a measurable physical quantity, one has to take *a posterior* numerical differentiation to obtain useful velocity components, and thus accuracy is reduced by one order. In the computer age this is unnecessary. The least-squares method based on first-order system can directly give an accurate solution of the velocity components.

If the original governing equations are of higher order, one can always reduce them to a first-order system by introducing some new variables, then use the least-squares method with C^0 elements.

4.6.2 Linear Triangular Element

For simplicity we assume that Ω is a polygonal domain in the x, y plane. We take the linear triangular element, i.e., the P_1 element, as an example. We divide Ω into a set $T_h = \{K_1, ..., K_m\}$ of non-overlapping triangles K_i,

$$\Omega = K_1 \cup K_2 ... \cup K_m,$$

such that no vertex (node) of one triangle lies on the edge of another triangle as illustrated in Fig. 4.3. We introduce the mesh parameter h as the maximum diameter of all circles circumscribing the triangles, and ρ as the minimum diameter of all circles inscribed in the triangles. We shall assume that the subdivision satisfies the standard regularity condition, i.e., there is a positive constant β independent of h such that

$$\frac{\rho}{h} \geq \beta. \tag{4.21}$$

This condition means that the triangles $K \in T_h$ are not allowed to be arbitrarily thin, or equivalently, the angles of the triangles K are not allowed to be arbitrarily small; the constant β is a measure of the smallest possible angle in any $K \in T_h$ for any triangulation T_h.

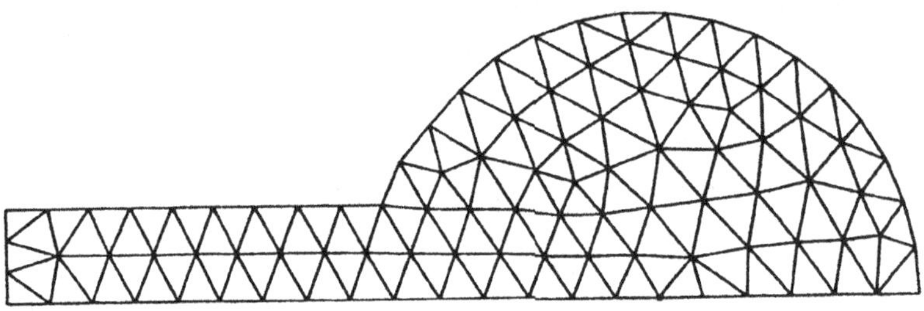

Fig. 4.3. Triangulation

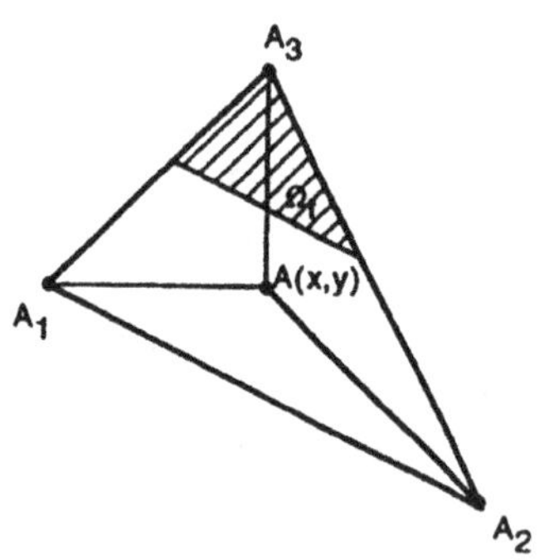

Fig. 4.4. Linear triangular element

Consider a triangle in the mesh (Fig. 4.4). The nodes of this triangle are the vertices $A_i (i = 1, 2, 3)$. Now we choose the linear interpolant

$$\Pi_h u(x, y) = a_0 + a_1 x + a_2 y. \tag{4.22}$$

To evaluate the three constants a_0, a_1 and a_2, we must provide three values of u, x and y at each of the three nodes, i.e.,

$$\begin{pmatrix} u_1 \\ u_2 \\ u_3 \end{pmatrix} = \begin{pmatrix} 1 & x_1 & y_1 \\ 1 & x_2 & y_2 \\ 1 & x_3 & y_3 \end{pmatrix} \begin{pmatrix} a_0 \\ a_1 \\ a_2 \end{pmatrix}.$$

Solving for the constants (a_0, a_1, a_2) and substituting them into (4.22) gives

$$\Pi_h u(x,y) = \psi_1(x,y)u_1 + \psi_2(x,y)u_2 + \psi_3(x,y)u_3, \tag{4.23}$$

where

$$\psi_1(x,y) = \frac{1}{D}\det\begin{pmatrix} 1 & x & y \\ 1 & x_2 & y_2 \\ 1 & x_3 & y_3 \end{pmatrix}, \tag{4.24a}$$

$$\psi_2(x,y) = \frac{1}{D}\det\begin{pmatrix} 1 & x_1 & y_1 \\ 1 & x & y \\ 1 & x_3 & y_3 \end{pmatrix}, \tag{4.24b}$$

$$\psi_3(x,y) = \frac{1}{D}\det\begin{pmatrix} 1 & x_1 & y_1 \\ 1 & x_2 & y_2 \\ 1 & x & y \end{pmatrix}, \tag{4.24c}$$

in which

$$D = \det\begin{pmatrix} 1 & x_1 & y_1 \\ 1 & x_2 & y_2 \\ 1 & x_3 & y_3 \end{pmatrix}. \tag{4.24d}$$

D is twice the area of the triangle $(A_1A_2A_3)$. When the node numbers 1, 2, 3 are assigned counterclockwise as in Fig. 4.4, $D > 0$; otherwise $D < 0$.

The shape functions (ψ_1, ψ_2, ψ_3) have a geometric explanation. We connect a point $A(x,y)$ in the triangle with three nodes to form three triangles. The ratio of the areas, ψ_i, called the area coordinates, are :

$$\psi_1 = \frac{\text{area}(AA_2A_3)}{\text{area}(A_1A_2A_3)},$$

$$\psi_2 = \frac{\text{area}(A_1AA_3)}{\text{area}(A_1A_2A_3)},$$

$$\psi_3 = \frac{\text{area}(A_1A_2A)}{\text{area}(A_1A_2A_3)}.$$

We see that

$$\sum_{i=1}^{3} \psi_i(x,y) = 1. \tag{4.25}$$

4.6.3 Interpolation Errors

We shall use this linear triangular element Ω_e (Fig. 4.4) as an example to estimate the interpolation errors $\|u - \Pi_h u\|_0$ and $\|u - \Pi_h u\|_1$. Readers may skip this section and just accept the general estimates (4.42) and (4.43) which are discussed in any book on the mathematical theory of finite elements. Here we follow the analysis presented by Ying (1988). This analysis has three steps. First, the nodal value of $u(A_i)$ is expressed by the Taylor expansion of u at any point $A(x,y)$. Next, the relation between the point-wise error $u(A) - \Pi_h u(A)$ and the remainders of the Taylor expansion is established. Finally, the L_2 and H_1 norms of the error are derived.

(1) We assume that $u \in H^2(\Omega)$. For any point $A(x,y)$ within the element, we have

$$u(A_i) = u(A) + \int_0^1 \frac{d}{dt} u(tx_i + (1-t)x,\ ty_i + (1-t)y)dt,\ i = 1,2,3. \quad (4.26)$$

Replacing $\int_0^1 (...)dt$ in (4.26) by $\int_0^1 (...)d(t-1)$ and integrating by parts we obtain

$$\begin{aligned} u(A_i) &= u(A) + \frac{\partial u(A)}{\partial x}(x_i - x) + \frac{\partial u(A)}{\partial y}(y_i - y) \\ &+ \int_0^1 (1-t)\frac{d^2}{dt^2} u(tx_i + (1-t)x,\ ty_i + (1-t)y)dt. \end{aligned} \quad (4.27)$$

Let

$$\xi_i = tx_i + (1-t)x, \qquad \eta_i = ty_i + (1-t)y,$$

then (4.27) can be written as

$$u(A_i) = u(A) + \frac{\partial u(A)}{\partial x}(x_i - x) + \frac{\partial u(A)}{\partial y}(y_i - y) + R_i(A), \quad (4.28)$$

in which

$$R_i(A) = \int_0^1 (1-t)\frac{d^2}{dt^2} u(\xi_i,\ \eta_i)dt. \quad (4.29)$$

In fact, (4.28) is a Taylor expansion with an integral remainder (4.29).

(2) Since the interpolant $\Pi_h u(A)$ can be expressed as (see (4.23))

$$\Pi_h u(A) = \sum_{i=1}^{3} u(A_i)\psi_i(A),$$

by virtue of (4.28) we have

$$\Pi_h u(A) = \sum_{i=1}^{3} \Big\{ u(A) + \frac{\partial u(A)}{\partial x}(x_i - x) + \frac{\partial u(A)}{\partial y}(y_i - y) + R_i(A) \Big\} \psi_i(A). \quad (4.30)$$

Now let us define a linear function l such that

$$l(A) = u(A),$$

$$\frac{\partial l(A)}{\partial x} = \frac{\partial u(A)}{\partial x}, \qquad \frac{\partial l(A)}{\partial y} = \frac{\partial u(A)}{\partial y}.$$

Since the second derivatives of l are equal to zero, by virtue of (4.30) we have

$$\Pi_h l(A) = \sum_{i=1}^{3} \Big\{ u(A) + \frac{\partial u(A)}{\partial x}(x_i - x) + \frac{\partial u(A)}{\partial y}(y_i - y) \Big\} \psi_i(A).$$

The interpolant of a linear function is equal to itself, therefore

$$\Pi_h l(A) = l(A) = u(A),$$

and

$$\sum_{i=1}^{3} \Big\{ \frac{\partial u(A)}{\partial x}(x_i - x) + \frac{\partial u(A)}{\partial y}(y_i - y) \Big\} \psi_i(A) = 0,$$

which together with (4.30) gives

$$\Pi_h u(A) - u(A) = \sum_{i=1}^{3} R_i(A)\psi_i(A). \tag{4.31}$$

Equation (4.31) is the expression for the point-wise interpolation error $\Pi_h u - u$.

Next, we shall give the derivatives of the point-wise error. Differentiation of (4.31) with respect to x leads to

$$\frac{\partial [\Pi_h u(A) - u(A)]}{\partial x} = \sum_{i=1}^{3} R_i(A) \frac{\partial \psi_i(A)}{\partial x} + \sum_{i=1}^{3} \frac{\partial R_i(A)}{\partial x} \psi_i(A). \tag{4.32}$$

Now we show that the last term in (4.32) is equal to zero. We differentiate (4.28) with respect to x to obtain

$$\frac{\partial^2 u(A)}{\partial x^2}(x_i - x) + \frac{\partial^2 u(A)}{\partial x \partial y}(y_i - y) + \frac{\partial R_i(A)}{\partial x} = 0,$$

therefore from the above equation we obtain

$$\sum_{i=1}^{3} \frac{\partial R_i(A)}{\partial x} \psi_i(A) = -\sum_{i=1}^{3} \Big\{ \frac{\partial^2 u(A)}{\partial x^2}(x_i - x) + \frac{\partial^2 u(A)}{\partial x \partial y}(y_i - y) \Big\} \psi_i(A). \tag{4.33}$$

Define a linear function λ such that

$$\lambda(A) = 0,$$

$$\frac{\partial \lambda(A)}{\partial x} = -\frac{\partial^2 u(A)}{\partial x^2}, \qquad \frac{\partial \lambda(A)}{\partial y} = -\frac{\partial^2 u(A)}{\partial x \partial y}.$$

According to (4.30) we have

$$\Pi_h\lambda(A) = -\sum_{i=1}^{3}\Big\{\frac{\partial^2 u(A)}{\partial x^2}(x_i - x) + \frac{\partial^2 u(A)}{\partial x\partial y}(y_i - y)\Big\}\psi_i(A). \tag{4.34}$$

But

$$\Pi_h\lambda(A) = \lambda(A) = 0,$$

which implies the right-hand sides of (4.34) and (4.33) equal zero, and hence

$$\sum_{i=1}^{3}\frac{\partial R_i(A)}{\partial x}\psi_i(A) = 0.$$

Therefore from (4.32) we obtain

$$\frac{\partial[\Pi_h u(A) - u(A)]}{\partial x} = \sum_{i=1}^{3} R_i(A)\frac{\partial\psi_i(A)}{\partial x}. \tag{4.35}$$

Similarly,

$$\frac{\partial[\Pi_h u(A) - u(A)]}{\partial y} = \sum_{i=1}^{3} R_i(A)\frac{\partial\psi_i(A)}{\partial y}. \tag{4.36}$$

(3) Using (4.31) we can estimate $\|u(A) - \Pi_h u(A)\|_{0,\Omega_e}$. To this end, we use (4.31) and $|\psi_i(A)| \le 1$ to obtain

$$\begin{aligned}\|u(A) - \Pi_h u(A)\|_{0,\Omega_e}^2 &\le \int_{\Omega_e}\Big(\sum_{i=1}^{3}|R_i(A)|\Big)^2 dxdy\\ &\le 3\sum_{i=1}^{3}\int_{\Omega_e} R_i^2(A)dxdy.\end{aligned}$$

Recall that

$$|x_i - x| \le h, \qquad |y_i - y| \le h,$$

which together with (4.29) give

$$\begin{aligned}R_i^2(A) &= \Big\{\int_0^1 (1-t)\Big[\frac{\partial^2 u(\xi_i,\eta_i)}{\partial\xi_i^2}(x_i - x)^2\\ &+ 2\frac{\partial^2 u(\xi_i,\eta_i)}{\partial\xi_i\partial\eta_i}(x_i - x)(y_i - y)\\ &+ \frac{\partial^2 u(\xi_i,\eta_i)}{\partial\eta_i^2}(y_i - y)^2\Big]dt\Big\}^2\\ &\le Ch^4\Big\{\int_0^1 (1-t)\Big(\Big|\frac{\partial^2 u}{\partial\xi_i^2}\Big| + 2\Big|\frac{\partial^2 u}{\partial\xi_i\partial\eta_i}\Big| + \Big|\frac{\partial^2 u}{\partial\eta_i^2}\Big|\Big)dt\Big\}^2.\end{aligned}$$

Using the Schwarz inequality we obtain

$$R_i^2(A) \le Ch^4 \int_0^1 (1-t)^2 \Big(\Big|\frac{\partial^2 u}{\partial \xi_i^2}\Big|^2 + 2\Big|\frac{\partial^2 u}{\partial \xi_i \partial \eta_i}\Big|^2 + \Big|\frac{\partial^2 u}{\partial \eta_i^2}\Big|^2 \Big) dt.$$

Therefore

$$\begin{aligned} \|u - \Pi_h u\|_{0,\Omega_e}^2 \quad &\le \quad Ch^4 \sum_{i=1}^3 \int_0^1 (1-t)^2 \int_{\Omega_e} \Big(\Big|\frac{\partial^2 u}{\partial \xi_i^2}\Big|^2 + 2\Big|\frac{\partial^2 u}{\partial \xi_i \partial \eta_i}\Big|^2 \\ &+ \quad \Big|\frac{\partial^2 u}{\partial \eta_i^2}\Big|^2 \Big) dx dy dt. \end{aligned}$$

Taking ξ_i, η_i as the integration variables and noting that

$$dx = \frac{d\xi_i}{1-t}, \qquad dy = \frac{d\eta_i}{1-t},$$

we have

$$\begin{aligned} \|u - \Pi_h u\|_{0,\Omega_e}^2 \quad &\le \quad Ch^4 \sum_{i=1}^3 \int_0^1 \int_{\Omega_t} \Big(\Big|\frac{\partial^2 u}{\partial \xi_i^2}\Big|^2 + 2\Big|\frac{\partial^2 u}{\partial \xi_i \partial \eta_i}\Big|^2 \\ &+ \quad \Big|\frac{\partial^2 u}{\partial \eta_i^2}\Big|^2 \Big) d\xi_i d\eta_i dt, \end{aligned} \tag{4.37}$$

in which the domain Ω_t is a subdomain of Ω_e, as illustrated in Fig. 4.4 for $i = 3$. From (4.37) we finally obtain the estimate

$$\|u - \Pi_h u\|_{0,\Omega_e}^2 \le Ch^4 |u|_{2,\Omega_e}^2. \tag{4.38}$$

Using (4.35) and (4.36) and following the same steps as above we can estimate $\|u - \Pi_h u\|_{1,\Omega_e}$. The only additional consideration is that we need to estimate the upper bounds of $|\partial\psi_i/\partial x|$ and $|\partial\psi_i/\partial y|$. We differentiate (4.24a) with respect to x to obtain

$$\frac{\partial \psi_1}{\partial x} = \frac{1}{D}(y_2 - y_3) \le \frac{h}{D} \le \frac{h}{2\frac{\rho}{2}h} = \frac{1}{\rho}.$$

Similarly we have

$$\Big|\frac{\partial \psi_i}{\partial x}\Big| \le \frac{1}{\rho}, \quad \Big|\frac{\partial \psi_i}{\partial y}\Big| \le \frac{1}{\rho}, \qquad i = 1, 2, 3.$$

Therefore we can derive

$$\|u - \Pi_h u\|_{1,\Omega_e}^2 \le C\frac{h^4}{\rho^2} |u|_{2,\Omega_e}^2 \le Ch^2 |u|_{2,\Omega_e}^2, \tag{4.39}$$

in which the constant β from (4.21) is absorbed into C. In fact, we have proved the following theorem by summing (4.38) and (4.39) over all elements:

Theorem 4.5 (Interpolation Errors) Let $T_h = K$ be a regular family of triangulations of a polygon Ω, let $u \in H^2(\Omega)$ and let the finite element subspaces $V_h = \{v \in C^0(\Omega) : v \in P_1(K)\}$. Then there exists a constant $C > 0$ such that the following estimates hold

$$\|u - \Pi_h u\|_{0,\Omega} \le Ch^2 |u|_{2,\Omega}, \tag{4.40}$$

$$\|u - \Pi_h u\|_{1,\Omega} \le Ch |u|_{2,\Omega}. \tag{4.41}$$

The estimates (4.40) and (4.41) are typical examples of estimates for the interpolation error $u - \Pi_h u$ with piecewise linear functions. If we work with piecewise polynomials of order $r \ge 1$ on the triangulation T_h satisfying (4.21), and u is smooth enough, we have the following estimates:

$$\|u - \Pi_h u\|_{0,\Omega} \le Ch^{r+1} |u|_{r+1,\Omega}, \tag{4.42}$$

$$\|u - \Pi_h u\|_{1,\Omega} \le Ch^r |u|_{r+1,\Omega}. \tag{4.43}$$

For a proof of the above estimates, see, e.g., Ciarlet (1991) and Oden and Carey (1983).

4.7 First-Order System

The least-squares method studied in this book is based on minimizing the L_2 norm of the residuals in the first-order differential equations. Almost all problems arising in fluid dynamics, solid mechanics, heat transfer, electromagnetics and other disciplines can be expressed by first-order systems.

Let us consider the first-order system of partial differential equations:

$$\sum_{k=1}^{n} \sum_{j=1}^{m} a_{ij,k} \frac{\partial u_j}{\partial x_k} + \sum_{j=1}^{m} a_{ij} u_j = f_i \qquad i = 1, ..., m. \tag{4.44}$$

The above system can be written in matrix form:

$$\boldsymbol{A u} \equiv \sum_{k=1}^{n} \boldsymbol{A}_k \frac{\partial \boldsymbol{u}}{\partial x_k} + \boldsymbol{A}_0 \boldsymbol{u} = \boldsymbol{f}, \tag{4.45}$$

where $\boldsymbol{u}^T = (u_1, u_2, ... u_m)$ is a vector of m unknown functions of $\boldsymbol{x} = (x_1, x_2, ..., x_n)$, $\boldsymbol{A}_k = (a_{ij,k})$ and $\boldsymbol{A}_0 = (a_{ij})$ are $m \times m$ matrices, and $\boldsymbol{f}$ is a given vector-valued function of $\boldsymbol{x}$. We emphasize here that for time-dependent problems the time variable t is included in $\boldsymbol{x}$.

(1) System (4.44) is called a quasi-linear system, if $a_{ij,k}$ and a_{ij} are functions of $\boldsymbol{x}$ and $\boldsymbol{u}$;

(2) It is called an almost-linear system, if $a_{ij,k}$ are functions depending only on $\boldsymbol{x}$, and a_{ij} are functions as in (1);

(3) It is called a linear system, if both $a_{ij,k}$ and a_{ij} are functions of $\boldsymbol{x}$ only.

For convenience, we shall restrict our treatment initially to linear problems. It is straightforward to extend the treatment to quasi-linear problems by using successive substitution or Newton's linearization (see Sects. 7.2 and 8.3).

Linear first-order systems of partial differential equations can be classified as elliptic, parabolic, hyperbolic and mixed. Different types of partial differential equations describe different physical phenomena and have different mathematical features. Now let us classify linear first-order systems.

Assume that the solution $\boldsymbol{u}$ and all its first-order derivatives have already been determined on a smooth surface S:

$$\varphi(x_1, ..., x_n) = 0.$$

We wish to extend the solution off S. To replace the independent variables $x_1, ..., x_n$ we introduce new independent variables $\xi_1, \xi_2, ..., \xi_n$, such that $\xi_n = \varphi$, and if $\xi_n = 0$ there exists a relationship:

$$x_i = g_i(\xi_1, ..., \xi_{n-1}) \qquad i = 1, ..., n$$

which is the parametric expression for the surface S. $(\xi_1, ..., \xi_{n-1})$ are the so called internal variables of the surface S. The derivatives of $\boldsymbol{u}$ with respect to these internal variables are called internal derivatives. In other words, the tangential derivatives of $\boldsymbol{u}$ along the surface S are defined as internal derivatives. Correspondingly, if the direction $\boldsymbol{\ell}$ is not tangential to the surface S, then $\partial\boldsymbol{u}/\partial\boldsymbol{\ell}$ is called an external derivative. If the function $\boldsymbol{u}$ on the surface S is given, then all its internal first-order derivatives on S are known. However, its first-order external derivative cannot be determined only by the value of $\boldsymbol{u}$ on S. Now the question is: Can the equations (4.45) help us uniquely determine the external derivatives? To answer this question we use the transformation of independent variables:

$$x_i = x_i(\xi_1, ..., \xi_n) \qquad i = 1, ..., n$$

under which the first-order derivative $\partial\boldsymbol{u}/\partial x_k$ becomes

$$\sum_{j=1}^{n} \frac{\partial \boldsymbol{u}}{\partial \xi_j} \frac{\partial \xi_j}{\partial x_k},$$

and therefore system (4.45) becomes

$$\Big(\sum_{k=1}^{n} \boldsymbol{A}_k \frac{\partial \xi_n}{\partial x_k}\Big) \frac{\partial \boldsymbol{u}}{\partial \xi_n} + ... = \boldsymbol{f}$$

in which, except for the first term, all terms depend only on the value of $\boldsymbol{u}$ and its internal derivatives on the surface S, and thus are known. If $\partial\boldsymbol{u}/\partial\xi_n$ can be determined, then the solution can be extended off S. This depends on whether the matrix $\sum_{k=1}^{n} \boldsymbol{A}_k \partial\xi_n/\partial x_k$ is of full rank or not. If it is of full rank, then the external derivative $\partial\boldsymbol{u}/\partial\xi_n$ can be determined. Noting that $\xi_n = \varphi$, if

$$\det\Big(\sum_{k=1}^{n} \boldsymbol{A}_k \frac{\partial \varphi}{\partial x_k}\Big) = 0 \tag{4.46}$$

is valid on the surface S, then the external derivatives on S cannot be determined uniquely. In this case the solution $\boldsymbol{u}$ may be discontinuous across the surface S. If there exists a direction $\boldsymbol{\ell}(\alpha_1, ..., \alpha_n)$ such that

$$\det\Big(\sum_{k=1}^{n} \boldsymbol{A}_k \alpha_k\Big) = 0, \tag{4.47}$$

then $\boldsymbol{\ell}$ is called the characteristic direction, and (4.47) is called the characteristic equation. The surface S which satisfies (4.47) is called the characteristic surface. Since $(\partial\varphi/\partial x_1, ..., \partial\varphi/\partial x_n)$ represents the normal direction of the surface $\varphi = 0$, the characteristic surface is a surface such that at every point the normal is in the characteristic direction.

The classification of the system (4.45) depends on the solution of (4.47):
(1) If $\alpha_1, ..., \alpha_{n-1}$ are arbitrarily given real numbers,
there exist m real roots α_n satisfying (4.47), the system is hyperbolic;
(2) If $\det(\sum_{k=1}^{n} \boldsymbol{A}_k \alpha_k) \neq 0$ for all non-zero real $\alpha_1, ..., \alpha_n$, it is elliptic.

For large systems some roots may be complex and some may be real; this gives a mixed system.

In the elliptic case all roots are complex; since complex roots occur as pairs, the number of equations in an elliptic system must be even. In other words a system with an odd number of equations cannot be elliptic.

We remark that the classification discussed here is in the ordinary sense. A system which is not elliptic in the sense discussed here may still be elliptic in the Agmon–Douglis–Nirenberg theory (Agmon et al. 1964).

4.8 General Formulation of LSFEM

In this section we shall give a general formulation of the least-squares finite element method for steady-state boundary-value problems. For time-dependent problems we always can use an appropriate finite difference method in the time domain to discretize time-derivative terms, so that at each time-step the problems are converted into boundary-value problems.

Now let us consider the linear boundary-value problem:

$$\boldsymbol{Au} = \boldsymbol{f} \qquad \text{in } \Omega, \tag{4.48a}$$

$$\boldsymbol{Bu} = \boldsymbol{g} \qquad \text{on } \Gamma, \tag{4.48b}$$

where $\boldsymbol{A}$ is a first-order partial differential operator:

$$\boldsymbol{Au} = \sum_{i=1}^{n_d} \boldsymbol{A}_i \frac{\partial \boldsymbol{u}}{\partial x_i} + \boldsymbol{A}_0 \boldsymbol{u}, \tag{4.49}$$

in which $\Omega \in \mathbb{R}^{n_d}$ is a bounded domain with a piecewise smooth boundary Γ, $n_d = 2$ or 3 represents the number of space dimensions, $\boldsymbol{u}^T = (u_1, u_2, ...u_m)$ is a vector of m unknown functions of $\boldsymbol{x} = (x_1, ..., x_{n_d})$, $\boldsymbol{A}_i$ and $\boldsymbol{A}_0$ are $N_{eq} \times m$

matrices which continuously depend on $\boldsymbol{x}$, $\boldsymbol{f}$ is a given vector-valued function, $\boldsymbol{B}$ is a boundary algebraic operator, and $\boldsymbol{g}$ is a given vector-valued function on the boundary. Without loss of generality we assume that the vector $\boldsymbol{g}$ is null. We should mention that the number of equations N_{eq} in the system (4.48a) must be greater than or equal to the number of unknowns m. In the classification discussed in Sect. 4.7, a first-order system is required to have square coefficient matrices. This is not necessary for the least-squares method in which the number of equations N_{eq} can be greater than the number of unknowns m. We will address this point in the next chapter.

Suppose that $\boldsymbol{f} \in \boldsymbol{L}_2(\Omega)$. We choose an appropriate subspace $\boldsymbol{V}$ of the Hilbert space $\boldsymbol{L}_2(\Omega)$. The functions in $\boldsymbol{V}$ satisfy the boundary condition:

$$\boldsymbol{B}\boldsymbol{v} = \boldsymbol{0} \qquad \text{on } \Gamma. \tag{4.50}$$

The first-order differential operator $\boldsymbol{A}$ maps the subspace $\boldsymbol{V}$ into $\boldsymbol{L}_2(\Omega)$:

$$\boldsymbol{A} : \boldsymbol{V} \to \boldsymbol{L}_2(\Omega). \tag{4.51}$$

For an arbitrary trial function $\boldsymbol{v} \in \boldsymbol{V}$, we define the residual function:

$$\boldsymbol{R} = \boldsymbol{A}\boldsymbol{v} - \boldsymbol{f} \qquad \text{in } \Omega. \tag{4.52}$$

In general, the residual $\boldsymbol{R}$ is not equal to zero, unless $\boldsymbol{v}$ is equal to the exact solution $\boldsymbol{u}$. The squared distance between $\boldsymbol{A}\boldsymbol{v}$ and $\boldsymbol{f}$ will be nonnegative:

$$\|\boldsymbol{R}\|_0^2 = \int_\Omega (\boldsymbol{A}\boldsymbol{v} - \boldsymbol{f})^2 d\Omega \geq 0. \tag{4.53}$$

A solution $\boldsymbol{u}$ to the problem (4.48) can thus be interpreted as a member of $\boldsymbol{V}$ that minimizes the squared distance between $\boldsymbol{A}\boldsymbol{v}$ and $\boldsymbol{f}$:

$$0 = \|\boldsymbol{R}(\boldsymbol{u})\|_0^2 \leq \|\boldsymbol{R}(\boldsymbol{v})\|_0^2 \qquad \forall \boldsymbol{v} \in \boldsymbol{V}. \tag{4.54}$$

The least-squares method amounts to seeking a minimizer of the squared distance $\|\boldsymbol{A}\boldsymbol{v} - \boldsymbol{f}\|_0^2$ in $\boldsymbol{V}$. We write the quadratic functional in (4.53) as

$$I(\boldsymbol{v}) = \|\boldsymbol{A}\boldsymbol{v} - \boldsymbol{f}\|_0^2 = (\boldsymbol{A}\boldsymbol{v} - \boldsymbol{f}, \boldsymbol{A}\boldsymbol{v} - \boldsymbol{f}). \tag{4.55}$$

A necessary condition that $\boldsymbol{u} \in \boldsymbol{V}$ be a minimizer of the functional I in (4.55) is that its first variation vanishes at $\boldsymbol{u}$, that is,

$$\lim_{t \to 0} \frac{d}{dt} I(\boldsymbol{u} + t\boldsymbol{v}) \equiv 2 \int_\Omega (\boldsymbol{A}\boldsymbol{v})^T (\boldsymbol{A}\boldsymbol{u} - \boldsymbol{f}) d\Omega = \boldsymbol{0} \qquad \forall \boldsymbol{v} \in \boldsymbol{V}.$$

Thus, the least-squares method leads us to the variational formulation: Find $\boldsymbol{u} \in \boldsymbol{V}$ such that

$$B(\boldsymbol{u}, \boldsymbol{v}) = F(\boldsymbol{v}) \qquad \forall \boldsymbol{v} \in \boldsymbol{V}, \tag{4.56a}$$

where

$$B(\boldsymbol{u}, \boldsymbol{v}) \equiv (\boldsymbol{A}\boldsymbol{u}, \boldsymbol{A}\boldsymbol{v}), \tag{4.56b}$$

$$F(\boldsymbol{v}) \equiv (\boldsymbol{f}, \boldsymbol{A}\boldsymbol{v}). \tag{4.56c}$$

We note that the bilinear form in (4.56) is symmetric. For a well-posed problem (4.48), the operator $\boldsymbol{A}$ is bounded below. As a consequence, when discretized, (4.56) always lead to a symmetric positive-definite matrix.

In finite element analysis, we first subdivide the domain into a union of finite elements and then introduce an appropriate finite element basis. Let N_n denote the number of nodes for one element and ψ_j denote the element shape functions. If equal-order finite elements are employed, that is, for all unknown variables the same finite element is used, we can write the expansion in each element

$$\boldsymbol{u}_h^e(\boldsymbol{x}) = \sum_{j=1}^{N_n} \psi_j(\boldsymbol{x}) \begin{pmatrix} u_1 \\ u_2 \\ \vdots \\ u_m \end{pmatrix}_j, \tag{4.57}$$

where $(u_1, u_2, ..., u_m)_j$ are the nodal values at the jth node, and h denotes the mesh parameter.

Introducing the finite element approximation defined in (4.57) into the variational statement (4.56), we have the linear algebraic equations

$$\boldsymbol{KU} = \boldsymbol{F}, \tag{4.58}$$

where the $\boldsymbol{U}$ is the global vector of nodal values. The global matrix $\boldsymbol{K}$ is assembled from the element matrices

$$\boldsymbol{K}_e = \int_{\Omega_e} (\boldsymbol{A}\psi_1, \boldsymbol{A}\psi_2, ..., \boldsymbol{A}\psi_{N_n})^T (\boldsymbol{A}\psi_1, \boldsymbol{A}\psi_2, ..., \boldsymbol{A}\psi_{N_n}) d\Omega, \tag{4.59}$$

in which $\Omega_e \subset \Omega$ is the domain of the eth element, and T denotes the transpose, and the vector $\boldsymbol{F}$ is assembled from the element vectors

$$\boldsymbol{F}_e = \int_{\Omega_e} (\boldsymbol{A}\psi_1, \boldsymbol{A}\psi_2, ..., \boldsymbol{A}\psi_{N_n})^T \boldsymbol{f} d\Omega, \tag{4.60}$$

in (4.59) and (4.60)

$$\boldsymbol{A}\psi_j = \sum_{i=1}^{n_d} \frac{\partial \psi_j}{\partial x_i} \boldsymbol{A}_i + \psi_j \boldsymbol{A}_0. \tag{4.61}$$

We remark that the boundary conditions (4.48b) can also be included into the quadratic functional (4.55). In this approach, no boundary conditions are imposed on the subspace V. This is another advantage of the least-squares method.

4.9 The Euler–Lagrange Equation

To give an alternative interpolation of the least-squares method let us derive the Euler–Lagrange equation corresponding to the variational formulation (4.56). Upon integration by parts or application of Green's formula from (4.56) we obtain

$$(\boldsymbol{A}^*\boldsymbol{A}\boldsymbol{u} - \boldsymbol{A}^*\boldsymbol{f}, \boldsymbol{v}) + \langle \boldsymbol{A}\boldsymbol{u} - \boldsymbol{f}, \boldsymbol{v}\rangle_\Gamma = 0 \qquad \forall \boldsymbol{v} \in \boldsymbol{V}, \tag{4.62}$$

where $\boldsymbol{A}^*$ with range in $L_2(\Omega)$ is the adjoint operator of $\boldsymbol{A}$, and $\langle\cdot,\cdot\rangle_\Gamma$ is the so called bilinear concomitant: the form that contains boundary terms generated by the intergrations by parts. Hence we have the Euler–Lagrange equation

$$\boldsymbol{A}^*\boldsymbol{A}\boldsymbol{u} = \boldsymbol{A}^*\boldsymbol{f} \qquad \text{in } \Omega. \tag{4.63}$$

The original boundary condition $\boldsymbol{B}\boldsymbol{v} = \boldsymbol{0}$ serves as an essential boundary condition for (4.63). The natural boundary condition can be obtained from the boundary term $\langle \boldsymbol{A}\boldsymbol{u} - \boldsymbol{f}, \boldsymbol{v}\rangle_\Gamma = 0$ for all admissible $\boldsymbol{v}$, that is, the original first-order equation serves as a natural boundary condition. Therefore, the least-squares method for first-order system is formally equivalent to the Galerkin method for second-order system with the differential equations given in (4.63).

Observe that the operator $\boldsymbol{A}^*\boldsymbol{A}$ in (4.62) is formally self-adjoint, even though $\boldsymbol{A}$ itself is not self-adjoint. Now we understand the advantage of the least-squares method: it converts a difficult non-self-adjoint first-order system into a relatively easy self-adjoint second-order system. For this reason, it is not surprising that the condition number of the resulting matrix produced by LSFEM is $O(h^{-2})$ which is the same as that for the Galerkin method for second-order equations. This can be easily proved for a general multi-dimensional problems just by following the steps introduced in Sect. 2.7.4.

4.10 Error Estimates for LSFEM

4.10.1 General Problems

Now we apply the bounded inverse theorem introduced in Sect. 4.3 to a special case – problem (4.48). Suppose that the system of first-order partial differential equations (4.48) has a unique solution $\boldsymbol{u}$ in a subspace $\boldsymbol{V}$ of the Hilbert space $\boldsymbol{L}_2(\Omega)$ for any $\boldsymbol{f} \in \boldsymbol{L}_2(\Omega)$. By virtue of Theorem 4.1, we have positive constants α such that

$$\alpha\|\boldsymbol{u}\|_V \leq \|\boldsymbol{A}\boldsymbol{u}\|_0 \qquad \forall \boldsymbol{u} \in \boldsymbol{V}, \tag{4.64}$$

and thus

$$\|\boldsymbol{u}\|_V \leq \frac{1}{\alpha}\|\boldsymbol{f}\|_0. \tag{4.65}$$

For a given particular problem (4.48) we often can identify the subspace $\boldsymbol{V}$ and its norm, see examples in Sect. 8.2.2. For a general problem (4.48) without further information about the operator $\boldsymbol{A}$, we do not know all details of the space $\boldsymbol{V}$ and its associated norm $\|\boldsymbol{u}\|_V$. But fortunately, if the operator equation (4.48) is well-posed, we can at least know some thing about the relations between $\|\boldsymbol{A}\boldsymbol{u}\|_0$ and $\|\boldsymbol{u}\|_0$.

Theorem 4.6 If the linear first-order differential equations $\boldsymbol{A}\boldsymbol{u} = \boldsymbol{f}$ has a unique solution that continuously depends on the data $\boldsymbol{f} \in \boldsymbol{L}_2(\Omega)$, in other words if the inverse operator $\boldsymbol{A}^{-1}$ exists, then there exists a positive constant α_1 such that

$$\alpha_1\|\boldsymbol{u}\|_0 \leq \|\boldsymbol{A}\boldsymbol{u}\|_0. \tag{4.66a}$$

Moreover, if the solution $\boldsymbol{u} \in \boldsymbol{H}^1(\Omega)$, then there exists a positive constant M_1 such that

$$\|\boldsymbol{A}\boldsymbol{u}\|_0 \leq M_1\|\boldsymbol{u}\|_1. \tag{4.66b}$$

Proof: Since the first-order linear differential equation $\boldsymbol{A}\boldsymbol{u} = \boldsymbol{f}$ has a unique solution in $\boldsymbol{V}$ which is a subspace of $\boldsymbol{L}_2(\Omega)$, we must have

$$\alpha_1\|\boldsymbol{u}\|_0 \leq \alpha\|\boldsymbol{u}\|_V \leq \|\boldsymbol{A}\boldsymbol{u}\|_0.$$

Otherwise, $\|\boldsymbol{A}\boldsymbol{u}\|_0 = 0$ will not guarantee that $\|\boldsymbol{u}\|_0 = 0$, or in other words, system (4.48) will not have a unique solution.

Since $\boldsymbol{u} \in \boldsymbol{H}^1(\Omega)$ and $\boldsymbol{A}$ is a continuous first-order linear differential operator, we must have inequality (4.66b). □

We now turn to error estimates of LSFEM. Suppose that $\boldsymbol{V}_h$ is a finite element subspace of $\boldsymbol{V}$ which consists of piecewise complete polynomials of order r, and that the equation $\boldsymbol{A}\boldsymbol{u} = \boldsymbol{f}$ is well-posed. From Sect. 4.8 we know that $\boldsymbol{u} \in \boldsymbol{V}$ and $\boldsymbol{u}_h \in \boldsymbol{V}_h$ are the solutions of the following problems respectively:

$$(\boldsymbol{A}\boldsymbol{u}, \boldsymbol{A}\boldsymbol{v}) = (\boldsymbol{f}, \boldsymbol{A}\boldsymbol{v}) \qquad \forall \boldsymbol{v} \in \boldsymbol{V}, \tag{4.67}$$

$$(\boldsymbol{A}\boldsymbol{u}_h, \boldsymbol{A}\boldsymbol{v}_h) = (\boldsymbol{f}, \boldsymbol{A}\boldsymbol{v}_h) \qquad \forall \boldsymbol{v}_h \in \boldsymbol{V}_h. \tag{4.68}$$

Since $\boldsymbol{V}_h \subset \boldsymbol{V}$, we have from (4.67) in particular

$$(\boldsymbol{A}\boldsymbol{u}, \boldsymbol{A}\boldsymbol{v}_h) = (\boldsymbol{f}, \boldsymbol{A}\boldsymbol{v}_h) \qquad \forall \boldsymbol{v}_h \in \boldsymbol{V}_h. \tag{4.69}$$

Subtracting (4.68) from (4.69) we obtain the orthogonal condition:

$$(\boldsymbol{A}(\boldsymbol{u} - \boldsymbol{u}_h), \boldsymbol{A}\boldsymbol{v}_h) = 0 \qquad \forall \boldsymbol{v}_h \in \boldsymbol{V}_h. \tag{4.70}$$

Let $\Pi_h\boldsymbol{u} \in \boldsymbol{V}_h$ be the equal-order interpolant of $\boldsymbol{u}$, by virtue of (4.70) we have

$$\begin{aligned}\|A(u-u_h)\|_0^2 &= (A(u-u_h), A(u-u_h)) \\ &= (A(u-u_h), A(u-\Pi_h u)) \\ &+ (A(u-u_h), A(\Pi_h u-u_h)) \\ &= (A(u-u_h), A(u-\Pi_h u)) \\ &\leq \|A(u-u_h)\|_0 \, \|A(u-\Pi_h u)\|_0\end{aligned}$$

where the last inequality follows from the Schwarz inequality (4.1).

Dividing by $\|A(u-u_h)\|_0$ and using Theorem 4.6, we obtain

$$\alpha_1\|u-u_h\|_0 \leq \|A(u-u_h)\|_0 \leq \|A(u-\Pi_h u)\|_0 \leq M_1\|u-\Pi_h u\|_1. \tag{4.71}$$

Recalling that the residual $R(u_h) = Au_h - f = -A(u-u_h)$, and considering (4.71) and (4.43), we conclude that

Theorem 4.7 (Errors of LSFEM for General Problems) If the linear first-order differential system (4.48) is well-posed, and its solution is smooth enough, then the following error estimates hold for the approximate solution by LSFEM:

$$\|R(u_h)\|_0 \leq C_1 h^r |u|_{r+1}, \tag{4.72}$$

$$\|u-u_h\|_0 \leq C_2 h^r |u|_{r+1}. \tag{4.73}$$

This theorem is the foundation of LSFEM for general linear first-order systems. It does not matter what the type of the first-order system is, the LSFEM gives a convergent solution, provided that the problem has a unique and smooth solution. Especially important, this theorem explains why the simple LSFEM without any special treatment works for hyperbolic and mixed-type problems.

4.10.2 Elliptic Problems

The estimate (4.73) is not optimal. For most elliptic problems we are able to obtain an improved result. If the linear first-order differential operator A is elliptic, the following boundedness (below) or coerciveness of A can often be proved:

$$\alpha\|v\|_1 \leq \|Av\|_0 \qquad \forall v \in V = \{v \in [H^1(\Omega)]^m : Bv = 0 \text{ on } \Gamma\}. \tag{4.74}$$

In the following chapters we shall prove that the coerciveness (4.74) holds for many important problems. The boundedness (below) from (4.74) and the following obvious boundedness (above) of A:

$$\|Av\|_0 \leq M\|v\|_1 \qquad \forall v \in V \tag{4.75}$$

indicate that for coercive (strictly elliptic) operators A the norm $\|v\|_1$ and $\|Av\|_0$ are equivalent for $v \in V$. By virtue of the bounded inverse theorem, the solution of (4.48) uniquely exists and

$$\|\boldsymbol{u}\|_1 \leq \frac{1}{\alpha}\|\boldsymbol{f}\|_0. \tag{4.76}$$

Moreover, the least-squares finite element method, which seeks a minimizer of $I(\boldsymbol{v})$ in finite element subspaces $\boldsymbol{V}_h$ of the Hilbert space $\boldsymbol{V}$, has an optimal rate of convergence, i.e., the following theorem holds:

Theorem 4.8 (Errors of LSFEM for Elliptic Problems) Suppose that $\boldsymbol{u}_h$ is the least-squares finite element solution of the strictly elliptic system (4.48) with piecewise complete polynomials of order r, then there exist constants C_1 and C_2 independent of $\boldsymbol{u}$ and h such that

$$\|\boldsymbol{u} - \boldsymbol{u}_h\|_1 \leq C_1 h^r |\boldsymbol{u}|_{r+1}, \tag{4.77}$$

$$\|\boldsymbol{u} - \boldsymbol{u}_h\|_0 \leq C_2 h^{r+1} |\boldsymbol{u}|_{r+1}. \tag{4.78}$$

The proof of (4.77) follows from similar steps in the proof of Theorem 4.6. Using the Aubin–Nitsche trick as in Sect. 2.7.3 we can further obtain the error estimate (4.78).

This theorem implies that for strictly elliptic problems the rate of convergence of the least-squares method with equal-order finite elements is optimal for all variables, since the rate predicted is as high as possible for a given order of interpolation.

It should be noted that for strictly elliptic problems we do not assume in advance that the solution uniquely exists. The coerciveness or the boundedness below in the H^1 norm guarantees the existence and uniqueness of the solution. For many problems in mathematical physics, the existence and uniqueness of the solutions are often proved by using the Galerkin method. The least-squares technique introduced in this section not only has practical value in numerical solution, but can also be used for investigating the existence, uniqueness and stability of the solutions for some difficult problems, such as the solvability of the incompressible Navier–Stokes equations with non-standard boundary conditions (see Chap. 8).

4.11 Implementation of LSFEM

In the implementation of any finite element method, a key procedure is the derivation of the element matrices. Usually, the coefficients of the partial differential equations are not constant, and the shape of the element domain is distorted, making the exact evaluation of the integrals in the element matrices almost impossible. For such cases, an approximate numerical evaluation is necessary; the integral, such as that in (4.59), is commonly replaced by the Gaussian quadrature.

Our purpose in the present section is to show that in the evaluation of element matrices, the number of sampling points in the Gaussian quadrature is of great importance for the success of LSFEM, and from this knowledge to establish a principle for choosing a proper order of Gaussian quadrature.

We begin by studying the least-squares method for linear algebraic equations.

4.11.1 The Least-Squares Solution to Linear Algebraic Equations

Consider the system of linear algebraic equations

$$\begin{pmatrix} k_{11} & k_{12} & \dots & k_{1m} \\ k_{21} & k_{22} & \dots & k_{2m} \\ \vdots & \vdots & \ddots & \vdots \\ k_{n1} & k_{n2} & \dots & k_{nm} \end{pmatrix} \begin{pmatrix} u_1 \\ u_2 \\ \vdots \\ u_m \end{pmatrix} = \begin{pmatrix} f_1 \\ f_2 \\ \vdots \\ f_n \end{pmatrix}, \tag{4.79}$$

or simply

$$\boldsymbol{K}\boldsymbol{u} = \boldsymbol{f}, \tag{4.80}$$

where $\boldsymbol{K}$, $\boldsymbol{u}$ and $\boldsymbol{f}$ denote the corresponding matrices in system (4.79). We note that in system (4.79) the number of equations n may not be equal to the number of unknowns m. We define the augmented matrix as

$$\begin{pmatrix} k_{11} & k_{12} & \dots & k_{1m} & f_1 \\ k_{21} & k_{22} & \dots & k_{2m} & f_2 \\ \vdots & \vdots & \ddots & \vdots & \vdots \\ k_{n1} & k_{n2} & \dots & k_{nm} & f_n \end{pmatrix} \tag{4.81}$$

and the rank of a matrix as the order of the largest square array within that matrix, formed by deleting certain rows and columns, whose determinant does not vanish. Obviously, if $n \geq m$, the possible maximum rank of the augmented matrix (4.81) is $m+1$.

Depending on the properties of both the coefficient matrix $\boldsymbol{K}$ and the augmented matrix, system (4.79) may have a unique solution, an infinite number of solutions, or no solution at all. We shall investigate the conditions under which the least-squares method can be used to find the solution to system (4.79).

The least-squares method amounts to minimizing the following summation of the weighted squared residuals:

$$\begin{aligned} I(\boldsymbol{u}) = \quad & W_1(k_{11}u_1 + k_{12}u_2 + \ldots + k_{1m}u_m - f_1)^2 \\ + \quad & W_2(k_{21}u_1 + k_{22}u_2 + \ldots + k_{2m}u_m - f_2)^2 \\ + \quad & \ldots \\ + \quad & W_n(k_{n1}u_1 + k_{n2}u_2 + \ldots + k_{nm}u_m - f_n)^2, \end{aligned} \tag{4.82}$$

where ($W_i > 0,\ i = 1, n$) are the weighting factors. The minimizer of I is the solution of the normal equation:

$$\boldsymbol{K}^T\boldsymbol{W}\boldsymbol{K}\boldsymbol{u} = \boldsymbol{K}^T\boldsymbol{W}\boldsymbol{f}, \tag{4.83}$$

where $\boldsymbol{W}$ is an $n \times n$ diagonal matrix whose diagonal entries consist of W_i, and T denotes the transpose.

If the number of equations is less than the number of unknowns, i.e., $n < m$, in other words if system (4.79) is underdetermined, then this system has an infinite number of solutions or no solution. For illustration, consider the cases

$$\begin{aligned} x_1 + x_2 + x_3 &= 3, \\ 2x_1 + x_2 + x_3 &= 6, \end{aligned} \tag{4.84}$$

and

$$\begin{aligned} x_1 + x_2 + x_3 &= 3, \\ x_1 + x_2 + x_3 &= 6. \end{aligned} \tag{4.85}$$

System (4.84) has an infinite number of solutions, since for any given x_3 one can find the solution x_1 and x_2; there is no solution to system (4.85). It is easy to verify that in both cases the matrix $\boldsymbol{K}^T\boldsymbol{W}\boldsymbol{K}$ is singular, i.e., its determinant is equal to zero. This means that neither (4.84) nor (4.85) can be solved by using the least-squares method. Obviously, the underdetermined system of linear algebraic equations is not well-posed, and thus cannot be solved by any method including the least-squares method.

If $\boldsymbol{K}$ is an $n \times n$ matrix and $\det\boldsymbol{K} \neq 0$, then the solution to system (4.79) exists and is unique. It is well-known that in this case system (4.79) can be solved by the least-squares method. No matter what the weighting matrix $\boldsymbol{W}$ is, the least-squares solution obtained by solving the normal equation (4.83) is identical to the solution of system (4.79). In this case, if the least-squares solution is substituted into system (4.79), then all equation residuals are equal to zero.

If the number of equations is greater than the number of unknowns, i.e., $n > m$, and the rank of the augmented matrix is equal to $m + 1$ (rank condition), then system (4.79) is overdetermined. In this case, although system (4.79) has no solution, the least-squares solution to the normal equation (4.83) exists uniquely. That is, for an overdetermined system we always can find a solution in the least-squares sense. In this case, different weighting matrices lead to different least-squares solutions; the larger the weighting factor, the closer the satisfaction of the corresponding equation; besides, the equation residuals for the least-squares solution in general are not equal to zero.

To illustrate the least-squares method for overdetermined systems, let us consider the following example:

$$\begin{aligned} x_1 + x_2 &= 3, \\ 2x_1 + x_2 &= 4, \\ x_1 + 0x_2 &= 2, \\ 2x_1 + 2x_2 &= 6. \end{aligned} \tag{4.86}$$

This system has four equations and two unknowns. Moreover, the rank of

$$\begin{pmatrix} 1 & 1 & 3 \\ 2 & 1 & 4 \\ 1 & 0 & 2 \\ 2 & 2 & 6 \end{pmatrix}$$

is three. Hence system (4.86) is overdetermined and has no solution. If ($W_i = 1, i = 1, n$) are chosen, the normal equation corresponding to (4.86) is

$$\begin{pmatrix} 1 & 2 & 1 & 2 \\ 1 & 1 & 0 & 2 \end{pmatrix} \begin{pmatrix} 1 & 1 \\ 2 & 1 \\ 1 & 0 \\ 2 & 2 \end{pmatrix} \begin{pmatrix} x_1 \\ x_2 \end{pmatrix} = \begin{pmatrix} 1 & 2 & 1 & 2 \\ 1 & 1 & 0 & 2 \end{pmatrix} \begin{pmatrix} 3 \\ 4 \\ 2 \\ 6 \end{pmatrix},$$

or

$$\begin{pmatrix} 10 & 7 \\ 7 & 6 \end{pmatrix} \begin{pmatrix} x_1 \\ x_2 \end{pmatrix} = \begin{pmatrix} 25 \\ 19 \end{pmatrix}. \tag{4.87}$$

The least-squares solution to system (4.86), i.e., the solution to the normal equation (4.87) is $(1.545, 1.364)$.

We emphasize that the satisfaction of the rank condition is important for classifying a system as overdetermined. The rank condition guarantees that $m + 1$ linearly independent equations can be found in the system. In example (4.86), the first three equations are linearly independent, i.e., the third equation cannot be obtained by a linear combination of the first two equations.

If $n > m$, but the rank of the augmented matrix equals m, then the system is determined. For example, the rank of the augmented matrix of

$$\begin{aligned} x_1 + x_2 &= 3, \\ 2x_1 + x_2 &= 4, \\ 4x_1 + 2x_2 &= 8, \\ 2x_1 + 2x_2 &= 6 \end{aligned} \tag{4.88}$$

is $2 = m$, therefore (4.88) is determined. The rank of the augmented matrix of

$$\begin{aligned} x_1 + x_2 &= 3, \\ 2x_1 + 2x_2 &= 6, \\ 3x_1 + 3x_2 &= 9, \\ 4x_1 + 4x_2 &= 12 \end{aligned} \tag{4.89}$$

is $1 < m = 2$, therefore system (4.89) is underdetermined.

In summary, the least-squares method can be used only for the solution of determined and overdetermined systems of linear algebraic equations for which $n \geq m$ and $\text{rank}_{\text{aug}} = m$ or $m + 1$.

4.11.2 The Least-Squares Finite Element Collocation Method

The finite element methods that we have considered so far have been based on the Galerkin and least-squares variational principles. In this section we introduce finite element methods based on the concept of collocation and then consider the least-squares finite element collocation method in which both the collocation and the least-squares ideas are combined.

In the finite element collocation method, an approximate solution is constructed so that it satisfies boundary conditions in advance, and the nodal values in the finite element expansion are then determined so that the approximation satisfies the differential equation at a number of distinct points in the domain. For simplicity, we shall describe the method as it applies to the general two-dimensional first-order partial differential equations:

$$\boldsymbol{A}_1\frac{\partial \boldsymbol{u}}{\partial x} + \boldsymbol{A}_2\frac{\partial \boldsymbol{u}}{\partial y} + \boldsymbol{A}_0\boldsymbol{u} = \boldsymbol{f} \tag{4.90}$$

with an appropriate boundary condition. The notations in (4.90) have been defined in Sect. 4.8. Using the finite element expansion (4.57), the element residuals may be defined on the element:

$$\boldsymbol{R}(x,y) = \sum_{j=1}^{N_n}\left(\frac{\partial \psi_j}{\partial x}\boldsymbol{A}_1 + \frac{\partial \psi_j}{\partial y}\boldsymbol{A}_2 + \psi_j\boldsymbol{A}_0\right)\begin{pmatrix} u_1 \\ u_2 \\ \vdots \\ u_m \end{pmatrix}_j - \boldsymbol{f}. \tag{4.91}$$

Usually, in the finite element computation the evaluation of element matrices is performed in the master element with the local ξ and η coordinates, see e.g., Carey and Oden (1983a). Collocating at an interior point (ξ_c, η_c) the collocation equations on the element are

$$\boldsymbol{R}(\xi_c, \eta_c) = \boldsymbol{0},$$

or more clearly

$$\left\{\sum_{j=1}^{N_n}\left(\frac{\partial \psi_j}{\partial x}\boldsymbol{A}_1 + \frac{\partial \psi_j}{\partial y}\boldsymbol{A}_2 + \psi_j\boldsymbol{A}_0\right)\begin{pmatrix} u_1 \\ u_2 \\ \vdots \\ u_m \end{pmatrix}_j - \boldsymbol{f}\right\}_{(\xi_c,\eta_c)} = \boldsymbol{0}. \tag{4.92}$$

The unknown nodal values can be found by solving the resulting system of linear algebraic equations (4.92). In order to obtain a determined system, we should choose a proper number of interior collocation points in each element such that the total number of equations is equal to the total number of unknown nodal values. Since the resulting system of algebraic equations (4.92) by the collocation method is sparse but not symmetric, it is difficult to find its solution.

Instead of solving the residual equations (4.92) directly, we may use the least-squares method discussed in the above section to obtain the solution.

We now collocate at N_{col} interior points in each element and minimize the summation of weighted, squared residuals:

$$I(\boldsymbol{v}_h) = \sum_{i=1}^{N_{\text{elem}}} \left\{ \sum_{l=1}^{N_{\text{col}}} W_l \boldsymbol{R}^2(\xi_l, \eta_l) \right\}, \tag{4.93}$$

where N_{elem} is the number of elements in the finite element discretization. The method based on the minimization of (4.93) is called the least-squares finite element collocation method.

4.11.3 Importance of the Order of Gaussian Quadrature

Now let us investigate the relationship between the LSFEM introduced in Sect. 4.8 and the least-squares finite element collocation method. As mentioned in the beginning of this section, in the LSFEM computation we use Gaussian quadrature to evaluate the element matrices. This is equivalent to minimizing the following summation of weighted squared residuals:

$$I(\boldsymbol{v}_h) = \sum_{i=1}^{N_{\text{elem}}} \left\{ \sum_{l=1}^{N_{\text{Gauss}}} w_l \boldsymbol{R}^2(\xi_l, \eta_l) |J(\xi_l, \eta_l)| \right\}, \tag{4.94}$$

where N_{Gauss} is the number of Gaussian points, w_l is the weighting factor in the Gaussian quadrature, and (ξ_l, η_l) is the location of Gaussian points in the master element, and J is the determinant of the Jacobian matrix of the coordinate transformation.

Comparing (4.93) with (4.94) we realize that the LSFEM based on Gaussian quadrature is equivalent to the least-squares finite element collocation method in which Gaussian points are chosen as the collocation points.

As discussed in Sect. 4.11.1, in order for this least-squares finite element collocation method to make sense, we must solve a determined or an overdetermined set of residual equations, that is, the total number of residual equations must be equal to or greater than the total number of unknown nodal values. This requirement can be expressed as

$$N_{\text{elem}} \times N_{\text{Gauss}} \times N_{\text{eq}} \geq N_{\text{node}} \times m - N_{\text{bc}}, \tag{4.95}$$

in which N_{eq} is the number of equations in (4.90), N_{node} is the total number of nodes, m is the number of components of $\boldsymbol{u}$ or the number of degrees of freedom at each node, N_{bc} is the total number of given nodal values at boundary nodes.

Requirement (4.95) is very important in the implementation of LSFEM. The satisfaction of requirement (4.95) depends on the chosen number of Gaussian points. If the order of Gaussian quadrature is too low, then this condition is not satisfied; consequently the global (stiffness) matrix will be singular and thus cannot be inverted by any direct solver. Further, in this case iterative solutions do not converge. However, the order of Gaussian quadrature should not be too high either; otherwise, the least-squares method amounts to solving an extremely overdetermined system. This implies that one intends forcing too many residual equations to be zero at Gaussian points with too few adjustable unknowns. Obviously this is impossible. In this case, the solution is inaccurate and often underestimated.

We also can use requirement (4.95) to explain why in general the linear triangle and the linear tetrahedron are not suitable for LSFEM. For example, consider a set of two first-order equations in a square domain. Assume that this problem involves two unknown functions, and has two partial differential equations in the domain and one boundary condition on each boundary. The domain is divided into $(i \times i)$ small squares, and then each small square is divided into two triangles. We choose one-point quadrature. This is the minimum number of Gauss points one can choose for triangles. For this problem, $N_{\text{elem}} = 2i^2$, $N_{\text{node}} = (i+1)^2$, $N_{\text{bc}} = 4(i+1)$ and $N_{\text{Gauss}} = 1$. The total number of unknown nodal values N_{unknown} is equal to $N_{\text{node}} \times 2 - N_{\text{bc}} = 2i^2 - 2$. The total number of residual equations is equal to $4i^2$ which is almost two-times greater than N_{unknown}, thus this least-squares solution will not be satisfactory. We note that this trouble is not unique to LSFEM. For the same reason, the linear triangle and the linear tetrahedron also should be avoided for Rayleigh–Ritz finite element methods. We are familiar with this situation in elasticity: the linear triangle is too stiff, and thus the solution is underestimated.

4.12 Concluding Remarks

In this chapter we have established that the mathematical foundation of the least-squares method for the solution of linear operator equations is simply the bounded inverse theorem. As long as the linear operator equation is well-posed, or in other words, the linear operator is bounded below, it is guaranteed that minimization of a certain norm of the residual will produce a convergent approximate solution. We have also shown that LSFEM based on first-order systems has an optimal rate of convergence for strictly elliptic problems and a suboptimal rate for general well-posed problems.

It has already become standard practice to employ different numerical schemes to deal with each type of differential equation. For example, central differences are used for elliptic problems; upwinding methods are designed for hyperbolic problems; for mixed-type problems, in which the governing equation is elliptic in a part of the region and hyperbolic in another part, one has to distinguish between them in advance and then use different schemes in different regions. Some problems, such as the incompressible Euler equations governing incompressible rotational flows, are neither elliptic nor hyperbolic, and thus are very difficult to solve by conventional methods. We show in this chapter that the standard LSFEM with one formulation (4.56) is suitable for all kinds of boundary value problems of partial differential equations, and is valid for the whole problem domain. This is the reason why various problems in engineering and physics can be simulated within a unified LSFEM framework. Special treatments, particularly upwinding, are totally unnecessary.

The least-squares formulation (4.56) in terms of the bilinear form and the linear forms offers great advantages. First, it is a unified variational formulation exclusively accepted in all modern mathematical literature. Based on this formulation one can further conduct theoretical analyses by using the bounded inverse theorem or the Lax–Milgram theorem (see Appendix D). Second, based on formulation (4.56) one can develop a general purpose least-squares finite element program to solve any first-order system of partial differential equations. When a new problem is given, one needs only to supply the entries of the matrices $\boldsymbol{A}_i$, $\boldsymbol{A}_0$ and the vector $\boldsymbol{f}$, and then let computers do the rest of the work. For example, one may write a symbolic language code to generate the FORTRAN subroutine for formation of the matrix $\boldsymbol{K}$, if a direct solver is employed; or for multiplication of the element matrix with the element vector, if a matrix-free, element-by-element iterative solver is used (see Chap. 15).

To avoid underestimated solutions, (1) the use of linear triangle and tetrahedron elements in LSFEM is not recommended; (2) the order of Gaussian quadrature in the computation of matrices should be carefully chosen such that the criterion (4.95) is just marginally satisfied, that corresponds to solving a determined or slightly overdetermined algebraic equations.

In some literature, e.g., Eason (1976), the least-squares method is formulated as:

$$\frac{\partial}{\partial a_j}\|\boldsymbol{A}\boldsymbol{v}_h - \boldsymbol{f}\|_0^2 = 0, \qquad j = 1, \ldots, N_h$$

where a_j are the degrees of freedom of the approximate solution and N_h is the total number of degrees of freedom. Some authors even directly use the least-squares method to the non-linear governing equations, then linearize the resulting non-linear algebraic equations. In all these approaches one has to derive a corresponding algebraic system of equations for each particular problem by hand. Clearly, it is very tedious and completely unnecessary. We must mention that the least-squares method based on linear operators leads to the minimization problems of quadratic functionals which have been well studied, and the bounded below theorem is applicable only for linear operators. Currently, no simple and general mathematical theory is available for higher order functionals. For these reasons, we prefer linearizing the differential equations first, then using the least-squares method.

5. Div–Curl System

The div–curl system is an important class of first-order partial differential equations. This system governs, for example, static electromagnetic fields, and incompressible irrotational fluid flows. The div–curl system is also fundamental from a theoretical point of view, since the Stokes equations and the incompressible Navier–Stokes equations written in the first-order velocity–pressure–vorticity formulation, as well as the Maxwell equations consist of two div–curl systems. The three-dimensional div–curl system is traditionally considered as "overdetermined" or "overspecified", because it has four equations involving only three unknowns. For this reason, it is not easy to solve by using conventional numerical methods. In this chapter, we will prove that the div–curl system is really well determined and strongly elliptic by introducing a dummy variable, and explain that for the well-posedness the div–curl system should have two algebraic boundary conditions. We will also show that the LSFEM is the best choice for numerical solution of the div–curl system.

In fluid dynamics and electromagnetics there are some good reasons why other higher-order versions of the Navier–Stokes equations and the Maxwell equations are often useful (see the discussion in Chaps. 8 and 14). In this chapter, we take the div–curl system as an example to show how to use the div–curl approach and the least-squares method to derive equivalent second-order equations and their boundary conditions.

5.1 Basic Theorems

In order to study the div–curl system and its boundary conditions, to establish a rigorous method for deriving equivalent higher-order equations, and to analyze the least-squares method, we need some basic theorems and inequalities which are related to the div, curl and grad operators.

We assume that $\Omega \subset \mathbb{R}^3$ is an open bounded domain with a piecewise smooth (as discussed in Chap. 4) boundary $\Gamma = \Gamma_1 \cup \Gamma_2$. Either Γ_1 or Γ_2, not both, may be empty. If both Γ_1 and Γ_2 are not empty, they are required to have at least one common point. $\boldsymbol{x}$ or (x, y, z) denotes a point in Ω; $\boldsymbol{n}$ a unit outward normal vector on the boundary, and $\boldsymbol{\tau}$ a tangential vector to Γ at a boundary point, respectively. $\boldsymbol{\tau}_1$ and $\boldsymbol{\tau}_2$ represent two orthogonal vectors tangential to Γ at a boundary point. In order to emphasize the basic ideas and maintain simplicity, we often further assume that the domain Ω is convex or simply connected, although these restrictions are not necessary in many cases.

In this book, we use the symbols $\nabla, \nabla\cdot, \nabla\times$ and Δ to denote the gradient, divergence, curl and Laplacian operators, respectively. We also use the following notations:

$$\langle u, v\rangle_\Gamma = \int_\Gamma uv d\Gamma,$$

$$\langle \boldsymbol{u}, \boldsymbol{v}\rangle_\Gamma = \int_\Gamma \boldsymbol{u}\cdot\boldsymbol{v} d\Gamma.$$

When there is no confusion, we will often omit the measure Γ from these inner products.

Lemma 5.1 Let Ω be a bounded open subset of $\mathbb{R}^3$ with a piecewise smooth boundary Γ. Then every function $\boldsymbol{u}$ of $[H^1(\Omega)]^3$ with $\boldsymbol{n}\times\boldsymbol{u} = \boldsymbol{0}$ on Γ satisfies

$$|\boldsymbol{u}|_1^2 + \int_\Gamma \Big(\frac{1}{R_1} + \frac{1}{R_2}\Big)\boldsymbol{u}\cdot\boldsymbol{u} d\Gamma = \|\nabla\cdot\boldsymbol{u}\|_0^2 + \|\nabla\times\boldsymbol{u}\|_0^2, \tag{5.1}$$

where R_1 and R_2 denote the principal radii of curvature for Γ.

Proof: Using Green's formulae (B.2), (B.3) and (B.7), and Equality (A.4), we have

$$\begin{aligned}\|\nabla\cdot\boldsymbol{u}\|_0^2 + \|\nabla\times\boldsymbol{u}\|_0^2 &= (\nabla\cdot\boldsymbol{u},\ \nabla\cdot\boldsymbol{u}) + (\nabla\times\boldsymbol{u},\ \nabla\times\boldsymbol{u})\\ &= \Big(\boldsymbol{u},\ -\nabla(\nabla\cdot\boldsymbol{u})\Big) + \langle\nabla\cdot\boldsymbol{u},\ \boldsymbol{n}\cdot\boldsymbol{u}\rangle\\ &\quad + \Big(\boldsymbol{u},\ \nabla(\nabla\cdot\boldsymbol{u}) - \Delta\boldsymbol{u}\Big) + \langle\nabla\times\boldsymbol{u},\ \boldsymbol{n}\times\boldsymbol{u}\rangle\\ &= (\nabla\boldsymbol{u},\ \nabla\boldsymbol{u}) + \langle\nabla\cdot\boldsymbol{u},\ \boldsymbol{n}\cdot\boldsymbol{u}\rangle + \langle\nabla\times\boldsymbol{u},\ \boldsymbol{n}\times\boldsymbol{u}\rangle - \Big\langle\frac{\partial\boldsymbol{u}}{\partial n},\ \boldsymbol{u}\Big\rangle. \end{aligned}\tag{5.2}$$

Obviously

$$(\nabla\boldsymbol{u},\ \nabla\boldsymbol{u}) = |\boldsymbol{u}|_1^2.$$

Now we turn to the boundary integral terms. $\boldsymbol{n}\times\boldsymbol{u} = \boldsymbol{0}$ on Γ implies that

$$\boldsymbol{u} = U\boldsymbol{n} \qquad \text{on } \Gamma,$$

where U is a scalar function whose value depends on its location on the boundary surface. By using (A.1), the boundary integral terms in (5.2) can be written as follows:

$$\int_\Gamma \left\{ U\nabla \cdot (U\boldsymbol{n}) - U\frac{\partial U}{\partial n} \right\} d\Gamma$$

$$= \int_\Gamma \Big\{ U(U\nabla \cdot \boldsymbol{n} + \boldsymbol{n} \cdot \nabla U) - U\nabla U \cdot \boldsymbol{n} \Big\} d\Gamma = \int_\Gamma U^2 \nabla \cdot \boldsymbol{n} d\Gamma.$$

It can be verified that on a smooth curved surface

$$\nabla \cdot \boldsymbol{n} = \frac{1}{R_1} + \frac{1}{R_2} = 2\kappa,$$

where κ is the mean curvature. □

Lemma 5.2 Let Ω be a bounded open subset of $\mathbb{R}^3$ with a piecewise smooth boundary Γ. Then every function $\boldsymbol{u}$ of $[H^1(\Omega)]^3$ with $\boldsymbol{n} \cdot \boldsymbol{u} = 0$ on Γ satisfies

$$|\boldsymbol{u}|_1^2 + \int_\Gamma \frac{1}{R} \boldsymbol{u} \cdot \boldsymbol{u} d\Gamma = \|\nabla \cdot \boldsymbol{u}\|_0^2 + \|\nabla \times \boldsymbol{u}\|_0^2, \tag{5.3}$$

where $R_1 \leq R \leq R_2$, in which R_1 and R_2 denote the principal radii of curvature for Γ.

Proof: In this case we still have (5.2). Since $\boldsymbol{n} \cdot \boldsymbol{u} = 0$, we may assume that

$$\boldsymbol{u} = U\boldsymbol{\tau} \qquad \text{on } \Gamma.$$

By virtue of the triple scalar product

$$(\nabla \times \boldsymbol{u}) \cdot (\boldsymbol{n} \times \boldsymbol{u}) = \boldsymbol{n} \cdot \big(\boldsymbol{u} \times (\nabla \times \boldsymbol{u})\big)$$

and using (A.5), the boundary integral terms in (5.2) can be written as

$$\int_\Gamma \left\{ \boldsymbol{n} \cdot (\boldsymbol{u} \times \nabla \times \boldsymbol{u}) - \frac{1}{2}\frac{\partial \boldsymbol{u}^2}{\partial n} \right\} d\Gamma$$

$$= \int_\Gamma \left\{ \boldsymbol{n} \cdot \Big(\nabla\Big(\frac{1}{2}\boldsymbol{u}^2\Big) - (\boldsymbol{u} \cdot \nabla)\boldsymbol{u} \Big) - \frac{1}{2}\frac{\partial \boldsymbol{u}^2}{\partial n} \right\} d\Gamma = \int_\Gamma \{ \boldsymbol{n} \cdot (-(\boldsymbol{u} \cdot \nabla)\boldsymbol{u}) \} d\Gamma$$

$$= \int_\Gamma \{ U^2 \boldsymbol{n} \cdot (-(\boldsymbol{\tau} \cdot \nabla)\boldsymbol{\tau}) \} d\Gamma = \int_\Gamma \Big\{ U^2 \boldsymbol{n} \cdot \Big(\frac{1}{R}\boldsymbol{n} \Big) \Big\} d\Gamma.$$

So the Lemma is proved. Here R is the radius of curvature of the boundary surface in the direction of $\boldsymbol{\tau}$. □

Since the curvature $1/R$ is always positive when the boundary surface is convex, we derive immediately the following theorems:

Theorem 5.1 Let Ω be a bounded and convex open subset of $\mathbb{R}^3$. Then every function $\boldsymbol{u}$ of $[H^1(\Omega)]^3$ with $\boldsymbol{n} \cdot \boldsymbol{u} = 0$ on Γ_1 and $\boldsymbol{n} \times \boldsymbol{u} = \boldsymbol{0}$ on Γ_2 satisfies:

$$|\boldsymbol{u}|_1^2 \leq \|\nabla \cdot \boldsymbol{u}\|_0^2 + \|\nabla \times \boldsymbol{u}\|_0^2. \tag{5.4}$$

Theorem 5.2 (The Div–Curl Theorem) Suppose that Ω is a (1) bounded and convex, or (2) bounded and simply connected subset of $\mathbb{R}^3$. If $\boldsymbol{u} \in [H^1(\Omega)]^3$ satisfies

$$\nabla \cdot \boldsymbol{u} = 0 \quad \text{in } \Omega,$$

$$\nabla \times \boldsymbol{u} = \mathbf{0} \quad \text{in } \Omega,$$

$$\boldsymbol{n} \cdot \boldsymbol{u} = 0 \quad \text{on } \Gamma_1,$$

$$\boldsymbol{n} \times \boldsymbol{u} = \mathbf{0} \quad \text{on } \Gamma_2,$$

then

$$\boldsymbol{u} \equiv \mathbf{0} \quad \text{in } \Omega.$$

Proof: Case (1). From Theorem 5.1, we have

$$|\boldsymbol{u}|_1^2 \leq 0,$$

that is, $\boldsymbol{u}$ must be a constant vector in Ω. If Γ_2 is empty, we always can find three different points on the boundary with three different normal directions. Due to the boundary condition, the components of this constant vector along these three directions must be zero, therefore this constant vector must be zero. This conclusion is also true for the case $\Gamma_1 = 0$. If both Γ_1 and Γ_2 exist, at the common point of Γ_1 and Γ_2 we have $\boldsymbol{u} = \mathbf{0}$, therefore this constant vector must be zero.

Case (2). Since the domain Ω is simply connected, we can introduce the potential ϕ such that $\boldsymbol{u} = \nabla\phi$, and show $\phi =$ constant. □

Remark. We distinguish between case (1) and case (2) only for convenience. In fact, case (2) includes case (1), because a convex domain must be simply connected.

Theorem 5.3 (The Friedrichs First Div–Curl Inequality) Suppose that Ω is a (1) bounded and convex, or (2) bounded and simply connected subset of $\mathbb{R}^3$. Then every function $\boldsymbol{u}$ of $[H^1(\Omega)]^3$ with $\boldsymbol{n} \cdot \boldsymbol{u} = 0$ on Γ_1 and $\boldsymbol{n} \times \boldsymbol{u} = \mathbf{0}$ on Γ_2 satisfies:

$$\|\boldsymbol{u}\|_1^2 \leq C(\|\nabla \cdot \boldsymbol{u}\|_0^2 + \|\nabla \times \boldsymbol{u}\|_0^2), \tag{5.5}$$

where the constant $C > 0$ depends only on Ω.

Proof: Case (1). Theorem 5.2 indicates that the homogeneous div–curl system with homogeneous boundary conditions has only a trivial solution. In other words the div–curl system under the given boundary conditions has a unique solution. Therefore, by virtue of Theorem 4.6, there exists a positive constant α such that

$$\alpha\|\boldsymbol{u}\|_0^2 \le (\|\nabla\cdot\boldsymbol{u}\|_0^2 + \|\nabla\times\boldsymbol{u}\|_0^2). \tag{5.6}$$

Combining (5.6) with (5.4) leads to (5.5).

Case (2). In this case (5.6) is still valid, but (5.4) cannot be guaranteed. However, we always can multiply (5.6) by a positive constant M and add it to (5.1) or (5.3) to eliminate the possible negative boundary integral term, and obtain

$$|\boldsymbol{u}|_1^2 \le (M+1)(\|\nabla\cdot\boldsymbol{u}\|_0^2 + \|\nabla\times\boldsymbol{u}\|_0^2). \tag{5.7}$$

Combining (5.7) with (5.6) leads to (5.5). □

This theorem indicates the fact that for the function space

$$\boldsymbol{H} = \{[H^1(\Omega)]^3 : \boldsymbol{n}\cdot\boldsymbol{u} = 0 \text{ on } \Gamma_1, \boldsymbol{n}\times\boldsymbol{u} = \boldsymbol{0} \text{ on } \Gamma_2\}$$

$\|\boldsymbol{u}\|_1^2$ and $(\|\nabla\cdot\boldsymbol{u}\|_0^2 + \|\nabla\times\boldsymbol{u}\|_0^2)$ are equivalent norms for an appropriate domain Ω.

The proof of Theorem 5.3 can also be based on the use of contradiction arguments together with Theorem 5.1, see e.g., Saranen (1982) and the references therein. In the two-dimensional case, a direct proof is available (Křížek and Neittaanmäki 1984a or 1990).

Theorem 5.4 (The Gradient Theorem). If $g \in H^1(\Omega)$ satisfies

$$\nabla g = \boldsymbol{0} \qquad \text{in } \Omega,$$

$$g = 0 \qquad \text{on } \Gamma_1 \ne 0 \text{ (or on } \Gamma_2 \ne 0),$$

then

$$g \equiv 0 \qquad \text{in } \Omega.$$

Proof: Since $g \in H^1(\Omega)$, g must be continuous in Ω. From $\nabla g = \boldsymbol{0}$ in Ω, we know that g is a constant. Due to the boundary condition, g must be identical to zero. □

In fact, $g = 0$ needs to be specified only at any point in the domain or on the boundary. This theorem will be used to derive the higher-order equations which are equivalent to a scalar equation.

Theorem 5.5 If $\boldsymbol{u} \in [H^1(\Omega)]^3$ and $\boldsymbol{n}\times\boldsymbol{u} = \boldsymbol{0}$ on $\Gamma_2 \ne 0$, then $\boldsymbol{n}\cdot\nabla\times\boldsymbol{u} = 0$ on Γ_2.

Proof: Assume the contrary, say, $\boldsymbol{n}\cdot(\nabla\times\boldsymbol{u}) > 0$ at a point P on Γ_2, then in a neighborhood surface σ of P we have

$$\boldsymbol{n}\cdot(\nabla\times\boldsymbol{u}) > \varepsilon > 0,$$

in which ε is a small positive constant. Considering that $\boldsymbol{u}$ on Γ_2 is in the normal direction $\boldsymbol{n}$, i.e., $\boldsymbol{u}$ is orthogonal to any tangential direction on Γ_2, and from the Stokes theorem we have a contradiction:

$$0 = \oint_c \boldsymbol{u} \cdot d\boldsymbol{s} = \int_\sigma (\nabla \times \boldsymbol{u}) \cdot \boldsymbol{n} d\sigma > 0,$$

where c is the boundary contour of σ. □

The proof of Theorem 5.5 can also be found in Pironneau (1989, p.53).

5.2 Determinacy and Ellipticity

Let us consider the following three-dimensional div–curl system:

$$\nabla \times \boldsymbol{u} = \boldsymbol{\omega} \qquad \text{in } \Omega, \tag{5.8a}$$

$$\nabla \cdot \boldsymbol{u} = \rho \qquad \text{in } \Omega, \tag{5.8b}$$

$$\boldsymbol{n} \cdot \boldsymbol{u} = 0 \qquad \text{on } \Gamma_1, \tag{5.8c}$$

$$\boldsymbol{n} \times \boldsymbol{u} = \boldsymbol{0} \qquad \text{on } \Gamma_2, \tag{5.8d}$$

where the given vector function $\boldsymbol{\omega} \in [L_2(\Omega)]^3$ cannot be arbitrary; it must satisfy the following solvability conditions:

$$\nabla \cdot \boldsymbol{\omega} = 0 \qquad \text{in } \Omega, \tag{5.9a}$$

$$\boldsymbol{n} \cdot \boldsymbol{\omega} = 0 \qquad \text{on } \Gamma_2, \tag{5.9b}$$

$$\int_\Gamma \boldsymbol{n} \cdot \boldsymbol{\omega} d\Gamma = 0. \tag{5.9c}$$

If Γ_2 is empty, then the given scalar function $\rho \in L_2(\Omega)$ must satisfy the solvability condition:

$$\int_\Omega \rho d\Omega = 0. \tag{5.9d}$$

At first glance, system (5.8) seems "overdetermined" or "overspecified", since there are four equations involving only three unknowns. However, after careful investigation we shall find that system (5.8) is properly determined and elliptic.

By introducing a dummy variable ϑ, system (5.8) can be written as

$$\nabla \vartheta + \nabla \times \boldsymbol{u} = \boldsymbol{\omega} \qquad \text{in } \Omega, \tag{5.10a}$$

$$\nabla \cdot \boldsymbol{u} = \rho \qquad \text{in } \Omega, \tag{5.10b}$$

$$\boldsymbol{n} \cdot \boldsymbol{u} = 0 \qquad \text{on } \Gamma_1, \tag{5.10c}$$

$$\vartheta = 0 \qquad \text{on } \Gamma_1, \tag{5.10d}$$

$$\boldsymbol{n} \times \boldsymbol{u} = \boldsymbol{0} \qquad \text{on } \Gamma_2. \tag{5.10e}$$

Notice that we impose $\vartheta = 0$ on Γ_1, and do not specify any boundary condition for the dummy variable ϑ on Γ_2.

By virtue of Theorem 5.2, the vector equation (5.10a) is equivalent to the following equations and boundary conditions:

$$\nabla \times (\nabla\vartheta + \nabla \times \boldsymbol{u} - \boldsymbol{\omega}) = \boldsymbol{0} \qquad \text{in } \Omega, \tag{5.11a}$$

$$\nabla \cdot (\nabla\vartheta + \nabla \times \boldsymbol{u} - \boldsymbol{\omega}) = 0 \qquad \text{in } \Omega, \tag{5.11b}$$

$$\boldsymbol{n} \times (\nabla\vartheta + \nabla \times \boldsymbol{u} - \boldsymbol{\omega}) = \boldsymbol{0} \qquad \text{on } \Gamma_1, \tag{5.11c}$$

$$\boldsymbol{n} \cdot (\nabla\vartheta + \nabla \times \boldsymbol{u} - \boldsymbol{\omega}) = 0 \qquad \text{on } \Gamma_2. \tag{5.11d}$$

Taking into account the solvability conditions (5.9a) and (5.9b), the boundary condition (5.10e) and Theorem 5.5, from (5.11b), (5.10d) and (5.11d) we have

$$\Delta\vartheta = 0 \qquad \text{in } \Omega, \tag{5.12a}$$

$$\vartheta = 0 \qquad \text{on } \Gamma_1, \tag{5.12b}$$

$$\frac{\partial\vartheta}{\partial n} = 0 \qquad \text{on } \Gamma_2. \tag{5.12c}$$

From (5.12) we know that $\vartheta \equiv 0$ in Ω. That is, the introduction of ϑ into (5.8) does not change anything, and thus system (5.10) with four equations and four unknowns is indeed equivalent to system (5.8).

We cannot classify system (5.8), since the classification introduced in Sect. 4.7 requires that the coefficient matrices of a first-order system be square. But now we can classify system (5.10). In Cartesian coordinates the equations in system (5.10) are given as

$$\begin{aligned}
\frac{\partial\vartheta}{\partial x} + \frac{\partial w}{\partial y} - \frac{\partial v}{\partial z} &= \omega_x, \\
\frac{\partial\vartheta}{\partial y} + \frac{\partial u}{\partial z} - \frac{\partial w}{\partial x} &= \omega_y, \\
\frac{\partial\vartheta}{\partial z} + \frac{\partial v}{\partial x} - \frac{\partial u}{\partial y} &= \omega_z, \\
\frac{\partial u}{\partial x} + \frac{\partial v}{\partial y} + \frac{\partial w}{\partial z} &= \rho.
\end{aligned} \tag{5.13}$$

We may write system (5.13) in standard matrix form:

$$\boldsymbol{A}_1 \frac{\partial \underline{\boldsymbol{u}}}{\partial x} + \boldsymbol{A}_2 \frac{\partial \underline{\boldsymbol{u}}}{\partial y} + \boldsymbol{A}_3 \frac{\partial \underline{\boldsymbol{u}}}{\partial z} + \boldsymbol{A}_0 \underline{\boldsymbol{u}} = \boldsymbol{f}, \tag{5.14}$$

in which

$$\boldsymbol{A}_1 = \begin{pmatrix} 0 & 0 & 0 & 1 \\ 0 & 0 & -1 & 0 \\ 0 & 1 & 0 & 0 \\ 1 & 0 & 0 & 0 \end{pmatrix}, \quad \boldsymbol{A}_2 = \begin{pmatrix} 0 & 0 & 1 & 0 \\ 0 & 0 & 0 & 1 \\ -1 & 0 & 0 & 0 \\ 0 & 1 & 0 & 0 \end{pmatrix},$$

$$\boldsymbol{A}_3 = \begin{pmatrix} 0 & -1 & 0 & 0 \\ 1 & 0 & 0 & 0 \\ 0 & 0 & 0 & 1 \\ 0 & 0 & 1 & 0 \end{pmatrix}, \quad \boldsymbol{A}_0 = \begin{pmatrix} 0 & 0 & 0 & 0 \\ 0 & 0 & 0 & 0 \\ 0 & 0 & 0 & 0 \\ 0 & 0 & 0 & 0 \end{pmatrix},$$

$$\boldsymbol{f} = \begin{pmatrix} \omega_x \\ \omega_y \\ \omega_z \\ \rho \end{pmatrix}, \qquad \underline{\boldsymbol{u}} = \begin{pmatrix} u \\ v \\ w \\ \vartheta \end{pmatrix}.$$

The characteristic polynomial associated with system (5.13) is

$$\det(\boldsymbol{A}_1\xi + \boldsymbol{A}_2\eta + \boldsymbol{A}_3\zeta) = \det \begin{pmatrix} 0 & -\zeta & \eta & \xi \\ \zeta & 0 & -\xi & \eta \\ -\eta & \xi & 0 & \zeta \\ \xi & \eta & \zeta & 0 \end{pmatrix}$$

$$= (\xi^2 + \eta^2 + \zeta^2)^2 \neq 0$$

for all nonzero real triplets (ξ, η, ζ), system (5.10) is thus elliptic and properly determined.

The first-order elliptic system (5.10) has four equations in four unknowns, so two boundary conditions on each boundary are needed to make system (5.10) well-posed. Here $\vartheta = 0$ and $\boldsymbol{n} \cdot \boldsymbol{u} = 0$ serve as two boundary conditions on Γ_1; while $\boldsymbol{n} \times \boldsymbol{u} = \boldsymbol{0}$ implies that two tangential components of $\boldsymbol{u}$ are zero on Γ_2.

Since system (5.8) is equivalent to system (5.10), and system (5.10) is elliptic and properly determined, so is system (5.8).

Remark. In fact, the solvability conditions (5.9a), (5.9b) can be obtained by applying the div–curl method (see the next section) to (5.8a).

5.3 The Div–Curl Method

Let us derive a second-order system which is equivalent to the div–curl system (5.8). We further assume that $\boldsymbol{u}$, $\boldsymbol{\omega}$ and ρ are sufficiently smooth. By virtue of Theorem 5.2, system (5.8) is equivalent to the following system:

$$\nabla \times (\nabla \times \boldsymbol{u} - \boldsymbol{\omega}) = \boldsymbol{0} \qquad \text{in } \Omega, \tag{5.15a}$$

$$\nabla \cdot (\nabla \times \boldsymbol{u} - \boldsymbol{\omega}) = 0 \qquad \text{in } \Omega, \tag{5.15b}$$

$$\boldsymbol{n} \times (\nabla \times \boldsymbol{u} - \boldsymbol{\omega}) = \boldsymbol{0} \qquad \text{on } \Gamma_1, \tag{5.15c}$$

$$\boldsymbol{n} \cdot (\nabla \times \boldsymbol{u} - \boldsymbol{\omega}) = 0 \qquad \text{on } \Gamma_2. \tag{5.15d}$$

$$\nabla \cdot \boldsymbol{u} = \rho \qquad \text{in } \Omega, \tag{5.15e}$$

$$\boldsymbol{n} \cdot \boldsymbol{u} = 0 \qquad \text{on } \Gamma_1, \tag{5.15f}$$

$$\boldsymbol{n} \times \boldsymbol{u} = \boldsymbol{0} \qquad \text{on } \Gamma_2. \tag{5.15g}$$

Equations (5.15b) and (5.15d) are satisfied, due to the solvability conditions (5.9a), (5.9b), the boundary condition (5.15g) and Theorem 5.5. Therefore, system (5.15) can be simplified as

$$\nabla \times (\nabla \times \boldsymbol{u}) = \nabla \times \boldsymbol{\omega} \qquad \text{in } \Omega, \tag{5.16a}$$

$$\nabla \cdot \boldsymbol{u} = \rho \qquad \text{in } \Omega, \tag{5.16b}$$

$$\boldsymbol{n} \cdot \boldsymbol{u} = 0 \qquad \text{on } \Gamma_1, \tag{5.16c}$$

$$\boldsymbol{n} \times (\nabla \times \boldsymbol{u}) = \boldsymbol{n} \times \boldsymbol{\omega} \qquad \text{on } \Gamma_1, \tag{5.16d}$$

$$\boldsymbol{n} \times \boldsymbol{u} = \boldsymbol{0} \qquad \text{on } \Gamma_2. \tag{5.16e}$$

Now at least one thing is made clear by this div–curl procedure. That is, the curl–curl equation (5.16a) cannot stand alone; it must go with the divergence equation (5.16b) and the additional Neumann boundary condition (5.16d).

It is still difficult to solve the second-order curl–curl equation and the first-order divergence equation together. To avoid this difficulty we further simplify system (5.16). By virtue of Theorem 5.4, (5.16b) is equivalent to the following system of equations and boundary condition (assuming that $\Gamma_2 \neq 0$):

$$\nabla(\nabla \cdot \boldsymbol{u} - \rho) = \boldsymbol{0} \qquad \text{in } \Omega, \tag{5.17a}$$

$$\nabla \cdot \boldsymbol{u} = \rho \qquad \text{on } \Gamma_2. \tag{5.17b}$$

Taking into account (5.17) and the following vector identity:

$$\nabla \times \nabla \times \boldsymbol{u} = \nabla(\nabla \cdot \boldsymbol{u}) - \Delta \boldsymbol{u}, \tag{5.18}$$

system (5.16) can be reduced as

$$\Delta \boldsymbol{u} = -\nabla \times \boldsymbol{\omega} + \nabla \rho \qquad \text{in } \Omega, \tag{5.19a}$$

$$\nabla(\nabla \cdot \boldsymbol{u} - \rho) = \boldsymbol{0} \qquad \text{in } \Omega, \tag{5.19b}$$

$$\boldsymbol{n} \cdot \boldsymbol{u} = 0 \qquad \text{on } \Gamma_1, \tag{5.19c}$$

$$\boldsymbol{n} \times (\nabla \times \boldsymbol{u}) = \boldsymbol{n} \times \boldsymbol{\omega} \qquad \text{on } \Gamma_1, \tag{5.19d}$$

$$\boldsymbol{n} \times \boldsymbol{u} = \boldsymbol{0} \qquad \text{on } \Gamma_2, \tag{5.19e}$$

$$\nabla \cdot \boldsymbol{u} = \rho \qquad \text{on } \Gamma_2. \tag{5.19f}$$

The solution of the derived second-order system (5.16) or (5.19) is completely identical to the solution of the original div–curl system (5.8), therefore no spurious solutions will be produced by systems (5.16) or (5.19). Moreover, the divergence equation (5.19b) in system (5.19) can be deleted. That is, the divergence equation is implicitly satisfied by (5.19a) and boundary conditions (5.19c)–(5.19f). A rigorous proof of this statement will be given by using the least-squares method in Sect. 5.5. Here we give a simple explanation. Let us consider a slightly different problem:

$$\Delta \boldsymbol{u} = -\nabla \times \boldsymbol{\omega} + \nabla \rho \qquad \text{in } \Omega, \tag{5.20a}$$

$$\boldsymbol{n} \cdot \boldsymbol{u} = 0 \qquad \text{on } \Gamma_1, \tag{5.20b}$$

$$\boldsymbol{n} \times (\nabla \times \boldsymbol{u}) = \boldsymbol{n} \times \boldsymbol{\omega} \qquad \text{on } \Gamma_1, \tag{5.20c}$$

$$\boldsymbol{n} \times \boldsymbol{u} = \boldsymbol{0} \qquad \text{on } \Gamma_2, \tag{5.20d}$$

$$\nabla \cdot \boldsymbol{u} - \rho = 0 \qquad \text{on } \Gamma. \tag{5.20e}$$

That is, we let the divergence equation be satisfied on the whole boundary. Although this condition needs to be specified only on Γ_2, it is not wrong for it to be enforced on Γ. By taking the divergence of (5.20a) we obtain a Poisson equation of $\phi = \nabla \cdot \boldsymbol{u} - \rho$:

$$\Delta \phi = 0 \qquad \text{in } \Omega. \tag{5.21}$$

Since $\phi = 0$ on the whole boundary, ϕ must be equal to zero in the domain, i.e., the divergence equation is implicitly satisfied in system (5.20).

Now we have shown that when $\boldsymbol{u}$ is sufficiently smooth, the three-dimensional div–curl system can have three equivalent differential forms: (1) the first-order system (5.8); (2) the curl–curl equation (5.16a) which must be accompanied by the divergence equation (5.16b) and the additional Neumann boundary condition (5.16d); (3) three uncoupled Poisson equations (5.20a) with additional Neumann boundary conditions (5.20c) and (5.20e) provided by the original first-order system.

5.4 The Least-Squares Method

Let us introduce a powerful and systematic method, the least-squares method, to solve system (5.8) and to derive an equivalent second-order system which is free of spurious solutions. We construct the following quadratic functional:

$$I : \boldsymbol{V} \longrightarrow \mathbb{R},$$

$$I(\boldsymbol{u}) = \|\nabla \times \boldsymbol{u} - \boldsymbol{\omega}\|_0^2 + \|\nabla \cdot \boldsymbol{u} - \rho\|_0^2,$$

where $\boldsymbol{V} = \{\boldsymbol{u} \in [H^1(\Omega)]^3 : \boldsymbol{n} \cdot \boldsymbol{u} = 0 \text{ on } \Gamma_1, \boldsymbol{n} \times \boldsymbol{u} = 0 \text{ on } \Gamma_2\}$. We note that the introduction of a dummy variable ϑ in Sect. 5.2 is purely for verification of the determinacy, and it is not required in the least-squares functional I. Taking the variation of I with respect to $\boldsymbol{u}$, and letting $\delta\boldsymbol{u} = \boldsymbol{v}$ and $\delta I = 0$, we obtain a least-squares variational formulation of the following type: Find $\boldsymbol{u} \in \boldsymbol{V}$ such that

$$(\boldsymbol{A}\boldsymbol{u}, \boldsymbol{A}\boldsymbol{v}) = (\boldsymbol{f}, \boldsymbol{A}\boldsymbol{v}) \qquad \forall \boldsymbol{v} \in \boldsymbol{V}, \tag{5.22}$$

where $(\boldsymbol{A}\boldsymbol{u}, \boldsymbol{A}\boldsymbol{v})$ is a bilinear form of the type

$$(\boldsymbol{A}\boldsymbol{u}, \boldsymbol{A}\boldsymbol{v}) = (\nabla \times \boldsymbol{u}, \nabla \times \boldsymbol{v}) + (\nabla \cdot \boldsymbol{u}, \nabla \cdot \boldsymbol{v}),$$

and $(\boldsymbol{f}, \boldsymbol{A}\boldsymbol{v})$ is a linear form of the type

$$(\boldsymbol{f}, \boldsymbol{A}\boldsymbol{v}) = (\boldsymbol{\omega}, \nabla \times \boldsymbol{v}) + (\rho, \nabla \cdot \boldsymbol{v}).$$

The boundedness (below) or coerciveness of $\boldsymbol{A}$ is due to Theorem 5.3. Therefore, we immediately have

$$\frac{1}{C}\|\boldsymbol{u}\|_1^2 \leq (\boldsymbol{A}\boldsymbol{u}, \boldsymbol{A}\boldsymbol{u}) = (\boldsymbol{f}, \boldsymbol{A}\boldsymbol{u}) \leq \|\boldsymbol{u}\|_1(\|\boldsymbol{\omega}\|_0 + \|\rho\|_0).$$

From the above inequality we obtain the following stability result.

Theorem 5.6 The solution of (5.8) or (5.22) uniquely exists and satisfies:

$$\|\boldsymbol{u}\|_1 \leq C(\|\boldsymbol{\omega}\|_0 + \|\rho\|_0). \tag{5.23}$$

To consider the LSFEM, we introduce the finite element subspace $\boldsymbol{V}_h \subset \boldsymbol{V}$, i.e., $\boldsymbol{V}_h$ is the space of continuous piecewise polynomial functions of order $r \geq 1$. The following theorem about error estimates of LSFEM is a direct consequence of Theorems 4.7 and 4.8.

Theorem 5.7 The LSFEM based on (5.22) has an optimal rate of convergence and an optimal satisfaction of the divergence equation:

$$\|\boldsymbol{u} - \boldsymbol{u}_h\|_0 \leq C_1 h^{r+1} \|\boldsymbol{u}\|_{r+1}, \tag{5.24a}$$

$$\|\nabla \cdot \boldsymbol{u}_h - \rho\|_0 \leq C_2 h^r \|\boldsymbol{u}\|_{r+1}, \tag{5.24b}$$

where $\boldsymbol{u}_h \in \boldsymbol{V}_h$ is the finite element solution.

5.5 The Euler–Lagrange Equation

In order to further understand the least-squares method, we derive the Euler–Lagrange equations associated with the least-squares variational formulation (5.22) which can be rewritten as: Find $\boldsymbol{u} \in \boldsymbol{V}$ such that

$$(\nabla \times \boldsymbol{u} - \boldsymbol{\omega}, \nabla \times \boldsymbol{v}) + (\nabla \cdot \boldsymbol{u} - \rho, \nabla \cdot \boldsymbol{v}) = 0 \qquad \forall \boldsymbol{v} \in \boldsymbol{V}. \tag{5.25}$$

Suppose that $\boldsymbol{u}$, $\boldsymbol{\omega}$ and ρ are sufficiently smooth. By using Green's formulae (B.3) and (B.5), (5.25) can be written as

$$\Big(\nabla\times(\nabla\times\boldsymbol{u}-\boldsymbol{\omega}),\boldsymbol{v}\Big)+\langle(\nabla\times\boldsymbol{u}-\boldsymbol{\omega}),\boldsymbol{n}\times\boldsymbol{v}\rangle_{\Gamma}$$
$$-\Big(\nabla(\nabla\cdot\boldsymbol{u}-\rho),\boldsymbol{v}\Big)+\langle(\nabla\cdot\boldsymbol{u}-\rho),\boldsymbol{n}\cdot\boldsymbol{v}\rangle_{\Gamma}=0 \qquad \forall\boldsymbol{v}\in\boldsymbol{V}. \quad (5.26)$$

Taking into account the vector identity (5.18) and that $\boldsymbol{v}$ satisfies $\boldsymbol{n}\cdot\boldsymbol{v}=0$ on Γ_1 and $\boldsymbol{n}\times\boldsymbol{v}=\boldsymbol{0}$ on Γ_2, we obtain from (5.26)

$$(-\Delta\boldsymbol{u}-\nabla\times\boldsymbol{\omega}+\nabla\rho,\boldsymbol{v})$$
$$-\langle\boldsymbol{n}\times(\nabla\times\boldsymbol{u}-\boldsymbol{\omega}),\boldsymbol{v}\rangle_{\Gamma_1}+\langle(\nabla\cdot\boldsymbol{u}-\rho),\boldsymbol{n}\cdot\boldsymbol{v}\rangle_{\Gamma_2}=0 \qquad (5.27)$$

for all admissible $\boldsymbol{v}\in\boldsymbol{V}$, hence we have the Euler–Lagrange equation and boundary conditions:

$$\Delta\boldsymbol{u}=-\nabla\times\boldsymbol{\omega}+\nabla\rho \qquad \text{in } \Omega, \qquad (5.28a)$$

$$\boldsymbol{n}\cdot\boldsymbol{u}=0 \qquad \text{on } \Gamma_1, \qquad (5.28b)$$

$$\boldsymbol{n}\times(\nabla\times\boldsymbol{u})=\boldsymbol{n}\times\boldsymbol{\omega} \qquad \text{on } \Gamma_1, \qquad (5.28c)$$

$$\boldsymbol{n}\times\boldsymbol{u}=\boldsymbol{0} \qquad \text{on } \Gamma_2, \qquad (5.28d)$$

$$\nabla\cdot\boldsymbol{u}=\rho \qquad \text{on } \Gamma_2. \qquad (5.28e)$$

We remark that included in the first boundary integral term in (5.27) is a triple vector product of $\boldsymbol{n}$, $(\nabla\times\boldsymbol{u}-\boldsymbol{\omega})$ and $\boldsymbol{v}$. Since we already know that $\boldsymbol{v}$ is orthogonal to the normal $\boldsymbol{n}$ on Γ_1, to make the triple vector product zero requires only that $(\nabla\times\boldsymbol{u}-\boldsymbol{\omega})$ is parallel to $\boldsymbol{n}$ on Γ_1, which is represented algebraically by (5.28c). It is not necessary to require that $(\nabla\times\boldsymbol{u}-\boldsymbol{\omega})=\boldsymbol{0}$ on Γ_1. Of course, in a practical finite difference implementation one may just take $\nabla\times\boldsymbol{u}=\boldsymbol{\omega}$ on Γ_1 instead of (5.28c) as one of the additional boundary conditions.

We note that in system (5.28) the divergence equation does not appear in the domain. In fact, we have rigorously proved that the solution of the uncoupled Poisson equations (5.28a) under additional boundary conditions (5.28c) and (5.28e) automatically satisfies the divergence equation. We also remark that if Γ_2 is empty, the divergence equation does not even appear on the boundary. The attraction of using the second-order system (5.28) now becomes apparent: one avoids dealing with the divergence condition (5.8b) which is implicitly satisfied; instead, one deals with three Poisson equations that are preferable. However, we should mention that if one chooses the finite difference method to solve (5.28a), additional natural boundary conditions (5.28c) and (5.28e) must be supplemented.

Now it is clear that when $\boldsymbol{u}$ is sufficiently smooth, the following four formulations are equivalent to each other: (1) the first-order div–curl system

(5.8); (2) the least-squares variational formulation (5.22); (3) the uncoupled Poisson equations (5.28); and (4) the Galerkin formulation (5.27).

It turns out that the least-squares method (5.22) for the div–curl equations (5.8) corresponds to using the Galerkin method (5.27) to solve system (5.28) which consists of three independent second-order Poisson equations (5.28a) and three coupled boundary conditions on each boundary, where the original first-order equations (5.28c) and (5.28e) serve as the natural boundary conditions, and (5.28b) and (5.28d) as the essential boundary conditions.

The least-squares method (5.22) is the simplest approach among these equivalent methods, because (1) the sources $\boldsymbol{\omega}$ and ρ need only to be square integrable; (2) it does not need any additional boundary conditions; the trial function $\boldsymbol{u}$ and the test function $\boldsymbol{v}$ need to satisfy only the original essential boundary conditions; (3) the corresponding finite element method has an optimal rate of convergence and leads to a symmetric positive-definite matrix. These are the reasons why we strongly recommend the least-squares method.

5.6 The Friedrichs Second Div–Curl Inequality

In order to study problems with inhomogeneous boundary conditions in Chap. 8, we need the Friedrichs second div–curl inequality which can be proved by using the following theorem:

Theorem 5.8 (The Orthogonal Decomposition) Every vector $\boldsymbol{u} \in [H^1(\Omega)]^3$ has the decomposition:

$$\boldsymbol{u} = \nabla q + \nabla \times \boldsymbol{\psi} \tag{5.29}$$

satisfying the orthogonal condition $(\nabla q, \nabla \times \boldsymbol{\psi}) = 0$ in which $q \in H^2(\Omega)/\mathbb{R}$ and $\boldsymbol{\psi} \in [H^2(\Omega)]^3$. Here the notation $/\mathbb{R}$ stands for the fact that q is determinable only to within an arbitrary constant.

Proof: If we can actually find q and $\boldsymbol{\psi}$, this theorem is proved. By virtue of Theorem 5.2, (5.29) is equivalent to the following equations and boundary condition:

$$\nabla \cdot (\nabla q + \nabla \times \boldsymbol{\psi} - \boldsymbol{u}) = 0 \quad \text{in } \Omega, \tag{5.30a}$$

$$\nabla \times (\nabla q + \nabla \times \boldsymbol{\psi} - \boldsymbol{u}) = \boldsymbol{0} \quad \text{in } \Omega, \tag{5.30b}$$

$$\boldsymbol{n} \cdot (\nabla q + \nabla \times \boldsymbol{\psi} - \boldsymbol{u}) = 0 \quad \text{on } \Gamma. \tag{5.30c}$$

Taking into account $\nabla \cdot \nabla \times \boldsymbol{\psi} = \boldsymbol{0}$ and $\nabla \times \nabla q = 0$, system (5.30) can be written as follows:

$$\Delta q = \nabla \cdot \boldsymbol{u} \quad \text{in } \Omega, \tag{5.31a}$$

$$\nabla \times (\nabla \times \boldsymbol{\psi}) = \nabla \times \boldsymbol{u} \quad \text{in } \Omega, \tag{5.31b}$$

$$\boldsymbol{n}\cdot(\nabla q+\nabla\times\boldsymbol{\psi})=\boldsymbol{n}\cdot\boldsymbol{u}\qquad\text{on } \Gamma. \tag{5.31c}$$

To obtain q we may solve the following Poisson equation with the Neumann boundary condition:

$$\Delta q=\nabla\cdot\boldsymbol{u}\qquad\text{in } \Omega, \tag{5.32a}$$

$$\boldsymbol{n}\cdot\nabla q=\boldsymbol{n}\cdot\boldsymbol{u}\qquad\text{on } \Gamma, \tag{5.32b}$$

where the boundary condition (5.32b) is additionally supplied. Although q is not unique, i.e., an arbitrary constant can be added into q, ∇q is uniquely determined.

Now $\boldsymbol{\psi}$ should satisfy

$$\nabla\cdot(\nabla\times\boldsymbol{\psi})=0\qquad\text{in } \Omega, \tag{5.33a}$$

$$\nabla\times(\nabla\times\boldsymbol{\psi})=\nabla\times\boldsymbol{u}\qquad\text{in } \Omega, \tag{5.33b}$$

$$\boldsymbol{n}\cdot(\nabla\times\boldsymbol{\psi})=0\qquad\text{on } \Gamma. \tag{5.33c}$$

$\nabla\times\boldsymbol{\psi}$ in system (5.33) may be considered as an unknown vector that can be uniquely determined by the least-squares method described in Sect. 5.4.

Finally, we can solve the following div–curl system to obtain $\boldsymbol{\psi}$:

$$\nabla\cdot\boldsymbol{\psi}=0\qquad\text{in } \Omega, \tag{5.34a}$$

$$\nabla\times\boldsymbol{\psi}=\text{given}\qquad\text{in } \Omega, \tag{5.34b}$$

$$\boldsymbol{n}\times\boldsymbol{\psi}=0\qquad\text{on } \Gamma. \tag{5.34c}$$

Therefore the validation of the decomposition (5.29) is proved.

Using Appendix (B.4) and (5.33c) we find

$$(\nabla q,\nabla\times\boldsymbol{\psi})=\langle\boldsymbol{n}\cdot(\nabla\times\boldsymbol{\psi}),q\rangle_\Gamma=0,$$

that is, ∇q and $\nabla\times\boldsymbol{\psi}$ are orthogonal. □

Since q is the solution to the Neumann problem of Poisson equation (5.32a), we have the following regularity result (e.g., see Lions and Magenes 1972 or Oden and Reddy 1976):

$$|q|_2^2\le C\{\|\nabla\cdot\boldsymbol{u}\|_0^2+\|\boldsymbol{n}\cdot\boldsymbol{u}\|_{1/2,\Gamma}^2\}. \tag{5.35}$$

Since $\nabla\times\boldsymbol{\psi}$ is the least-squares solution of (5.33), by virtue of (5.23) we have

$$|\nabla\times\boldsymbol{\psi}|_1\le C\|\nabla\times\boldsymbol{u}\|_0. \tag{5.36}$$

Hence by using (5.29), (5.35) and (5.36) we obtain

$$\begin{aligned}\|\boldsymbol{u}\|_1^2 &\le C(\|\boldsymbol{u}\|_0^2+|\nabla q|_1^2+|\nabla\times\boldsymbol{\psi}|_1^2)\\ &\le C(\|\boldsymbol{u}\|_{0,\Omega}^2+\|\nabla\cdot\boldsymbol{u}\|_{0,\Omega}^2+\|\boldsymbol{n}\cdot\boldsymbol{u}\|_{1/2,\Gamma}^2+\|\nabla\times\boldsymbol{u}\|_{0,\Omega}^2).\end{aligned}$$

In fact, we have proved the following theorem.

Theorem 5.9 (The Friedrichs Second Div–Curl Inequality) Let Ω be a bounded and convex (or bounded and simply connected) open region of $\mathbb{R}^3$ with a smooth boundary Γ. Every $\boldsymbol{u} \in [H^1(\Omega)]^3$ satisfies

$$\|\boldsymbol{u}\|_{1,\Omega}^2 \leq C(\|\boldsymbol{u}\|_{0,\Omega}^2 + \|\nabla \cdot \boldsymbol{u}\|_{0,\Omega}^2 + \|\nabla \times \boldsymbol{u}\|_{0,\Omega}^2 + \|\boldsymbol{n} \cdot \boldsymbol{u}\|_{1/2,\Gamma}^2). \quad (5.37)$$

We remark that Theorem 5.9 is valid also for bounded and multiply connected domains (Girault and Raviart 1986).

5.7 Concluding Remarks

By introducing a dummy variable $\vartheta \equiv 0$, the three-dimensional div–curl system is proved to be properly determined and strictly elliptic. For well-posedness the div–curl system needs two algebraic boundary conditions: either ($\boldsymbol{n} \cdot \boldsymbol{u} =$ given and $\vartheta = 0$) or $\boldsymbol{n} \times \boldsymbol{u} =$ given.

Among various methods for solving the div–curl system, the LSFEM is the simplest one and has an optimal rate of convergence.

In order to derive an equivalent higher-order version of vector differential equations without spurious solutions, one should either apply the div–curl method, that is, the curl operation and the div operation must act together with appropriate boundary conditions, or apply the least-squares method for more rigorous results. This idea as well as the least-squares method for the general div–curl system were originally proposed in Jiang et al. (1994b).

The numerical solution of the div–curl system by LSFEM first appeared in Jiang and Chai (1980). The dummy variable was first introduced by Chang and Gunzburger (1987). Early works on the convergence of LSFEM for the div–curl system with a particular boundary condition can be found in Neittaanmäki and Saranen (1981a,b), Křížek and Neittaanmäki (1984b), Fix and Rose (1985), Jiang (1986), Chang and Gunzburger (1987), Křížek and Neittaanmäki (1990) and the references therein.

6. Div–Curl–Grad System

In this chapter we study and compare different finite element methods for the solution of two- and three-dimensional diffusion-type problems usually governed by second-order elliptic partial differential equations. In the mixed Galerkin method, the second-order scalar equation is decomposed into a first-order grad–div system by introducing additional variables (the fluxes). The conventional least-squares finite element method is also based on the same grad–div system. We shall present theoretical analysis and numerical results to show that this simple procedure of reduction destroys ellipticity and thus the conventional LSFEM is not optimal, that is, the rate of convergence for the fluxes is one order lower than optimal. In order to have an optimal LSFEM, the div–curl–grad system should be employed.

6.1 A Model Problem

Let $\Omega \subset \mathbb{R}^{n_d}$ ($n_d = 2$ *or* 3) be an open bounded convex (or bounded simply connected) domain with a piecewise smooth boundary Γ, $\boldsymbol{x} = (x, y, z)$ be a point in Ω, and $\boldsymbol{n} = (n_1, n_2, n_3)$ be a unit outward normal vector on the boundary. We consider the following second-order elliptic boundary-value problem:

$$-\nabla \cdot \nabla \phi = f(\boldsymbol{x}) \qquad \text{in } \Omega, \tag{6.1a}$$

$$\phi = g(\boldsymbol{x}) \qquad \text{on } \Gamma, \tag{6.1b}$$

where $f(\boldsymbol{x}) \in L_2(\Omega)$ and $g(\boldsymbol{x})$ are given functions. Without loss of generality, we shall hereafter consider only the homogeneous boundary condition for simplicity, that is, we shall take $g(\boldsymbol{x}) \equiv 0$. The primal variable ϕ can be, for instance, temperature for heat conduction, potential for incompressible and irrotational flows, or electric potential for electrostatics, etc.

We shall work with the following function spaces:

$$H = H_0^1(\Omega) = \{\psi \in H^1(\Omega) : \psi = 0 \text{ on } \Gamma\},$$

$$S = \{\boldsymbol{v} \in [H^1(\Omega)]^{n_d} : \boldsymbol{n} \times \boldsymbol{v} = \boldsymbol{0} \text{ on } \Gamma\},$$

$$W = \{\boldsymbol{v} \in [L_2(\Omega)]^{n_d}\},$$

$$M = H(\text{div}, \Omega) = \{\boldsymbol{v} \in [L_2(\Omega)]^{n_d} : \nabla \cdot \boldsymbol{v} \in L_2(\Omega)\},$$

where $H(\text{div}, \Omega)$ is a Hilbert space equipped with the norm

$$\|\boldsymbol{v}\|_{H(\text{div},\Omega)} = \{\|\boldsymbol{v}\|_{0,\Omega}^2 + \|\nabla \cdot \boldsymbol{v}\|_{0,\Omega}^2\}^{1/2}.$$

The basic properties of the space $H(\text{div}, \Omega)$ can be found in Girault and Raviart (1986). We also introduce the corresponding finite element subspaces H_h, S_h and W_h, i.e., H_h and S_h are the spaces of continuous piecewise polynomial functions of order $r \geq 1$, and W_h is the space of piecewise polynomial functions of order $r-1$. One may choose a Raviart–Thomas (1977) space for M_h, in which continuity requirements are weaker. For simplicity, we choose M_h to consist of continuous piecewise polynomial functions of order r.

By the finite element interpolation theory (see Sect. 4.6.3), we have:
(1) Given a function $\phi \in H^{r+1}(\Omega)$ and a function $\boldsymbol{u} \in [H^{r+1}(\Omega)]^{n_d}$, there exist interpolants $\Pi_h\phi \in H_h$ and $\Pi_h\boldsymbol{u} \in S_h$ such that

$$\|\phi - \Pi_h\phi\|_0 \leq Ch^{r+1}|\phi|_{r+1}, \tag{6.2a}$$

$$\|\phi - \Pi_h\phi\|_1 \leq Ch^{r}|\phi|_{r+1}, \tag{6.2b}$$

$$\|\boldsymbol{u} - \Pi_h\boldsymbol{u}\|_0 \leq Ch^{r+1}|\boldsymbol{u}|_{r+1}, \tag{6.2c}$$

$$\|\boldsymbol{u} - \Pi_h\boldsymbol{u}\|_1 \leq Ch^{r}|\boldsymbol{u}|_{r+1}, \tag{6.2d}$$

(2) Given a function $\boldsymbol{u} \in [H^r(\Omega)]^{n_d}$, there exists an interpolant $\Pi_h\boldsymbol{u} \in W_h$ such that

$$\|\boldsymbol{u} - \Pi_h\boldsymbol{u}\|_0 \leq Ch^{r}|\boldsymbol{u}|_{r}, \tag{6.2e}$$

(3) Given a function $\boldsymbol{u} \in [H^{r+1}(\Omega)]^{n_d}$, there exists an $\Pi_h\boldsymbol{u} \in M_h$ such that

$$\|\boldsymbol{u} - \Pi_h\boldsymbol{u}\|_{H(\text{div},\Omega)} \leq Ch^{r}|\boldsymbol{u}|_{r+1}, \tag{6.2f}$$

where C here and below denotes a constant independent of the mesh parameter h with generally different values in each appearance.

6.2 The Mixed Galerkin Method

The most commonly used method for problem (6.1) is the Rayleigh–Ritz method. However, as indicated in Sect. 3.2 for one-dimensional problems, *a posteriori* numerical differentiation is required to obtain the dual variables (flux for heat transfer; velocity for fluid flows; or electric field intensity for electrostatics) which are often of most interest. In general, the accuracy of so-computed dual variables is one order lower than that of the primal variable. Moreover, the computed dual variables are not continuous across the element boundary.

As mentioned in Sect. 3.3, the mixed Galerkin method is an alternative choice for solving problem (6.1) (see, e.g., Oden and Carey 1983, Roberts and Thomas 1991, or Brezzi and Fortin 1991). All conclusions about the mixed method drawn for one-dimensional problems in Sect. 3.3 are still valid for multi-dimensional problems, so here we only give a brief review of the mixed method.

In the mixed method, (6.1a) is decomposed into an equivalent first-order grad–div system:

$$\boldsymbol{u} - \nabla\phi = 0 \qquad \text{in } \Omega, \tag{6.3a}$$

$$\nabla \cdot \boldsymbol{u} = -f \qquad \text{in } \Omega, \tag{6.3b}$$

$$\phi = 0 \qquad \text{on } \Gamma. \tag{6.3c}$$

Now multiplying (6.3a) by $\boldsymbol{v} \in W$ and integrating, multiplying (6.3b) by $\psi \in H$ and integrating by parts, we are led to the mixed Galerkin variational statement: Find a pair $\{\phi, \boldsymbol{u}\} \in H \times W$ such that

$$(\boldsymbol{u}, \boldsymbol{v}) - (\nabla\phi, \boldsymbol{v}) = 0 \qquad \forall \boldsymbol{v} \in W, \tag{6.4a}$$

$$-(\nabla\psi, \boldsymbol{u}) = -(f, \psi) \qquad \forall \psi \in H. \tag{6.4b}$$

The formulation (6.4) is a primal mixed method. Another (dual) formulation is also possible (Roberts and Thomas 1991).

Problem (6.4) corresponds to a saddle-point variational problem, and thus in order to guarantee the existence of the solution, the following LBB condition must be satisfied:

$$\sup_{0 \neq \phi \in H} \left(\int_\Omega \nabla\phi \cdot \boldsymbol{u} d\Omega \right) (|\phi|_1)^{-1} \geq \gamma \|\boldsymbol{u}\|_0 \qquad \forall \boldsymbol{u} \in W, \tag{6.5}$$

where the constant $\gamma > 0$. The LBB condition precludes the application of simple equal-order finite elements. It can be proved that the finite element spaces H_h and W_h satisfy the discrete LBB condition (6.5), and if the solution $(\phi, \boldsymbol{u})$ of (6.3) belongs to $H^{r+1}(\Omega) \times [H^r(\Omega)]^{n_d}$, we have the following error estimate (see, e.g., Roberts and Thomas 1991, p.578):

$$|\phi - \phi_h|_1 + \|\boldsymbol{u} - \boldsymbol{u}_h\|_0 \leq Ch^r(|\phi|_{r+1} + \|\boldsymbol{u}\|_r). \tag{6.6}$$

The estimate (6.6) shows that in this mixed method, the accuracy of the flux $\boldsymbol{u}$ is always one order lower than that for the primal variable ϕ.

Inspecting equation (6.4) shows that the matrix associated with the mixed method is non-positive-definite. This makes the use of iterative methods to solve large-scale problems relatively difficult.

6.3 The Conventional LSFEM

The conventional LSFEM is also based on the grad–div system (6.3). For two-dimensional problems, the first-order system (6.3) consists of three equations and three unknown functions. In Sect. 4.7 we have pointed out that a first-order system with an odd number of unknowns and an odd number of equations cannot form an elliptic system in the ordinary sense. For three-dimensional problems, although the first-order system (6.3) has four unknowns and four equations, it is easy to verify that this system is also not elliptic in the ordinary sense. Therefore, the linear operator in system (6.3) cannot be bounded below in the H^1 norm. The fact that the rate of convergence of the conventional LSFEM based on system (6.3) is not optimal due to this destruction of the H^1 ellipticity.

Now let us analyze the conventional LSFEM that minimizes the following functional:

$$I : H \times M \to \mathbb{R},$$

$$I(\phi, \boldsymbol{u}) = \|\nabla \cdot \boldsymbol{u} + f\|_0^2 + \|\nabla\phi - \boldsymbol{u}\|_0^2. \tag{6.7}$$

Taking the variation of I with respect to ϕ and $\boldsymbol{u}$, and letting $\delta I = 0$, $\delta\phi = \psi$ and $\delta\boldsymbol{u} = \boldsymbol{v}$ lead to a variational statement: Find $U = \{\phi, \boldsymbol{u}\} \in H \times M$, such that

$$(\boldsymbol{A}U, \boldsymbol{A}V) = (\boldsymbol{f}, \boldsymbol{A}V) \qquad \forall V = \{\psi, \boldsymbol{v}\} \in H \times M, \tag{6.8}$$

where

$$(\boldsymbol{A}U, \boldsymbol{A}V) = (\nabla \cdot \boldsymbol{u}, \nabla \cdot \boldsymbol{v}) + (\nabla\phi - \boldsymbol{u}, \nabla\psi - \boldsymbol{v}),$$

$$(\boldsymbol{f}, \boldsymbol{A}V) = (-f, \nabla \cdot \boldsymbol{v}).$$

The corresponding finite element problem is then to find $U_h = \{\phi_h, \boldsymbol{u}_h\} \in H_h \times M_h$, such that

$$(\boldsymbol{A}U_h, \boldsymbol{A}V_h) = (\boldsymbol{f}, \boldsymbol{A}V_h) \qquad \forall V_h = \{\psi_h, \boldsymbol{v}_h\} \in H_h \times M_h, \tag{6.9}$$

where

$$(\boldsymbol{A}U_h, \boldsymbol{A}V_h) = (\nabla \cdot \boldsymbol{u}_h, \nabla \cdot \boldsymbol{v}_h) + (\nabla\phi_h - \boldsymbol{u}_h, \nabla\psi_h - \boldsymbol{v}_h),$$

$$(\boldsymbol{f}, \boldsymbol{A}V_h) = (-f, \nabla \cdot \boldsymbol{v}_h).$$

It is easy to verify that

$$\|\boldsymbol{A}V\|_0 \le C\|V\|, \tag{6.10}$$

where $\|V\|^2 = \|\phi\|_1^2 + \|\boldsymbol{u}\|_{H(\mathrm{div},\Omega)}^2 = \|\phi\|_1^2 + \|\boldsymbol{u}\|_0^2 + \|\nabla \cdot \boldsymbol{u}\|_0^2$. Thus, the operator $\boldsymbol{A}$ is continuous on $H \times M$. We note that the norm associated with $H \times M$ is not an H^1 norm. We need to show that the operator $\boldsymbol{A}$ is bounded below, that is, there exists a constant $\alpha > 0$ such that for $V \in H \times M$

$$\|\boldsymbol{A}\boldsymbol{V}\|_0^2 \geq \alpha \|V\|^2. \tag{6.11}$$

Let us prove (6.11). We know that

$$\|\boldsymbol{A}\boldsymbol{V}\|_0^2 = \|\nabla \cdot \boldsymbol{v}\|_0^2 + \|\nabla \psi - \boldsymbol{v}\|_0^2. \tag{6.12}$$

Consequently,

$$\|\boldsymbol{A}\boldsymbol{V}\|_0^2 \geq \|\nabla \cdot \boldsymbol{v}\|_0^2, \tag{6.13}$$

$$\|\boldsymbol{A}\boldsymbol{V}\|_0^2 \geq \|\nabla \psi - \boldsymbol{v}\|_0^2. \tag{6.14}$$

From (6.14) and by using Green's formula (B.1), the boundary condition (6.3c) and the Friedrichs inequality (4.14), we have

$$\begin{aligned} \|\boldsymbol{A}\boldsymbol{V}\|_0^2 &\geq \|\nabla \psi - \boldsymbol{v}\|_0^2 = \|\nabla \psi\|_0^2 + \|\boldsymbol{v}\|_0^2 - 2(\boldsymbol{v}, \nabla \psi) \\ &\geq C\|\psi\|_0^2 + \|\boldsymbol{v}\|_0^2 + 2(\nabla \cdot \boldsymbol{v}, \psi). \end{aligned} \tag{6.15}$$

From (6.13) we have

$$\frac{1}{C}\|\boldsymbol{A}\boldsymbol{V}\|_0^2 \geq \frac{1}{C}\|\nabla \cdot \boldsymbol{v}\|_0^2. \tag{6.16}$$

The combination of (6.15) and (6.16) yields

$$\begin{aligned} (1+\frac{1}{C})\|\boldsymbol{A}\boldsymbol{V}\|_0^2 &\geq \frac{1}{C}\|\nabla \cdot \boldsymbol{v}\|_0^2 + C\|\psi\|_0^2 + \|\boldsymbol{v}\|_0^2 + 2(\nabla \cdot \boldsymbol{v}, \psi) \\ &= \|\frac{1}{\sqrt{C}}\nabla \cdot \boldsymbol{v} + \sqrt{C}\psi\|_0^2 + \|\boldsymbol{v}\|_0^2 \geq \|\boldsymbol{v}\|_0^2 \end{aligned}$$

or

$$C_1\|\boldsymbol{A}\boldsymbol{V}\|_0^2 \geq \|\boldsymbol{v}\|_0^2. \tag{6.17}$$

By combining (6.14) and (6.17), we have

$$C_2\|\boldsymbol{A}\boldsymbol{V}\|_0 \geq \|\nabla \psi - \boldsymbol{v}\|_0 + \|\boldsymbol{v}\|_0 \geq \|\nabla \psi\|_0,$$

that is

$$C_3\|\boldsymbol{A}\boldsymbol{V}\|_0^2 \geq \|\nabla \psi\|_0^2. \tag{6.18}$$

By virtue of the Friedrichs inequality (4.14) and (6.18) we have

$$C_4\|\boldsymbol{A}\boldsymbol{V}\|_0^2 \geq \|\psi\|_0^2. \tag{6.19}$$

By combining (6.13), (6.17), (6.18) and (6.19), we obtain the boundedness (below):

$$\|\boldsymbol{A}\boldsymbol{V}\|_0^2 \geq C_5(\|\nabla \psi\|_0^2 + \|\psi\|_0^2 + \|\nabla \cdot \boldsymbol{v}\|_0^2 + \|\boldsymbol{v}\|_0^2),$$

or

$$\|\boldsymbol{A}\boldsymbol{V}\|_0^2 \geq C_5(\|\psi\|_1^2 + \|\boldsymbol{v}\|_{H(\mathrm{div},\Omega)}^2). \tag{6.20}$$

Therefore, the following theorem about the rate of convergence of the conventional least-squares solutions with equal-order finite elements can be derived (see Sect. 4.10).

Theorem 6.1 Assume that $f(\boldsymbol{x}) \in L_2(\Omega)$, the solution $(\phi, \boldsymbol{u})$ of (6.3) belongs to $H^{r+1}(\Omega) \times [H^{r+1}(\Omega)]^{n_d}$, and the finite element interpolation estimates (6.2a), (6.2b) and (6.2f) hold. Then for the approximate solution associated with (6.9), we have the error estimate:

$$\|\phi - \phi_h\|_1 + \|\boldsymbol{u} - \boldsymbol{u}_h\|_{H(\mathrm{div},\Omega)} \leq Ch^r(\|\phi\|_{r+1} + \|\boldsymbol{u}\|_{r+1}). \tag{6.21}$$

Although the conventional LSFEM has significant advantages over the mixed Galerkin method, namely, that the conventional LSFEM is not subject to the LBB condition and thus can accommodate simple equal-order elements, we don't recommend its use due to the following considerations: (1) Theorem 6.1 shows that the accuracy of the flux $\boldsymbol{u}$ for the conventional LSFEM with equal-order finite elements is lower than optimal; (2) linear elements with one-point quadrature rule (in general, the reduced integration) will result in a singular global matrix; the reason is that the LSFEM with Gaussian quadrature is equivalent to the discrete least-squares collocation method (see Sect. 4.11), and in this case one solves an underdetermined algebraic system; (3) the squares of the residuals at Gaussian points may not decrease monotonically with mesh refinement, this will cause a problem for adaptive refinement using the residual as an error indicator.

6.4 The Optimal LSFEM

The optimal LSFEM is based on the following first-order div–curl–grad system:

$$\nabla \cdot \boldsymbol{u} = -f \quad \text{in } \Omega, \tag{6.22a}$$

$$\nabla \times \boldsymbol{u} = 0 \quad \text{in } \Omega, \tag{6.22b}$$

$$\nabla \phi - \boldsymbol{u} = 0 \quad \text{in } \Omega, \tag{6.22c}$$

$$\phi = 0 \quad \text{on } \Gamma, \tag{6.22d}$$

$$\boldsymbol{n} \times \boldsymbol{u} = 0 \quad \text{on } \Gamma. \tag{6.22e}$$

This system differs from (6.3) in that the compatibility equation (6.22b) and the boundary compatibility condition (6.22e) are added. For irrotational flow problems, $\boldsymbol{u}$ is the velocity vector, ϕ is the potential, (6.22a) represents mass conservation, (6.22b) irrotationality, and (6.22c) constitutive relation.

The curl-free equation (6.22b) and the boundary condition (6.22e) seem 'redundant', since (6.22b) can be obtained by taking the curl operation on

(6.22c), and (6.22e) can be derived by (6.22d). We shall show that the inclusion of these relations is important.

We note that the calculation of ϕ and $\boldsymbol{u}$ in (6.22) can be decoupled. We may solve the div–curl system of (6.22a) and (6.22b) with the boundary condition (6.22e) to obtain $\boldsymbol{u}$, and then (if necessary) to find ϕ by using numerical integration or by solving (6.22c) and (6.22d) in terms of LSFEM. For more general problems such as

$$-\nabla \cdot \nabla\phi + \phi = f(\boldsymbol{x}) \qquad \text{in } \Omega,$$

$$\frac{\partial \phi}{\partial n} = 0 \qquad \text{on } \Gamma,$$

this decoupling is impossible (Jiang and Povinelli 1993).

6.4.1 Two-Dimensional Case

For two-dimensional problems, the first-order system (6.22) consists of four equations involving three unknowns. At first glance, one may think that this is an 'overdetermined' system. We now show that system (6.22) is properly determined and elliptic by introducing a dummy variable ϑ. We have used this trick in Sect. 5.2. We rewrite system (6.22) as

$$\nabla \cdot \boldsymbol{u} = -f \qquad \text{in } \Omega, \tag{6.23a}$$

$$\text{rot } \boldsymbol{u} = 0 \qquad \text{in } \Omega, \tag{6.23b}$$

$$-\textbf{curl } \vartheta + \nabla\phi - \boldsymbol{u} = \boldsymbol{0} \qquad \text{in } \Omega, \tag{6.23c}$$

$$\phi = 0 \qquad \text{on } \Gamma, \tag{6.23d}$$

$$\boldsymbol{n} \times \boldsymbol{u} = 0 \qquad \text{on } \Gamma, \tag{6.23e}$$

in which $\boldsymbol{u} = (u, v)$, and the following particular notations for the two-dimensional case are adopted:

$$\text{rot } \boldsymbol{u} = \frac{\partial v}{\partial x} - \frac{\partial u}{\partial y},$$

$$\textbf{curl } \vartheta = \begin{pmatrix} \frac{\partial \vartheta}{\partial y} \\ -\frac{\partial \vartheta}{\partial x} \end{pmatrix}.$$

In Cartesian Coordinates, system (6.23) can be written as

$$\frac{\partial u}{\partial x} + \frac{\partial v}{\partial y} = -f \qquad \text{in } \Omega, \tag{6.24a}$$

$$-\frac{\partial u}{\partial y} + \frac{\partial v}{\partial x} = 0 \qquad \text{in } \Omega, \tag{6.24b}$$

$$\frac{\partial \phi}{\partial x} - \frac{\partial \vartheta}{\partial y} - u = 0 \qquad \text{in } \Omega, \tag{6.24c}$$

$$\frac{\partial \phi}{\partial y} + \frac{\partial \vartheta}{\partial x} - v = 0 \qquad \text{in } \Omega, \tag{6.24d}$$

$$\phi = 0 \qquad \text{on } \Gamma, \tag{6.24e}$$

$$un_2 - vn_1 = 0 \qquad \text{on } \Gamma. \tag{6.24f}$$

Recalling the div–curl method introduced in Sect. 5.3 we know that (6.23c) is equivalent to the following equations and boundary condition

$$\nabla \cdot (-\mathbf{curl}\,\vartheta + \nabla\phi - \boldsymbol{u}) = 0 \qquad \text{in } \Omega,$$

$$\operatorname{rot}(-\mathbf{curl}\,\vartheta + \nabla\phi - \boldsymbol{u}) = 0 \qquad \text{in } \Omega,$$

$$\boldsymbol{n} \times (-\mathbf{curl}\,\vartheta + \nabla\phi - \boldsymbol{u}) = 0 \qquad \text{on } \Gamma.$$

After simplification the second and third above equations become

$$\Delta\vartheta = 0 \qquad \text{in } \Omega,$$

$$\frac{\partial \vartheta}{\partial n} = \boldsymbol{n} \times \nabla\phi = 0 \qquad \text{on } \Gamma.$$

Therefore, ϑ is a constant. That is, the introduction of ϑ does not change anything. This means that system (6.24) of four equations with four unknowns is indeed equivalent to the two-dimensional version of system (6.22) of four equations with only three unknowns.

Now we write (6.24) in standard matrix form:

$$\boldsymbol{A}_1 \frac{\partial \underline{\boldsymbol{u}}}{\partial x} + \boldsymbol{A}_2 \frac{\partial \underline{\boldsymbol{u}}}{\partial y} + \boldsymbol{A}_0 \underline{\boldsymbol{u}} = \boldsymbol{f} \qquad \text{in } \Omega, \tag{6.25}$$

in which

$$\boldsymbol{A}_1 = \begin{pmatrix} 1 & 0 & 0 & 0 \\ 0 & 1 & 0 & 0 \\ 0 & 0 & 1 & 0 \\ 0 & 0 & 0 & 1 \end{pmatrix}, \qquad \boldsymbol{A}_2 = \begin{pmatrix} 0 & 1 & 0 & 0 \\ -1 & 0 & 0 & 0 \\ 0 & 0 & 0 & -1 \\ 0 & 0 & 1 & 0 \end{pmatrix},$$

$$\boldsymbol{A}_0 = \begin{pmatrix} 0 & 0 & 0 & 0 \\ 0 & 0 & 0 & 0 \\ -1 & 0 & 0 & 0 \\ 0 & -1 & 0 & 0 \end{pmatrix}, \qquad \boldsymbol{f} = \begin{pmatrix} -f \\ 0 \\ 0 \\ 0 \end{pmatrix}, \qquad \underline{\boldsymbol{u}} = \begin{pmatrix} u \\ v \\ \phi \\ \vartheta \end{pmatrix}.$$

Since

$$\det(\boldsymbol{A}_1\xi + \boldsymbol{A}_2\eta) = \det \begin{pmatrix} \xi & \eta & 0 & 0 \\ -\eta & \xi & 0 & 0 \\ 0 & 0 & \xi & -\eta \\ 0 & 0 & \eta & \xi \end{pmatrix} = (\xi^2 + \eta^2)^2 \neq 0$$

for all nonzero real pairs (ξ, η), system (6.24) and thus the two-dimensional version of system (6.22) is determined and elliptic, as contended. The elliptic first-order system (6.24) has four equations involving four unknowns, so two boundary conditions (6.24e) and (6.24f) are required.

Remark. In this chapter we analyze only the boundary condition (6.1b). If equation (6.1a) is supplemented by the boundary condition

$$\frac{\partial \phi}{\partial n} = 0 \qquad \text{on } \Gamma,$$

then the following boundary conditions

$$\vartheta = 0 \qquad \text{on } \Gamma, \tag{6.26a}$$

$$\boldsymbol{n} \cdot \boldsymbol{u} = 0 \qquad \text{on } \Gamma, \tag{6.26b}$$

should be specified for (6.23a)–(6.23c).

6.4.2 Three-Dimensional Case

In the three-dimensional case, we consider the following first-order system:

$$\nabla \cdot \boldsymbol{u} = -f \qquad \text{in } \Omega, \tag{6.27a}$$

$$\nabla \vartheta + \nabla \times \boldsymbol{u} = \boldsymbol{0} \qquad \text{in } \Omega, \tag{6.27b}$$

$$\nabla \times \boldsymbol{\chi} + \nabla \phi - \boldsymbol{u} = \boldsymbol{0} \qquad \text{in } \Omega, \tag{6.27c}$$

$$\nabla \cdot \boldsymbol{\chi} = 0 \qquad \text{in } \Omega, \tag{6.27d}$$

$$\phi = 0 \qquad \text{on } \Gamma, \tag{6.27e}$$

$$\boldsymbol{n} \cdot \boldsymbol{\chi} = 0 \qquad \text{on } \Gamma, \tag{6.27f}$$

$$\boldsymbol{n} \times \boldsymbol{u} = \boldsymbol{0} \qquad \text{on } \Gamma, \tag{6.27g}$$

where $\boldsymbol{u} = (u, v, w)$; while ϑ and $\boldsymbol{\chi} = \{\chi_1, \chi_2, \chi_3\}$ are dummy variables. In system (6.27) there are eight equations and four algebraic boundary conditions involving eight unknowns.

In the following we use the div–curl method discussed in Sect. 5.3 to show that the dummy variables are identical to zero. For simplicity, some details have been omitted. By applying the divergence operation to (6.27b) and considering $\nabla \cdot \nabla \times \boldsymbol{u} = 0$ we have

$$\Delta \vartheta = 0 \qquad \text{in } \Omega.$$

By applying the $\boldsymbol{n}\cdot$ operation to (6.27b) on the boundary and considering (6.27g) and Theorem 5.5, we have

$$\boldsymbol{n} \cdot \nabla \vartheta = 0 \qquad \text{on } \Gamma.$$

Therefore ϑ = constant, and thus (6.27b) is not different from (6.22b). By applying the curl operation to (6.27c) and considering that $\nabla \times \nabla\phi = \mathbf{0}$ and $\nabla \times \boldsymbol{u} = \mathbf{0}$, we have

$$\nabla \times (\nabla \times \boldsymbol{\chi}) = \mathbf{0} \qquad \text{in } \Omega. \tag{6.28}$$

We also know that

$$\nabla \cdot (\nabla \times \boldsymbol{\chi}) = 0 \qquad \text{in } \Omega. \tag{6.29}$$

Furthermore, by applying $\boldsymbol{n}\times$ operation to (6.27c) on Γ and considering the boundary conditions (6.27g) and (6.27e) (the consequence of which is $\boldsymbol{n} \times \nabla\phi = 0$ on Γ) we have

$$\boldsymbol{n} \times (\nabla \times \boldsymbol{\chi}) = \mathbf{0} \qquad \text{on } \Gamma. \tag{6.30}$$

From (6.28), (6.29) and (6.30) by virtue of Theorem 5.2 , we have

$$\nabla \times \boldsymbol{\chi} = \mathbf{0} \qquad \text{in } \Omega. \tag{6.31}$$

Combining (6.27d), (6.31) and (6.27f) and using Theorem 5.2, we finally find that $\boldsymbol{\chi} \equiv \mathbf{0}$. Thus we have proved that system (6.27) of eight equations involving eight unknowns is indeed equivalent to system (6.22) of four equations with four unknowns.

In Cartesian Coordinates, system (6.27) has the following form:

$$\begin{aligned}
\frac{\partial u}{\partial x}+\frac{\partial v}{\partial y}+\frac{\partial w}{\partial z} &= -f && \text{in } \Omega,\\
-\frac{\partial v}{\partial z}+\frac{\partial w}{\partial y}+\frac{\partial \vartheta}{\partial x} &= 0 && \text{in } \Omega,\\
-\frac{\partial w}{\partial x}+\frac{\partial u}{\partial z}+\frac{\partial \vartheta}{\partial y} &= 0 && \text{in } \Omega,\\
-\frac{\partial u}{\partial y}+\frac{\partial v}{\partial x}+\frac{\partial \vartheta}{\partial z} &= 0 && \text{in } \Omega,\\
-\frac{\partial \chi_2}{\partial z}+\frac{\partial \chi_3}{\partial y}+\frac{\partial \phi}{\partial x}-u &= 0 && \text{in } \Omega,\\
-\frac{\partial \chi_3}{\partial x}+\frac{\partial \chi_1}{\partial z}+\frac{\partial \phi}{\partial y}-v &= 0 && \text{in } \Omega,\\
-\frac{\partial \chi_1}{\partial y}+\frac{\partial \chi_2}{\partial x}+\frac{\partial \phi}{\partial z}-w &= 0 && \text{in } \Omega,\\
\frac{\partial \chi_1}{\partial x}+\frac{\partial \chi_2}{\partial y}+\frac{\partial \chi_3}{\partial z} &= 0 && \text{in } \Omega.
\end{aligned} \tag{6.32}$$

We may write system (6.32) in standard matrix form:

$$\boldsymbol{A}_1\frac{\partial \underline{\boldsymbol{u}}}{\partial x}+\boldsymbol{A}_2\frac{\partial \underline{\boldsymbol{u}}}{\partial y}+\boldsymbol{A}_3\frac{\partial \underline{\boldsymbol{u}}}{\partial z}+\boldsymbol{A}_0\underline{\boldsymbol{u}}=\boldsymbol{f} \qquad \text{in } \Omega, \tag{6.33}$$

in which

$$
\boldsymbol{A}_1 = \begin{pmatrix}
1 & 0 & 0 & 0 & 0 & 0 & 0 & 0 \\
0 & 0 & 0 & 0 & 1 & 0 & 0 & 0 \\
0 & 0 & -1 & 0 & 0 & 0 & 0 & 0 \\
0 & 1 & 0 & 0 & 0 & 0 & 0 & 0 \\
0 & 0 & 0 & 1 & 0 & 0 & 0 & 0 \\
0 & 0 & 0 & 0 & 0 & 0 & 0 & -1 \\
0 & 0 & 0 & 0 & 0 & 0 & 1 & 0 \\
0 & 0 & 0 & 0 & 0 & 1 & 0 & 0
\end{pmatrix},
$$

$$
\boldsymbol{A}_2 = \begin{pmatrix}
0 & 1 & 0 & 0 & 0 & 0 & 0 & 0 \\
0 & 0 & 1 & 0 & 0 & 0 & 0 & 0 \\
0 & 0 & 0 & 0 & 1 & 0 & 0 & 0 \\
-1 & 0 & 0 & 0 & 0 & 0 & 0 & 0 \\
0 & 0 & 0 & 0 & 0 & 0 & 0 & 1 \\
0 & 0 & 0 & 1 & 0 & 0 & 0 & 0 \\
0 & 0 & 0 & 0 & 0 & -1 & 0 & 0 \\
0 & 0 & 0 & 0 & 0 & 0 & 1 & 0
\end{pmatrix},
$$

$$
\boldsymbol{A}_3 = \begin{pmatrix}
0 & 0 & 1 & 0 & 0 & 0 & 0 & 0 \\
0 & -1 & 0 & 0 & 0 & 0 & 0 & 0 \\
1 & 0 & 0 & 0 & 0 & 0 & 0 & 0 \\
0 & 0 & 0 & 0 & 1 & 0 & 0 & 0 \\
0 & 0 & 0 & 0 & 0 & 0 & -1 & 0 \\
0 & 0 & 0 & 0 & 0 & 1 & 0 & 0 \\
0 & 0 & 0 & 1 & 0 & 0 & 0 & 0 \\
0 & 0 & 0 & 0 & 0 & 0 & 0 & 1
\end{pmatrix},
$$

$$
\boldsymbol{A}_0 = \begin{pmatrix}
0 & 0 & 0 & 0 & 0 & 0 & 0 & 0 \\
0 & 0 & 0 & 0 & 0 & 0 & 0 & 0 \\
0 & 0 & 0 & 0 & 0 & 0 & 0 & 0 \\
0 & 0 & 0 & 0 & 0 & 0 & 0 & 0 \\
-1 & 0 & 0 & 0 & 0 & 0 & 0 & 0 \\
0 & -1 & 0 & 0 & 0 & 0 & 0 & 0 \\
0 & 0 & -1 & 0 & 0 & 0 & 0 & 0 \\
0 & 0 & 0 & 0 & 0 & 0 & 0 & 0
\end{pmatrix},
$$

$$
\boldsymbol{f} = \begin{pmatrix} -f \\ 0 \\ 0 \\ 0 \\ 0 \\ 0 \\ 0 \\ 0 \end{pmatrix}, \qquad \underline{\boldsymbol{u}} = \begin{pmatrix} u \\ v \\ w \\ \phi \\ \vartheta \\ \chi_1 \\ \chi_2 \\ \chi_3 \end{pmatrix}.
$$

Since

$$\det(\boldsymbol{A}_1\xi + \boldsymbol{A}_2\eta + \boldsymbol{A}_3\zeta) = \det \begin{pmatrix} \xi & \eta & \zeta & 0 & 0 & 0 & 0 & 0 \\ 0 & -\zeta & \eta & 0 & \xi & 0 & 0 & 0 \\ \zeta & 0 & -\xi & 0 & \eta & 0 & 0 & 0 \\ -\eta & \xi & 0 & 0 & \zeta & 0 & 0 & 0 \\ 0 & 0 & 0 & \xi & 0 & 0 & -\zeta & \eta \\ 0 & 0 & 0 & \eta & 0 & \zeta & 0 & -\xi \\ 0 & 0 & 0 & \zeta & 0 & -\eta & \xi & 0 \\ 0 & 0 & 0 & 0 & 0 & \xi & \eta & \zeta \end{pmatrix}$$

$$= -(\xi^2 + \eta^2 + \zeta^2)^4 \neq 0$$

for all nonzero real triplets (ξ, η, ζ), systems (6.32), (6.27) and thus system (6.22) is properly determined and elliptic. The elliptic first-order system (6.27) has eight unknowns, so that four boundary conditions are needed. Considering that two algebraic conditions are included in (6.27g), this requirement is satisfied in (6.27).

Remark. If equation (6.1a) is supplemented by the Neumann boundary condition:

$$\frac{\partial \phi}{\partial n} = 0 \qquad \text{on } \Gamma,$$

then for three-dimensional problems the following boundary conditions

$$\vartheta = 0 \qquad \text{on } \Gamma, \tag{6.34a}$$

$$\boldsymbol{n} \times \boldsymbol{\chi} = \boldsymbol{0} \qquad \text{on } \Gamma, \tag{6.34b}$$

$$\boldsymbol{n} \cdot \boldsymbol{u} = 0 \qquad \text{on } \Gamma, \tag{6.34c}$$

should be given to (6.27a)–(6.27d). If the original second-order problem (6.1) has a mixed boundary condition:

$$\phi + \beta\frac{\partial \phi}{\partial n} = 0 \qquad \text{on } \Gamma,$$

then (6.34c) should be replaced by

$$\phi + \beta \boldsymbol{n} \cdot \boldsymbol{u} = 0 \qquad \text{on } \Gamma.$$

6.4.3 Error Analysis

The optimal LSFEM minimizes the following functional:

$$I : H \times S \to \mathbb{R},$$

$$I(\phi, \boldsymbol{u}) = \|\nabla \cdot \boldsymbol{u} + f\|_0^2 + \|\nabla \times \boldsymbol{u}\|_0^2 + \|\nabla\phi - \boldsymbol{u}\|_0^2. \tag{6.35}$$

We emphasize again that the introduction of the dummy variables is purely for proving the determinacy and ellipticity of the div–curl–grad system, and it has nothing to do with the numerical computation.

Taking the variation of I with respect to ϕ and $\boldsymbol{u}$, and letting $\delta I = 0$, $\delta\phi = \psi$ and $\delta\boldsymbol{u} = \boldsymbol{v}$ lead to a least-squares formulation: Find $\boldsymbol{U} = (\phi, \boldsymbol{u}) \in H \times S$, such that

$$(\boldsymbol{AU}, \boldsymbol{AV}) = (\boldsymbol{f}, \boldsymbol{AV}) \qquad \forall \boldsymbol{V} = (\psi, \boldsymbol{v}) \in H \times S, \tag{6.36}$$

in which

$$(\boldsymbol{AU}, \boldsymbol{AV}) = (\nabla \cdot \boldsymbol{u}, \nabla \cdot \boldsymbol{v}) + (\nabla \times \boldsymbol{u}, \nabla \times \boldsymbol{v}) + (\nabla\phi - \boldsymbol{u}, \nabla\psi - \boldsymbol{v}),$$

$$(\boldsymbol{f}, \boldsymbol{AV}) = (-f, \nabla \cdot \boldsymbol{v}).$$

The corresponding finite element problem is then to find $U_h = (\phi_h, \boldsymbol{u}_h) \in H_h \times S_h$, such that

$$(\boldsymbol{AU}_h, \boldsymbol{AV}_h) = (\boldsymbol{f}, \boldsymbol{AV}_h) \qquad \forall \boldsymbol{V}_h = (\psi_h, \boldsymbol{v}_h) \in H_h \times S_h, \tag{6.37}$$

where

$$\begin{aligned}(\boldsymbol{AU}_h, \boldsymbol{AV}_h) = (\nabla \cdot \boldsymbol{u}_h, \nabla \cdot \boldsymbol{v}_h) &+ (\nabla \times \boldsymbol{u}_h, \nabla \times \boldsymbol{v}_h) \\ &+ (\nabla\phi_h - \boldsymbol{u}_h, \nabla\psi_h - \boldsymbol{v}_h),\end{aligned}$$

$$(\boldsymbol{f}, \boldsymbol{AV}_h) = (-f, \nabla \cdot \boldsymbol{v}_h).$$

It is easy to verify that

$$\|\boldsymbol{AV}\|_0 \le C\|\boldsymbol{V}\|_1, \tag{6.38}$$

where

$$\|\boldsymbol{V}\|_1^2 = \|\phi\|_1^2 + \|\boldsymbol{u}\|_1^2$$

is an H^1 norm. Thus, $\boldsymbol{A}$ is continuous on $H \times S$. We shall then prove that there exists a constant $\alpha > 0$ such that for $\boldsymbol{V} \in H \times S$

$$\|\boldsymbol{AV}\|_0^2 \ge \alpha\|\boldsymbol{V}\|_1^2. \tag{6.39}$$

Following the same argument as in Sect. 6.3, we obtain

$$\|\boldsymbol{AV}\|_0^2 \ge C(\|\nabla\psi\|_0^2 + \|\psi\|_0^2 + \|\nabla \cdot \boldsymbol{v}\|_0^2 + \|\nabla \times \boldsymbol{v}\|_0^2 + \|\boldsymbol{v}\|_0^2). \tag{6.40}$$

The combination of (6.40) and the Friedrichs first div–curl inequality (5.5) yields H^1 coerciveness (6.39).

Once the H^1 coerciveness is proved, the following theorem is a consequence of Theorem 4.8.

Theorem 6.2 Assume that $f(\boldsymbol{x}) \in L_2(\Omega)$, the solution $(\phi, \boldsymbol{u}) \in H^{r+1}(\Omega) \times [H^{r+1}(\Omega)]^{n_d}$ and the finite element interpolation estimates (6.2a)–(6.2d) hold. Then for the approximate solution associated with (6.37), we have the error estimate:

$$\|\phi - \phi_h\|_1 + \|\boldsymbol{u} - \boldsymbol{u}_h\|_1 \le Ch^r(\|\phi\|_{r+1} + \|\boldsymbol{u}\|_{r+1}). \tag{6.41}$$

This theorem implies that the rate of convergence (in the H^1 norm) of LSFEM based on the first-order div–curl–grad system (6.22) with equal-order finite elements is optimal for all variables. Optimal L_2 convergence can be obtained by using the Aubin–Nitsche trick. The optimality is attributed to the fact that the optimal LSFEM controls the errors in the full H^1 norm.

As discussed in Sect. 4.11, the LSFEM with numerical quadrature is equivalent to a weighted collocation least-squares method. We may use this idea to choose an appropriate number of Gaussian points. The conventional LSFEM with linear (Q_1) elements and one-point quadrature will lead to a singular global matrix, because it corresponds to solving an underdetermined algebraic system. In contrast, the optimal LSFEM with reduced integration works very well; although the computed nodal values of $\boldsymbol{u}$ may have some oscillations, the values of $\boldsymbol{u}$ at Gaussian points are very smooth. The use of reduced integration is important in practice, because it significantly reduces the computing work when the matrix-free element-by-element conjugate gradient method is employed for the solution of the resulting system of algebraic equations.

6.5 Numerical Results

As an example, we choose

$$f = (x^2 + 3x)e^x(y^2 - y) + (y^2 + 3y)e^y(x^2 - x) \qquad \text{in } \Omega,$$

where $\Omega = \{(x, y) \in \mathbb{R}^2 : 0 < x < 1,\ 0 < y < 1\}$ is the unit square with boundary Γ. The boundary conditions are

$$\begin{aligned}
\phi &= 0 \qquad \text{on } \Gamma,\\
v &= 0 \qquad \text{on } \Gamma_1 = \{(x, y) \in \Gamma : x = 0\},\\
v &= 0 \qquad \text{on } \Gamma_3 = \{(x, y) \in \Gamma : x = 1\},\\
u &= 0 \qquad \text{on } \Gamma_2 = \{(x, y) \in \Gamma : y = 0\},\\
u &= 0 \qquad \text{on } \Gamma_4 = \{(x, y) \in \Gamma : y = 1\}.
\end{aligned}$$

The exact solution should be

$$\begin{aligned}
\phi &= (x^2 - x)e^x(y^2 - y)e^y,\\
u &= (x^2 + x - 1)e^x(y^2 - y)e^y,\\
v &= (y^2 + y - 1)e^y(x^2 - x)e^x.
\end{aligned}$$

Numerical experiments were carried out using bilinear (Q_1) elements on uniform meshes with $1/h = 4, 9, 20, 29$. We calculated the L_2 errors for ϕ and $\boldsymbol{u}$:

$$e_\phi = \|\phi - \phi_h\|_0, \qquad e_u = (\|u - u_h\|_0^2 + \|v - v_h\|_0^2)^{\frac{1}{2}}.$$

The numerical results on the rates of convergence are given in Fig. 6.1. As expected, the rate of convergence of the flux $\boldsymbol{u}$ for the conventional LSFEM is lower than $O(h^2)$. The rates of convergence are $O(h^2)$ for both the primal variable ϕ and the dual variables $\boldsymbol{u}$ for the optimal LSFEM.

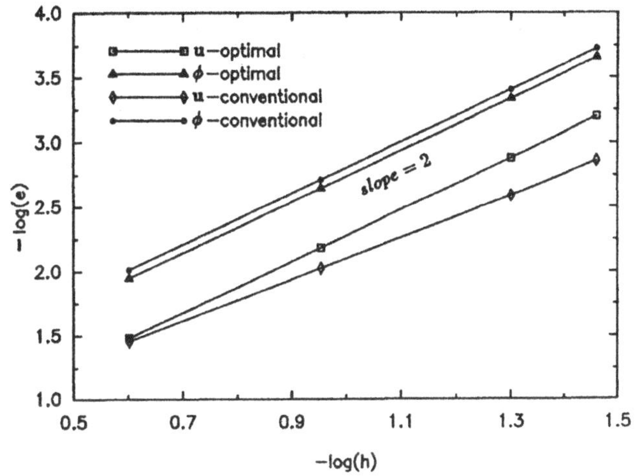

Fig. 6.1. Computed rates of convergence

Here we mention that in all our calculations, 2×2 Gaussian quadrature was used for finite element solutions, and 3×3 Gaussian quadrature was used for error evaluation.

6.6 Concluding Remarks

For the numerical solution of two- and three-dimensional diffusion-type second-order elliptic partial differential equations, the Rayleigh–Ritz method leads to symmetric and positive-definite matrix and has an optimal order of convergence. However, it involves difficulties related to the loss of accuracy and the discontinuity of the dual variables (gradients or fluxes) computed from the primal variable by numerical differentiation.

The mixed Galerkin method is based on the grad–div system deduced from the usual second-order scalar differential equation. The mixed Galerkin method can directly yield an approximation of the fluxes. But it requires the use of different finite elements for the primal and dual variables. Moreover, the resulting matrix is non-positive-definite and thus difficult to solve.

The conventional LSFEM is also based on the grad–div system. The conventional LSFEM has a significant advantage over the mixed Galerkin method in that it can accommodate equal-order elements. This method, which represents the beginning of LSFEM for first-order systems, was proposed in the pioneering papers of Lynn and Arya (1973) and Zienkiewicz et al. (1974) and later developed and investigated by Fix and Gunzburger (1978), Fix et al. (1979), and Cox and Fix (1984). This approach is still adopted by some people (e.g., Kececioglu and Rubinsky 1989, Bentley and Pinder 1992, and Winterscheidt and Surana 1993). However, theoretical and numerical studies have shown that, in the conventional LSFEM method, the rate of convergence for dual variables (fluxes) is lower than optimal, since the reduction of second-order elliptic problems to the first-order grad–div system destroys the full H^1 ellipticity.

In order to (1) achieve an optimal rate of convergence for both primal and dual variables; (2) allow the use of reduced intergration; (3) allow the use of the element residual as an error indicator for adaptive refinement; (4) gain fast convergence for iterative solvers; the LSFEM should be based on the div–curl–grad formulation. Usually, the div–curl–grad system should be supplemented by one of the following boundary data: (1) ϕ and $\boldsymbol{n} \times \boldsymbol{u}$; (2) $\boldsymbol{n} \cdot \boldsymbol{u}$; (3) $\phi + \beta \boldsymbol{n} \cdot \boldsymbol{u}$.

The idea of including the curl equation in the LSFEM can be found in Jiang and Chai (1980), Chen (1986) and Jiang (1986). The optimal LSFEM and its analysis described in this chapter was first developed by Jiang and Povinelli (1993). Following the same basic idea, Cai et al. (1994, 1997) gave a more general and rigorous analysis including a fast multigrid solver. The treatment of this subject by the Agmon–Douglis–Nirenberg theory can be found in Chang (1992).

Part III

LSFEM in Fluid Dynamics

7. Inviscid Irrotational Flows

In this chapter the LSFEM is applied to the solution of inviscid irrotational flow problems. Both incompressible and subsonic compressible flows are considered. The flow of incompressible, irrotational fluids is governed by the div–curl system that has been studied in Chap. 5. From understanding of the physical principles underlying compressible flows, we derive the associated governing equations in the form of a first-order system appropriate for LSFEM computation. In aerodynamic applications, uniform flow past a profile is of particular interest. Boundary conditions at the profile and in the far field, and the Kutta condition for lifting profiles are described. Applications of LSFEM to airfoil problems are presented.

7.1 Incompressible Irrotational Flow

Let us consider stationary incompressible, irrotational flows. The velocity $\boldsymbol{u}$ of the fluid flow is governed by the continuity equation:

$$\nabla \cdot \boldsymbol{u} = 0 \qquad \text{in } \Omega, \tag{7.1a}$$

and the condition of zero vorticity:

$$\nabla \times \boldsymbol{u} = \boldsymbol{0} \qquad \text{in } \Omega. \tag{7.1b}$$

On the boundary, in general, the normal component of the velocity is given, that is,

$$\boldsymbol{n} \cdot \boldsymbol{u} = g \qquad \text{on } \Gamma. \tag{7.1c}$$

System (7.1) has a typical div–curl structure. We may write system (7.1) in the standard matrix form (4.48), and use the LSFEM described in Sect. 4.8 to obtain the solution for a particular problem.

In order to verify the accuracy of LSFEM, we have numerically determined rates of convergence for the two-dimensional model problem in the unit square $0 \leq x, y \leq 1$:

$$\frac{\partial u}{\partial x} + \frac{\partial v}{\partial y} = 0, \tag{7.2a}$$

$$\frac{\partial v}{\partial x} - \frac{\partial u}{\partial y} = 0, \tag{7.2b}$$

with the boundary conditions

$$\begin{aligned} v &= \sin(x) \quad &&\text{on } y = 0, \\ u &= e^y \cos(1) \quad &&\text{on } x = 1, \\ v &= e \sin(x) \quad &&\text{on } y = 1\ , \\ u &= e^y \quad &&\text{on } x = 0. \end{aligned} \tag{7.3}$$

Combining the equations in (7.2), we see that this example corresponds to solving the Laplace equation for the potential ϕ in the unit square. For example, this problem describes incompressible potential flow in the square domain with the normal derivative $\partial\phi/\partial n$ prescribed on the boundary. The analytic solution to the model problem is

$$u = e^y \cos(x), \quad v = e^y \sin(x). \tag{7.4}$$

This is now used with the approximate solution to calculate the norm of the error in a sequence of mesh refinement studies.

The domain is divided into uniform bilinear elements. In the computation we are faced with the problem of evaluating an integral of the form (4.59) for the element matrices. The integrand consists of the squares of the first-order derivatives of the shape functions. Since bilinear elements are used, the integrand is a polynomial of degree 2, and hence the integral can be evaluated exactly with 2×2 Gaussian quadrature. However, as explained in Sect. 4.11.2, the LSFEM with Gaussian quadrature is equivalent to the least-squares finite element collocation method, and hence here one-point Gaussian quadrature should be employed. From this example, we understand that in LSFEM computations, in general, reduced integration should be used instead of full (exact) integration. For comparison purposes we have tested both one-point and 2×2 Gaussian quadrature in the computation.

The L_2 norm of the error is plotted against the mesh size h in a log-log plot in Fig. 7.1. We see that the error is $O(h^2)$, which is consistent with the optimal rate predicted by the estimate (5.24a) with this choice of elements. We remark that, although the estimate (5.24a) is derived for the case with homogeneous boundary conditions, it is not very difficult to show its validity for inhomogeneous boundary conditions.

Both full integration with 2×2 Gaussian quadrature and the reduced integration with one-point Gaussian quadrature produced the same rates, as shown in Fig. 7.1. The reason is as follows: first, the coefficients in system (7.2) are constants; second, we used uniform square elements which are not distorted and thus the Jacobian of the coordinate transformation in (4.94) is a constant; and more important, there are no zero-order terms in system (7.2), i.e., system (7.2) is of homogeneous first-order type. The consequence is that the equation can be satisfied everywhere in an element, and hence an increase in the order of quadrature does not mean the inclusion of more independent residual equations in the equivalent least-squares finite element collocation computation. Therefore, for this particular example reduced and full integration produced almost the same results. But in general when the elements are distorted, the coefficients in the partial differential equations are not constant, and the equations have zero-order terms, reduced integration becomes necessary.

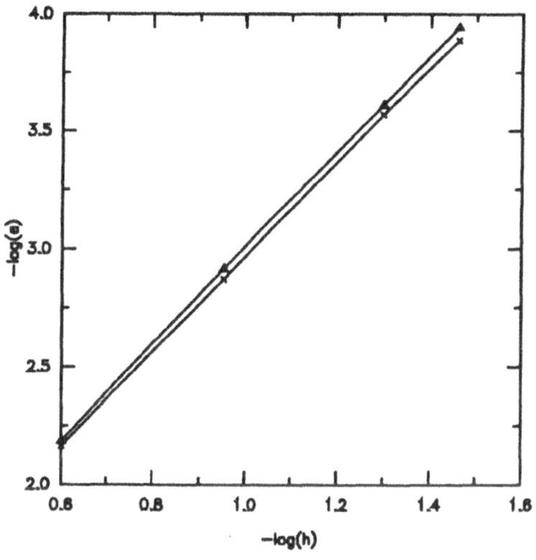

Fig. 7.1. Experimental rates of convergence of LSFEM for the model problem: $\triangle$ 2×2 Gaussian quadrature; $\times$ one-point Gaussian quadrature

Flow Around NACA0012 Airfoil. Consider the NACA0012 airfoil in incompressible flow with incident velocity at an angle of attack α relative to the airfoil chord. The flow is considered to be uniform at upstream infinity with velocity components $(U_\infty \cos\alpha, U_\infty \sin\alpha)$, where the subscript "$\infty$" denotes conditions for the incident flow at infinity. For our computation, the flow domain is approximated by a large but finite flow region. The outer boundary of this computational domain is 10.5 chords away from the center of the airfoil chord. The mesh consists of 600 bilinear elements and 660 nodes, see Fig. 7.2. The angles of incidence α are 2 and 4 degrees, respectively. The normal velocity component on the airfoil surface is prescribed to be zero. The condition $\boldsymbol{u} = (U_\infty \cos\alpha, U_\infty \sin\alpha)$ is specified on the outer boundary. The NACA0012 airfoil has a sharp trailing edge. The Kutta condition is the requirement that the flow leaves smoothly from the trailing edge in the direction of the bisector of the angle between the upper surface and the lower surface. For symmetrical airfoils, such as NACA0012, the Kutta condition requires that the vertical velocity component be zero at the trailing edge of the airfoil.

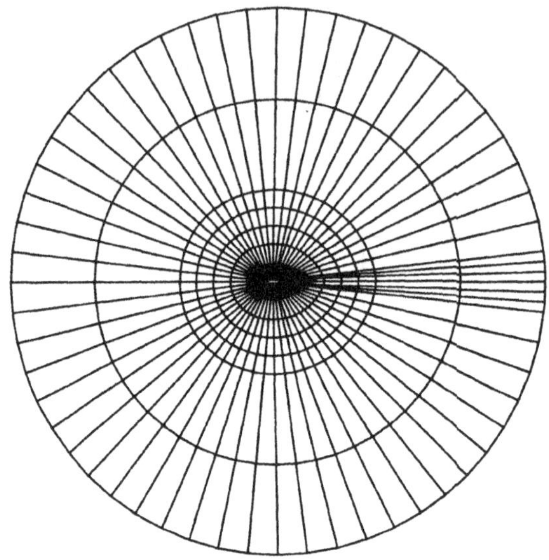

Fig. 7.2. The mesh for flow around NACA0012 airfoil

For the methods based on the potential, establishing a single-valued potential for a lifting body such as an inclined airfoil requires a branch cut in the domain and iteration of the solution to satisfy the Kutta conditions. These allow the potential solution to have a jump discontinuity across the cut. Since the flow velocity is continuous, the branch cut and iteration are not needed for LSFEM based on the velocity, and hence the satisfaction of the Kutta condition becomes very easy.

Once the velocity components at the nodes on the airfoil surface are obtained, the pressure coefficient C_p follows from the Bernoulli equation:

$$C_p = \frac{p - p_\infty}{0.5\rho U_\infty^2} = 1 - \frac{q^2}{U_\infty^2}, \tag{7.5}$$

where ρ is the density, and $q = \sqrt{u^2 + v^2 + w^2}$ is the magnitude of the flow velocity. The LSFEM solutions for the pressure coefficient distribution on the airfoil surface for $\alpha = 2°$ and $\alpha = 4°$ are shown in Fig. 7.3 and Fig. 7.4, respectively. In the computation, reduced integration with one-point Gaussian quadrature is used for evaluating the element matrices. As expected 2×2 Gaussian quadrature does not give good results.

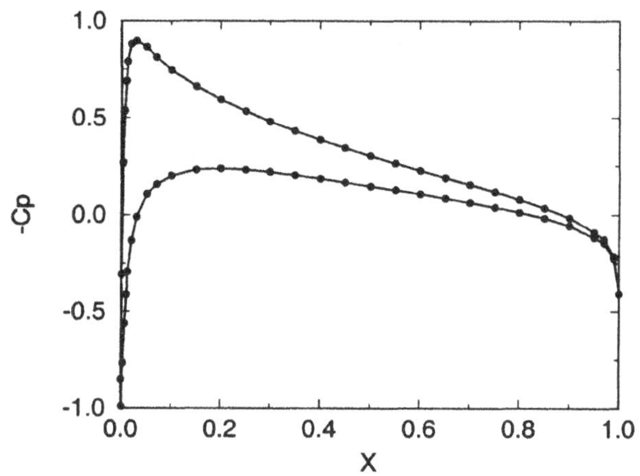

Fig. 7.3. Pressure coefficient distribution for NACA0012 airfoil at $\alpha = 2°$

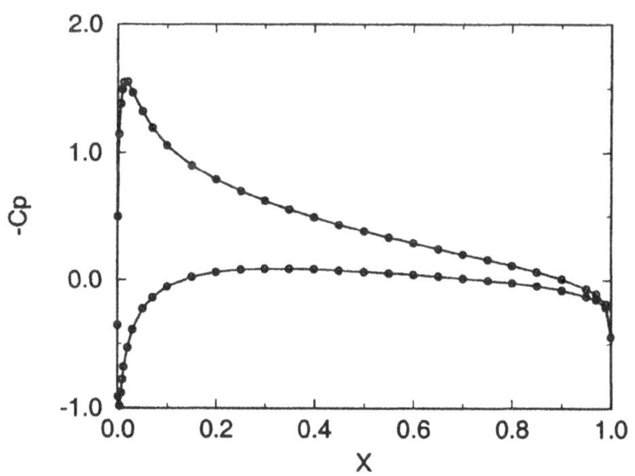

Fig. 7.4. Pressure coefficient distribution for NACA0012 airfoil at $\alpha = 4°$

7.2 Subsonic Compressible Irrotational Flow

7.2.1 The First-Order Governing Equations

In aerodynamics applications, as long as the flow velocity is sufficiently small, the density of the fluid is approximately constant and the classic theory of incompressible flow studied in the previous section adequately describes the flow field. As the incident flow velocity is increased, compressibility of the gas in the vicinity of the profile becomes more important, and the assumption of incompressibility is no longer applicable. Introducing the local Mach number $M = q/a$, in which a is the local speed of sound, we may characterize compressible flow as subsonic when $M < 1$ and supersonic when $M > 1$.

Let us consider steady subsonic inviscid, isothermal fluid flow with density $\rho(\boldsymbol{x})$. In addition, the fluid is assumed to be barotropic (elastic) and irrotational. Then there is a one-to-one relation between pressure and density, and this implies that the flow is isentropic.

The principle of conservation of mass leads to the continuity equation:

$$\frac{\partial \rho u}{\partial x} + \frac{\partial \rho v}{\partial y} + \frac{\partial \rho w}{\partial z} = 0. \tag{7.6a}$$

Since the flow is also irrotational, the velocity components are related by

$$\frac{\partial w}{\partial y} - \frac{\partial v}{\partial z} = 0, \quad \frac{\partial u}{\partial z} - \frac{\partial w}{\partial x} = 0, \quad \frac{\partial v}{\partial x} - \frac{\partial u}{\partial y} = 0. \tag{7.6b}$$

For isentropic flows, we have

$$p = k\rho^{\gamma}, \tag{7.6c}$$

where $\gamma > 1$ is the ratio of specific heats ($\gamma = C_p/C_v$) for calorically perfect gas and k is a constant. Using the first law of thermodynamics, we derive the energy equation for steady flow:

$$\frac{\gamma}{\gamma - 1}\frac{p}{\rho} + \frac{q^2}{2} = \frac{\gamma}{\gamma - 1}\frac{p_\infty}{\rho_\infty} + \frac{q_\infty^2}{2}. \tag{7.6d}$$

To obtain the governing equations in terms of velocity components, we eliminate pressure and density from the system of equations (7.6). First, eliminating p from (7.6c) and (7.6d), we obtain

$$\frac{\gamma}{\gamma - 1}\rho^{\gamma-1}\frac{p_\infty}{\rho_\infty^{\gamma}} + \frac{q^2}{2} = \frac{\gamma}{\gamma - 1}\frac{p_\infty}{\rho_\infty} + \frac{q_\infty^2}{2}. \tag{7.7}$$

Differentiating (7.7) with respect to x,

$$\gamma\rho^{\gamma-2}\frac{\partial \rho}{\partial x}\frac{p_\infty}{\rho_\infty^{\gamma}} + \frac{\partial}{\partial x}\left(\frac{q^2}{2}\right) = 0,$$

solving for $\partial\rho/\partial x$ and utilizing $\gamma p/\rho = a^2$, we have

$$\frac{\partial \rho}{\partial x} = -\rho\frac{\partial}{\partial x}\left(\frac{q^2}{2}\right)/a^2. \tag{7.8a}$$

Similarly we may obtain

$$\frac{\partial \rho}{\partial y} = -\rho\frac{\partial}{\partial y}\left(\frac{q^2}{2}\right)/a^2, \tag{7.8b}$$

$$\frac{\partial \rho}{\partial z} = -\rho\frac{\partial}{\partial z}\left(\frac{q^2}{2}\right)/a^2. \tag{7.8c}$$

Substituting the above three equations into (7.6a), we obtain

$$\rho\Big(\frac{\partial u}{\partial x}+\frac{\partial v}{\partial y}+\frac{\partial w}{\partial z}\Big)-\frac{\rho}{a^2}\Big[u\frac{\partial}{\partial x}\Big(\frac{q^2}{2}\Big)+v\frac{\partial}{\partial y}\Big(\frac{q^2}{2}\Big)+w\frac{\partial}{\partial z}\Big(\frac{q^2}{2}\Big)\Big]=0,$$

and hence the first-order equations in the velocity components:

$$\Big(1-\frac{u^2}{a^2}\Big)\frac{\partial u}{\partial x}+\Big(1-\frac{v^2}{a^2}\Big)\frac{\partial v}{\partial y}+\Big(1-\frac{w^2}{a^2}\Big)\frac{\partial w}{\partial z}$$

$$-\frac{uv}{a^2}\Big(\frac{\partial u}{\partial y}+\frac{\partial v}{\partial x}\Big)-\frac{vw}{a^2}\Big(\frac{\partial v}{\partial z}+\frac{\partial w}{\partial y}\Big)-\frac{wu}{a^2}\Big(\frac{\partial w}{\partial x}+\frac{\partial u}{\partial z}\Big)=0, \tag{7.9a}$$

where the local isentropic speed of sound a is given by (7.6d) as

$$\frac{a^2}{\gamma-1}+\frac{q^2}{2}=\frac{a_\infty^2}{\gamma-1}+\frac{q_\infty^2}{2},$$

that is,

$$a^2=a_\infty^2+\frac{\gamma-1}{2}(q_\infty^2-q^2). \tag{7.9b}$$

The above (7.9a) and (7.9b) together with the irrotational condition (7.6b) constitute the first-order system of equations governing compressible flows.

We introduce dimensionless variables of space and velocity by dividing the physical variables by the characteristic length scale L and a_∞ respectively. The governing equations are then rewritten as

$$\Big(1-\frac{\bar u^2}{\bar a^2}\Big)\frac{\partial \bar u}{\partial \bar x}+\Big(1-\frac{\bar v^2}{\bar a^2}\Big)\frac{\partial \bar v}{\partial \bar y}+\Big(1-\frac{\bar w^2}{\bar a^2}\Big)\frac{\partial \bar w}{\partial \bar z}$$

$$-\frac{\bar u\bar v}{\bar a^2}\Big(\frac{\partial \bar u}{\partial \bar y}+\frac{\partial \bar v}{\partial \bar x}\Big)-\frac{\bar v\bar w}{\bar a^2}\Big(\frac{\partial \bar v}{\partial \bar z}+\frac{\partial \bar w}{\partial \bar y}\Big)-\frac{\bar w\bar u}{\bar a^2}\Big(\frac{\partial \bar w}{\partial \bar x}+\frac{\partial \bar u}{\partial \bar z}\Big)=0, \tag{7.10a}$$

$$\frac{\partial \bar v}{\partial \bar x}-\frac{\partial \bar u}{\partial \bar y}=0,\quad \frac{\partial \bar v}{\partial \bar x}-\frac{\partial \bar u}{\partial \bar y}=0,\quad \frac{\partial \bar v}{\partial \bar x}-\frac{\partial \bar u}{\partial \bar y}=0, \tag{7.10b}$$

where

$$\bar a^2=1+\frac{\gamma-1}{2}({M_\infty}^2-\bar q^2), \tag{7.10c}$$

in which the free stream Mach number $M_\infty=q_\infty/a_\infty$.

It is also appropriate here to introduce some useful local dimensionless parameters for the fluid flow. Using $a_\infty^2=\gamma p_\infty/\rho_\infty$ and $\bar\rho=\rho/\rho_\infty$ and (7.9b), we obtain the dimensionless local density

$$\bar\rho=\Big[1+\frac{\gamma-1}{2}({M_\infty}^2-\bar q^2)\Big]^{1/(\gamma-1)}. \tag{7.11}$$

By virtue of (7.10c), the local Mach number can be expressed as

$$M=\frac{q}{a}=\frac{q/a_\infty}{a/a_\infty}=\frac{\bar q}{[1+\frac{\gamma-1}{2}({M_\infty}^2-\bar q^2)]^{1/2}}. \tag{7.12}$$

The pressure coefficient C_p is defined by

$$C_p=\frac{p-p_\infty}{\frac{\rho_\infty q_\infty^2}{2}}=\frac{2}{\gamma M_\infty^2}\Big\{\Big[1+\frac{\gamma-1}{2}(M_\infty^2-\bar q^2)\Big]^{\gamma/(\gamma-1)}-1\Big\}. \tag{7.13}$$

7.2.2 Application of LSFEM

Now let us apply the LSFEM to the solution of plane steady compressible flow problems. The fundamental dimensionless equations for two-dimensional steady subsonic irrotational flows are

$$\left(1-\frac{u^2}{a^2}\right)\frac{\partial u}{\partial x}-\frac{uv}{a^2}\left(\frac{\partial u}{\partial y}+\frac{\partial v}{\partial x}\right)+\left(1-\frac{v^2}{a^2}\right)\frac{\partial v}{\partial y}=0 \qquad \text{in } \Omega, \tag{7.14a}$$

$$\frac{\partial v}{\partial x}-\frac{\partial u}{\partial y}=0 \qquad \text{in } \Omega, \tag{7.14b}$$

$$a^2=1+\frac{\gamma-1}{2}[M_\infty^2-(u^2+v^2)], \tag{7.14c}$$

where a is the local sound speed normalized by the sound speed of the free stream, and for convenience the dimensionless bar notation is suppressed. Usually, on the solid surface the normal component of the velocity is zero, that is,

$$\boldsymbol{n}\cdot\boldsymbol{u}=0 \qquad \text{on } \Gamma. \tag{7.15}$$

The uniform condition $\boldsymbol{u}=(U_\infty\cos\alpha, U_\infty\sin\alpha)$ is prescribed on the outer boundary.

The quasi-linear first-order partial differential equations (7.14) may be linearized, for example, by using the successive substitution method. Let

$$A=\left(1-\frac{u^2}{a^2}\right),\ B=\left(1-\frac{v^2}{a^2}\right),\ C=-\frac{uv}{a^2}. \tag{7.16}$$

Coefficients A, B, C are regarded as known quantities and are determined by the velocity field solved in the preceding iteration step. Then system (7.14) can be rewritten as

$$A\frac{\partial u}{\partial x}+C\left(\frac{\partial u}{\partial y}+\frac{\partial v}{\partial x}\right)+B\frac{\partial v}{\partial y}=0 \qquad \text{in } \Omega, \tag{7.17a}$$

$$\frac{\partial v}{\partial x}-\frac{\partial u}{\partial y}=0 \qquad \text{in } \Omega. \tag{7.17b}$$

Now we write system (7.17) in standard matrix form:

$$\boldsymbol{A}_1\frac{\partial \boldsymbol{u}}{\partial x}+\boldsymbol{A}_2\frac{\partial \boldsymbol{u}}{\partial y}+\boldsymbol{A}_0\boldsymbol{u}=\boldsymbol{f} \qquad \text{in } \Omega, \tag{7.18}$$

in which

$$\boldsymbol{A}_1=\begin{pmatrix} A & C \\ 0 & 1 \end{pmatrix}, \qquad \boldsymbol{A}_2=\begin{pmatrix} C & B \\ -1 & 0 \end{pmatrix},$$

$$\boldsymbol{A}_0=\begin{pmatrix} 0 & 0 \\ 0 & 0 \end{pmatrix}, \qquad \boldsymbol{f}=\begin{pmatrix} 0 \\ 0 \end{pmatrix}, \qquad \boldsymbol{u}=\begin{pmatrix} u \\ v \end{pmatrix}.$$

Since

$$\det(\boldsymbol{A}_1\xi + \boldsymbol{A}_2\eta) = \det\begin{pmatrix} A\xi + C\eta & C\xi + B\eta \\ -\eta & \xi \end{pmatrix}$$
$$= A\xi^2 + 2C\xi\eta + B\eta^2 \neq 0$$

for all nonzero real pairs (ξ, η), if

$$C^2 - 4AB = \frac{u^2 + v^2}{a^2} - 1 < 0.$$

It means that for subsonic flows, (7.14) is elliptic.

At each iteration step, the LSFEM described in Sect. 4.8 can be used to solve the linearized equation system (7.18). By using (4.59) we can derive the element matrices which consist of the following (2×2) sub-matrix:

$$[\boldsymbol{K}_e]_{ij} =$$
$$\int_{\Omega_e} \begin{pmatrix} \begin{array}{c} A^2\psi_{i,x}\psi_{j,x} \\ +AC(\psi_{i,x}\psi_{j,y} + \psi_{i,y}\psi_{j,x}) \\ +(C^2+1)\psi_{i,y}\psi_{j,y} \end{array} & \vdots & \begin{array}{c} AB\psi_{i,x}\psi_{j,y} \\ +AC\psi_{i,x}\psi_{j,x} + BC\psi_{i,y}\psi_{j,y} \\ +(C^2-1)\psi_{i,y}\psi_{j,x} \end{array} \\ \cdots\cdots & \vdots & \cdots\cdots \\ \begin{array}{c} AB\psi_{i,y}\psi_{j,x} \\ +AC\psi_{i,x}\psi_{j,x} + BC\psi_{i,y}\psi_{j,y} \\ +(C^2-1)\psi_{i,x}\psi_{j,y} \end{array} & \vdots & \begin{array}{c} B^2\psi_{i,y}\psi_{j,y} \\ +BC(\psi_{i,y}\psi_{j,x} + \psi_{i,x}\psi_{j,y}) \\ +(C^2+1)\psi_{i,x}\psi_{j,x} \end{array} \end{pmatrix} dxdy. \tag{7.19}$$

Now we explain how to treat the boundary condition (7.15) on the inclined solid surface. Assume that at a surface node i the normal direction is $(\cos\beta, \sin\beta)$. At this node the normal component and the tangential component of the velocity $\boldsymbol{u} = (u, v)$ are denoted by u_n and u_t respectively. Then we have the following relations between (u, v) and (u_n, u_t):

$$\begin{pmatrix} u_n \\ u_t \end{pmatrix} = \boldsymbol{C}\begin{pmatrix} u \\ v \end{pmatrix}, \tag{7.20a}$$

$$\begin{pmatrix} u \\ v \end{pmatrix} = \boldsymbol{C}\begin{pmatrix} u_n \\ u_t \end{pmatrix}, \tag{7.20b}$$

where $\boldsymbol{C}$ is the coordinate transformation matrix:

$$\boldsymbol{C} = \begin{pmatrix} \cos\beta & \sin\beta \\ \sin\beta & -\cos\beta \end{pmatrix}. \tag{7.20c}$$

Taking nodal values of the velocity components in global rectangular coordinates as the unknowns, the global algebraic equations have the following form

$$\begin{pmatrix} \boldsymbol{K}_{11} & \boldsymbol{K}_{12} & \dots & \boldsymbol{K}_{1i} & \dots & \boldsymbol{K}_{1n} \\ \boldsymbol{K}_{21} & \boldsymbol{K}_{22} & \dots & \boldsymbol{K}_{2i} & \dots & \boldsymbol{K}_{2n} \\ \vdots & \vdots & \ddots & \vdots & \ddots & \vdots \\ \boldsymbol{K}_{i1} & \boldsymbol{K}_{i2} & \dots & \boldsymbol{K}_{ii} & \dots & \boldsymbol{K}_{in} \\ \vdots & \vdots & \ddots & \vdots & \ddots & \vdots \\ \boldsymbol{K}_{n1} & \boldsymbol{K}_{n2} & \dots & \boldsymbol{K}_{ni} & \dots & \boldsymbol{K}_{nn} \end{pmatrix} \begin{pmatrix} \boldsymbol{U}_1 \\ \boldsymbol{U}_2 \\ \vdots \\ \boldsymbol{U}_i \\ \vdots \\ \boldsymbol{U}_n \end{pmatrix} = \begin{pmatrix} \boldsymbol{F}_1 \\ \boldsymbol{F}_2 \\ \vdots \\ \boldsymbol{F}_i \\ \vdots \\ \boldsymbol{F}_n \end{pmatrix}, \tag{7.21}$$

in which $\boldsymbol{K}_{11}$, $\boldsymbol{K}_{12}$, ... are 2×2 matrices, and

$$\boldsymbol{U}_i = \begin{pmatrix} u \\ v \end{pmatrix}_i, \qquad \boldsymbol{F}_i = \begin{pmatrix} F_x \\ F_y \end{pmatrix}_i$$

are the velocity components and the components of the 'force' vector at the ith node respectively. For the homogeneous equations (7.14), the components of the 'force' vectors in system (7.21) are zero. In general, they are not zero.

For nodes on the inclined solid surface, the normal and the tangential velocity components are taken as the unknowns. By considering (7.20b), we can rewrite system (7.21) as

$$\begin{pmatrix} \boldsymbol{K}_{11} & \boldsymbol{K}_{12} & \dots & \boldsymbol{K}_{1i}\boldsymbol{C} & \dots & \boldsymbol{K}_{1n} \\ \boldsymbol{K}_{21} & \boldsymbol{K}_{22} & \dots & \boldsymbol{K}_{2i}\boldsymbol{C} & \dots & \boldsymbol{K}_{2n} \\ \vdots & \vdots & \ddots & \vdots & \ddots & \vdots \\ \boldsymbol{K}_{i1} & \boldsymbol{K}_{i2} & \dots & \boldsymbol{K}_{ii}\boldsymbol{C} & \dots & \boldsymbol{K}_{in} \\ \vdots & \vdots & \ddots & \vdots & \ddots & \vdots \\ \boldsymbol{K}_{n1} & \boldsymbol{K}_{n2} & \dots & \boldsymbol{K}_{ni}\boldsymbol{C} & \dots & \boldsymbol{K}_{nn} \end{pmatrix} \begin{pmatrix} \boldsymbol{U}_1 \\ \boldsymbol{U}_2 \\ \vdots \\ \boldsymbol{U}_i^* \\ \vdots \\ \boldsymbol{U}_n \end{pmatrix} = \begin{pmatrix} \boldsymbol{F}_1 \\ \boldsymbol{F}_2 \\ \vdots \\ \boldsymbol{F}_i \\ \vdots \\ \boldsymbol{F}_n \end{pmatrix}, \tag{7.22}$$

in which

$$\boldsymbol{U}_i^* = \begin{pmatrix} u_n \\ u_t \end{pmatrix}_i .$$

In order to obtain a symmetric global matrix we multiply the ith equation in system (7.22) by $\boldsymbol{C}$ and finally obtain

$$\begin{pmatrix} \boldsymbol{K}_{11} & \boldsymbol{K}_{12} & \dots & \boldsymbol{K}_{1i}\boldsymbol{C} & \dots & \boldsymbol{K}_{1n} \\ \boldsymbol{K}_{21} & \boldsymbol{K}_{22} & \dots & \boldsymbol{K}_{2i}\boldsymbol{C} & \dots & \boldsymbol{K}_{2n} \\ \vdots & \vdots & \ddots & \vdots & \ddots & \vdots \\ \boldsymbol{C}\boldsymbol{K}_{i1} & \boldsymbol{C}\boldsymbol{K}_{i2} & \dots & \boldsymbol{C}\boldsymbol{K}_{ii}\boldsymbol{C} & \dots & \boldsymbol{C}\boldsymbol{K}_{in} \\ \vdots & \vdots & \ddots & \vdots & \ddots & \vdots \\ \boldsymbol{K}_{n1} & \boldsymbol{K}_{n2} & \dots & \boldsymbol{K}_{ni}\boldsymbol{C} & \dots & \boldsymbol{K}_{nn} \end{pmatrix} \begin{pmatrix} \boldsymbol{U}_1 \\ \boldsymbol{U}_2 \\ \vdots \\ \boldsymbol{U}_i^* \\ \vdots \\ \boldsymbol{U}_n \end{pmatrix} = \begin{pmatrix} \boldsymbol{F}_1 \\ \boldsymbol{F}_2 \\ \vdots \\ \boldsymbol{C}\boldsymbol{F}_i \\ \vdots \\ \boldsymbol{F}_n \end{pmatrix}. \tag{7.23}$$

Using given boundary data we can get the solution to system (7.23) by using a direct solver. If the problem is very large, one may use the conjugate gradient method to obtain the solution. In that case, it is neither necessary to form the element matrices, nor to assemble the global matrix and to transfer the coordinates for the nodes on the inclined solid surface (see Chap. 15).

7.2.3 Examples

In the following two numerical examples, the flow field obtained by solving the corresponding incompressible problem was taken as an initial field. At all nodal points of the field, the solution was considered convergent if the relative variation of two successive values of the dimensionless density $\rho = [1 + \frac{\gamma-1}{2}(M_\infty^2 - (u^2 + v^2))]^{1/(\gamma-1)}$ is less than a given small quantity ϵ.

Flow Past Circular Cylinder. The flow condition at seven times the radius of the cylinder was considered undisturbed from the free stream. One quarter of the circular cylinder was divided into 4×5 eight-node quadrilateral elements. Reduced integration with 2×2 Gaussian quadrature was used. ϵ was taken as 0.0009. The solutions at various M_∞ including the critical Mach number 0.42 depicted in Fig. 7.5 were obtained in $2 \sim 3$ linearizations.

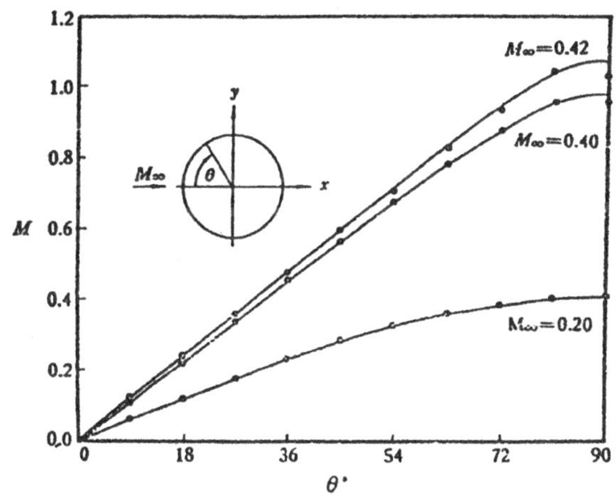

Fig. 7.5. Mach number on the surface of the circular cylinder at upstream Mach number of 0.42, o LSFEM, — results from Imai (1941) (from Jiang and Cai 1980)

Flow Past NACA0012 Airfoil. The flow around the symmetrical NACA 0012 airfoil with zero angle of attack was calculated. The upper half region of the airfoil was divided into 5×12 eight-node quadrilateral elements and $\epsilon = 0.0001$ was taken as the convergence criterion. In the computation reduced integration with 2×2 Gaussian quadrature was used. The solution at the critical Mach number 0.72 was obtained after six linearizations and is shown in Fig. 7.6.

The LSFEM has also been tested for lifting cases. Figure 7.7 and Figure 7.8 show the pressure distribution for the subsonic flow on the NACA0012 airfoil at 3.5° incidence and $M_\infty = 0.5$, and 10° incidence and $M_\infty = 0.5$,

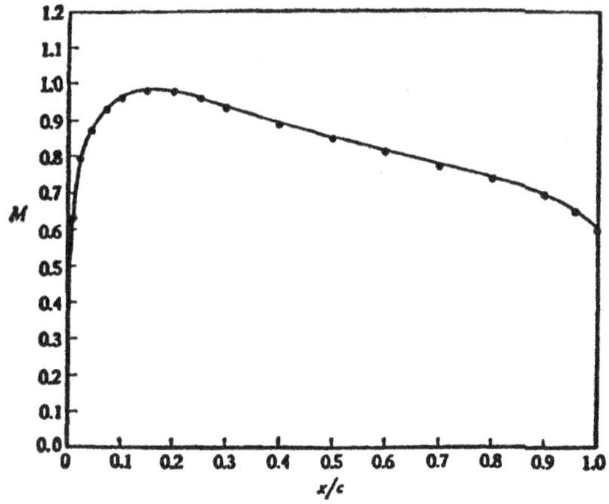

Fig. 7.6. Mach number on the surface of the NACA0012 airfoil surface at 0° incidence and upstream Mach number of 0.72, o LSFEM, — results from Lock (1970) (from Jiang and Cai 1980)

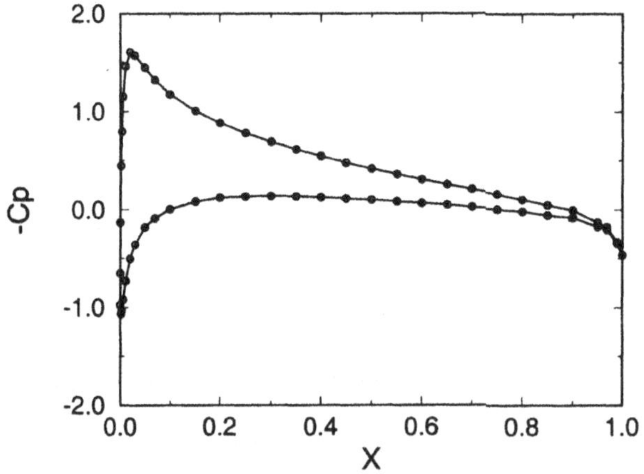

Fig. 7.7. Pressure distribution on the NACA0012 airfoil surface at 3.5° incidence and upstream Mach number of 0.5

respectively. In these computations, the flow domain and mesh, the number of integration points, the boundary conditions and the Kutta condition are the same as those used for the incompressible case. Converged solutions are obtained after four linearizations.

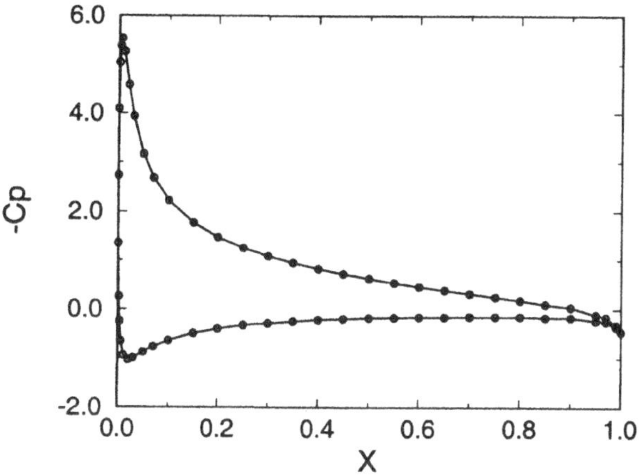

Fig. 7.8. Pressure distribution on the NACA0012 airfoil surface at 10° incidence and upstream Mach number of 0.3

7.3 Concluding Remarks

Irrotational flow problems are usually solved by using methods based on the potential. The finite element solution for the potential has optimal accuracy. For example, if linear elements are used, the accuracy of the potential is $O(h^2)$. However, numerical differentiation is needed to obtain the velocity from the potential. As a result, the velocity derived from the potential has lower accuracy, viz., $O(h)$, and is discontinuous across element boundaries. Moreover, for multi-connected domains, such as lifting airfoils, potential methods need a branch cut. Due to the continuity of velocity, branch cut is not necessary for LSFEM. This makes the satisfaction of the Kutta condition trivial in LSFEM.

The LSFEM and numerical results based on the first-order div–curl system (7.14) in terms of the velocity components for subsonic irrotational flows first appeared in Jiang and Chai (1980). More numerical tests including adaptive refinement were presented in Jiang (1986). Similar formulations can also be found in Chen (1986) and Chen and Fix (1986).

Fletcher (1979) first used a LSFEM based on the compressible Euler equations to solve subsonic flow problems. The feature of his formulation is the designation of groups of variables rather than single variables.

8. Incompressible Viscous Flows

The incompressible Navier–Stokes equations which govern the motion of viscous fluids are among the most important partial differential equations in engineering. The Navier–Stokes equations are derived from the mass and momentum conservation laws and the linear stress–strain relation. While the physical model leading to the Navier–Stokes equations is simple, the mathematical theory and the numerical solution are challenging. The major difficulties in the study and the solution of these equations are related to incompressibility and nonlinearity.

In the last two decades various methods including finite elements, finite differences, and finite volumes based on different formulations have been used for solving the Navier–Stokes equations. Usually the Galerkin method based on the velocity–pressure formulation is employed for the theoretical analysis and the finite element solution of the incompressible Navier–Stokes equations. In the beginning of this chapter we will show that the Galerkin method leads to a saddle-point problem, and thus a difficult Ladyzhenskaya–Babuška–Brezzi (LBB) condition is invoked. Then we will briefly show that all modified Galerkin-type methods have some restrictions.

In this chapter attention will be focussed on the least-squares method based on the first-order velocity–pressure–vorticity formulation. The nonlinear convective terms in the Navier–Stokes equations have no effect on the classification (see Sect. 8.4); also according to the theory in Brezzi et al. (1980), error analysis for the nonlinear Navier–Stokes equations is essentially the same as that for the linear Stokes equations, at least away from singular points. Thus, the main concern of the mathematical analysis in this chapter is the verification of the well-posedness of the Stokes equations with different boundary conditions.

In the investigation of existence and uniqueness of the incompressible Stokes and Navier–Stokes equations, mathematicians mainly confined themselves to the standard case in which velocity components are prescribed on the boundary. On the other hand, physicists and engineers apply the non-standard boundary conditions often based on physical intuition. The permissibility of non-standard boundary conditions applied to the Navier–Stokes equations has not been systematically studied, and thus has been the subject of constant controversy.

The principal part of the Stokes equations in the first-order velocity–pressure–vorticity formulation consists of two generalized div–curl systems, so the techniques learned in Chap. 5 about a div–curl system can play an essential role in the analysis. We will use this knowledge to prove that the velocity–pressure–vorticity formulation is elliptic and not "overdetermined", to derive all permissible combinations of non-standard boundary conditions for the Stokes equations, and show the optimality or sub-optimality of the corresponding LSFEM.

In this chapter we will demonstrate how to use the LSFEM based on the first-order velocity–pressure–vorticity formulation to solve various problems related to incompressible viscous flows.

Finally we will apply the least-squares method and the div–curl method to rigorously derive the pressure Poisson equation, the second-order velocity–vorticity formulation of the Stokes and Navier–Stokes equations and their additional boundary conditions.

8.1 The Stokes Equations in the $\boldsymbol{u} - p$ Formulation

8.1.1 The Mixed Galerkin Method

Consider the Stokes equations: Find the velocity $\boldsymbol{u} = (u_1, u_2, u_3) = (u, v, w)$ and the pressure p such that

$$-\nu \Delta \boldsymbol{u} + \nabla p = \boldsymbol{f} \qquad \text{in } \Omega, \tag{8.1a}$$

$$\nabla \cdot \boldsymbol{u} = 0 \qquad \text{in } \Omega, \tag{8.1b}$$

$$\boldsymbol{u} = \boldsymbol{0} \qquad \text{on } \Gamma, \tag{8.1c}$$

where ν is the kinematic viscosity, and $\boldsymbol{f} = (f_x, f_y, f_z) \in [L_2(\Omega)]^3$ the body force. Since the viscosity ν can be absorbed into the pressure and the body force terms for the Stokes equations, we need to consider only case $\nu = 1$.

The most popular finite element method for the Stokes equations is based on the mixed Galerkin formulation, see e.g., Oden and Carey (1983), Girault and Raviart (1986), Gunzburger (1989) and Pironneau (1989): Find $\boldsymbol{u} \in [H_0^1(\Omega)]^3$ and $p \in S(\Omega)$ such that

$$a(\boldsymbol{u}, \boldsymbol{v}) - (\nabla \cdot \boldsymbol{v}, p) = (\boldsymbol{f}, \boldsymbol{v}) \qquad \forall \boldsymbol{v} \in [H_0^1(\Omega)]^3, \tag{8.2a}$$

$$(\nabla \cdot \boldsymbol{u}, q) = 0 \qquad \forall q \in S(\Omega), \tag{8.2b}$$

where

$$H_0^1(\Omega) = \{u \in H^1(\Omega) : u = 0 \text{ on } \Gamma\}, \tag{8.2c}$$

$$S(\Omega) = \{q \in L_2(\Omega) : \int_\Omega q dx = 0\}, \tag{8.2d}$$

$$a(\boldsymbol{u},\boldsymbol{v}) = \sum_{i=1}^{3}(\nabla u_i, \nabla v_i). \tag{8.2e}$$

The variational statement (8.2) can be obtained directly from the stationary condition for the Lagrangian

$$L(\boldsymbol{u},p) = \frac{1}{2}a(\boldsymbol{u},\boldsymbol{u}) - (\boldsymbol{f},\boldsymbol{u}) - (\nabla\cdot\boldsymbol{u},p). \tag{8.3}$$

The solution $\{\boldsymbol{u},p\}$ defines a saddle-point of this functional.

Similar to the situation discussed in Sect. 3.3, the existence of a finite element approximate solution to (8.2) depends on choosing a pair of spaces $\boldsymbol{H}_h \subset [H_0^1(\Omega)]^3$ and $S_h \subset S(\Omega)$ such that the following LBB "inf-sup" condition holds:

$$\inf_{q\in S_h}\sup_{\boldsymbol{v}\in \boldsymbol{H}_h}\frac{(\nabla\cdot\boldsymbol{v},q)}{\|\boldsymbol{v}\|_1\|q\|_0} \geq \alpha > 0, \tag{8.4}$$

where α is independent of the mesh size h. This condition precludes the use of equal-order finite elements. Although for two-dimensional problems quite a few convergent pairs of velocity and pressure elements have been developed, most of these combinations employ some basis functions that are inconvenient to implement. For three-dimensional problems, this difficulty becomes more severe and only elaborate constructions can pass the LBB test. The other basic difficulty associated with the mixed method is the non-positiveness of the resulting linear algebraic system.

Yet another important choice is the penalty method which pre-eliminates the pressure variable from the Stokes equations by penalizing the continuity equation, see, e.g., Reddy and Gartling (1994). The penalty functional is

$$J_\varepsilon(\boldsymbol{u}) = \frac{1}{2}a(\boldsymbol{u},\boldsymbol{u}) - (\boldsymbol{f},\boldsymbol{u}) + \frac{1}{2\varepsilon}(\nabla\cdot\boldsymbol{u},\nabla\cdot\boldsymbol{u}), \tag{8.5}$$

where ε is a positive penalty parameter, and the variational problem is to find the velocity $\boldsymbol{u}_\varepsilon$ making J_ε stationary over $[H_0^1(\Omega)]^3$. Since it involves only velocities, the penalty method is efficient. However, pressure is calculated from the computed velocity field and thus the accuracy of pressure is lower than that of the velocity. Another disadvantage is that the penalty parameter ε causes loss of accuracy for small values and for too large values prevents convergence to the solution. Furthermore, the resulting algebraic system is ill-conditioned. Therefore, this system cannot normally be solved by iterative techniques.

8.1.2 The Mixed Galerkin/Least-Squares Method

For the Stokes equations, Hughes and his colleagues put additional least-squares terms into the mixed Galerkin functional (8.3):

$$L_{M/L}(\boldsymbol{u},p) = \frac{1}{2}a(\boldsymbol{u},\boldsymbol{u}) - (\boldsymbol{f},\boldsymbol{u}) - (\nabla\cdot\boldsymbol{u},p)$$
$$-\frac{\alpha}{2}h^2(-\Delta\boldsymbol{u}+\nabla p-\boldsymbol{f},-\Delta\boldsymbol{u}+\nabla p-\boldsymbol{f})_{\tilde{\Omega}} - \frac{\beta}{2}h([\![p]\!],[\![p]\!])_{\tilde{\Gamma}}, \tag{8.6}$$

where $\tilde{\Omega}$ denotes element interiors, $\tilde{\Gamma}$ denotes element interfaces, $[\![\cdot]\!]$ is the jump operator. α and β are nondimensional stability parameters, such that for small parameters this new functional (8.6) has the desired coerciveness over the finite element space $\boldsymbol{H}_h \times S_h$ in the sense that as $\|p\|_0 \to \infty$ and $\|\boldsymbol{u}\|_1 \to \infty$, $L_{M/L}(\boldsymbol{u},p) = +\infty$. This coerciveness implies unique solvability of the problem without imposition of the LBB condition. Because it involves least-squares terms, this method is named the mixed Galerkin/least-squares method (Franca et al. 1989). This method is attractive because one obtains convergence for arbitrary combinations of velocity and pressure interpolations. However, when it is extended to solve the Navier–Stokes equations, the resulting matrix is nonsymmetric and thus difficult to solve by iterative methods for large-scale problems.

8.2 The Stokes Equations in the $\boldsymbol{u}-p-\boldsymbol{\omega}$ Formulation

The difficulties associated with the Galerkin method prompted us to consider the least-squares method. Since the momentum equation (8.1a) involves second-order derivatives of velocity, application of the least-squares method to (8.1) requires the use of impractical continuously differentiable shape functions. In order to avoid this problem, one has to consider the governing equations of incompressible flow in the form of a first-order system. Therefore, we introduce the vorticity $\boldsymbol{\omega} = (\omega_x, \omega_y, \omega_z) = \nabla\times\boldsymbol{u}$ as an additional independent unknown vector, and rewrite the incompressible Stokes equations in the following first-order velocity–pressure–vorticity formulation:

$$\nabla p + \nabla\times\boldsymbol{\omega} = \boldsymbol{f} \quad \text{in } \Omega, \tag{8.7a}$$

$$\nabla\cdot\boldsymbol{\omega} = 0 \quad \text{in } \Omega, \tag{8.7b}$$

$$\boldsymbol{\omega} - \nabla\times\boldsymbol{u} = \boldsymbol{0} \quad \text{in } \Omega, \tag{8.7c}$$

$$\nabla\cdot\boldsymbol{u} = 0 \quad \text{in } \Omega. \tag{8.7d}$$

For simplicity the domain Ω is assumed to be open, bounded, and simply connected with a piece-wise smooth boundary Γ.

At first glance, one may think that the compatibility constraint condition (8.7b), i.e., the divergence of the vorticity vector equals zero, is 'redundant'. In (8.7) there are eight equations involving only seven unknown variables, i.e., three velocity components u, v, w, the pressure p, and three vorticity components $\omega_x, \omega_y, \omega_z$; also the definition of vorticity (8.7c) implies satisfaction of (8.7b), and inclusion of (8.7b) appears to make system (8.7) "overdetermined".

8.2.1 Determinacy and Ellipticity

We shall show that system (8.7) is really properly determined and elliptic by using the same technique as discussed in Sect. 5.2. To proceed we may consider the following first-order system for the Stokes equations:

$$\nabla p + \nabla \times \boldsymbol{\omega} = \boldsymbol{f} \quad \text{in } \Omega, \tag{8.8a}$$

$$\nabla \cdot \boldsymbol{\omega} = 0 \quad \text{in } \Omega, \tag{8.8b}$$

$$-\boldsymbol{\omega} + \nabla \phi + \nabla \times \boldsymbol{u} = \boldsymbol{0} \quad \text{in } \Omega, \tag{8.8c}$$

$$\nabla \cdot \boldsymbol{u} = 0 \quad \text{in } \Omega. \tag{8.8d}$$

Here we have introduced a dummy variable ϕ in (8.8c) , which satisfies the boundary condition $\phi = 0$ on Γ. Substituting (8.8c) into (8.8b) yields $\Delta\phi = 0$, thus $\phi \equiv 0$ in Ω. It means that the introduction of ϕ does not change anything. However, now there are eight unknowns and eight equations in (8.8). System (8.8), and hence system (8.7) is indeed properly determined.

We note that in some cases specification of the boundary condition for the dummy variable ϕ is unnecessary, and $\nabla\phi \equiv \boldsymbol{0}$ can also be guaranteed, see Sect. 8.2.2.

Now let us classify system (8.8). In Cartesian coordinates, system (8.8) is given as

$$\begin{aligned}
&\frac{\partial p}{\partial x} + \frac{\partial \omega_z}{\partial y} - \frac{\partial \omega_y}{\partial z} = f_x, \\
&\frac{\partial p}{\partial y} + \frac{\partial \omega_x}{\partial z} - \frac{\partial \omega_z}{\partial x} = f_y, \\
&\frac{\partial p}{\partial z} + \frac{\partial \omega_y}{\partial x} - \frac{\partial \omega_x}{\partial y} = f_z, \\
&\frac{\partial \omega_x}{\partial x} + \frac{\partial \omega_y}{\partial y} + \frac{\partial \omega_z}{\partial z} = 0, \\
&-\omega_x + \frac{\partial \phi}{\partial x} + \frac{\partial w}{\partial y} - \frac{\partial v}{\partial z} = 0, \\
&-\omega_y + \frac{\partial \phi}{\partial y} + \frac{\partial u}{\partial z} - \frac{\partial w}{\partial x} = 0, \\
&-\omega_z + \frac{\partial \phi}{\partial z} + \frac{\partial v}{\partial x} - \frac{\partial u}{\partial y} = 0, \\
&\frac{\partial u}{\partial x} + \frac{\partial v}{\partial y} + \frac{\partial w}{\partial z} = 0.
\end{aligned} \tag{8.9}$$

We may write (8.9) in standard matrix form:

$$\boldsymbol{A}_1 \frac{\partial \underline{\boldsymbol{u}}}{\partial x} + \boldsymbol{A}_2 \frac{\partial \underline{\boldsymbol{u}}}{\partial y} + \boldsymbol{A}_3 \frac{\partial \underline{\boldsymbol{u}}}{\partial z} + \boldsymbol{A}\underline{\boldsymbol{u}} = \underline{\boldsymbol{f}}, \tag{8.10}$$

in which

$$\boldsymbol{A}_1 = \begin{pmatrix} 0 & 0 & 0 & 0 & 0 & 0 & 0 & 1 \\ 0 & 0 & 0 & 0 & 0 & 0 & -1 & 0 \\ 0 & 0 & 0 & 0 & 0 & 1 & 0 & 0 \\ 0 & 0 & 0 & 0 & 1 & 0 & 0 & 0 \\ 0 & 0 & 0 & 1 & 0 & 0 & 0 & 0 \\ 0 & 0 & -1 & 0 & 0 & 0 & 0 & 0 \\ 0 & 1 & 0 & 0 & 0 & 0 & 0 & 0 \\ 1 & 0 & 0 & 0 & 0 & 0 & 0 & 0 \end{pmatrix},$$

$$\boldsymbol{A}_2 = \begin{pmatrix} 0 & 0 & 0 & 0 & 0 & 0 & 1 & 0 \\ 0 & 0 & 0 & 0 & 0 & 0 & 0 & 1 \\ 0 & 0 & 0 & 0 & -1 & 0 & 0 & 0 \\ 0 & 0 & 0 & 0 & 0 & 1 & 0 & 0 \\ 0 & 0 & 1 & 0 & 0 & 0 & 0 & 0 \\ 0 & 0 & 0 & 1 & 0 & 0 & 0 & 0 \\ -1 & 0 & 0 & 0 & 0 & 0 & 0 & 0 \\ 0 & 1 & 0 & 0 & 0 & 0 & 0 & 0 \end{pmatrix},$$

$$\boldsymbol{A}_3 = \begin{pmatrix} 0 & 0 & 0 & 0 & 0 & -1 & 0 & 0 \\ 0 & 0 & 0 & 0 & 1 & 0 & 0 & 0 \\ 0 & 0 & 0 & 0 & 0 & 0 & 0 & 1 \\ 0 & 0 & 0 & 0 & 0 & 0 & 1 & 0 \\ 0 & -1 & 0 & 0 & 0 & 0 & 0 & 0 \\ 1 & 0 & 0 & 0 & 0 & 0 & 0 & 0 \\ 0 & 0 & 0 & 1 & 0 & 0 & 0 & 0 \\ 0 & 0 & 1 & 0 & 0 & 0 & 0 & 0 \end{pmatrix},$$

$$\boldsymbol{A} = \begin{pmatrix} 0 & 0 & 0 & 0 & 0 & 0 & 0 & 0 \\ 0 & 0 & 0 & 0 & 0 & 0 & 0 & 0 \\ 0 & 0 & 0 & 0 & 0 & 0 & 0 & 0 \\ 0 & 0 & 0 & 0 & 0 & 0 & 0 & 0 \\ 0 & 0 & 0 & 0 & -1 & 0 & 0 & 0 \\ 0 & 0 & 0 & 0 & 0 & -1 & 0 & 0 \\ 0 & 0 & 0 & 0 & 0 & 0 & -1 & 0 \\ 0 & 0 & 0 & 0 & 0 & 0 & 0 & 0 \end{pmatrix},$$

$$\underline{f} = \begin{pmatrix} f_x \\ f_y \\ f_z \\ 0 \\ 0 \\ 0 \\ 0 \\ 0 \end{pmatrix}, \qquad \underline{u} = \begin{pmatrix} u \\ v \\ w \\ \phi \\ \omega_x \\ \omega_y \\ \omega_z \\ p \end{pmatrix}.$$

The characteristic polynomial associated with system (8.10) is

$$\det(\boldsymbol{A}_1\xi + \boldsymbol{A}_2\eta + \boldsymbol{A}_3\zeta) = \det \begin{pmatrix} 0 & 0 & 0 & 0 & 0 & -\zeta & \eta & \xi \\ 0 & 0 & 0 & 0 & \zeta & 0 & -\xi & \eta \\ 0 & 0 & 0 & 0 & -\eta & \xi & 0 & \zeta \\ 0 & 0 & 0 & 0 & \xi & \eta & \zeta & 0 \\ 0 & -\zeta & \eta & \xi & 0 & 0 & 0 & 0 \\ \zeta & 0 & -\xi & \eta & 0 & 0 & 0 & 0 \\ -\eta & \xi & 0 & \zeta & 0 & 0 & 0 & 0 \\ \xi & \eta & \zeta & 0 & 0 & 0 & 0 & 0 \end{pmatrix}$$

$$= (\xi^2 + \eta^2 + \zeta^2)^4 \neq 0$$

for all nonzero triplet (ξ, η, ζ). System (8.10) or (8.8) and thus system (8.7) is indeed elliptic. Ellipticity can be easily understood in the following way: (8.8a) and (8.8b) constitute a div–curl system for vorticity, and (8.8c) and (8.8d) constitute a div–curl system for velocity. These two div–curl systems are coupled through the vorticity in (8.8c) and form the Stokes equations. In other words, the principal part of the Stokes operator consists of two identical elliptic div–curl operators.

The classification here is based on an ordinary method (see Sect. 4.7), so (8.7b) is needed to guarantee the ellipticity.

8.2.2 Boundary Conditions

Before investigating boundary conditions, we first discuss the equivalence between the velocity–pressure formulation (8.1) and the velocity–pressure–vorticity formulation (8.7). System (8.7) is obtained from (8.1) by introducing the vorticity $\boldsymbol{\omega}$ which is a linear combination of derivatives of the velocity $\boldsymbol{u}$, and thus reducing the order of differential operator. Reducing the order of the differential equations does not generate spurious solutions. It means that system (8.7) is equivalent to system (8.1), that is, the solution of (8.7) is the solution of (8.1). Conversely, system (8.1) can be deduced from system (8.7) by substituting the definition of the vorticity (8.7c) into (8.7a) and (8.7b). This type of substitution and combination does not generate spurious solutions either. Therefore, the velocity–pressure formulation (8.1) and the velocity–pressure–vorticity formulation (8.7) are mutually equivalent. This equivalence implies that the permissible boundary conditions for system (8.7) must be the permissible boundary conditions for system (8.1). The reverse is also true. Consequently, it is not possible that some boundary conditions that have been shown to be legitimate for system (8.7) might not be so for system (8.1), and vice versa.

As will be pointed out in Sect. 8.4, the nonlinear convective term has no effect on the classification of the Navier–Stokes equations, hence the boundary conditions for the Stokes equations are valid for the Navier–Stokes equations. So we need only analyze the Stokes equations (8.7) or (8.8). Since system (8.8) is of first order, the boundary conditions do not involve derivatives of unknowns. In other words, there are only algebraic boundary conditions for the solution of first-order partial differential equations. This fact precludes the mathematical legitimacy of specifying the derivative of pressure on the boundary for system (8.7) or (8.1).

For convenience we rewrite the Stokes equations (8.8) as the following two coupled generalized div–curl systems:

$$\nabla p + \nabla \times \boldsymbol{\omega} = \boldsymbol{f} \qquad \text{in } \Omega, \tag{8.11a}$$

$$\nabla \cdot \boldsymbol{\omega} = 0 \qquad \text{in } \Omega, \tag{8.11b}$$

and

$$\nabla \phi + \nabla \times \boldsymbol{u} = \boldsymbol{\omega} \qquad \text{in } \Omega, \tag{8.12a}$$

$$\nabla \cdot \boldsymbol{u} = 0 \qquad \text{in } \Omega. \tag{8.12b}$$

If the vorticity $\boldsymbol{\omega}$ in (8.12a) is known, the structure of these two generalized div–curl systems would be identical. In Chap. 5 we have thoroughly investigated this type of div–curl system. In order to solve (8.11) to obtain p and $\boldsymbol{\omega}$, on the boundary we should specify

$$(p \text{ and } \boldsymbol{n} \cdot \boldsymbol{\omega}) \qquad \text{or} \qquad \boldsymbol{n} \times \boldsymbol{\omega},$$

where $\boldsymbol{n} \times \boldsymbol{\omega}$ is counted as two conditions.

When $\boldsymbol{\omega}$ is known, to obtain ϕ and $\boldsymbol{u}$, we may solve (8.12) with the boundary conditions specified as

$$(\phi \text{ and } \boldsymbol{n} \cdot \boldsymbol{u}) \qquad \text{or} \qquad \boldsymbol{n} \times \boldsymbol{u}.$$

From the above considerations we can immediately list four permissible combinations of boundary conditions ((1)–(4) in Table 8.1) for the Stokes equations. In Table 8.1 for three-dimensional problems, we don't explicitly include the boundary condition $\phi = 0$ on Γ. If we understand that $\phi = 0$ always goes with the condition of $\boldsymbol{n} \cdot \boldsymbol{u}$ and include this condition, then the total number of boundary conditions is four for three-dimensional problems. Since there are eight equations and eight unknowns in the first-order elliptic system (8.8), we indeed need four boundary conditions on each fixed boundary. In (5) and (6) of Table 8.1 we list other possible choices which are not obtained from the above arguments but still satisfy the requirement of four conditions on the boundary. For two-dimensional problems, in the first-order velocity–pressure–vorticity formulation there are four unknowns u, v, p, ω, and four equations, viz., two momentum equations, the definition of the vorticity and the incompressibility equation; no dummy variable is involved. Therefore, two boundary conditions are needed for two-dimensional problems. In Table 8.1 we also list corresponding boundary conditions for two-dimensional problems.

Table 8.1. Combination of boundary conditions

Boundary Conditions	3D	2D	Remarks
(1) Normal velocity	$\boldsymbol{n}\cdot\boldsymbol{u}$	$\boldsymbol{n}\cdot\boldsymbol{u}$	symmetric plane, inflow
Tangential vorticity	$\boldsymbol{n}\times\boldsymbol{\omega}$	ω	
(2) Pressure	p	p	inflow
Normal vorticity	$\boldsymbol{n}\cdot\boldsymbol{\omega}$		
Normal velocity	$\boldsymbol{n}\cdot\boldsymbol{u}$	$\boldsymbol{n}\cdot\boldsymbol{u}$	
(3) Pressure	p	p	outflow
Normal vorticity	$\boldsymbol{n}\cdot\boldsymbol{\omega}$		
Tangential velocity	$\boldsymbol{n}\times\boldsymbol{u}$	$\boldsymbol{n}\times\boldsymbol{u}$	
(4) Tangential velocity	$\boldsymbol{n}\times\boldsymbol{u}$	$\boldsymbol{n}\times\boldsymbol{u}$	uniform outflow
Tangential vorticity	$\boldsymbol{n}\times\boldsymbol{\omega}$	ω	
(5) Velocity	$\boldsymbol{n}\cdot\boldsymbol{u}$	$\boldsymbol{n}\cdot\boldsymbol{u}$	wall, inflow
	$\boldsymbol{n}\times\boldsymbol{u}$	$\boldsymbol{n}\times\boldsymbol{u}$	
(6) Pressure	p	p	outflow on a part of Γ
Normal vorticity	$\boldsymbol{n}\cdot\boldsymbol{\omega}$		
Tangential vorticity	$\boldsymbol{n}\times\boldsymbol{\omega}$	ω	

We shall rigorously prove the well-posedness of system (8.7) under the boundary conditions (1)–(5) listed in Table 8.1. These boundary conditions can be used on the entire boundary or on a part of the boundary Γ. For simplicity we consider only one kind of homogeneous boundary conditions specified on the entire boundary. The results can be extended to mixed and inhomogeneous cases without difficulty.

Given the elliptic differential operator, the question of well-posedness reduces to verification of the permissibility of the boundary conditions. In this section we try to identify the permissible boundary conditions by using an elementary treatment. The mathematical tools used are the bounded inverse theorem, Green's formulae (integration by parts) and the Friedrichs inequalities for the div and curl operators established in Chap. 5. An elliptic system with certain boundary conditions is considered to be well-posed, if one can prove that the differential operator is bounded below, or equivalently, the corresponding least-squares method leads to a coercive bilinear form. In the following subsections, we discuss these boundary conditions case by case. We choose case (1) as an example to show the technique for proving coerciveness. The proof of coerciveness for cases (2)–(4) is similar to that for case (1), the details can be found in Jiang et al. (1994b), and thus are omitted here. Case (5) needs special attention.

(1) Symmetric Plane. The boundary conditions $u_n = 0$, $\omega_{\tau 1} = 0$, $\omega_{\tau 2} = 0$ ($\boldsymbol{n} \cdot \boldsymbol{u} = 0$, $\boldsymbol{n} \times \boldsymbol{\omega} = \boldsymbol{0}$) on Γ may be used for symmetric planes. The inhomogeneous version may be used for inflow boundaries where the normal component of the velocity and two tangential components of the vorticity are prescribed. These conditions correspond to those in the velocity–pressure formulation, in which the normal velocity and the tangential stresses are given. For example, let us consider a piece of boundary with $\boldsymbol{n} = (1,0,0)$. We have that

$$u = 0,$$

$$\omega_y = \frac{\partial u}{\partial z} - \frac{\partial w}{\partial x} = 0,$$

$$\omega_z = \frac{\partial v}{\partial x} - \frac{\partial u}{\partial y} = 0.$$

From $u = 0$ on this boundary we know that

$$\frac{\partial u}{\partial z} = 0, \qquad \frac{\partial u}{\partial y} = 0.$$

Therefore,

$$\frac{\partial w}{\partial x} = 0, \qquad \frac{\partial v}{\partial x} = 0.$$

This implies that

$$\gamma_{xz} = \frac{\partial u}{\partial z} + \frac{\partial w}{\partial x} = 0,$$

$$\gamma_{xy} = \frac{\partial v}{\partial x} + \frac{\partial u}{\partial y} = 0.$$

That is, the tangential strains and thus the tangential stresses are equal to zero.

In order to guarantee the uniqueness of the solution of pressure p, we require that the pressure has a zero mean over Ω as indicated in (8.2d), i.e.,

$$\int_\Omega p dx = 0. \tag{8.13}$$

For the Stokes equations (8.7), $\|\boldsymbol{A}\boldsymbol{U}\|_0$ is defined as

$$\|\boldsymbol{A}\boldsymbol{U}\|_0^2 = \|\nabla p + \nabla \times \boldsymbol{\omega}\|_0^2 + \|\nabla \cdot \boldsymbol{\omega}\|_0^2 + \|\boldsymbol{\omega} - \nabla \times \boldsymbol{u}\|_0^2 + \|\nabla \cdot \boldsymbol{u}\|_0^2, \tag{8.14}$$

where $\boldsymbol{U} = (\boldsymbol{u}, p, \boldsymbol{\omega}) \in [H^1(\Omega)]^7$, and $\boldsymbol{U}$ satisfies the corresponding homogeneous boundary conditions on Γ.

If we can prove that $\|\boldsymbol{A}\boldsymbol{U}\|_0$ is coercive, i.e., bounded below in the H^1 norm, then the existence and uniqueness of the solution to the Stokes problem follow from the bounded inverse theorem (Theorem 4.1) in a standard manner. Consequently, the corresponding finite element method has an optimal rate of convergence for all unknowns by virtue of Theorem 4.8.

Now we examine the coerciveness of $\|\boldsymbol{AU}\|_0$. Let us expand the first term in (8.14). Since $\boldsymbol{n}\times\boldsymbol{\omega}=\boldsymbol{0}$ on Γ, using Green's formula (B.6), we have

$$(\nabla p, \nabla\times\boldsymbol{\omega}) = \langle\nabla p,\ \boldsymbol{n}\times\boldsymbol{\omega}\rangle = 0, \tag{8.15}$$

and thus

$$\begin{aligned}\|\nabla p+\nabla\times\boldsymbol{\omega}\|_0^2 &= \|\nabla p\|_0^2+\|\nabla\times\boldsymbol{\omega}\|_0^2+2(\nabla p,\nabla\times\boldsymbol{\omega})\\ &= \|\nabla p\|_0^2+\|\nabla\times\boldsymbol{\omega}\|_0^2.\end{aligned} \tag{8.16}$$

By virtue of (8.16) we have

$$\|\boldsymbol{AU}\|_0^2 = \|\nabla p\|_0^2+\|\nabla\times\boldsymbol{\omega}\|_0^2+\|\nabla\cdot\boldsymbol{\omega}\|_0^2+\|\boldsymbol{\omega}-\nabla\times\boldsymbol{u}\|_0^2+\|\nabla\cdot\boldsymbol{u}\|_0^2. \tag{8.17}$$

From (8.17) we have

$$\|\boldsymbol{AU}\|_0^2 \geq \|\nabla p\|_0^2 = |p|_1^2. \tag{8.18}$$

Since p satisfies the zero mean constraint (8.13), from the Poincaré inequality (C.1) or (4.18) we have

$$C|p|_1^2 \geq \|p\|_1^2. \tag{8.19}$$

The combination of (8.18) and (8.19) yields

$$C\|\boldsymbol{AU}\|_0^2 \geq \|p\|_1^2. \tag{8.20}$$

From (8.17) we also have

$$\|\boldsymbol{AU}\|_0^2 \geq \|\nabla\times\boldsymbol{\omega}\|_0^2+\|\nabla\cdot\boldsymbol{\omega}\|_0^2, \tag{8.21}$$

$$\|\boldsymbol{AU}\|_0^2 \geq \|\boldsymbol{\omega}-\nabla\times\boldsymbol{u}\|_0^2, \tag{8.22}$$

$$\|\boldsymbol{AU}\|_0^2 \geq \|\nabla\cdot\boldsymbol{u}\|_0^2. \tag{8.23}$$

Since $\boldsymbol{n}\times\boldsymbol{\omega}=\boldsymbol{0}$ on Γ, from the Friedrichs first div–curl inequality (Theorem 5.3) we have the inequality:

$$C(\|\nabla\times\boldsymbol{\omega}\|_0^2+\|\nabla\cdot\boldsymbol{\omega}\|_0^2) \geq \|\boldsymbol{\omega}\|_1^2 \geq \|\boldsymbol{\omega}\|_0^2. \tag{8.24}$$

The combination of (8.21) and (8.24) yields

$$C\|\boldsymbol{AU}\|_0^2 \geq \|\boldsymbol{\omega}\|_1^2, \tag{8.25}$$

and

$$C\|\boldsymbol{AU}\|_0^2 \geq \|\boldsymbol{\omega}\|_0^2$$

or

$$C(\|\boldsymbol{AU}\|_0^2)^{\frac{1}{2}} \geq \|\boldsymbol{\omega}\|_0. \tag{8.26}$$

From (8.22) we have

$$(\|\boldsymbol{A}\boldsymbol{U}\|_0^2)^{\frac{1}{2}} \geq \| -\boldsymbol{\omega} + \nabla \times \boldsymbol{u}\|_0. \tag{8.27}$$

Combining (8.26) with (8.27) and using the triangle inequality we have

$$C(\|\boldsymbol{A}\boldsymbol{U}\|_0^2)^{\frac{1}{2}} \geq \|\boldsymbol{\omega}\|_0 + \| -\boldsymbol{\omega} + \nabla \times \boldsymbol{u}\|_0 \geq \|\nabla \times \boldsymbol{u}\|_0,$$

that is

$$C(\|\boldsymbol{A}\boldsymbol{U}\|_0^2) \geq \|\nabla \times \boldsymbol{u}\|_0^2. \tag{8.28}$$

The combination of (8.23) and (8.28) leads to

$$C\|\boldsymbol{A}\boldsymbol{U}\|_0^2 \geq \|\nabla \times \boldsymbol{u}\|_0^2 + \|\nabla \cdot \boldsymbol{u}\|_0^2. \tag{8.29}$$

Since $\boldsymbol{n} \cdot \boldsymbol{u} = 0$ on Γ, again due to Theorem 5.3 we have the inequality:

$$C(\|\nabla \times \boldsymbol{u}\|_0^2 + \|\nabla \cdot \boldsymbol{u}\|_0^2) \geq \|\boldsymbol{u}\|_1^2. \tag{8.30}$$

The combination of (8.29) and (8.30) yields

$$C\|\boldsymbol{A}\boldsymbol{U}\|_0^2 \geq \|\boldsymbol{u}\|_1^2. \tag{8.31}$$

Bringing (8.20), (8.25) and (8.31) together we finally obtain

$$C\|\boldsymbol{A}\boldsymbol{U}\|_0^2 \geq \|\boldsymbol{u}\|_1^2 + \|p\|_1^2 + \|\boldsymbol{\omega}\|_1^2. \tag{8.32}$$

This shows that $\|\boldsymbol{A}\boldsymbol{U}\|_0^2$ is indeed bounded below in the H^1 norm and thus coercive. Consequently, it is trivial to prove that this problem has a unique solution that satisfies the following desired bound:

$$\|\boldsymbol{u}\|_1 + \|p\|_1 + \|\boldsymbol{\omega}\|_1 \leq C\|\boldsymbol{f}\|_0. \tag{8.33}$$

Also by the virtue of Theorem 4.8 the corresponding LSFEM solution based on equal-order finite elements has an optimal rate of convergence for all variables.

(2) Inflow. The boundary conditions are $p = 0,\ u_n = 0,\ \omega_n = 0$ ($p = 0,\ \boldsymbol{n} \cdot \boldsymbol{u} = 0,\ \boldsymbol{n} \cdot \boldsymbol{\omega} = 0$) on Γ. The related inhomogeneous case represents, for example, the well developed inflow boundary, on which the normal velocity is given, and the normal vorticity as well as the pressure are prescribed to be zero. In two-dimensional cases, only u_n and p are prescribed. The validity of this group of boundary conditions would be difficult to demonstrate by the Galerkin method.

By using $p = 0$ on Γ and (B.4) we can show the validity of (8.16). Due to the fact that $u_n = 0$ and $\omega_n = 0$, the Friedrichs first div–curl inequality is valid. Therefore, coerciveness of $\|\boldsymbol{A}\boldsymbol{U}\|_0^2$ can be proved in the same way as in case (1).

(3) Outflow. The boundary conditions $p = 0,\ u_{\tau 1} = 0,\ u_{\tau 2} = 0,\ \omega_n = 0$ $(p = 0, \boldsymbol{n} \times \boldsymbol{u} = \boldsymbol{0},\ \boldsymbol{n} \cdot \boldsymbol{\omega} = 0)$ on Γ may be employed for a well developed exit boundary. Here four boundary conditions are prescribed. As shown in Theorem 5.5, $\boldsymbol{n} \times \boldsymbol{u} = \boldsymbol{0}$ on Γ analytically implies that

$$\boldsymbol{n} \cdot (\nabla \times \boldsymbol{u}) = 0 \qquad \text{on } \Gamma. \tag{8.34}$$

It appears that there are too many boundary conditions. In the previous cases, for the dummy variable ϕ the boundary condition $\phi = 0$ on Γ is specified in advance, and hence only three boundary conditions are needed. In the following we show that in the present case no boundary condition is needed for the dummy variable ϕ, so it is appropriate to specify four conditions.

By virtue of Theorem 5.2, (8.8c) is equivalent to the following equations and boundary condition:

$$\nabla \times (-\boldsymbol{\omega} + \nabla\phi + \nabla \times \boldsymbol{u}) = \boldsymbol{0} \qquad \text{in } \Omega, \tag{8.35a}$$

$$\nabla \cdot (-\boldsymbol{\omega} + \nabla\phi + \nabla \times \boldsymbol{u}) = 0 \qquad \text{in } \Omega, \tag{8.35b}$$

$$\boldsymbol{n} \cdot (-\boldsymbol{\omega} + \nabla\phi + \nabla \times \boldsymbol{u}) = 0 \qquad \text{on } \Gamma. \tag{8.35c}$$

Taking into account (8.8b) and $\nabla \cdot \nabla \times \boldsymbol{u} = 0$, (8.35b) becomes

$$\Delta\phi = 0 \qquad \text{in } \Omega. \tag{8.36a}$$

Considering $\omega_n = 0$ on Γ and (8.34), boundary condition (8.35c) becomes

$$\frac{\partial \phi}{\partial n} = 0 \qquad \text{on } \Gamma. \tag{8.36b}$$

Equations (8.36a) and (8.36b) imply that ϕ is a constant or $\nabla\phi \equiv \boldsymbol{0}$ in Ω. Therefore, four conditions in the present case automatically guarantee that the dummy variable ϕ can be eliminated in (8.8c).

The boundary conditions in this case correspond to those in the velocity–pressure formulation, in which the tangential velocity components and the normal stress are prescribed. To show this let us consider, for example, the surface with $\boldsymbol{n} = (1, 0, 0)$. Since $v = w = 0$, we have

$$\frac{\partial v}{\partial y} = 0, \qquad \frac{\partial w}{\partial z} = 0.$$

Hence from the continuity of velocity we know that

$$\frac{\partial u}{\partial x} = 0.$$

Therefore,

$$\sigma_x = p + 2\nu \frac{\partial u}{\partial x} = 0,$$

that is, the normal stress is equal to zero.

By using $p = 0$ on Γ and (B.4) we can show the validity of (8.16). Due to the fact that $\boldsymbol{n} \times \boldsymbol{u} = \boldsymbol{0}$ and $\omega_n = 0$, the Friedrichs first div–curl inequality is valid. Therefore, the coerciveness of $\|\boldsymbol{AU}\|_0^2$ can be proved in the same way as in case (1).

If we really specify that the dummy variable ϕ be zero on Γ in advance, then only three boundary conditions are needed, and $\omega_n = 0$ can be imposed in a weak sense. In this case, the least-squares method minimizes the following functional:

$$\begin{aligned} J(\boldsymbol{U}) &= \|\nabla p + \nabla \times \boldsymbol{\omega}\|_0^2 + \|\nabla \cdot \boldsymbol{\omega}\|_0^2 + \|\boldsymbol{\omega} - \nabla \times \boldsymbol{u}\|_0^2 + \|\nabla \cdot \boldsymbol{u}\|_0^2 \\ &+ \|\boldsymbol{n} \cdot \boldsymbol{\omega}\|_{1/2,\Gamma}^2, \end{aligned} \tag{8.37}$$

and coerciveness can still be proved. For details see Jiang et al. (1994b).

(4) Uniform Outflow. The boundary conditions are $u_{\tau 1} = 0,\ u_{\tau 2} = 0,\ \omega_{\tau 1} = 0,\ \omega_{\tau 2} = 0$ $(\boldsymbol{n} \times \boldsymbol{u} = \boldsymbol{0},\ \boldsymbol{n} \times \boldsymbol{\omega} = \boldsymbol{0})$ on Γ. For the same reason as explained in the previous case, $\phi \equiv 0$ is guaranteed even if no boundary condition for ϕ is given. The coerciveness of $\boldsymbol{AU}$ can be proved by just following the steps in case (1).

(5) Wall. The boundary conditions $u_n = 0,\ u_{\tau 1} = 0,\ u_{\tau 2} = 0$ $(\boldsymbol{n} \cdot \boldsymbol{u} = 0,\ \boldsymbol{n} \times \boldsymbol{u} = \boldsymbol{0})$ on Γ are the so called no slip conditions on the wall. This is a standard case of the permissible velocity boundary conditions. Bochev and Gunzburger (1994) pointed out that in this case the full H^1 coerciveness of $\boldsymbol{AU}$ does not hold. To confirm this, they consider a two-dimensional situation: $\boldsymbol{u} = \boldsymbol{0}$, $p = \sin(nx)e^{ny}$ and $\omega = \omega_z = -\cos(nx)e^{ny}$. Since $\mathbf{curl}\ \omega + \nabla p = \boldsymbol{0}$ (see the definition of **curl** in Sect. 6.4.1),

$$\|\boldsymbol{AU}\|_0^2 = \|\omega\|_0^2 = O(1).$$

However, $\|\omega\|_1^2 = O(n^2)$. Thus it is not true that

$$C\|\boldsymbol{AU}\|_0^2 \geq \|\omega\|_1^2.$$

Therefore, we shall try to find a proper norm to bound $\|\boldsymbol{AU}\|_0$ from below. It is well known that the Stokes equations (8.1) have a unique solution under the standard velocity boundary condition. This can be proved by using the Galerkin method, see e.g., Girault and Raviart (1986). The first-order version of the Stokes equations (8.7) is equivalent to (8.1), hence (8.7) must have a unique solution in the subspace of $[L_2(\Omega)]^3 \times L_2(\Omega) \times [L_2(\Omega)]^3$ under the velocity boundary conditions. By virtue of Theorem 4.6, we have

$$C\|\boldsymbol{AU}\|_0^2 \geq \|\boldsymbol{u}\|_0^2 + \|p\|_0^2 + \|\boldsymbol{\omega}\|_0^2, \tag{8.38}$$

and particularly

$$C\|\boldsymbol{AU}\|_0^2 \geq \|\boldsymbol{\omega}\|_0^2. \tag{8.39}$$

On the other hand, we have the definition of $\|\boldsymbol{AU}\|_0$:

$$\|\boldsymbol{AU}\|_0^2 = \|\nabla p + \nabla \times \boldsymbol{\omega}\|_0^2 + \|\nabla \cdot \boldsymbol{\omega}\|_0^2 + \|\boldsymbol{\omega} - \nabla \times \boldsymbol{u}\|_0^2 + \|\nabla \cdot \boldsymbol{u}\|_0^2, \tag{8.14}$$

and therefore

$$\|\boldsymbol{AU}\|_0^2 \geq \|\boldsymbol{\omega} - \nabla \times \boldsymbol{u}\|_0^2, \tag{8.40}$$

$$\|\boldsymbol{AU}\|_0^2 \geq \|\nabla \cdot \boldsymbol{u}\|_0^2. \tag{8.41}$$

Using (8.39) and (8.40) we can show that

$$C\|\boldsymbol{AU}\|_0^2 \geq \|\nabla \times \boldsymbol{u}\|_0^2. \tag{8.42}$$

Combining (8.41) and (8.42) and using Theorem 5.3 we have

$$C\|\boldsymbol{AU}\|_0^2 \geq \|\boldsymbol{u}\|_1^2. \tag{8.43}$$

The combination of (8.38) with (8.43) yields

$$C\|\boldsymbol{AU}\|_0^2 \geq \|\boldsymbol{u}\|_1^2 + \|p\|_0^2 + \|\boldsymbol{\omega}\|_0^2. \tag{8.44}$$

Consequently, the solution satisfies the following bound:

$$\|\boldsymbol{u}\|_1 + \|p\|_0 + \|\boldsymbol{\omega}\|_0 \leq C\|\boldsymbol{f}\|_0. \tag{8.45}$$

The inequality (8.44) implies that $\boldsymbol{AU}$ is bounded below in a less desired norm, that is, the pressure and the vorticity are measured in a L_2 norm. For this reason, the corresponding LSFEM with equal interpolations has an optimal convergence for the velocity, and is suboptimal for the pressure and vorticity.

(6) Pressure and Vorticity. In this case $p=0,\ \omega_n=0,\ \omega_{\tau 1}=0,\ \omega_{\tau 2}=0$ ($p=0,\ \boldsymbol{n}\cdot\boldsymbol{\omega}=0,\ \boldsymbol{n}\times\boldsymbol{\omega}=\boldsymbol{0}$) are given on Γ. Using the boundary conditions $p=0,\ \omega_n=0$ one can solve (8.11) to obtain p and $\boldsymbol{\omega}$. However, $\boldsymbol{u}$ cannot be uniquely determined by (8.12). Therefore, this combination can only be used on a part of the boundary. In this situation, at least on a part of the boundary $\boldsymbol{n}\cdot\boldsymbol{u}$ or $\boldsymbol{n}\times\boldsymbol{u}$ should be given.

8.2.3 Application of LSFEM

Let us apply the least-squares method to the two-dimensional Stokes equations:

$$\frac{\partial u}{\partial x} + \frac{\partial v}{\partial y} = 0 \quad \text{in} \quad \Omega, \tag{8.46a}$$

$$\frac{\partial p}{\partial x} + \frac{\partial \omega}{\partial y} = f_x \quad \text{in} \quad \Omega, \tag{8.46b}$$

$$\frac{\partial p}{\partial y} - \frac{\partial \omega}{\partial x} = f_y \quad \text{in} \quad \Omega, \tag{8.46c}$$

$$\omega + \frac{\partial u}{\partial y} - \frac{\partial v}{\partial x} = 0 \quad \text{in} \quad \Omega, \tag{8.46d}$$

where ω denotes the z-component of $\boldsymbol{\omega}$, and Ω is a bounded domain in $\mathbb{R}^2$ with piecewise smooth boundary Γ. The boundary conditions listed in Table 8.1 should be supplemented to complete the definition of the boundary-value problem. They may be written as:

(1) $u_n,\ \omega$ given,

(2) $p,\ u_n$ given,

(3) $p,\ u_t$ given,

(4) $u_t,\ \omega$ given,

(5) $u_n,\ u_t$ given,

(6) $p,\ \omega$ given.

We can write (8.46) in the general form of the first-order system:

$$\boldsymbol{A}\underline{\boldsymbol{u}} = \underline{\boldsymbol{f}}$$

or

$$\boldsymbol{A}_1\frac{\partial \underline{\boldsymbol{u}}}{\partial x} + \boldsymbol{A}_2\frac{\partial \underline{\boldsymbol{u}}}{\partial y} + \boldsymbol{A}_0\underline{\boldsymbol{u}} = \underline{\boldsymbol{f}}, \tag{8.47}$$

where

$$\boldsymbol{A}_1 = \begin{pmatrix} 1 & 0 & 0 & 0 \\ 0 & 0 & 1 & 0 \\ 0 & 0 & 0 & -1 \\ 0 & -1 & 0 & 0 \end{pmatrix}, \qquad \boldsymbol{A}_2 = \begin{pmatrix} 0 & 1 & 0 & 0 \\ 0 & 0 & 0 & 1 \\ 0 & 0 & 1 & 0 \\ 1 & 0 & 0 & 0 \end{pmatrix},$$

$$\boldsymbol{A}_0 = \begin{pmatrix} 0 & 0 & 0 & 0 \\ 0 & 0 & 0 & 0 \\ 0 & 0 & 0 & 0 \\ 0 & 0 & 0 & 1 \end{pmatrix}, \qquad \underline{\boldsymbol{f}} = \begin{pmatrix} 0 \\ f_x \\ f_y \\ 0 \end{pmatrix}, \qquad \underline{\boldsymbol{u}} = \begin{pmatrix} u \\ v \\ p \\ \omega \end{pmatrix}.$$

The LSFEM formulation for equal-order interpolations is given in Sect. 4.8. For the Stokes equations, the element matrices and vectors are computed from the following submatrices:

$$\boldsymbol{A}\psi_i = \psi_{i,x}\boldsymbol{A}_1 + \psi_{i,y}\boldsymbol{A}_2 + \psi_i\boldsymbol{A}_0 = \begin{pmatrix} \psi_{i,x} & \psi_{i,y} & 0 & 0 \\ 0 & 0 & \psi_{i,x} & \psi_{i,y} \\ 0 & 0 & \psi_{i,y} & -\psi_{i,x} \\ \psi_{i,y} & -\psi_{i,x} & 0 & \psi \end{pmatrix}. \tag{8.48}$$

Usually we use Gaussian quadrature to evaluate the coefficients of $\boldsymbol{K}_e$ and $\boldsymbol{F}_e$. As discussed in Sect. 4.11.3, the number of Gaussian points required for the solution is of importance. The least-squares method with Gaussian quadrature is equivalent to the least-squares collocation method. Therefore, the number of collocation points (Gaussian points) should be compatible with the number of unknowns to get good results. In general, exact integration in the LSFEM computation leads to underestimated and inaccurate solutions which even violate local and global mass conservation. The trouble encountered by Nelson and Chang (1995) is typically caused by the use of full integration. According to the principle given in (4.95), it is easy to verify that the quadratic triangle with three-point quadrature used by them is not suitable for LSFEM.

The importance of the use of reduced integration can be explained by another point of view. Inspection of (8.46d) shows that the vorticity ω and the derivatives of velocity components u and v appear simultaneously. When equal-order interpolations are employed, it is impracticable to reduce the residual of this equation to zero throughout. For this reason, reduced integration is necessary (see also Zienkiewicz et al. 1974).

We remark that non-equal-order finite elements are also applicable for the solution of the Stokes equations. For example, one may use quadratic elements for velocity, and linear elements for pressure and vorticity. In this case, regular integration can be employed.

To verify the theoretical error estimate given in the preceding section, we consider a model problem in a unit square taken from Oden and Jacquotte (1984). The Stokes equations have an exact solution with the polynomial divergence-free velocity field

$$u(x,y) = x^2(1-x)^2(2y - 6y^2 + 4y^3),$$

$$v(x,y) = y^2(1-y)^2(-2x + 6x^2 - 4x^3),$$

the pressure field

$$p(x,y) = x^2 - y^2 - 0.25,$$

and the vorticity field

$$\omega(x,y) = -x^2(1-x)^2(2 - 12y + 12y^2) + y^2(1-y)^2(-2 + 12x - 12x^2)$$

with two groups of boundary conditions shown in Fig. 8.1. One group consists of mixed boundary conditions including the first four cases in Table 8.1. Another one has a pure velocity boundary condition.

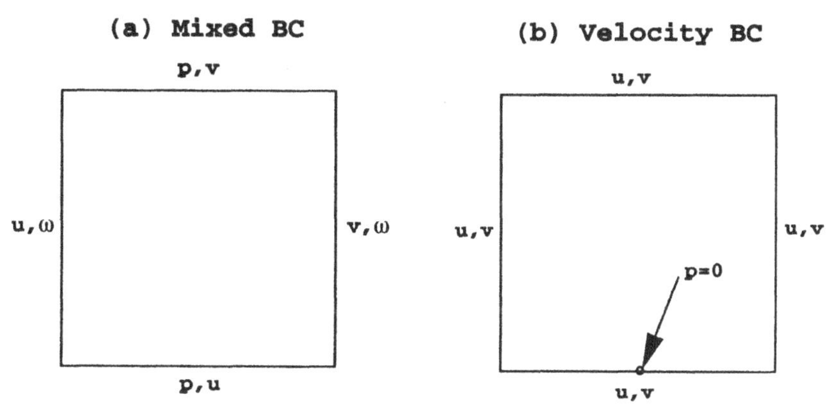

Fig. 8.1a,b. Boundary conditions for the Stokes equations (from Jiang 1998)

Uniform meshes with bilinear element are used. We employ both 2×2 and one-point Gaussian quadrature. The numerical results of the rate of convergence are shown in Fig. 8.2. We have confirmed that in the mixed case

$$\|e_{\mathbf{u}}\| \leq Ch^2, \quad \|e_p\| \leq Ch^2, \quad \|e_\omega\| \leq Ch^2,$$

that is, all variables u, v, p, and ω converge in L_2 norm at the optimal rate as predicted by the theory. From Fig. 8.2. we also infer that in the pure velocity boundary condition case, the rate of convergence for velocity is optimal, but that for pressure and vorticity is lower than optimal. This is consistent with the theory given in Sect. 8.2.2.

Driven-Cavity Flow. The least-squares method has also been tested by solving a driven-cavity flow (Jiang and Chang 1990). In this calculation, nonuniform 50×50 bilinear elements and one-point Gaussian quadrature are used. The no slip boundary condition $u = 0, v = 0$ is applied on all boundaries except at the upper boundary where $u = 1$ and $v = 0$. A pressure condition $p = 0$ is specified at the middle point of the lower boundary. Figure 8.3 shows the computed results of streamlines, pressure contours, vorticity contours and velocity vectors. We note that the two corner eddies are clearly shown.

We note that when reduced integration is used, the nodal values of pressure and vorticity have some oscillations. However, their solutions at the Gaussian points are smooth, therefore we reinterpolate the results at the Gaussian points to obtain the correct solutions at the nodal points by the method of area averaging. That is, the final nodal value is the area average of the values at the Gaussian points of all elements surrounding that node.

8.3 The Navier–Stokes Equations in the $u - p - \omega$ Formulation

The steady-state laminar incompressible Navier–Stokes equations are often written in the velocity–pressure formulation: Find the velocity $\boldsymbol{u} = (u, v, w)$ and the pressure p such that

$$\boldsymbol{u} \cdot \nabla \boldsymbol{u} + \nabla p - \frac{1}{\mathrm{Re}} \Delta \boldsymbol{u} = \boldsymbol{f} \qquad \text{in } \Omega, \tag{8.49a}$$

$$\nabla \cdot \boldsymbol{u} = 0 \qquad \text{in } \Omega. \tag{8.49b}$$

Here all variables are nondimensionalized, and Re denotes the Reynolds number, defined as

$$\mathrm{Re} = \frac{UL}{\nu},$$

where L is a reference length, U a reference velocity and ν the kinematic viscosity.

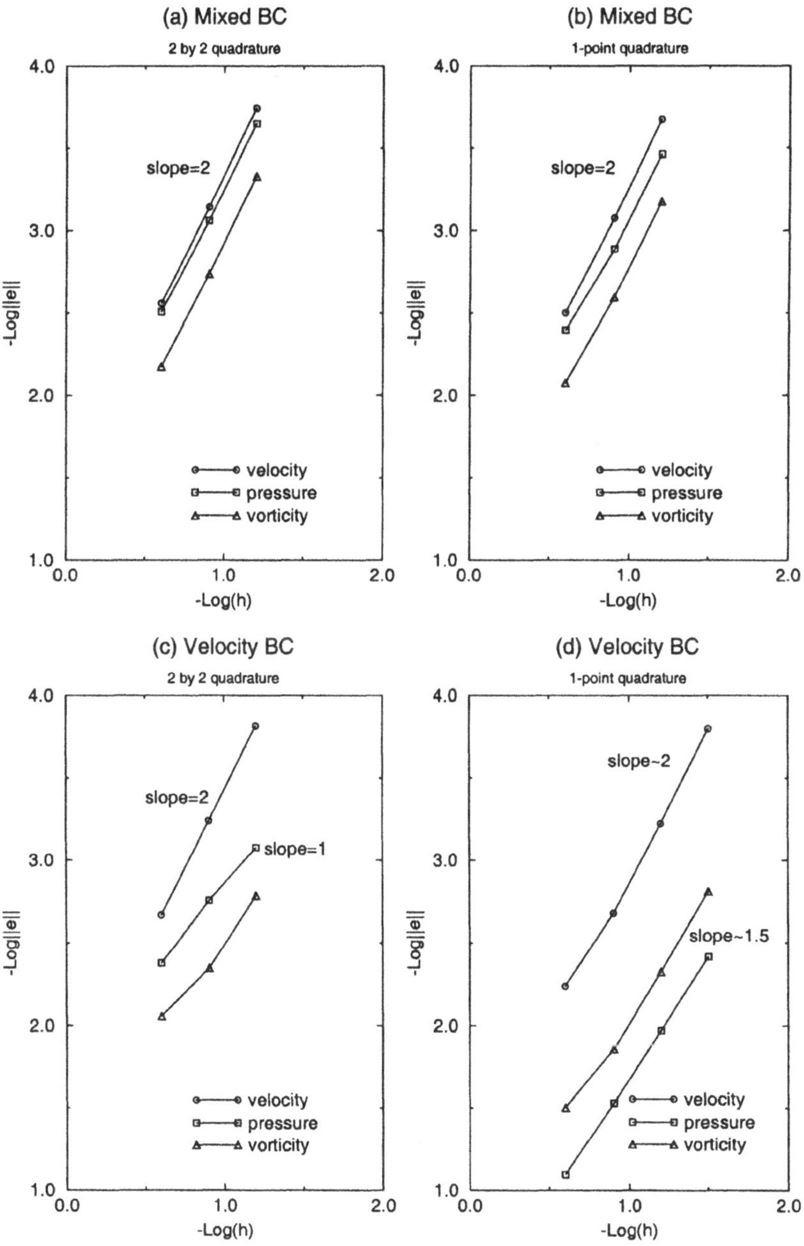

Fig. 8.2a–d. Computed convergence rate for the Stokes equations (from Jiang 1998)

Fig. 8.3. Computed results for the cavity Stokes flow: (**a**) Streamline (**b**) Pressure (**c**) Vorticity (**d**) Velocity (From Jiang and Chang 1990)

As explained in Sect. 8.2, in order to use the LSFEM to obtain an approximate solution, we introduce the vorticity $\boldsymbol{\omega} = \nabla \times \boldsymbol{u}$ as an independent unknown vector, and rewrite the incompressible Navier–Stokes equations (8.49) in the following first-order quasi-linear velocity–pressure–vorticity formulation:

$$\boldsymbol{u} \cdot \nabla \boldsymbol{u} + \nabla p + \frac{1}{\mathrm{Re}} \nabla \times \boldsymbol{\omega} = \boldsymbol{f} \qquad \text{in } \Omega, \tag{8.50a}$$

$$\nabla \cdot \boldsymbol{\omega} = 0 \qquad \text{in } \Omega, \tag{8.50b}$$

$$\boldsymbol{\omega} - \nabla \times \boldsymbol{u} = 0 \qquad \text{in } \Omega, \tag{8.50c}$$

$$\nabla \cdot \boldsymbol{u} = 0 \qquad \text{in } \Omega. \tag{8.50d}$$

These equations need to be supplemented by boundary conditions to complete the definition of the boundary value problem. According to the explanation in Sect. 8.4, the non-linear convective term $\boldsymbol{u} \cdot \nabla \boldsymbol{u}$ in the Navier–Stokes equations has no effect on the classification, and thus the boundary conditions required by the Navier–Stokes equations are exactly the same as for the Stokes equations. Therefore, the boundary conditions listed in Table 8.1 can be used for the Navier–Stokes equations (8.50) without any modification.

Before application of LSFEM, the convective term $\boldsymbol{u}\cdot\nabla\boldsymbol{u}$ in the momentum equation (8.50a) should be linearized by using successive substitution or Newton's method. In the successive substitution method, (8.50a) is linearized as

$$\boldsymbol{u}_0\cdot\nabla\boldsymbol{u}+\nabla p+\frac{1}{\mathrm{Re}}\nabla\times\boldsymbol{\omega}=\boldsymbol{f}\qquad\text{in }\Omega. \tag{8.51}$$

Here the superscript "0" indicates that the value of the corresponding variable is known from the previous calculation step. In Newton's method, (8.50a) is linearized as

$$\boldsymbol{u}_0\cdot\nabla\boldsymbol{u}+\boldsymbol{u}\cdot\nabla\boldsymbol{u}_0+\nabla p+\frac{1}{\mathrm{Re}}\nabla\times\boldsymbol{\omega}=\boldsymbol{f}+\boldsymbol{u}_0\cdot\nabla\boldsymbol{u}_0\qquad\text{in }\Omega. \tag{8.52}$$

Successive substitution has slow convergence and a large radius of convergence, and thus is often used for an initial linearization. Newton's method converges fast, but has a small radius of convergence, and hence is preferable when a close guess to the solution is available.

8.3.1 Two-Dimensional Case

The general first-order Navier–Stokes equations (8.50) take the following form for two-dimensional problems in rectangular coordinates:

$$\frac{\partial u}{\partial x}+\frac{\partial v}{\partial y}=0\qquad\text{in }\Omega, \tag{8.53a}$$

$$u\,\frac{\partial u}{\partial x}+v\,\frac{\partial u}{\partial y}+\frac{\partial p}{\partial x}+\frac{1}{\mathrm{Re}}\frac{\partial \omega}{\partial y}=f_x\qquad\text{in }\Omega, \tag{8.53b}$$

$$u\,\frac{\partial v}{\partial x}+v\,\frac{\partial v}{\partial y}+\frac{\partial p}{\partial y}-\frac{1}{\mathrm{Re}}\frac{\partial \omega}{\partial x}=f_y\qquad\text{in }\Omega, \tag{8.53c}$$

$$\omega+\frac{\partial u}{\partial y}-\frac{\partial v}{\partial x}=0\qquad\text{in }\Omega. \tag{8.53d}$$

By using, for example, Newton's method, the two-dimensional Navier–Stokes equations (8.53) can be linearized as:

$$\frac{\partial u}{\partial x}+\frac{\partial v}{\partial y}=0\qquad\text{in }\Omega, \tag{8.54a}$$

$$\begin{aligned}u_0\,\frac{\partial u}{\partial x}+v_0\,\frac{\partial u}{\partial y}+u\,\frac{\partial u_0}{\partial x}+v\,\frac{\partial u_0}{\partial y}+\frac{\partial p}{\partial x}+\frac{1}{\mathrm{Re}}\frac{\partial \omega}{\partial y}&\\ =f_x+u_0\,\frac{\partial u_0}{\partial x}+v_0\,\frac{\partial u_0}{\partial y}&\qquad\text{in }\Omega,\end{aligned} \tag{8.54b}$$

$$u_0 \frac{\partial v}{\partial x} + v_0 \frac{\partial v}{\partial y} + u \frac{\partial v_0}{\partial x} + v \frac{\partial v_0}{\partial y} + \frac{\partial p}{\partial y} - \frac{1}{\mathrm{Re}} \frac{\partial \omega}{\partial x}$$

$$= f_y + u_0 \frac{\partial v_0}{\partial x} + v_0 \frac{\partial v_0}{\partial y} \qquad \text{in } \Omega, \tag{8.54c}$$

$$\omega + \frac{\partial u}{\partial y} - \frac{\partial v}{\partial x} = 0 \qquad \text{in } \Omega. \tag{8.54d}$$

We can write (8.54) in standard matrix form (8.47) of the first-order system, in which

$$\boldsymbol{A}_1 = \begin{pmatrix} 1 & 0 & 0 & 0 \\ u_0 & 0 & 1 & 0 \\ 0 & u_0 & 0 & -\frac{1}{\mathrm{Re}} \\ 0 & -1 & 0 & 0 \end{pmatrix}, \qquad \boldsymbol{A}_2 = \begin{pmatrix} 0 & 1 & 0 & 0 \\ v_0 & 0 & 0 & \frac{1}{\mathrm{Re}} \\ 0 & v_0 & 1 & 0 \\ 1 & 0 & 0 & 0 \end{pmatrix},$$

$$\boldsymbol{A}_0 = \begin{pmatrix} 0 & 0 & 0 & 0 \\ \frac{\partial u_0}{\partial x} & \frac{\partial u_0}{\partial y} & 0 & 0 \\ \frac{\partial v_0}{\partial x} & \frac{\partial v_0}{\partial y} & 0 & 0 \\ 0 & 0 & 0 & 1 \end{pmatrix},$$

$$\underline{f} = \begin{pmatrix} 0 \\ f_x + u_0 \frac{\partial u_0}{\partial x} + v_0 \frac{\partial u_0}{\partial y} \\ f_y + u_0 \frac{\partial v_0}{\partial x} + v_0 \frac{\partial v_0}{\partial y} \\ 0 \end{pmatrix}, \qquad \underline{\boldsymbol{u}} = \begin{pmatrix} u \\ v \\ p \\ \omega \end{pmatrix}. \tag{8.55}$$

Two-Dimensional Driven-Cavity Flow. Now we are ready to use the LSFEM given in Sect. 4.8. The LSFEM has been tested by solving the two-dimensional driven-cavity and the backward-facing step flow problems (Jiang 1992). In these numerical experiments, a grid of 50×50 nonuniform bilinear elements with one-point quadrature is employed for the driven-cavity problem. The numerical results at $\mathrm{Re} = 10\,000$ are in good agreement with the fine mesh 257×257 results of Ghia et al. (1982).

A large-scale LSFEM computation has also been performed for the two-dimensional driven-cavity problem. The definition of driven-cavity problem is as follows: the boundary conditions for (u, v) are $u = v = 0$ on all solid walls and $u = 1, v = 0$ on the top lid. Also $p = 0$ is specified at the center of the bottom. No boundary conditions are given for the vorticity. At first we divide the cavity into 50×50 uniform bilinear elements with the size $h = 0.02$. In order to take care of the singularity at the top corners where the absolute value of the vorticity goes to infinite, we again divide the top layer of elements into two thin layers with thickness 0.005 and 0.015 units, respectively.

The matrix-free Jacobi preconditioned conjugate gradient method, see Chap. 15 for details, is employed for the solution of the resulting linear algebraic equations. One-point Gaussian quadrature is used in all computation. At first the calculation is for the flow at Re = 10 000 with this 50 × 51 mesh. $u = v = 0$ is used as an initial guess for Re = 100, the converged solution for Re = 100 is used as an initial guess for Re = 200, and so on until Re = 10 000. By using this kind of continuation method, the computation could be done on a PC-386 with 4M bytes memory in a few hours. The mesh is then refined to 100×102 elements, and the previous solution is interpolated as the initial guess for the conjugate gradient method. This procedure is applied sequentially until the converged solution is obtained for the 400 × 408 mesh in which most elements have the size $h = 0.0025$, and in the top layer the thickness of the elements is 0.0005. The storage required for this problem is less than 8M words on a CRAY-YMP. The quality of the solution can be judged from the L_2 norm of the residual which is 6.26×10^{-7}.

Table 8.2. Velocity results for two-dimensional cavity at Re = 10 000; u-velocity along vertical line through geometric center, v-velocity along horizontal line through geometric center (From Jiang et al. 1994a)

y	u	x	v
1.0000	1.0000	1.0000	0.0000
0.9775	0.4884	0.9725	−0.5669
0.9700	0.4959	0.9675	−0.5777
0.9625	0.4996	0.9625	−0.5603
0.9525	0.4949	0.9600	−0.5463
0.8500	0.3494	0.9525	−0.5037
0.7300	0.2016	0.9450	−0.4752
0.6200	0.0868	0.9050	−0.4277
0.4500	−0.0727	0.8050	−0.3090
0.2800	−0.2312	0.5000	0.0083
0.1700	−0.3356	0.2500	0.2571
0.1000	−0.3969	0.1500	0.3645
0.0800	−0.4198	0.1000	0.4190
0.0700	−0.4363	0.0700	0.4567
0.0625	−0.4485	0.0600	0.4628
0.0525	−0.4554	0.0500	0.4567
0.0400	−0.4286	0.0400	0.4329
0.0300	−0.3704	0.0200	0.3278
0.0000	0.0000	0.0000	0.0000

The profile of the horizontal velocity along the vertical center line and the profile of the vertical velocity along the horizontal centerline of the driven-cavity for Re = 10 000 are illustrated in Fig. 8.4a and b, respectively. Overall, the profile compares well with that given by Ghia et al. (1982), except in the boundary layers. In our results, the boundary layer phenomena are more pronounced. Table 8.2 lists the numerical values of selected nodes corresponding

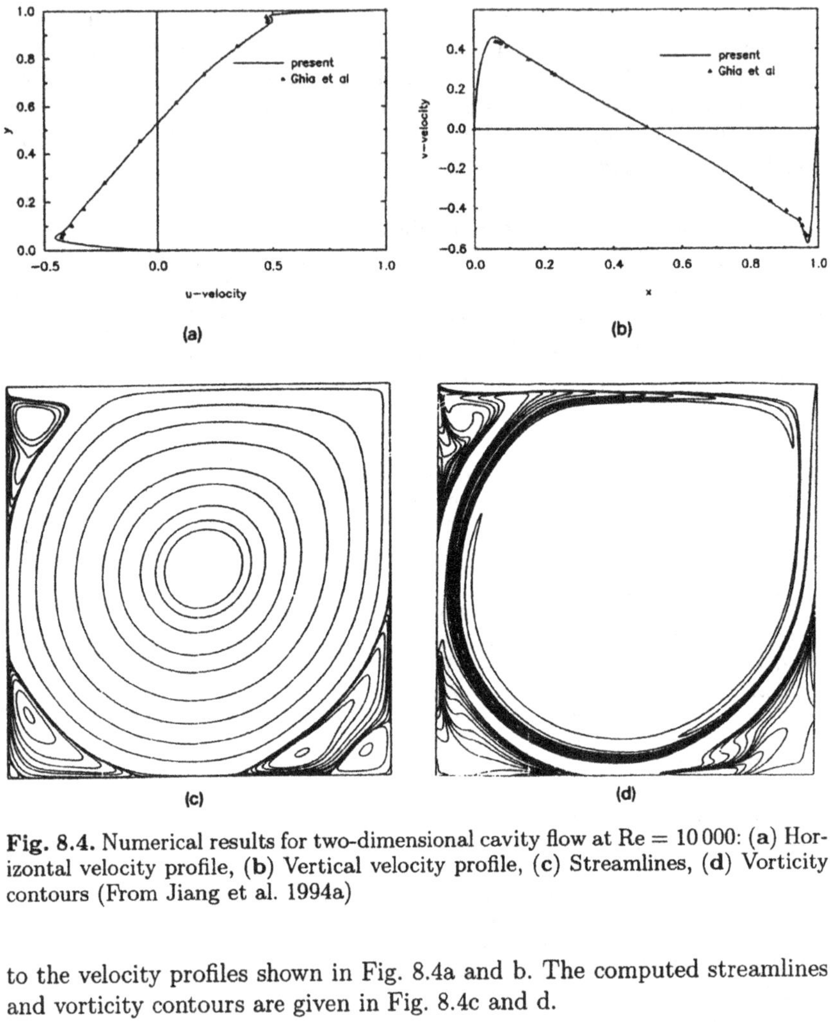

Fig. 8.4. Numerical results for two-dimensional cavity flow at Re = 10 000: (**a**) Horizontal velocity profile, (**b**) Vertical velocity profile, (**c**) Streamlines, (**d**) Vorticity contours (From Jiang et al. 1994a)

to the velocity profiles shown in Fig. 8.4a and b. The computed streamlines and vorticity contours are given in Fig. 8.4c and d.

8.3.2 Axisymmetric Case

The general, incompressible Navier–Stokes equations (8.50) can be applied to the study of axisymmetric fluid flows. The first-order system in cylindrical coordinates is given by:

$$\frac{\partial u}{\partial z} + \frac{\partial v}{\partial r} + \frac{v}{r} = 0 \qquad \text{in } \Omega, \tag{8.56a}$$

$$u\,\frac{\partial u}{\partial z} + v\,\frac{\partial u}{\partial r} + \frac{\partial p}{\partial z} + \frac{1}{\mathrm{Re}}\Big(\frac{\partial \omega}{\partial r} + \frac{\omega}{r}\Big) = f_z \qquad \text{in } \Omega, \tag{8.56b}$$

$$u\frac{\partial v}{\partial z}+v\frac{\partial v}{\partial r}+\frac{\partial p}{\partial r}-\frac{1}{\mathrm{Re}}\frac{\partial \omega}{\partial z}=f_r \quad \text{in } \Omega, \tag{8.56c}$$

$$\omega+\frac{\partial u}{\partial r}-\frac{\partial v}{\partial z}=0 \quad \text{in } \Omega, \tag{8.56d}$$

where z denote the axial coordinate, r the radial coordinate, u the velocity component in the axial direction, and v the velocity in the radial direction.

For example, when (8.56) is linearized by using simple substitution, the coefficient matrices in the standard first-order system form of (8.47) are

$$\boldsymbol{A}_1=\begin{pmatrix}1&0&0&0\\u_0&0&1&0\\0&u_0&0&-\frac{1}{\mathrm{Re}}\\0&-1&0&0\end{pmatrix},\qquad \boldsymbol{A}_2=\begin{pmatrix}0&1&0&0\\v_0&0&0&\frac{1}{\mathrm{Re}}\\0&v_0&1&0\\1&0&0&0\end{pmatrix},$$

$$\boldsymbol{A}_0=\begin{pmatrix}0&\frac{1}{r}&0&0\\0&0&0&\frac{1}{\mathrm{Re}}\frac{1}{r}\\0&0&0&0\\0&0&0&1\end{pmatrix},\quad \underline{\boldsymbol{f}}=\begin{pmatrix}0\\f_z\\f_r\\0\end{pmatrix},\quad \underline{\boldsymbol{u}}=\begin{pmatrix}u\\v\\p\\\omega\end{pmatrix}. \tag{8.57}$$

8.3.3 Three-Dimensional Case

In Cartesian coordinates, the three-dimensional incompressible Navier–Stokes equations (8.50) take the following form:

$$u\frac{\partial u}{\partial x}+v\frac{\partial u}{\partial y}+w\frac{\partial u}{\partial z}+\frac{\partial p}{\partial x}+\frac{1}{\mathrm{Re}}\left(\frac{\partial \omega_z}{\partial y}-\frac{\partial \omega_y}{\partial z}\right)=f_x \quad \text{in } \Omega, \tag{8.58a}$$

$$u\frac{\partial v}{\partial x}+v\frac{\partial v}{\partial y}+w\frac{\partial v}{\partial z}+\frac{\partial p}{\partial y}+\frac{1}{\mathrm{Re}}\left(\frac{\partial \omega_x}{\partial z}-\frac{\partial \omega_z}{\partial x}\right)=f_y \quad \text{in } \Omega, \tag{8.58b}$$

$$u\frac{\partial w}{\partial x}+v\frac{\partial w}{\partial y}+w\frac{\partial w}{\partial z}+\frac{\partial p}{\partial z}+\frac{1}{\mathrm{Re}}\left(\frac{\partial \omega_y}{\partial x}-\frac{\partial \omega_x}{\partial y}\right)=f_z \quad \text{in } \Omega, \tag{8.58c}$$

$$\frac{\partial \omega_x}{\partial x}+\frac{\partial \omega_y}{\partial y}+\frac{\partial \omega_z}{\partial z}=0 \quad \text{in } \Omega, \tag{8.58d}$$

$$\omega_x-\frac{\partial w}{\partial y}+\frac{\partial v}{\partial z}=0 \quad \text{in } \Omega, \tag{8.58e}$$

$$\omega_y-\frac{\partial u}{\partial z}+\frac{\partial w}{\partial x}=0 \quad \text{in } \Omega, \tag{8.58f}$$

$$\omega_z-\frac{\partial v}{\partial x}+\frac{\partial u}{\partial y}=0 \quad \text{in } \Omega, \tag{8.58g}$$

$$\frac{\partial u}{\partial x}+\frac{\partial v}{\partial y}+\frac{\partial w}{\partial z}=0 \quad \text{in } \Omega. \tag{8.58h}$$

In order to use the LSFEM given in Sect. 4.8, system (8.58) should be linearized, for example by Newton's method, and written in standard matrix form (8.10) in which:

$$\boldsymbol{A}_1 = \begin{pmatrix} u_0 & 0 & 0 & 0 & 0 & 0 & 1 \\ 0 & u_0 & 0 & 0 & 0 & -\frac{1}{\mathrm{Re}} & 0 \\ 0 & 0 & u_0 & 0 & \frac{1}{\mathrm{Re}} & 0 & 0 \\ 0 & 0 & 0 & 1 & 0 & 0 & 0 \\ 0 & 0 & 0 & 0 & 0 & 0 & 0 \\ 0 & 0 & -1 & 0 & 0 & 0 & 0 \\ 0 & 1 & 0 & 0 & 0 & 0 & 0 \\ 1 & 0 & 0 & 0 & 0 & 0 & 0 \end{pmatrix},$$

$$\boldsymbol{A}_2 = \begin{pmatrix} v_0 & 0 & 0 & 0 & 0 & \frac{1}{\mathrm{Re}} & 0 \\ 0 & v_0 & 0 & 0 & 0 & 0 & 1 \\ 0 & 0 & v_0 & -\frac{1}{\mathrm{Re}} & 0 & 0 & 0 \\ 0 & 0 & 0 & 0 & 1 & 0 & 0 \\ 0 & 0 & 1 & 0 & 0 & 0 & 0 \\ 0 & 0 & 0 & 0 & 0 & 0 & 0 \\ -1 & 0 & 0 & 0 & 0 & 0 & 0 \\ 0 & 1 & 0 & 0 & 0 & 0 & 0 \end{pmatrix},$$

$$\boldsymbol{A}_3 = \begin{pmatrix} w_0 & 0 & 0 & 0 & -\frac{1}{\mathrm{Re}} & 0 & 0 \\ 0 & w_0 & 0 & \frac{1}{\mathrm{Re}} & 0 & 0 & 0 \\ 0 & 0 & w_0 & 0 & 0 & 0 & 1 \\ 0 & 0 & 0 & 0 & 0 & 1 & 0 \\ 0 & -1 & 0 & 0 & 0 & 0 & 0 \\ 1 & 0 & 0 & 0 & 0 & 0 & 0 \\ 0 & 0 & 0 & 0 & 0 & 0 & 0 \\ 0 & 0 & 1 & 0 & 0 & 0 & 0 \end{pmatrix},$$

$$\boldsymbol{A} = \begin{pmatrix} \frac{\partial u_0}{\partial x} & \frac{\partial u_0}{\partial y} & \frac{\partial u_0}{\partial z} & 0 & 0 & 0 & 0 \\ \frac{\partial v_0}{\partial x} & \frac{\partial v_0}{\partial y} & \frac{\partial v_0}{\partial z} & 0 & 0 & 0 & 0 \\ \frac{\partial w_0}{\partial x} & \frac{\partial w_0}{\partial y} & \frac{\partial w_0}{\partial z} & 0 & 0 & 0 & 0 \\ 0 & 0 & 0 & 0 & 0 & 0 & 0 \\ 0 & 0 & 0 & -1 & 0 & 0 & 0 \\ 0 & 0 & 0 & 0 & -1 & 0 & 0 \\ 0 & 0 & 0 & 0 & 0 & -1 & 0 \\ 0 & 0 & 0 & 0 & 0 & 0 & 0 \end{pmatrix},$$

$$\underline{f} = \begin{pmatrix} f_x + u_0 \frac{\partial u_0}{\partial x} + v_0 \frac{\partial u_0}{\partial y} + w_0 \frac{\partial u_0}{\partial z} \\ f_y + u_0 \frac{\partial v_0}{\partial x} + v_0 \frac{\partial v_0}{\partial y} + w_0 \frac{\partial v_0}{\partial z} \\ f_z + u_0 \frac{\partial w_0}{\partial x} + v_0 \frac{\partial w_0}{\partial y} + w_0 \frac{\partial w_0}{\partial z} \\ 0 \\ 0 \\ 0 \\ 0 \end{pmatrix}, \qquad \underline{u} = \begin{pmatrix} u \\ v \\ w \\ \omega_x \\ \omega_y \\ \omega_z \\ p \end{pmatrix}. \tag{8.59}$$

Three-Dimensional Driven-Cavity Flow. The least-squares solution of the cubic driven-cavity problem was obtained by Jiang et al. (1994a) using formulation (8.59) and also by Tang et al. (1995) using the time marching method. Almost all previous researchers used time-marching methods, and claimed that steady-state solutions had been obtained for Re $= 2000$. Also no Taylor–Görtler-like vortices were reported for Re < 2000 by previous researchers.

Koseff and Street (1984a,b,c) conducted systematic experiments of flow in a driven-cavity. Their experiments reveal that the flow field is highly unsteady and possesses significant secondary motions (end-wall corner eddies and Taylor–Görtler-like vortices). They mention that the flow is not stable at Re $= 2000$ and show results for Re $= 3200$. However, a very important criterion, the Reynolds number at which the flow becomes unstable, had never been reported.

In a study conducted by Jiang et al. (1994a), the computation is carried out up to Re $= 3200$. The boundary conditions are as follows: $u = 1$, $v = w = 0$ are specified on the top driven surface ($y = 1.0$), and $u = v = w = 0$ on all solid walls (Fig. 8.5). At the center of the bottom ($x = 0.5$, $y = 0.0$, $z = 0.5$), $p = 0$ is specified. The mesh is based on $50 \times 50 \times 50$ uniform

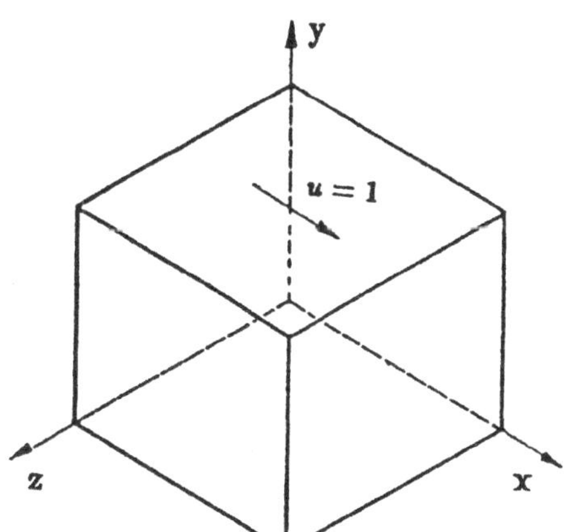

Fig. 8.5. Geometry and boundary conditions of cubic cavity flow

trilinear elements. In order to take care of corner singularities, two layers of thin elements are added close to the upper driven surface; so it had $50 \times 52 \times 50$ elements. One-point Gaussian quadrature is used for computing the product of the element matrix and the element vector in the matrix-free Jacobi preconditioned conjugate gradient method. Reduced integration not only guarantees the accuracy of the solution, but also saves a significant amount of computing time when an iterative solver is employed. Due to the use of the matrix-free conjugate gradient method, these problems are solved using less than 13M words of memory on a CRAY-YMP.

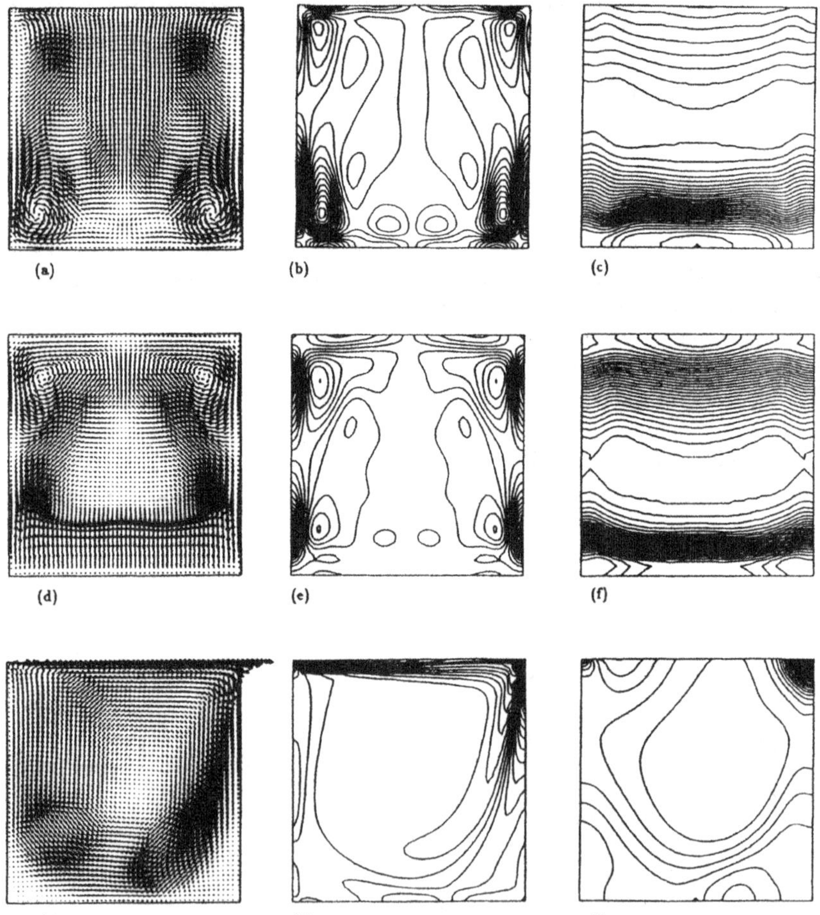

Fig. 8.6. Numerical results for three-dimensional cavity flow at Re = 1000: (**a**) Flow pattern, (**b**) Vorticity(ω_x) contours, (**c**) Pressure contour at x=0.5; (**d**) Flow pattern, (**e**) Vorticity(ω_x) contours , (**f**) Pressure contour at y=0.5; (**g**) Flow pattern, (**h**) Vorticity(ω_x) contours, (**i**) Pressure contour at z=0.5 (From Jiang et al. 1994a)

It is found that up to Re = 3200 the flow is symmetrical about the plane $z = 0.5$. This observation is consistent with the results of numerical simulation published by other researchers. It is also found that small eddies might appear near the center part of the cube. Therefore, an almost-uniform $50 \times 52 \times 25$ mesh for a half domain imposing the boundary conditions

$$w = 0, \ \omega_x = \omega_y = 0$$

on the symmetric plane $z = 0.5$ is chosen for the further computation.

The Reynolds numbers considered are 100, 400, and 1000. The solution of the Stokes equations is taken as an initial guess for the calculation at Re = 100 which is used as an initial guess for Re = 400, and so on. In order to control the solution time on the computer, the number of conjugate gradient iterations is fixed at between 3500 to 8000 in each linearized step. Since this is a nonlinear problem, it is not necessary to obtain a 'converged' solution in each Newton step.

Starting from the results at Re = 400, after ten Newton's steps, the results at Re = 1000 are obtained and comparable with the results of the fine stretched mesh $81 \times 81 \times 81$ by Iwatsu et al. (1989a,b). After twenty Newton's steps the results depicted in Fig. 8.6 are obtained which shows the corner vortices and the Taylor–Görtler-like vortices at the bottom region of the cavity at Re = 1000. At this moment, the L_2 norm of the residual is about 6.0×10^{-5}. Figure 8.7 displays the velocity profiles of the u component on the vertical centerline of the plane $z = 0.5$ for Re = 100, 400 and 1000. Table 8.3 lists some nodal values corresponding to the profiles of u velocity in Fig. 8.7.

The computation is continued for Re = 1000. After twenty-four Newton's steps, a pair of Taylor–Görtler-like vortices near the central part of the bot-

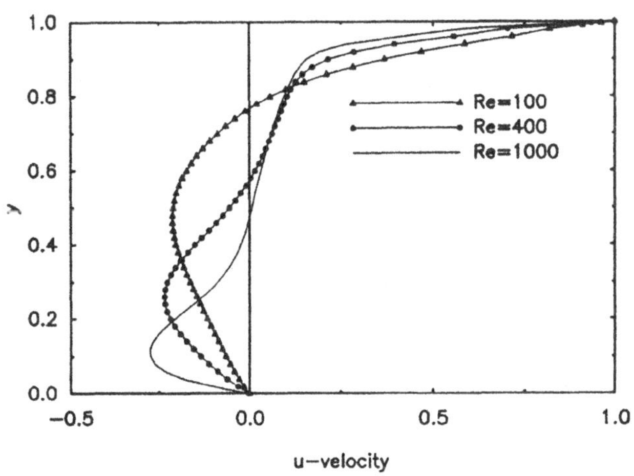

Fig. 8.7. Profiles of u velocity for three-dimensional cavity flow at Re = 100, 400 and 1000 (From Jiang et al. 1994a)

Table 8.3. u-velocity along vertical line through geometric center of three-dimensional cavity (From Jiang et al. 1994a)

y	Re 100	Re 400	Re 1000
1.0000	1.0000	1.0000	1.0000
0.9600	0.7159	0.5563	0.4119
0.9200	0.4704	0.2868	0.1912
0.8800	0.2841	0.1730	0.1389
0.8400	0.1516	0.1285	0.1148
0.8000	0.0577	0.1052	0.0959
0.7600	−0.0119	0.0875	0.0798
0.7200	−0.0660	0.0714	0.0661
0.6800	−0.1097	0.0550	0.0540
0.6400	−0.1452	0.0369	0.0432
0.6000	−0.1732	0.0159	0.0328
0.5600	−0.1937	−0.0092	0.0224
0.5200	−0.2067	−0.0391	0.0120
0.4800	−0.2120	−0.0734	0.0015
0.4400	−0.2100	−0.1113	−0.0099
0.4000	−0.2017	−0.1505	−0.0242
0.3600	−0.1886	−0.1872	−0.0435
0.3200	−0.1722	−0.2159	−0.0701
0.2800	−0.1539	−0.2319	−0.1063
0.2600	−0.1444	−0.2341	−0.1284
0.2400	−0.1349	−0.2324	−0.1528
0.2000	−0.1155	−0.2176	−0.2049
0.1600	−0.0955	−0.1903	−0.2512
0.1200	−0.0748	−0.1540	−0.2754
0.0800	−0.0529	−0.1111	−0.2611
0.0400	−0.0286	−0.0608	−0.1812
0.0000	0.0000	0.0000	0.0000

tom shown in Fig. 8.8 become more clearer and stronger. The L_2 norm of the residual remained around 6.0×10^{-5}. The maximum difference of the velocity magnitudes between the results of two sequential Newton's steps is around 9.0×10^{-3} and couldn't be further reduced. Newton's method employed here may be considered as an implicit time-marching method with a very large time step. The fact that the method cannot lead to a convergent solution may imply that at Re = 1000 the flow in the cavity is not stable. The appearance of the Taylor–Görtler-like vortices is consistent with this instability.

It appears that the results with LSFEM are highly accurate and also predict instability at a lower Reynolds number than shown by previous works in this field.

Three-Dimensional Backward-Facing Step Flow. The LSFEM based on (8.59) is further tested by solving the backward-facing step flow problem (Jiang et al. 1995). It is well known that for backward-facing step flows, the reattachment length obtained by two-dimensional calculations cannot

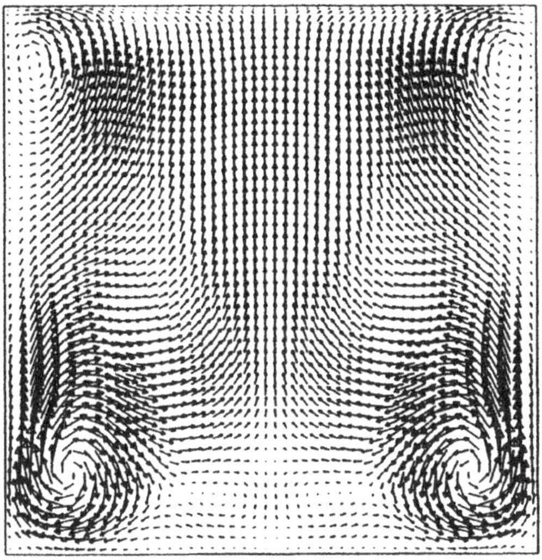

Fig. 8.8. Velocity vector at $x = 0.5$ for three-dimensional cavity flow at Re = 1000 (From Jiang et al. 1994a)

match experimental results for Re $\geq$ 450. The deviations of two-dimensional simulations from experimental results are mainly due to the growing effects of three-dimensionality as the Reynolds number increases.

Relatively few people have solved the three-dimensional backward-facing step flow problem as compared to the cavity problem. Ku et al (1987, 1989) applied a pseudospectral matrix element method to simulate the same three-dimensional problem only up to Re = 450 by employing the velocity–pressure formulation. A simplified marker-and-cell finite-difference scheme was applied by Ikohagi et al. (1992) for the same three-dimensional problem in curvilinear coordinates. However, no comparisons were made by them with the experimental data. The recent numerical simulations of the same problem presented by Williams and Baker (1997) are compared well with the experimental data. But the implementation of their Taylor–Galerkin finite element method might be cumbersome.

The geometry and boundary conditions for the three-dimensional model are shown in Fig. 8.9. Due to symmetry, only half of the problem is solved. The step has a height S=4.9 mm, and the inlet boundary is located 3.5 step heights upstream of this step. The channel downstream of the step has a height of 10.1 mm that provides an expansion ratio of 1 : 1.9423, and the half-span W = 90 mm that provides an aspect ratio for the test channel of 18 : 1.01. These data are taken from the experiment conducted by Armaly et al. (1982). The length L measured from the step to the end of the calculation domain is 45 step heights, which is 3.11 times the maximum experimentally

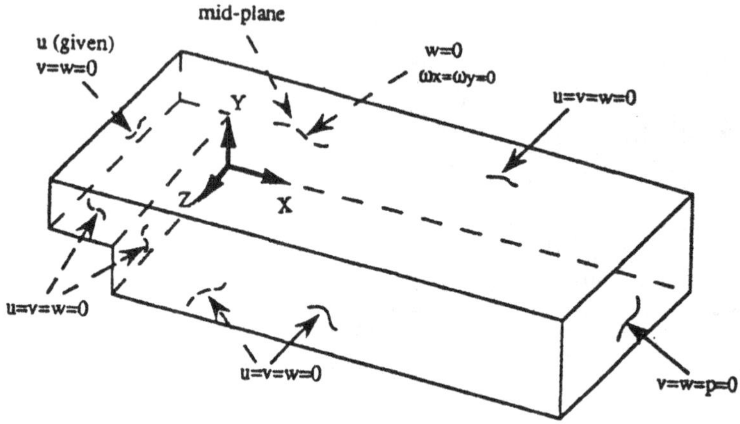

Fig. 8.9. Backward-facing step geometry and boundary conditions

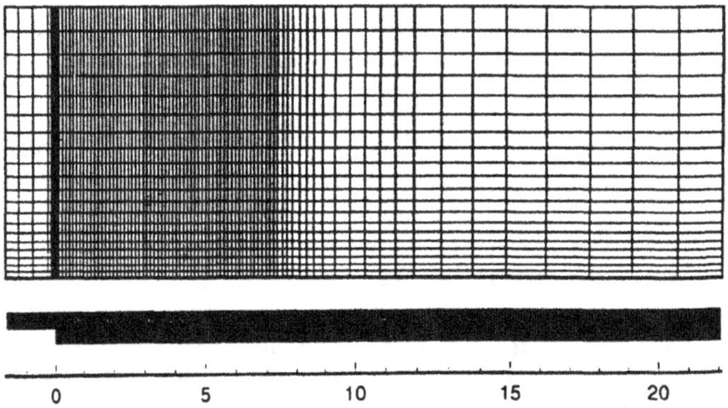

Fig. 8.10. Nonuniform mesh (xy plane and xz plane) (From Jiang et al. 1995)

measured reattachment length of the primary recirculation region for the Renolds numbers of interest. The Reynolds number $\mathrm{Re} = UD/\nu$ is based on the hydraulic diameter ($D = 10.4\,\mathrm{mm}$) of the inlet channel, and the average velocity is two-thirds of the maximum inlet velocity (normalized to unity) on the mid-span plane. The various Reynolds numbers are obtained by varying the kinematic viscosity ν. The velocity profiles obtained by solving the two-dimensional Poisson equation for a well-developed flow are imposed as the inflow boundary conditions. The outflow boundary conditions are defined such that a parallel flow and a constant pressure field exist.

The computation is performed on a mesh with 54 400 nonuniform trilinear elements ($6 \times 16 \times 20$ for the inlet channel and $82 \times 32 \times 20$ for the test channel). The smallest element has a size of 0.1251, 0.05714 and 0.421 steps

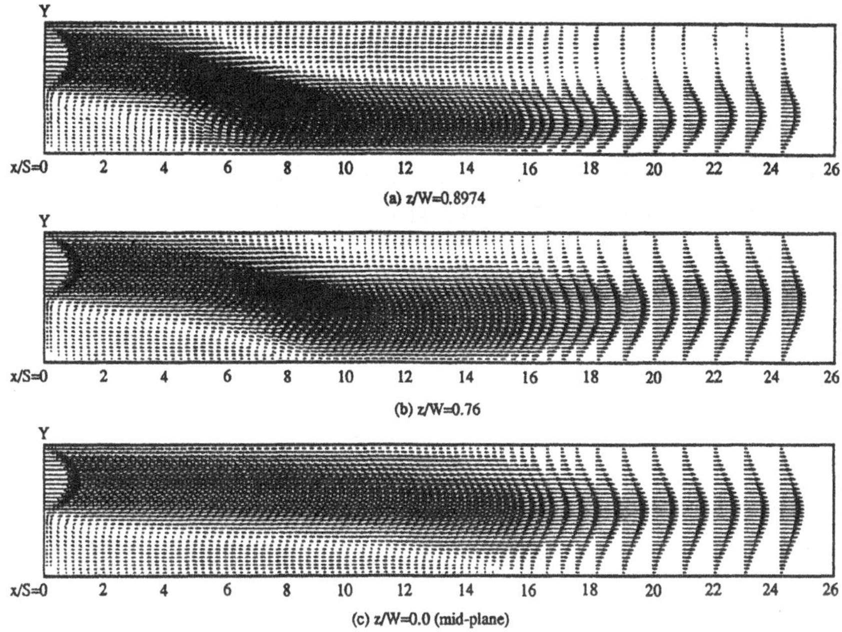

Fig. 8.11. Velocity vectors for Re=800 (From Jiang et al. 1995)

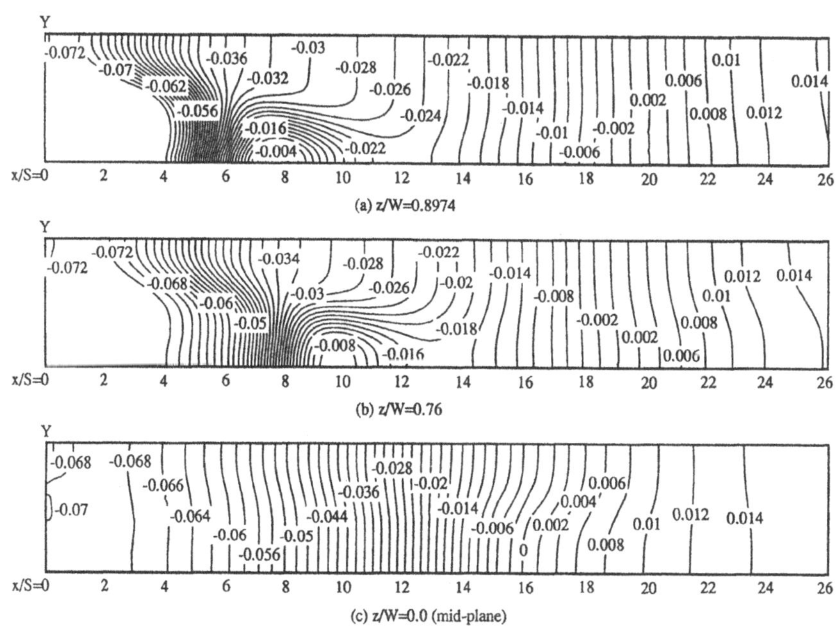

Fig. 8.12. Pressure contours for Re=800 (From Jiang et al. 1995)

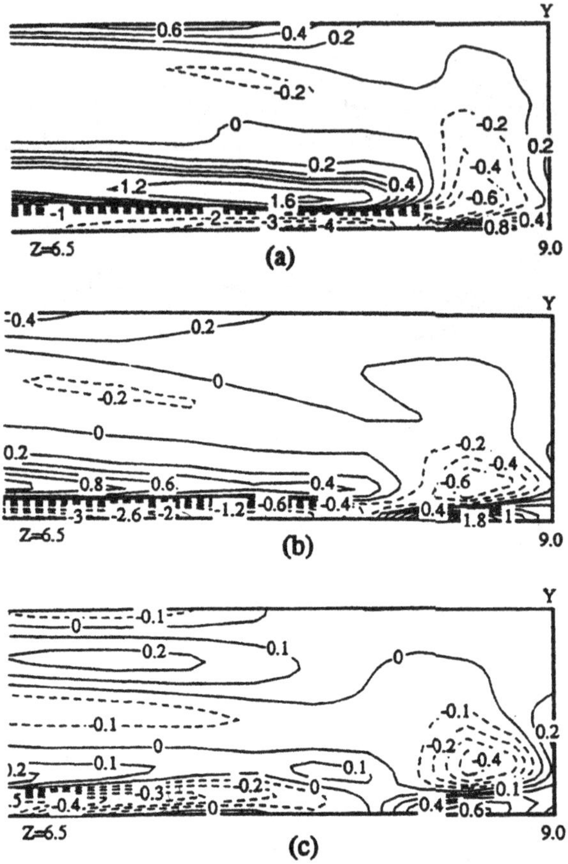

Fig. 8.13. ω_x contours for Re=800 at x/S=(**a**) 6.25, (**b**) 9.0, (**c**) 12.25 (From Jiang et al. 1995)

in the x, y and z direction individually, while the largest element has a size of 3.0, 0.06938 and 1.68367 steps respectively.

Figure 8.10 shows the nonuniform mesh which has more elements close to the sidewall, the floor and the roof in the test channel. The matrix-free Jacobi preconditioned conjugate gradient method is employed for the solution of the resulting algebraic equations. One-point Gaussian quadrature is used for the evaluation of element matrices. The solution of the Stokes equations is taken as an initial guess for the case of Re = 100, and the 'converged' solution is then used as an initial guess for higher Reynolds numbers. 'Converged' solutions are those whose L_2 norm of the residuals less than 1.0×10^{-4}. The problem is solved using about $5M$ words of memory on a CRAY-YMP.

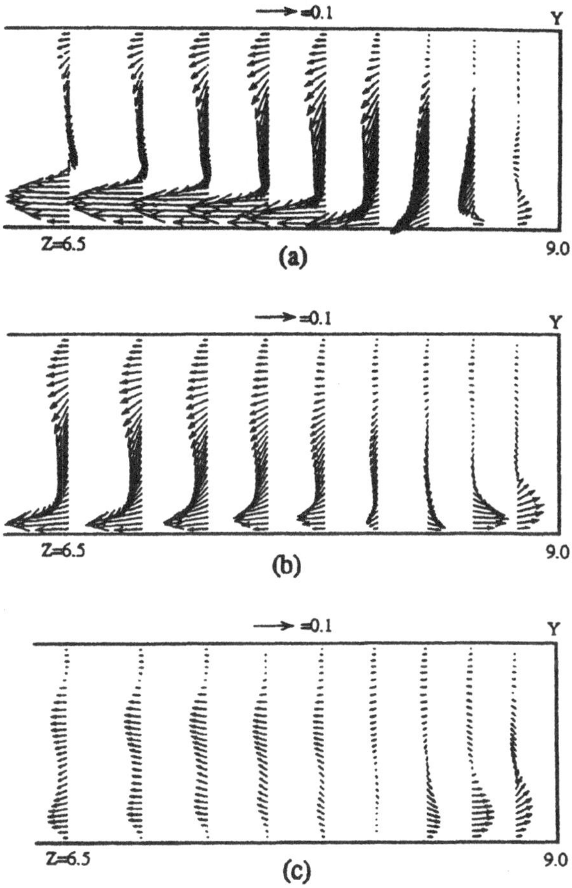

Fig. 8.14. Velocity vectors for Re=800 at x/S=(a) 6.25, (b) 9.0, (c) 12.25 (From Jiang et al. 1995)

Simulation of the three-dimensional model is performed from Re = 100 to 800.

This three-dimensional simulation is mainly for the demonstration of spanwise flow structure. The velocity vectors at three spanwise locations for Re = 800 are shown in Fig. 8.11. At Re = 800, the velocity vectors change significantly in the spanwise direction, and the reattachment length increases rapidly, moving towards the mid-plane. Figure 8.12 depicts the pressure contours at Re = 800. Again, the change in spanwise pressure distribution is significant. Most researchers predicted closely the reattachment length up to Re $\leq$ 450 with two-dimensional simulations, and assumed that two-dimensional phenomena are maintained up to Re $\approx$ 450. Numerical re-

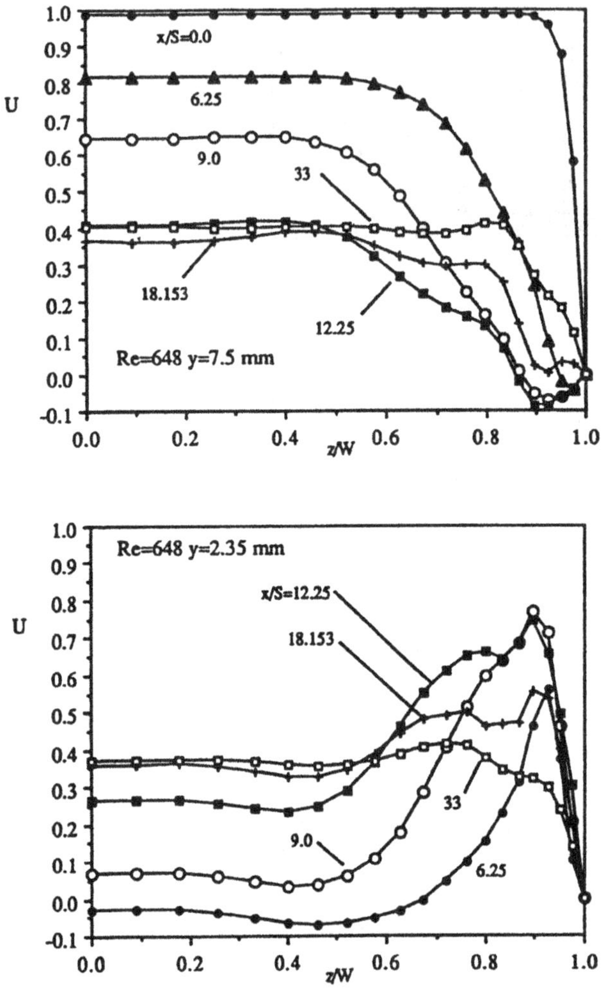

Fig. 8.15. Spanwise distributions of u-velocity profiles for Re=648 (From Jiang et al. 1995)

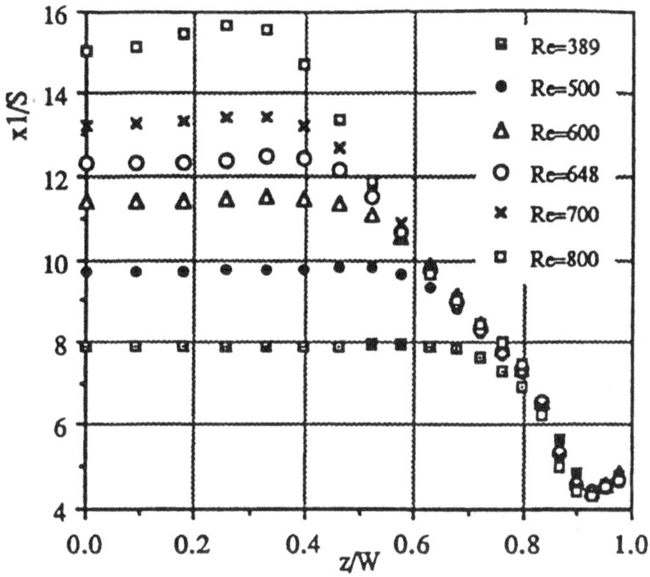

Fig. 8.16. Spanwise locations of reattachment line for various Reynolds numbers (From Jiang et al. 1995)

sults obtained by LSFEM, however, show that three-dimensional effects are significant even at low Reynolds numbers (e.g. Re = 277).

Figure 8.13 illustrates the contours of streamwise vorticity ω_x from $z =$ 6.5 cm to the sidewall ($z = 9.0$ cm) at Re = 800. The dashed lines indicate negative contour values. As shown in the figure, the vortex is stronger near the sidewall and has negligible influence on the region near the mid-plane. The study of velocity vectors along the span at different downstream locations provides a better view of the three-dimensional phenomena (see Fig. 8.14). The behavior of the inward flow toward the mid-span at the top roof and the outward flow toward the sidewall ($z/W = 1.0$) at the channel floor contributes to the three-dimensional phenomena. This three-dimensional behavior around the corner of the sidewall and the step can also be found at lower Reynolds numbers but is less significant.

It is found that except in the inlet channel and in the region which is very far downstream of the step, the three-dimensional effects are significant at the downstream in the vicinity of the step. A study of spanwise flow structure provides further details. The spanwise velocity vectors at various x-locations and at a fixed y-position for Re = 648 are shown in Fig. 8.15. At $y = 7.5$ mm, all velocity profiles close to the mid-plane are basically two-dimensional. The negative velocities in the figure indicate that a second separation bubble occurs on the ceiling, and its thickness grows as close to the sidewall. At $y =$ 2.35 mm, the x-component velocity, u, first increases rapidly in the boundary

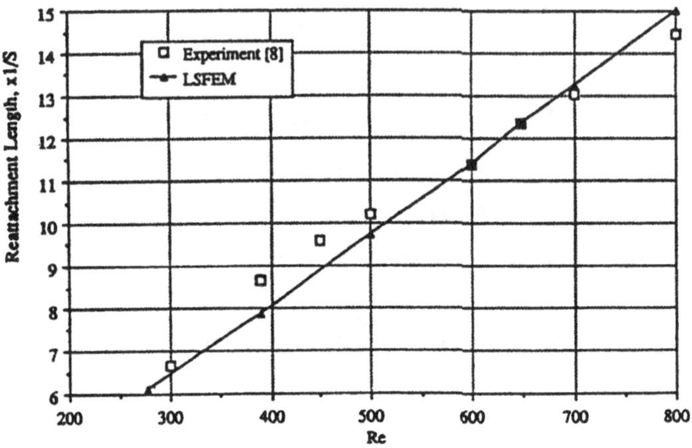

Fig. 8.17. Comparison of experimental and numerical results for primary reattachment length up to Re=800 (From Jiang et al. 1995)

layer, then drops and resumes two-dimensional flow from $z/W = 0.5$ to the mid-plane ($z/W = 0.0$).

Figure 8.16 demonstrates the three-dimensional effects by depicting the spanwise location of the reattachment length for the primary separated-flow region. Next to the sidewall, the reattachment length increases mainly due to the interaction of viscous stress between the channel floor and sidewall. Figure 8.17 shows the computed reattachment length of the primary recirculation zone, and compares well with the experimental results obtained by Armaly et al. (1982). In the Re range for which most simulations fail to predict the reattachment length because of the three-dimensional phenomenon (Re $= 450 \sim 800$), the calculated results by LSFEM agree very well with experimental results up to Re $= 800$.

As the Reynolds number increases, an additional separated-flow region occurs near the channel ceiling. Figure 8.18 illustrates the spanwise detachment (x_4/S) and reattachment (x_5/S) lines of this second eddy. The present results show that as the Reynolds number increases, this upper-wall eddy propagates toward the mid-plane and its length decreases in the spanwise direction. For example, as the Reynolds number increases from 600 to 800, it ranges from 17.5 to 22 step heights at $z/W = 0.97708$ and increases from 0.2 to 9.35 step heights at $z/W = 0.76$. The upper-wall eddy at the mid-plane is not observed.

From these comprehensive numerical results, it is seen that the LSFEM is a remarkably robust method even for three-dimensional flows at large Reynolds numbers.

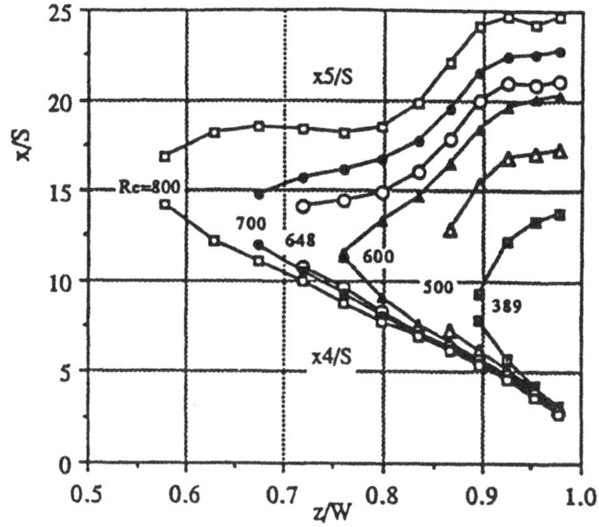

Fig. 8.18. Spanwise distributions of detachment and reattachment length for various Reynolds numbers (From Jiang et al. 1995)

8.4 The Navier–Stokes Equations in the $\boldsymbol{u} - b - \boldsymbol{\omega}$ Formulation

System (8.50) is not the only way to write a first-order formulation by using the vorticity as an independent unknown vector (Gresho 1991a,b, 1992). By introducing the Bernoulli pressure or the total pressure $b = p + (u^2+v^2+w^2)/2$ as an independent variable instead of the pressure p, we have the following Navier–Stokes problem: Find $\boldsymbol{u}$, b and $\boldsymbol{\omega}$ such that

$$-\boldsymbol{u} \times \boldsymbol{\omega} + \nabla b + \frac{1}{\mathrm{Re}} \nabla \times \boldsymbol{\omega} = \boldsymbol{f} \qquad \text{in } \Omega, \tag{8.60a}$$

$$\nabla \cdot \boldsymbol{\omega} = 0 \qquad \text{in } \Omega, \tag{8.60b}$$

$$\boldsymbol{\omega} - \nabla \times \boldsymbol{u} = 0 \qquad \text{in } \Omega, \tag{8.60c}$$

$$\nabla \cdot \boldsymbol{u} = 0 \qquad \text{in } \Omega. \tag{8.60d}$$

In the Cartesian coordinates, system (8.60) can be written as

$$w\omega_y - v\omega_z + \frac{\partial b}{\partial x} + \frac{1}{\mathrm{Re}}\left(\frac{\partial \omega_z}{\partial y} - \frac{\partial \omega_y}{\partial z}\right) = f_x \qquad \text{in } \Omega, \tag{8.61a}$$

$$u\omega_z - w\omega_x + \frac{\partial b}{\partial y} + \frac{1}{\mathrm{Re}}\left(\frac{\partial \omega_x}{\partial z} - \frac{\partial \omega_z}{\partial x}\right) = f_y \qquad \text{in } \Omega, \tag{8.61b}$$

$$v\omega_x - u\omega_y + \frac{\partial b}{\partial z} + \frac{1}{\mathrm{Re}}\left(\frac{\partial \omega_y}{\partial x} - \frac{\partial \omega_x}{\partial y}\right) = f_z \qquad \text{in } \Omega, \tag{8.61c}$$

$$\frac{\partial \omega_x}{\partial x} + \frac{\partial \omega_y}{\partial y} + \frac{\partial \omega_z}{\partial z} = 0 \qquad \text{in } \Omega, \tag{8.61d}$$

$$\omega_x - \frac{\partial w}{\partial y} + \frac{\partial v}{\partial z} = 0 \qquad \text{in } \Omega, \tag{8.61e}$$

$$\omega_y - \frac{\partial u}{\partial z} + \frac{\partial w}{\partial x} = 0 \qquad \text{in } \Omega, \tag{8.61f}$$

$$\omega_z - \frac{\partial v}{\partial x} + \frac{\partial u}{\partial y} = 0 \qquad \text{in } \Omega, \tag{8.61g}$$

$$\frac{\partial u}{\partial x} + \frac{\partial v}{\partial y} + \frac{\partial w}{\partial z} = 0 \qquad \text{in } \Omega. \tag{8.61h}$$

For two-dimensional problems, let $\omega = \omega_z$, we have

$$-v\omega + \frac{\partial b}{\partial x} + \frac{1}{\text{Re}}\frac{\partial \omega}{\partial y} = f_x \qquad \text{in } \Omega, \tag{8.62a}$$

$$u\omega + \frac{\partial b}{\partial y} - \frac{1}{\text{Re}}\frac{\partial \omega}{\partial x} = f_y \qquad \text{in } \Omega, \tag{8.62b}$$

$$\omega - \frac{\partial v}{\partial x} + \frac{\partial u}{\partial y} = 0 \qquad \text{in } \Omega, \tag{8.62c}$$

$$\frac{\partial u}{\partial x} + \frac{\partial v}{\partial y} = 0 \qquad \text{in } \Omega. \tag{8.62d}$$

We note that the first-order velocity–total pressure–vorticity formulation is almost-linear. In this formulation, a good portion of the nonlinearity has been absorbed into the Bernoulli pressure; the nonlinear term $\boldsymbol{u} \times \boldsymbol{\omega}$ is of zero order, and thus is not related to any derivatives; while the rest of the terms constitutes the linear Stokes equations. Therefore, the non-linear term has no effect on the classification of the system, and it does not matter how large the Reynolds number is, the whole system is elliptic. For this reason, the incompressible Navier–Stokes equations have the same boundary conditions as those listed in Table 8.1 for the Stokes equations.

The velocity–total pressure–vorticity formulation is also suitable for LSFEM. It is found that for tested problems the velocity–total pressure–vorticity formulation leads to slightly slow convergence (Jiang et al. 1994a).

8.5 The Navier–Stokes Equations in the $\boldsymbol{u} - p - \boldsymbol{\sigma}$ Formulation

The Navier–Stokes equations can also be written in the following first-order velocity–pressure–stress form:

$$\rho\Big(u\,\frac{\partial u}{\partial x}+v\,\frac{\partial u}{\partial y}+w\,\frac{\partial u}{\partial z}\Big)+\frac{\partial p}{\partial x}-\Big(\frac{\partial\sigma_{xx}}{\partial x}+\frac{\partial\sigma_{xy}}{\partial y}+\frac{\partial\sigma_{xz}}{\partial z}\Big)=f_x \quad \text{in } \Omega, \tag{8.63a}$$

$$\rho\Big(u\,\frac{\partial v}{\partial x}+v\,\frac{\partial v}{\partial y}+w\,\frac{\partial v}{\partial z}\Big)+\frac{\partial p}{\partial y}-\Big(\frac{\partial\sigma_{xy}}{\partial x}+\frac{\partial\sigma_{yy}}{\partial y}+\frac{\partial\sigma_{yz}}{\partial z}\Big)=f_y \quad \text{in } \Omega, \tag{8.63b}$$

$$\rho\Big(u\,\frac{\partial w}{\partial x}+v\,\frac{\partial w}{\partial y}+w\,\frac{\partial w}{\partial z}\Big)+\frac{\partial p}{\partial z}-\Big(\frac{\partial\sigma_{xz}}{\partial x}+\frac{\partial\sigma_{yz}}{\partial y}+\frac{\partial\sigma_{zz}}{\partial z}\Big)=f_z \quad \text{in } \Omega, \tag{8.63c}$$

$$\sigma_{xx}-2\mu\frac{\partial u}{\partial x}=0 \quad in\ \Omega, \tag{8.63d}$$

$$\sigma_{yy}-2\mu\frac{\partial v}{\partial y}=0 \quad in\ \Omega, \tag{8.63e}$$

$$\sigma_{zz}-2\mu\frac{\partial u}{\partial z}=0 \qquad \text{in } \Omega, \tag{8.63f}$$

$$\sigma_{xy}-\mu\Big(\frac{\partial u}{\partial y}+\frac{\partial v}{\partial x}\Big)=0 \qquad \text{in } \Omega, \tag{8.63g}$$

$$\sigma_{xz}-\mu\Big(\frac{\partial u}{\partial z}+\frac{\partial w}{\partial x}\Big)=0 \qquad \text{in } \Omega, \tag{8.63h}$$

$$\sigma_{yz}-\mu\Big(\frac{\partial v}{\partial z}+\frac{\partial w}{\partial y}\Big)=0 \qquad \text{in } \Omega, \tag{8.63i}$$

$$\frac{\partial u}{\partial x}+\frac{\partial v}{\partial y}+\frac{\partial w}{\partial z}=0 \qquad \text{in } \Omega, \tag{8.63j}$$

where ρ is the density, μ is the viscosity.

In (8.63) there are ten equations, that is, three momentum equations, six constitutive equations and one incompressibility equation, involving ten unknown variables $(u, v, w, p,\ \sigma_{xx}, \sigma_{yy}, \sigma_{zz}, \sigma_{xy}, \sigma_{xz}, \sigma_{yz})$. However, due to incompressibility, three normal stress $\sigma_{xx}, \sigma_{yy}, \sigma_{zz}$ are not algebraically independent, that is

$$\sigma_{zz}=-(\sigma_{xx}+\sigma_{yy}).$$

Therefore, in the velocity–pressure–stress formulation there are in fact only nine (odd) independent differential equations in nine (odd) independent unknown variables. As discussed before in Sect. 4.7, this kind of first-order system cannot be elliptic in the ordinary sense, and it is also difficult to specify an appropriate number of boundary conditions. Numerical experiments conducted by the author on the two-dimensional Stokes equations with this formulation reveal that the rate of convergence for the pressure and the stresses is one order lower than that of the velocity. Moreover, this formulation has too many unknowns. It is not profitable to do more computations to obtain directly the stress components which are not important in Newtonian fluid flows. In case someone is interested in the shear stresses on the boundary, the shear stresses can be algebraically expressed by the vorticity. For these reasons it is not recommended that the velocity–pressure–stress formulation be used for Newtonian fluid flows.

8.6 Time-Dependent Problems in the $\boldsymbol{u}-p-\boldsymbol{\omega}$ Formulation

We now turn our attention to time-dependent problems. We consider the LSFEM for the transient Navier–Stokes equations in the velocity–pressure–vorticity formulation: Find $\boldsymbol{u}(\boldsymbol{x},t)$, $p(\boldsymbol{x},t)$ and $\boldsymbol{\omega}(\boldsymbol{x},t)$ such that

$$\frac{\partial \boldsymbol{u}}{\partial t} + \boldsymbol{u}\cdot\nabla\boldsymbol{u} + \nabla p + \frac{1}{\mathrm{Re}}\nabla\times\boldsymbol{\omega} = \boldsymbol{f} \qquad \text{in } \Omega\times(0,T], \tag{8.64a}$$

$$\nabla\cdot\boldsymbol{\omega} = 0 \qquad \text{in } \Omega\times(0,T], \tag{8.64b}$$

$$\boldsymbol{\omega} - \nabla\times\boldsymbol{u} = 0 \qquad \text{in } \Omega\times(0,T], \tag{8.64c}$$

$$\nabla\cdot\boldsymbol{u} = 0 \qquad \text{in } \Omega\times(0,T], \tag{8.64d}$$

and

$$\boldsymbol{u}(\boldsymbol{x},0) = \boldsymbol{u}^0(\boldsymbol{x}) \qquad \text{in } \Omega, \tag{8.64e}$$

where T is a given positive constant, $\boldsymbol{f}(\boldsymbol{x},t)$ is a prescribed body force, and $\boldsymbol{u}^0(\boldsymbol{x})$ is a specified initial velocity. Note that due to the absence of any time derivatives of the pressure and vorticity, no initial conditions for the pressure and vorticity are needed. The boundary conditions required by the time dependent Navier–Stokes equations are the same as for the Stokes equations listed in Table 8.1.

The time discretization can be readily realized by using finite difference methods, for example, the θ method. Let the time step be $\Delta t = t^{n+1} - t^n$. Given $\boldsymbol{u}^n$ for the previous time step, the solution $(\boldsymbol{u},p,\boldsymbol{\omega})^{n+1}$ for the current time step is determined from

$$\frac{\boldsymbol{u}^{n+1}-\boldsymbol{u}^n}{\Delta t} + \theta\Big(\boldsymbol{u}^{n+1}\cdot\nabla\boldsymbol{u}^{n+1} + \nabla p^{n+1} + \frac{1}{\mathrm{Re}}\nabla\times\boldsymbol{\omega}^{n+1}\Big)$$
$$+(1-\theta)\Big(\boldsymbol{u}^n\cdot\nabla\boldsymbol{u}^n + \nabla p^n + \frac{1}{\mathrm{Re}}\nabla\times\boldsymbol{\omega}^n\Big) = \boldsymbol{f}^{n+1} \quad \text{in } \Omega\times(0,T], \tag{8.65a}$$

$$\nabla\cdot\boldsymbol{\omega}^{n+1} = 0 \qquad \text{in } \Omega\times(0,T], \tag{8.65b}$$

$$\boldsymbol{\omega}^{n+1} - \nabla\times\boldsymbol{u}^{n+1} = 0 \qquad \text{in } \Omega\times(0,T], \tag{8.65c}$$

$$\nabla\cdot\boldsymbol{u}^{n+1} = 0 \qquad \text{in } \Omega\times(0,T]. \tag{8.65d}$$

Usually, two choices can be made: $\theta = 1/2$ corresponds to the Crank–Nicolson scheme and provides second-order accuracy $O(\Delta t^2)$, and is used for true transient problems; and $\theta = 1$ represents the backward-Euler scheme and gives only first-order accuracy $O(\Delta t)$, and may be used for a time-marching approach to obtain steady-state solutions. Both schemes are unconditionally stable, and thus there is no limitation on the size of the time step. When a steady-state solution is of interest, one may use large time steps to accelerate the convergence. When a transient solution is sought, one may choose the time step in such a way that the largest element CFL (Courant–Friedrichs–Lewy) number, defined as $U\Delta t/h$, is close to unity to avoid large errors introduced by the time-discretization.

It is worth remarking that the accuracy of the finite difference discretization mentioned above is referred to the error of the Taylor expansion of the time derivative term. In other words, this is the error of the time-discretized equation (8.65a) compared to (8.64a). In general, this type of accuracy, which is widely used in the finite difference literature, does not guarantee the same solution accuracy measured in some norms.

Now let us employ the LSFEM to find the solution of (8.65). To give some details of the formulation, we consider the two-dimensional problem in Cartesian coordinates as an example:

$$\frac{\partial u^{n+1}}{\partial x} + \frac{\partial v^{n+1}}{\partial y} = 0 \qquad \text{in } \Omega, \tag{8.66a}$$

$$\begin{aligned}&\frac{u^{n+1}-u^n}{\Delta t} + \theta\Big(u^{n+1}\,\frac{\partial u^{n+1}}{\partial x} + v^{n+1}\,\frac{\partial u^{n+1}}{\partial y} + \frac{\partial p^{n+1}}{\partial x} + \frac{1}{\mathrm{Re}}\frac{\partial \omega^{n+1}}{\partial y}\Big)\\ &+(1-\theta)\Big(u^{n}\,\frac{\partial u^{n}}{\partial x} + v^{n}\,\frac{\partial u^{n}}{\partial y} + \frac{\partial p^{n}}{\partial x} + \frac{1}{\mathrm{Re}}\frac{\partial \omega^{n}}{\partial y}\Big) = f_x^{n+1} \quad \text{in } \Omega,\end{aligned} \tag{8.66b}$$

$$\begin{aligned}&\frac{v^{n+1}-v^n}{\Delta t} + \theta\Big(u^{n+1}\,\frac{\partial v^{n+1}}{\partial x} + v^{n+1}\,\frac{\partial v^{n+1}}{\partial y} + \frac{\partial p^{n+1}}{\partial y} - \frac{1}{\mathrm{Re}}\frac{\partial \omega^{n+1}}{\partial x}\Big)\\ &+(1-\theta)\Big(u^{n}\,\frac{\partial v^{n}}{\partial x} + v^{n}\,\frac{\partial v^{n}}{\partial y} + \frac{\partial p^{n}}{\partial y} - \frac{1}{\mathrm{Re}}\frac{\partial \omega^{n}}{\partial x}\Big) = f_y^{n+1} \quad \text{in } \Omega,\end{aligned} \tag{8.66c}$$

$$\omega^{n+1} + \frac{\partial u^{n+1}}{\partial y} - \frac{\partial v^{n+1}}{\partial x} = 0 \qquad \text{in } \Omega. \tag{8.66d}$$

The convective term in (8.66) can be linearized by using the simple substitution method or Newton's method as explained in Sect. 8.3. When a small time-step is used, one Newton's linearization is often enough at each time-step. Then, for example, $u^{n+1}\partial u^{n+1}/\partial x$ in (8.66b) can be approximated as

$$u^{n+1}\,\frac{\partial u^{n+1}}{\partial x} \approx u^{n}\,\frac{\partial u^{n+1}}{\partial x} + u^{n+1}\,\frac{\partial u^{n}}{\partial x} - u^{n}\,\frac{\partial u^{n}}{\partial x}.$$

System (8.66), after linearization, can be written in standard form (8.47) for first-order systems in which

$$\underline{u} = \begin{pmatrix} u \\ v \\ p \\ \omega \end{pmatrix}^{n+1}, \quad \boldsymbol{A}_0 = \begin{pmatrix} 0 & 0 & 0 & 0 \\ \frac{1}{\Delta t} + \theta \frac{\partial u^n}{\partial x} & \theta \frac{\partial u^n}{\partial y} & 0 & 0 \\ \theta \frac{\partial v^n}{\partial x} & \frac{1}{\Delta t} + \theta \frac{\partial v^n}{\partial y} & 0 & 0 \\ 0 & 0 & 0 & 1 \end{pmatrix},$$

$$\boldsymbol{A}_1 = \begin{pmatrix} 1 & 0 & 0 & 0 \\ \theta u^n & 0 & 1 & 0 \\ 0 & \theta u^n & 0 & -\theta \frac{1}{\mathrm{Re}} \\ 0 & -1 & 0 & 0 \end{pmatrix}, \quad \boldsymbol{A}_2 = \begin{pmatrix} 0 & 1 & 0 & 0 \\ \theta v^n & 0 & 0 & \theta \frac{1}{\mathrm{Re}} \\ 0 & \theta v^n & 1 & 0 \\ 1 & 0 & 0 & 0 \end{pmatrix}, \tag{8.67}$$

$$\underline{f} = \begin{pmatrix} 0 \\ f_x^{n+1} + \frac{u^n}{\Delta t} \\ f_y^{n+1} + \frac{v^n}{\Delta t} \\ 0 \end{pmatrix}$$

$$+ \begin{pmatrix} 0 \\ -(1-\theta)(u^n \frac{\partial u^n}{\partial x} + v^n \frac{\partial u^n}{\partial y} + \frac{\partial p^n}{\partial x} + \frac{1}{\mathrm{Re}} \frac{\partial \omega^n}{\partial y}) + \theta(u^n \frac{\partial u^n}{\partial x} + v^n \frac{\partial u^n}{\partial y}) \\ -(1-\theta)(u^n \frac{\partial v^n}{\partial x} + v^n \frac{\partial v^n}{\partial y} + \frac{\partial p^n}{\partial y} - \frac{1}{\mathrm{Re}} \frac{\partial \omega^n}{\partial x}) + \theta(u^n \frac{\partial v^n}{\partial x} + v^n \frac{\partial v^n}{\partial y}) \\ 0 \end{pmatrix}.$$

If more than one linearization is needed at each time-step, the formulation will be similar to that given in Sect. 10.3.

The first-order system is then discretized in space following the standard LSFEM procedure in Sect. 4.8. The matrix-free Jacobi preconditioned conjugate gradient method (see Chap. 15) is recommended for solving the resulting linear algebraic equations. The advantage of the conjugate gradient method is that the time step can be very small so that the largest element CFL number is close to unity. In this way, time accuracy is achieved without heavy expense of computer time. This is attributed to the fact that when the time step is small, the solution at the previous time step serves as a very good initial guess for the solution at the current time-step, thus the conjugate gradient method needs fewer iterations than when a larger time step is used. Therefore, although the time-marching LSFEM is an implicit method which has the advantage of unconditional stability, when combined with the conjugate gradient method, its implementation is similar to an explicit method, and thus is relatively inexpensive.

Vortex Shedding Behind Circular Cylinder. A circular cylinder is immersed in an infinite and uniform flow field with $u = 1, v = 0$. When $\mathrm{Re} \geq 40$, a periodic oscillatory flow occurs behind the cylinder. This is the well-known von Karman vortex street problem. The interest in this problem is due to its periodic flow pattern and the unknown outflow boundary conditions.

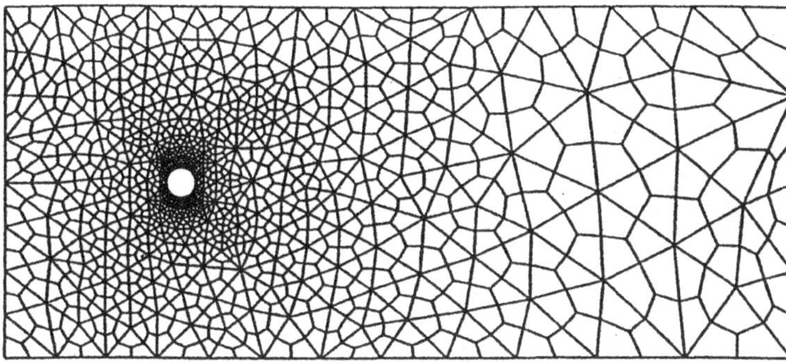

Fig. 8.19. Mesh grids for vortex shedding behind a circular cylinder (From Wu and Jiang 1995)

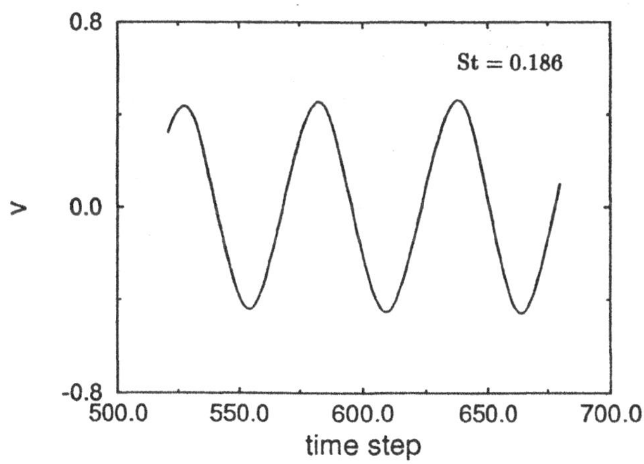

Fig. 8.20. Time history of velocity component v at point P (From Wu and Jiang 1995)

It is difficult and unnecessary to solve the problem in the infinite domain. The infinite domain is artificially truncated into a finite region ($\Omega : -6 \leq x \leq 20, -6 \leq y \leq 6$), as shown in Fig. 8.19. The circular cylinder with unit diameter is located at the point $(0,0)$. Figure 8.19 shows a mesh consisting of 1335 eight-noded quadrilateral elements and 4107 nodes. The boundary conditions are specified as $u = v = 0$ along the surface of the cylinder and $u = 1, v = 0$ on the inlet, top and bottom boundaries. No boundary conditions are given on the outflow boundary, except that the reference pressure $p = 0$ is specified at the center point of the outflow boundary. The initial condition consists of a uniform velocity distribution with $u = 1, v = 0$ everywhere,

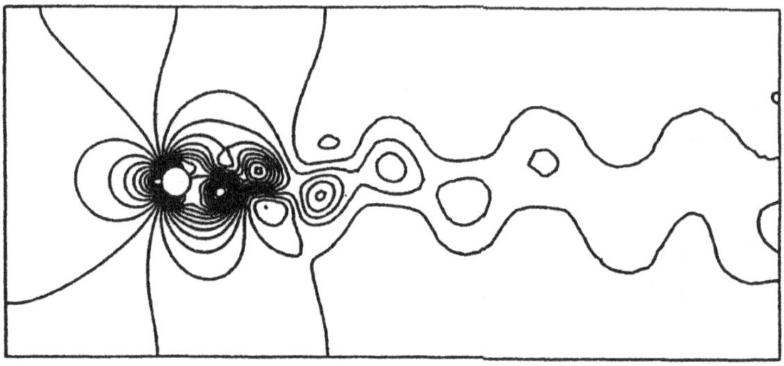

(a)

(b)

Fig. 8.21. Vortex shedding by a circular cylinder at Re = 200 at t = 150: (**a**) Pressure contours, (**b**) Streamlines (From Wu and Jiang 1995)

except at the surface of cylinder, where no-slip and impermeability conditions have been assumed.

The Reynolds number for the computation is 200 based on the diameter of the cylinder. The time step $\Delta t = 0.1$, and 2×2 Gaussian quadrature is used for the evaluation of the element matrices. A point P located about 1.5 diameters downstream from the center of the cylinder is chosen to trace the change of variables with time. The time history of the computed vertical component of velocity is presented in Fig. 8.20. The shedding period found from this numerical test is about $\tau = 5.95$; therefore the Strouhal number $\mathrm{St} = 0.186$ which is defined as

$$\mathrm{St} = \frac{D}{U\tau},$$

where D is the diameter of the cylinder and U the characteristic velocity. The computed Strouhal number falls within the range of experimental data given by Roshko (1953) and numerical results obtained by using other methods, see e.g., Engleman and Jamnia (1990). Figure 8.21 shows the instantaneous pressure contours and streamlines after the stable periodical solution is reached.

More details about the transient LSFEM solutions for this and other problems can also be found in Tang (1994).

8.7 Fluid-Thermal Coupling

We now turn from the isothermal flows that we have been investigating in previous sections to flows in which temperature variations are introduced. Consider a volume of fluid characterized by its density ρ_0, kinematic viscosity ν, thermal diffusivity κ, volumetric thermal expansion coefficient α, thermal coefficient γ of surface tension. The dynamics of the fluid is governed by the Navier–Stokes equations:

$$\frac{\partial \boldsymbol{u}}{\partial t} + \boldsymbol{u} \cdot \nabla \boldsymbol{u} + \frac{\nabla p}{\rho_0} - \nu \Delta \boldsymbol{u} - \frac{\boldsymbol{f}}{\rho_0} = \boldsymbol{0} \qquad \text{in } \Omega \times (0, T], \tag{8.68a}$$

$$\nabla \cdot \boldsymbol{u} = 0 \qquad \text{in } \Omega \times (0, T], \tag{8.68b}$$

and by the heat conduction equation:

$$\frac{\partial \theta}{\partial t} + \boldsymbol{u} \cdot \nabla \theta - \kappa \Delta \theta = 0 \qquad \text{in } \Omega \times (0, T], \tag{8.68c}$$

supplemented with appropriate initial and boundary conditions.

There are two mechanisms which can bring the fluid into motion, if the temperature $\theta(x, y, z)$ differs from some constant value θ_0. On the one hand, the buoyancy force

$$\boldsymbol{f} = -\rho_0 \alpha \boldsymbol{g}(\theta - \theta_0), \tag{8.69}$$

where $\boldsymbol{g}$ is the gravitational acceleration vector, is created due to the temperature dependence of the density in the framework of the Boussinesq approximation. On the other hand, surface forces, called thermal capillary force, are generated due to the temperature dependence of the surface tension, and can be expressed as

$$\sigma = \sigma_0 - \gamma(\theta - \theta_0). \tag{8.70}$$

The surface tension of a liquid decreases with increasing temperature. Consequently, if a part of the free surface should become locally hotter than the rest, then as a result of some small disturbance, fluid is drawn away from the region by the action of surface tension. Other hot fluid comes in from below the surface to replace it.

Introducing dimensionless variables of space, time, velocity, pressure and temperature obtained by dividing the physical variables by the reference scales L, L/U, U, $\rho_0 U^2$ and the temperature difference Θ, we obtain the following set of governing equations:

$$\frac{\partial \boldsymbol{u}}{\partial t} + \boldsymbol{u} \cdot \nabla \boldsymbol{u} + \nabla p - \frac{1}{\mathrm{Re}} \Delta \boldsymbol{u} + \frac{|\boldsymbol{g}| \alpha \Theta L^3}{U^2 L^2} \theta \frac{\boldsymbol{g}}{|\boldsymbol{g}|} = \boldsymbol{0} \quad \text{in } \Omega \times (0,T], \tag{8.71a}$$

$$\nabla \cdot \boldsymbol{u} = 0 \qquad \text{in } \Omega \times (0,T], \tag{8.71b}$$

$$\frac{\partial \theta}{\partial t} + \boldsymbol{u} \cdot \nabla \theta - \frac{1}{\mathrm{Pe}} \Delta \theta = 0 \qquad \text{in } \Omega \times (0,T], \tag{8.71c}$$

in which

$$\text{Reynolds number} \qquad \mathrm{Re} = \frac{UL}{\nu} \sim \frac{\text{inertial force}}{\text{viscous force}},$$

$$\text{Péclet number} \qquad \mathrm{Pe} = \frac{UL}{\kappa} \sim \frac{\text{advection of heat}}{\text{conduction of heat}}.$$

Depending on the characteristics of the fluid flows, the nondimensional governing equations (8.71) may be further simplified.

8.7.1 Natural Convection

Natural convection is caused by buoyancy forces. Temperature differences are introduced, for example, through boundaries maintained at different temperatures, and the consequent density differences induce the motion: hot fluid tends to rise, cold to fall.

We will confine attention to the case where the Péclet number Pe = 1 which may be interpreted as assigning the same importance to heat advection and conduction. In other words, in this case we can use the thermal diffusion speed $U = \kappa/L$ as the velocity scale. Thus the governing equations (8.71) become

$$\frac{\partial \boldsymbol{u}}{\partial t} + \boldsymbol{u} \cdot \nabla \boldsymbol{u} + \nabla p - \mathrm{Pr} \Delta \boldsymbol{u} + \mathrm{Pr}\mathrm{Ra} \theta \frac{\boldsymbol{g}}{|\boldsymbol{g}|} = \boldsymbol{0} \qquad \text{in } \Omega \times (0,T], \tag{8.72a}$$

$$\nabla \cdot \boldsymbol{u} = 0 \qquad \text{in } \Omega \times (0,T], \tag{8.72b}$$

$$\frac{\partial \theta}{\partial t} + \boldsymbol{u} \cdot \nabla \theta - \Delta \theta = 0 \qquad \text{in } \Omega \times (0,T], \tag{8.72c}$$

in which

$$\text{Prandtl number} \qquad \mathrm{Pr} = \frac{\nu}{\kappa},$$

$$\text{Rayleigh number} \qquad \mathrm{Ra} = \frac{|\boldsymbol{g}| \alpha L^3 \Theta}{\nu \kappa}.$$

The Prandtl number is the ratio of two diffusivities – ν being the diffusivity of momentum and vorticity and κ that of heat. Thus the Prandtl number is a property of the fluid, not of the particular flow. The Rayleigh number indicates the relative importance of the buoyancy force.

In order to apply LSFEM to obtain the approximate solution of (8.72), we need to introduce the vorticity $\boldsymbol{\omega}$ and the heat flux $\boldsymbol{q} = (q_x, q_y, q_z)$ as independent variables and transform system (8.72) into the following first-order $\boldsymbol{u} - p - \boldsymbol{\omega} - \theta - \boldsymbol{q}$ system:

$$\frac{\partial \boldsymbol{u}}{\partial t} + \boldsymbol{u} \cdot \nabla \boldsymbol{u} + \nabla p + \mathrm{Pr} \nabla \times \boldsymbol{\omega} + \mathrm{PrRa}\theta \frac{\boldsymbol{g}}{|\boldsymbol{g}|} = \boldsymbol{0} \quad \text{in } \Omega \times (0, T], \tag{8.73a}$$

$$\nabla \cdot \boldsymbol{\omega} = 0 \qquad \text{in } \Omega \times (0, T], \tag{8.73b}$$

$$\boldsymbol{\omega} - \nabla \times \boldsymbol{u} = \boldsymbol{0} \qquad \text{in } \Omega \times (0, T], \tag{8.73c}$$

$$\nabla \cdot \boldsymbol{u} = 0 \qquad \text{in } \Omega \times (0, T], \tag{8.73d}$$

$$\frac{\partial \theta}{\partial t} + \boldsymbol{u} \cdot \nabla \theta + \nabla \cdot \boldsymbol{q} = 0 \qquad \text{in } \Omega \times (0, T], \tag{8.73e}$$

$$\boldsymbol{q} + \nabla \theta = \boldsymbol{0} \qquad \text{in } \Omega \times (0, T], \tag{8.73f}$$

$$\nabla \times \boldsymbol{q} = \boldsymbol{0} \qquad \text{in } \Omega \times (0, T]. \tag{8.73g}$$

Equations (8.73e)–(8.73g) constitute a div–curl–grad system. As explained in Chap. 6, the inclusion of the curl equation (8.73g) is for more accurate results for the heat flux, and more important, for better performance of the conjugate gradient method. It is also worth mentioning that the addition of (8.73g) and (8.73b) does not increase the total number of degrees of freedom in LSFEM, but reduces the computational work when an iterative equation solver is used.

As an example of the application of the $\boldsymbol{u} - p - \boldsymbol{\omega} - \theta - \boldsymbol{q}$ formulation (8.73) for natural convection, we consider the steady-state, two-dimensional, thermally driven flow in a square cavity with insulated top and bottom walls but differentially heated vertical walls. The gravitational force is in the negative y direction. The boundary conditions are: $u = v = q_y = 0$ everywhere, $\theta = 1$ on the left wall and $\theta = 0$ on the right wall, the reference pressure $p = 0$ is specified at the center of the bottom. A 50×50 nonuniform mesh with bilinear elements is used. One-point Gaussian quadrature is employed for computing the element matrices. The LSFEM solution for the Prandtl number $\mathrm{Pr} = 0.71$ (air) and the Rayleigh number $\mathrm{Ra} = 10^6$ is given in Fig. 8.22. These results are obtained by taking the solution at $\mathrm{Ra} = 10^5$ as the initial guess, then using two successive substitution and ten Newton's linearization steps. In each linearization step, 1500 conjugate gradient iterations are employed.

8.7.2 Rayleigh–Bénard Convection

Consider a volume of fluid in a three-dimensional rectangular vessel with geometric aspect ratio $a_x : a_y : a_z$ where $a_x = l_x/d$, $a_y = l_y/d$, $a_z = 1$,

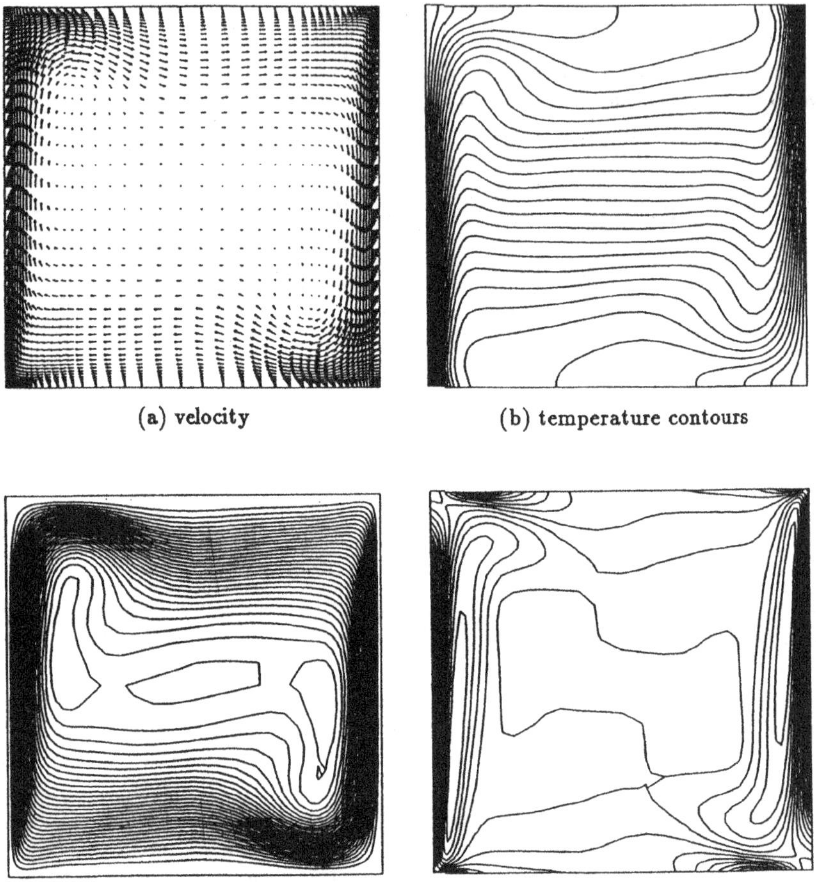

(a) velocity (b) temperature contours

(c) streamlines (d) vorticity contours

Fig. 8.22a–d. Numerical results for thermally driven-cavity flow at $\mathrm{Pr} = 0.71$ and $\mathrm{Ra} = 10^6$

and $d = 1$ is the thickness of the vessel, see Fig. 8.23. The gravitational force is in the negative z direction. The vertical walls are adiabatic. The bottom wall is heated isothermally to a higher temperature θ_h, and the top wall is maintained at a lower temperature θ_c. The fluid starts motion when the Rayleigh number Ra reaches the critical value. This is termed Rayleigh–Bénard convection.

For fluid between infinite parallel plates, the critical Rayleigh number $\mathrm{Ra}_c = 1708$. Theoretical (Davis 1967 and Catton 1972) and experimental (Stork and Müller 1972) results show that the critical value in small vessels

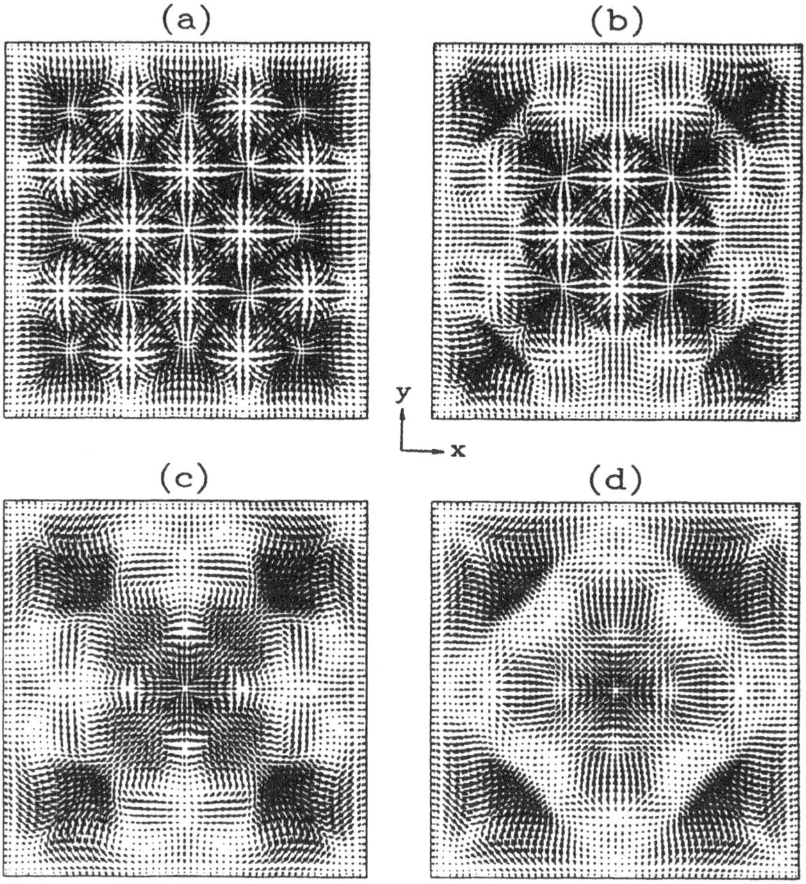

Fig. 8.23. Velocity vectors on the plane $z = 0.5$ at t = (**a**) 25, (**b**) 45, (**c**) 60 and (**d**) Steady state (From Tang and Tsang 1997)

is higher than in large one. In general, the critical Rayleigh number increases when the aspect ratio of the container decreases, because sidewalls have a stablizing effect on the onset of convection due to viscous drag. Moreover, the roll structures and thermal fields are decisively affected by the aspect ratio of the vessel, thermal properties of the sidewalls, and the physical properties of the fluid. The most interesting phenomenon of Rayleigh–Bénard convection is the pattern selection process. Due to the nonlinearity of the governing equations, several solutions are possible for the same set of parameters. The mechanisms on how different solutions to the problem evolve and compete and the processes by which a particular flow configuration undergoes change are broadly known as pattern selection. A numerical study using LSFEM that

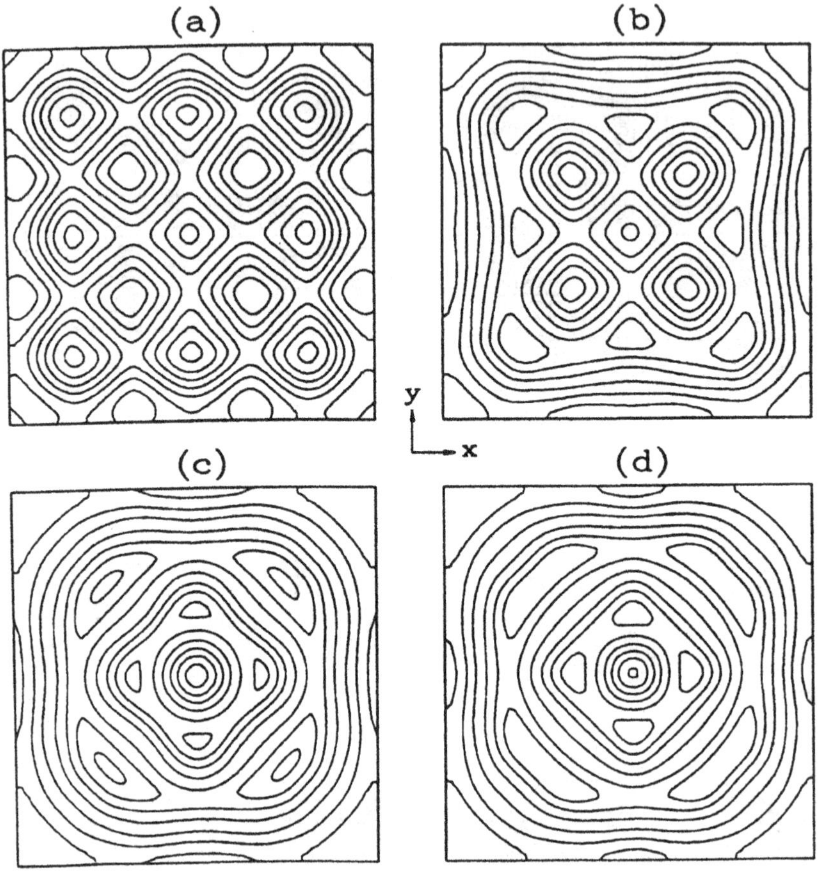

Fig. 8.24. Temperature contours on the plane $z = 0.5$ at t = (**a**) 25, (**b**) 45, (**c**) 60 and (**d**) Steady state (From Tang and Tsang 1997)

focuses on some issues relevant to pattern selection was conducted by Tang and Tsang (1997). The presentation of this subsection follows their work.

When nondimensionalized by using the buoyant speed $U = (|g|\alpha\Theta L)^{1/2}$ as the velocity scale and the length scale $L = d$, the first-order governing equations in the $\boldsymbol{u} - p - \boldsymbol{\omega} - \theta - \boldsymbol{q}$ formulation are

$$\frac{\partial \boldsymbol{u}}{\partial t} + \boldsymbol{u} \cdot \nabla \boldsymbol{u} + \nabla p + \frac{1}{\text{Re}} \nabla \times \boldsymbol{\omega} + \theta \frac{\boldsymbol{g}}{|\boldsymbol{g}|} = \boldsymbol{0} \qquad \text{in } \Omega \times (0, T], \tag{8.74a}$$

$$\nabla \cdot \boldsymbol{\omega} = 0 \qquad \text{in } \Omega \times (0, T], \tag{8.74b}$$

$$\boldsymbol{\omega} - \nabla \times \boldsymbol{u} = 0 \qquad \text{in } \Omega \times (0, T], \tag{8.74c}$$

$$\nabla \cdot \boldsymbol{u} = 0 \qquad \text{in } \Omega \times (0, T], \tag{8.74d}$$

$$\frac{\partial \theta}{\partial t} + \boldsymbol{u} \cdot \nabla \theta + \nabla \cdot \boldsymbol{q} = 0 \qquad \text{in } \Omega \times (0, T], \tag{8.74e}$$

$$\boldsymbol{q} + \frac{1}{\text{Pe}} \nabla \theta = \boldsymbol{0} \qquad \text{in } \Omega \times (0, T], \tag{8.74f}$$

$$\nabla \times \boldsymbol{q} = \boldsymbol{0} \qquad \text{in } \Omega \times (0, T], \tag{8.74g}$$

where $\text{Re} = (\text{Ra}/\text{Pr})^{1/2}$ and $\text{Pe} = (\text{Ra} \cdot \text{Pr})^{1/2}$.

Initially, it is assumed that the physical problem is at a low Rayleigh number $\text{Ra} < \text{Ra}_c$. The fluid remains motionless, the heat conduction is dominant and the temperature distribution is linear along the z direction:

$$\begin{aligned} \boldsymbol{u} &= \boldsymbol{0}, \\ \theta &= 1 - z \qquad \text{in } \Omega \text{ at } t = 0, \\ \boldsymbol{q} &= (0, 0, 1/\text{Pe}). \end{aligned}$$

At the very moment when $t > 0$, the Rayleigh number is suddenly increased to a certain value $\text{Ra} > \text{Ra}_c$ so that the fluid starts moving inside the domain. The boundary conditions consistent with the adiabatic and isothermal walls in dimensionless form are as follows:

$$\begin{aligned} \boldsymbol{u} &= \boldsymbol{0} && \text{on } \Gamma, \\ q_x &= 0 && \text{on } x = 0, a_x, \\ q_y &= 0 && \text{on } y = 0, a_y, \\ \theta &= 1 && \text{on } z = 0, \\ q_x &= 0 && \text{on } z = 0, \\ q_y &= 0 && \text{on } z = 0, \\ \theta &= 0 && \text{on } z = 1, \\ q_x &= 0 && \text{on } z = 1, \\ q_y &= 0 && \text{on } z = 1. \end{aligned}$$

The numerical technique is based on an implicit, fully coupled, and time accurate method, which consists of the Crank–Nicolson scheme for time integration, Newton's method for the convective terms with extensive linearization steps, and the LSFEM for space discretization. A matrix-free Jacobi preconditioned conjugate gradient method (see Chap. 15) is implemented to solve the resulting symmetric positive definite linear system of equations. For the solution illustrated in Fig. 8.23, $\text{Ra} = 8 \times 10^3$, $\text{Pr} = 0.71$ (air), the geometric aspect ratio is $a_x : a_y : a_z = 5 : 5 : 1$, and the mesh consists of $60 \times 60 \times 14$ trilinear elements. One-point Gaussian quadrature is used for the computation.

The velocity vectors and temperature contours on the plane $z = 0.5$ are shown in Fig. 8.23 and Fig. 8.24, respectively. Figure 8.25 shows the results on the plane y=0.5. When steady-state is reached, three vortex tubes, i.e.,

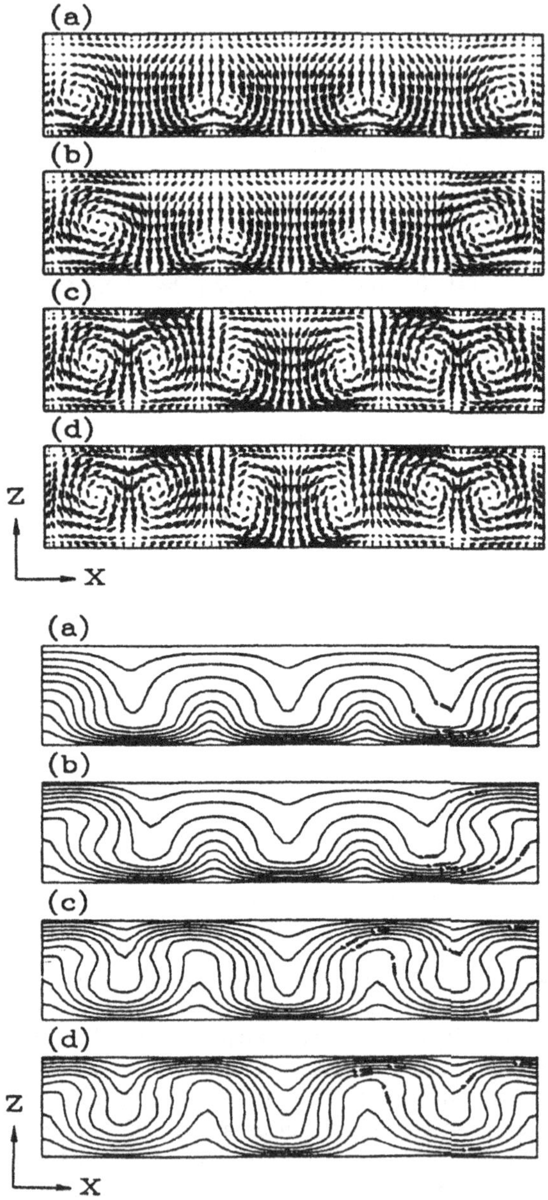

Fig. 8.25. Velocity vectors and temperature contours on the plane $y = 0.5$ at t = (**a**) 25, (**b**) 45, (**c**) 60 and (**d**) Steady state (From Tang and Tsang 1997)

six rolls in a vertical cross section, ring around the vertical center line of the vessel can be clearly observed. The flow structure is referred to as a multicellular flow pattern. More results for different cases can be found in Tang and Tsang (1997).

8.7.3 Surface-Tension-Driven Convection

In the previous two subsections we have focused on buoyancy-driven convection. This type of fluid flows is observed either in deep layers or in fluid layers enclosed between two plates with no free surface. When a temperature gradient is applied orthogonally to a thin planar liquid layer with a free surface, cellular convection occurs from an originally quiescent state. The onset of the convection is due to the combined effects of thermal stratification instability and thermal capillary effect. In particular, the temperature dependence of the surface tension on the free surface can destabilize the motionless fluid state to form regular convective cells. Usually, the diameters of these cells are of the same order of magnitude as compared to the depth of the fluid. This surface-tension-driven convection is often referred to as the Marangoni–Bénard instability due to the first reported observation of the flow phenomenon by Bénard.

In the past, extensive experimental studies of surface-tension-driven convection using silicon oil as the working fluid have been conducted by Koschmieder and his colleagues (1967, 1986, 1990, 1992). Theoretical studies of surface-tension-driven convection have been focused on stability analysis by using the linear theory (Pearson 1958, Nield 1964), the energy theory (Davis 1969) and the bifurcation theory (Scanlon and Segal 1967, Cloot and Lebon 1984). While these studies have significantly improved our understanding of the flow physics, numerical simulation of the flow phenomena is necessary for further investigation. Full flow equations can be numerically solved without assumptions and simplifications usually employed in stability analysis. The complications caused by buoyancy for experiments on Earth and the non-planar layer for experiments in Space can also be avoided. Moreover, direct simulation is an indispensable tool for investigating the regime of supercritical flows for which few studies have been conducted.

Only a few numerical studies exist for three-dimensional surface-tension-driven convection. By using a spectral method, Bestehorn (1993) presented numerical results for the planform evolution in containers with very large aspect ratios. Thess and Orszag (1995) reported numerical results of their pseudospectral method for an infinite and periodic layer, in which the flow motion is solely determined by the temperature distribution on the free surface so that the calculation was simplified.

Recently, Yu et al. (1996c) applied LSFEM to the full three-dimensional simulation of the surface-tension-driven convection in small square containers. In order to recapture the unusual cellular patterns observed by Koschmieder and Prahl (1990), the no-slip boundary condition on the peripheral walls must be used. In Table 8.4 the properties of the silicon oil used in their experiments are listed. According to these data, the Prandtl number of the silicon oil $\mathrm{Pr} = \nu/\kappa$ is about 1000, and the capillary number $\mathrm{Cc} = \rho_0\kappa\nu/\gamma d$ is about one thousandth. Note that d is the depth of the fluid layer and is in the order of millimeter. Since the flow motion is thermally driven, the Prandtl number is a measure of the sluggishness of the fluid: higher Prandtl number implies slower motion. On the other hand, in the absence of gravity, the capillary number Cc is a measure of the surface deflection. And smaller Cc implies higher surface tension that corresponds to a non-deflecting free surface.

Table 8.4. Properties of the silicon oil at 25°C

Symbol	Property	Unit	Value
ν	Viscosity	cm^2/s	1.0
ρ	Density	g/cm^3	0.968
κ	Thermal Diffusivity	cm^2/s	0.001095
γ	Surface Tension Coefficient	$dyne/\mathrm{cm}$	13.96

Accordingly Yu et al. (1996c) made two assumptions in their calculation: (1) the Prandtl number of the working fluid is large, and therefore the Stokes equations instead of the full Navier–Stokes equations can be used; (2) the capillary number is small and the free surface of liquid layer is flat. As a result, (8.68) of the fluid-thermal coupling with zero gravity become:

$$\frac{\nabla p}{\rho_0} - \nu\Delta \boldsymbol{u} = \boldsymbol{0} \qquad \text{in } \Omega \times (0, T], \tag{8.75a}$$

$$\nabla \cdot \boldsymbol{u} = 0 \qquad \text{in } \Omega \times (0, T], \tag{8.75b}$$

$$\frac{\partial \theta}{\partial t} + \boldsymbol{u} \cdot \nabla\theta - \kappa\Delta\theta = 0 \qquad \text{in } \Omega \times (0, T]. \tag{8.75c}$$

The boundary condition on the bottom and the side walls of the container is the no-slip condition:

$$u = v = w = 0. \tag{8.75d}$$

In addition, the bottom is heated to a constant temperature θ_1, and the side walls are insulated:

$$\theta = \theta_1 \qquad \text{on the bottom}, \tag{8.75e}$$

$$\boldsymbol{n} \cdot \nabla\theta = 0 \qquad \text{on the sidewalls}. \tag{8.75f}$$

On the top free surface, the Marangoni boundary conditions are applied:

$$\rho_0 \nu \frac{\partial u}{\partial z} = -\gamma \frac{\partial \theta}{\partial x}, \tag{8.75g}$$

$$\rho_0 \nu \frac{\partial v}{\partial z} = -\gamma \frac{\partial \theta}{\partial y}, \tag{8.75h}$$

$$w = 0. \tag{8.75i}$$

The Marangoni conditions represent the relationship between the flow shear stress and the tangential surface tension force across the free surface. Any inhomogeneity of the surface tension due to temperature variations creates a shear force on the free surface and therefore results in fluid motion. These Marangoni conditions are the driving force for Marangoni–Bénard convection.

The heat loss on the free surface is modeled by the usual heat transfer condition:

$$\rho_0 \kappa C_p \frac{\partial \theta}{\partial z} = -C_h(\theta - \theta_0), \tag{8.75j}$$

where C_h is a heat transfer coefficient, C_p is the constant pressure specific heat, and θ_0 is the prescribed cold temperature of the ambient air. The boundary heat transfer mechanism on the free surface could be conduction, convection and radiation, or combination of these effects.

For convenience, the energy equation is reformulated in terms of the temperature perturbation $\bar{\theta}$ defined via

$$\theta = \theta_1 - \frac{z}{d}(\theta_1 - \theta_0) + \bar{\theta}. \tag{8.76}$$

As a result, the energy equation (8.75c) becomes:

$$\frac{\partial \bar{\theta}}{\partial t} + \boldsymbol{u} \cdot \nabla \bar{\theta} - \frac{w\Theta}{d} - \kappa \Delta \bar{\theta} = 0 \qquad \text{in } \Omega \times (0, T], \tag{8.77}$$

in which w is the vertical component of the velocity vector, and $\Theta = \theta_1 - \theta_0$. In the following we suppress the bar notation for temperature perturbation for simplicity.

The governing equations and the boundary conditions are then nondimensionalized by the appropriate parameters. The reference parameters for the length, time, velocity, pressure and temperature are d, d^2/κ, κ/d, $\rho_0(\kappa/d)^2$ and Θ. As a result, the dimensionless governing equations are

$$\nabla p - \Delta \boldsymbol{u} = \boldsymbol{0} \qquad \text{in } \Omega \times (0, T], \tag{8.78a}$$

$$\nabla \cdot \boldsymbol{u} = 0 \qquad \text{in } \Omega \times (0, T], \tag{8.78b}$$

$$\frac{\partial \theta}{\partial t} + \boldsymbol{u} \cdot \nabla \theta - w - \kappa \Delta \theta = 0 \qquad \text{in } \Omega \times (0, T]. \tag{8.78c}$$

The dimensionless Marangoni boundary conditions are

$$\frac{\partial u}{\partial z} = -\mathrm{Ma} \frac{\partial \theta}{\partial x}, \tag{8.79a}$$

$$\frac{\partial v}{\partial z} = -\mathrm{Ma}\frac{\partial \theta}{\partial y}, \tag{8.79b}$$

where

$$\mathrm{Ma} = \frac{\gamma \Theta d}{\rho \nu \kappa} \tag{8.80}$$

is the Marangoni number.

Similarly, after the non-dimensionalization the heat convection condition on the free surface becomes

$$\frac{\partial \theta}{\partial z} + \mathrm{B_i}\theta = 0, \tag{8.81}$$

where

$$\mathrm{B_i} = \frac{C_h d}{\kappa \rho_0 C_p} \tag{8.82}$$

is the Biot number, which is a dimensionless measure of the heat loss on the free surface. Since all heat transfer on the free surface is mainly through the gradient of the linear distribution of the undisturbed temperature, it is reasonable to assume that $\mathrm{B_i} = 0$, which implies that heat transfer to the ambient air through the temperature fluctuation θ on the free surface is negligible. As a result, the boundary condition for the energy equation on the free surface becomes

$$\frac{\partial \theta}{\partial z} = 0. \tag{8.83}$$

By introducing the vorticity $\boldsymbol{\omega}$ and the heat flux $\boldsymbol{q}$ as independent variables the governing equations are reduced to the following first-order $\boldsymbol{u} - p - \boldsymbol{\omega} - \theta - \boldsymbol{q}$ system:

$$\nabla p + \nabla \times \boldsymbol{\omega} = \mathbf{0} \qquad \text{in } \Omega \times (0, T], \tag{8.84a}$$

$$\nabla \cdot \boldsymbol{\omega} = 0 \qquad \text{in } \Omega \times (0, T], \tag{8.84b}$$

$$\boldsymbol{\omega} - \nabla \times \boldsymbol{u} = \mathbf{0} \qquad \text{in } \Omega \times (0, T], \tag{8.84c}$$

$$\nabla \cdot \boldsymbol{u} = 0 \qquad \text{in } \Omega \times (0, T], \tag{8.84d}$$

$$\frac{\partial \theta}{\partial t} + \boldsymbol{u} \cdot \boldsymbol{q} - \nabla \cdot \boldsymbol{q} - w = 0 \qquad \text{in } \Omega \times (0, T], \tag{8.84e}$$

$$\boldsymbol{q} - \nabla \theta = \mathbf{0} \qquad \text{in } \Omega \times (0, T], \tag{8.84f}$$

$$\nabla \times \boldsymbol{q} = \mathbf{0} \qquad \text{in } \Omega \times (0, T]. \tag{8.84g}$$

Equation (8.84) consist of two div–curl systems and one div–curl–grad system, i.e., (8.84a) and (8.84b) constitute a div–curl system for the vorticity; (8.84c) and (8.84d) constitute a div–curl system for the velocity; and (8.84e)–(8.84f) constitute a div–curl–grad system for the temperature fluctuation and the heat flux.

Table 8.5. Boundary conditions of surface-tension-driven convection

Conditions	Flow Equations	Heat Equations
Heated Bottom	$u = v = w = 0$ $\omega_z = 0$	$\theta = 0,\ q_x = q_y = 0$
Insulated Walls	$u = v = w = 0$	$\boldsymbol{n} \cdot \boldsymbol{q} = 0$
Free Surface	$w = 0,$ $\omega_x = \mathrm{Ma} q_y,\ \omega_y = -\mathrm{Ma} q_x$	$q_z = 0$

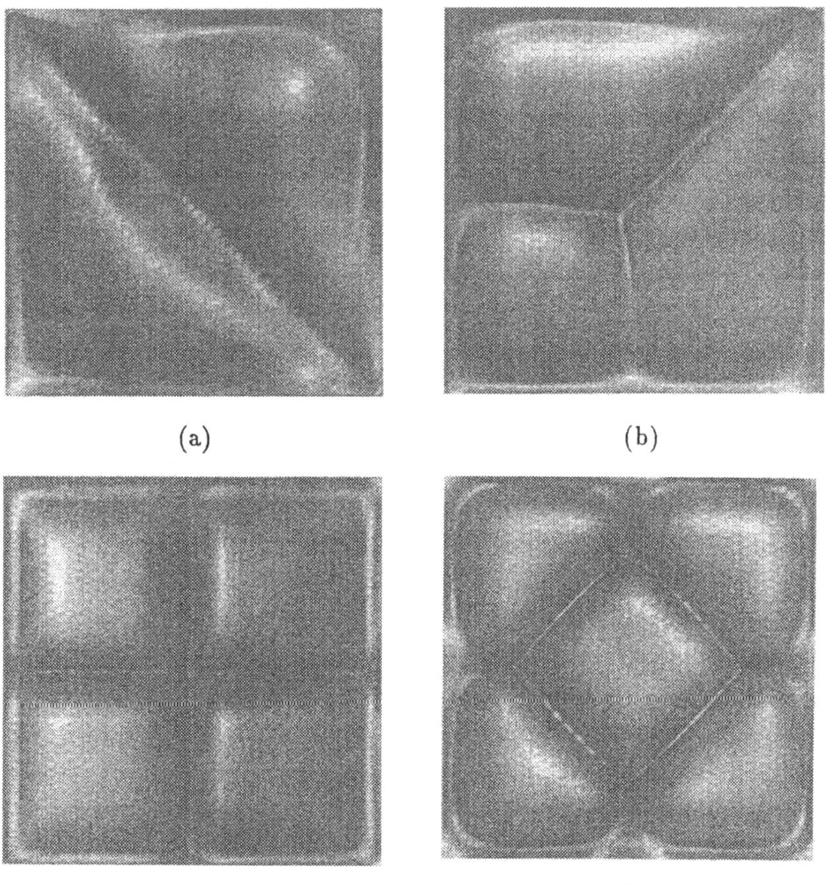

(a) (b) (c) (d)

Fig. 8.26. The computed patterns in the square container: (**a**) Two-cell, (**b**) Three-cell, (**c**) Four-cell, (**d**) Five-cell (From Yu et al. 1996c)

The dimensionless boundary conditions are listed in Table 8.5.

The initial temperature field is the same as that proposed by Thess and Orszag (1995):

$$\theta(x, y, z, 0) = \epsilon(x, y)z(2 - z). \tag{8.85}$$

The field $\epsilon(x, y)$ is the superposition of all Fourier modes supported by the numerical mesh. The magnitude of $\epsilon(x, y)$ is set to be one thousandth. All other flow variables are initialized to zero. As time evolves, the numerical procedure will pick up the most unstable mode and suppress others. In addition, the calculations are started with very low Ma which is gradually increased until the onset of convection.

The computation is for the flow in small containers with the aspect ratio $a \leq 9$ for which no previous numerical study has been reported. A mesh with $50 \times 50 \times 18$ trilinear elements is employed. The mesh is uniform in the x and y directions, and is clustered near the free surface in the z direction.

An implicit, fully coupled backward-Euler time marching method discussed in Sect. 8.6 is used. In each time step, three to five Newton linearization steps are performed for the convective term in the heat equation. For each Newton step about 300 iterations of the Jacobi preconditioned conjugate gradient method are taken to obtain the solution of the linear algebraic equations. Typically, it takes about 50 to 100 time steps to make the difference between the nodal values of two consecutive solutions converge to about 10^{-4}.

The LSFEM solution of the two, three, four and five-cell Marangoni–Bénard convection is shown in Fig. 8.26. The planform shown here is actually the smoothed contour plot of the velocity profile just beneath the free surface. The upward bulging portion of these shadowgraph images represents the rising flow motion, and the valley is the downward flow. All these computed patterns are the same as those observed in the laboratory by Koschmieder and Prahl (1990). Due to the existence of vertical walls, these patterns are quite different from the general conception of the hexagonal surface-tension driven convection pattern usually observed in the container with very large aspect ratio.

8.7.4 Double-Diffusive Convection

We turn now to situations in which, in addition to the temperature variation, there is a variable amount of some substance mixed with, dissolved in, or otherwise carried by the fluid. The variable salinity (salt concentration) of the oceans can be dynamically important; air pollution studies require understanding of the processes determining pollutant concentration; chemical engineering processes often involve the mixing of different substances. The interaction between concentration variations and a velocity field is closely analogous to that between temperature variations and a velocity field. The two situations are governed by very similar equations.

The dynamics of the fluid flow is governed by the Navier–Stokes equations (8.68a) and (8.68b), the heat conduction equation (8.68c) and the following mass transfer equation:

$$\frac{\partial c}{\partial t} + \boldsymbol{u} \cdot \nabla c - \kappa_c \Delta c = 0 \qquad \text{in } \Omega \times (0, T], \tag{8.86}$$

in which c is the concentration of the substance and κ_c is the binary mass diffusion coefficient. In the framework of the Boussinesq approximation, the driving force in (8.68a) is now generated due to the fact that both temperature and concentration depend on the density:

$$\boldsymbol{f} = -\rho_0 \alpha \boldsymbol{g}(\theta - \theta_0) - \rho_0 \alpha_c \boldsymbol{g} c, \tag{8.87}$$

where the density ρ_0 corresponds to the temperature θ_0 and $c = 0$; α_c is the coefficient of volumetric expansion due to the concentration.

By using the buoyant speed $U = (|\boldsymbol{g}|\alpha\Theta L)^{1/2}$ as the velocity scale, and the characteristic concentration difference Θ_c as the reference concentration, and by introducing the mass flux $\boldsymbol{J} = (J_x, J_y, J_z)$, the governing equations can be written in dimensionless first-order velocity–pressure–vorticity–temperature–heat flux–concentration–mass flux formulation:

$$\frac{\partial \boldsymbol{u}}{\partial t} + \boldsymbol{u} \cdot \nabla \boldsymbol{u} + \nabla p + \frac{1}{\sqrt{\mathrm{Gr}}} \nabla \times \boldsymbol{\omega} + (\theta + \chi c) \frac{\boldsymbol{g}}{|\boldsymbol{g}|} = \boldsymbol{0} \ \text{ in } \Omega \times (0, T], \tag{8.88a}$$

$$\nabla \cdot \boldsymbol{\omega} = 0 \qquad \text{in } \Omega \times (0, T], \tag{8.88b}$$

$$\boldsymbol{\omega} - \nabla \times \boldsymbol{u} = 0 \qquad \text{in } \Omega \times (0, T], \tag{8.88c}$$

$$\nabla \cdot \boldsymbol{u} = 0 \qquad \text{in } \Omega \times (0, T], \tag{8.88d}$$

$$\frac{\partial \theta}{\partial t} + \boldsymbol{u} \cdot \nabla \theta + \nabla \cdot \boldsymbol{q} = 0 \qquad \text{in } \Omega \times (0, T], \tag{8.88e}$$

$$\boldsymbol{q} + \frac{1}{\mathrm{Pr}\sqrt{\mathrm{Gr}}} \nabla \theta = \boldsymbol{0} \qquad \text{in } \Omega \times (0, T], \tag{8.88f}$$

$$\nabla \times \boldsymbol{q} = \boldsymbol{0} \qquad \text{in } \Omega \times (0, T], \tag{8.88g}$$

$$\frac{\partial c}{\partial t} + \boldsymbol{u} \cdot \nabla c + \nabla \cdot \boldsymbol{J} = 0 \qquad \text{in } \Omega \times (0, T], \tag{8.88h}$$

$$\boldsymbol{J} + \frac{1}{\mathrm{Sc}\sqrt{\mathrm{Gr}}} \nabla c = \boldsymbol{0} \qquad \text{in } \Omega \times (0, T], \tag{8.88i}$$

$$\nabla \times \boldsymbol{J} = \boldsymbol{0} \qquad \text{in } \Omega \times (0, T], \tag{8.88j}$$

where the dimensionless numbers are defined as

$$
\begin{aligned}
\text{Schmit number} \quad & \mathrm{Sc} &=& \ \frac{\nu}{\kappa_c}, \\
\text{Lewis number} \quad & \mathrm{Le} &=& \ \frac{\kappa}{\kappa_c} = \frac{\mathrm{Sc}}{\mathrm{Pr}}, \\
\text{Thermal Grashof number} \quad & \mathrm{Gr} &=& \ \frac{\alpha|\boldsymbol{g}|L^3\Theta}{\nu^2}, \\
\text{Solute Grashof number} \quad & \mathrm{Gr}_c &=& \ \frac{\alpha_c|\boldsymbol{g}|\kappa_c L^3\Theta_c}{\nu^2\kappa}, \\
\text{Buoyancy ratio} \quad & \chi &=& \ \frac{\alpha_c\Theta_c}{\alpha\Theta} = \frac{\mathrm{Gr}_c\cdot\mathrm{Sc}}{\mathrm{Gr}\cdot\mathrm{Pr}}, \\
\text{Rayleigh number} \quad & \mathrm{Ra} &=& \ \frac{\alpha|\boldsymbol{g}|L^3\Theta}{\kappa\nu} = \mathrm{Gr}\cdot\mathrm{Pr}.
\end{aligned}
$$

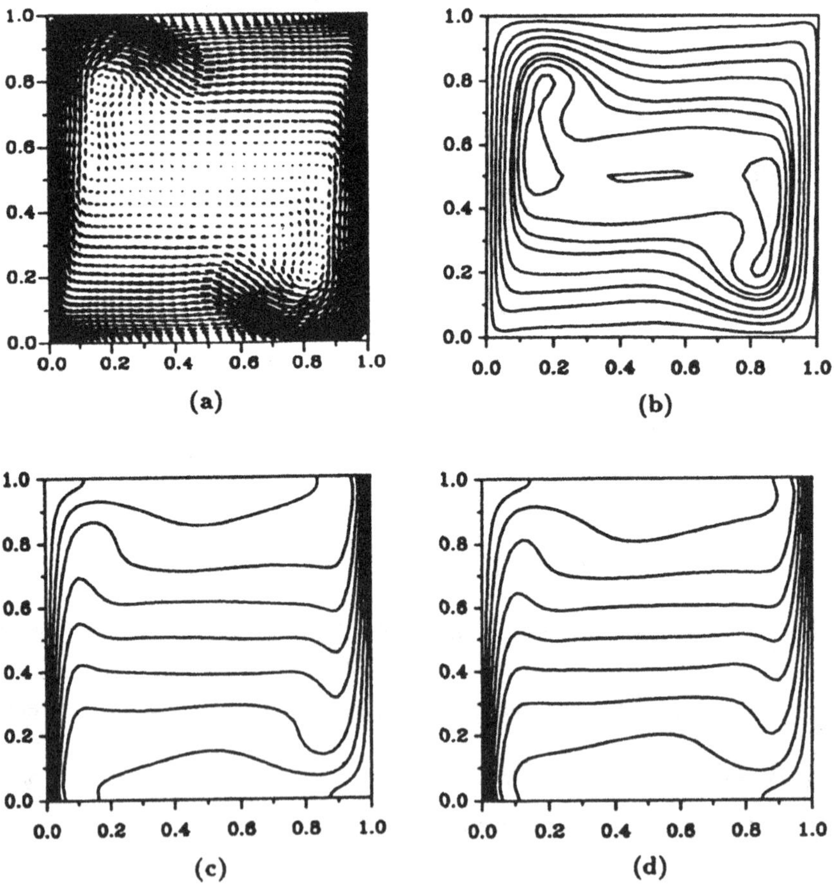

Fig. 8.27. Augmented flow of double-diffusive convection in a square cavity at $Gr = 10^6$ and $\chi = -1.0$: (**a**) Velocity vectors, (**b**) Stream functions, (**c**) Temperature contours, (**d**) Concentration contours (From Tang and Tsang 1994)

The LSFEM solution of the coupled system for heated side-walls and diffusion in a square cavity is given by Tang and Tsang (1994). The cavity has insulated and impermeable top and bottom walls but with heated and solute coated vertical walls. The fluid motion driven by temperature and solute gradients consists of two flow structures: augmenting and opposing flows. Each flow structure has two cases, $\mathrm{Gr}_c > 0$ and $\mathrm{Gr}_c < 0$. In their study, the solute is the heavier component. Therefore, the buoyancy ratio χ is always negative, i.e., $\mathrm{Gr}_c < 0$. Here we only discuss augmented flows. Results for opposing flows can be found in Tang and Tsang (1994).

Augmented flow is the flow driven vertically up by high temperature near the left wall and down by high concentration near the right wall. The fluid flow is always in the clockwise direction. The boundary conditions are: $u = v = q_y = J_y = 0$ everywhere; $\theta = 1,\ c = 0$ at the left wall and $\theta = 0,\ c = 1$ at the right wall. The reference pressure $p = 0$ is specified at the lower left corner. A non-uniform mesh of 50 × 50 bilinear elements is used. Since the steady-state solution is sought, the system is discretized in time by the backward-Euler differencing. A large time step $\Delta t = 1.0$ is applied. One-point Gaussian quadrature is used in the computation. The results for the case of $\mathrm{Gr} = 10^6$, $\chi = -1.0$, $\mathrm{Pr} = 0.7$ and $\mathrm{Sc} = 0.6$ are given in Fig. 8.27. These results compare well with Béghein et al. (1992) and Lin et al. (1990).

8.8 The Second-Order $\boldsymbol{u} - \boldsymbol{\omega}$ Formulation

In the previous sections we have briefly studied the velocity–pressure formulation, and have investigated in detail various first-order formulations of the Navier–Stokes equations. For incompressible viscous flows there are other useful higher-order versions of the Navier–Stokes equations (Gresho 1992). These are all derived from the "primitive" velocity–pressure equations by differentiation and often offer additional insights regarding fluid flow fields. They also serve as the starting point for devising alternative numerical schemes. A key issue is that a derived equation obtained by simple differentiation admits more solutions than do its progenitors, in other words, spurious solutions may be generated. With careful selection of additional equations and boundary conditions, the solution of the derived equations will also solve the original equations. Unfortunately, in the past, these additional conditions were often ignored or incorrect. In the best cases, they are discovered and/or justified mainly by numerical experience and heuristics.

In this section, we will show that the least-squares method provides a systematic and consistent way to derive higher-order versions of differential equations and corresponding boundary conditions without generating spurious solutions. That is, the Euler–Lagrange equations associated with the least-squares variational formulations are the most appropriate derived equations.

We will also use the div–curl method introduced in Sect. 5.3 to rigorously derive the popular second-order velocity–vorticity formulation, the pressure Poisson equation and their boundary conditions. We believe that many controversies over the permissibility of the boundary conditions for derived equations can be resolved definitively via this div–curl method.

8.8.1 The Stokes Equations

Consider the Stokes equations in the velocity–pressure–vorticity formulation (8.7) with any combination of permissible boundary conditions listed in Table 8.1. For the solution of the Stokes equations, the least-squares method minimizes the following quadratic functional:

$$I(\boldsymbol{u},p,\boldsymbol{\omega}) = \|\nabla p+\nabla\times\boldsymbol{\omega}-\boldsymbol{f}\|_0^2+\|\nabla\cdot\boldsymbol{\omega}\|_0^2+\|\boldsymbol{\omega}-\nabla\times\boldsymbol{u}\|_0^2+\|\nabla\cdot\boldsymbol{u}\|_0^2, \quad (8.89)$$

where $(\boldsymbol{u},p,\boldsymbol{\omega}) \in \boldsymbol{H} = \{[H^1(\Omega)]^7 : (\boldsymbol{u},p,\boldsymbol{\omega})$ satisfies the given boundary conditions on $\Gamma\}$. Letting the first variation of I vanish, we obtain the least-squares variational formulation: Find $(\boldsymbol{u},p,\boldsymbol{\omega}) \in \boldsymbol{U}$ such that

$$\begin{aligned}
&(\nabla p+\nabla\times\boldsymbol{\omega}-\boldsymbol{f},\ \nabla q)\\
&+(\nabla p+\nabla\times\boldsymbol{\omega}-\boldsymbol{f},\ \nabla\times\boldsymbol{\tau})\\
&+(\nabla\cdot\boldsymbol{\omega},\ \nabla\cdot\boldsymbol{\tau})\\
&+(\boldsymbol{\omega}-\nabla\times\boldsymbol{u},\ \boldsymbol{\tau})\\
&-(\boldsymbol{\omega}-\nabla\times\boldsymbol{u},\ \nabla\times\boldsymbol{v})\\
&+(\nabla\cdot\boldsymbol{u},\ \nabla\cdot\boldsymbol{v}) = 0 \qquad \forall(\boldsymbol{v},q,\boldsymbol{\tau}) \in \boldsymbol{H}_0.
\end{aligned} \quad (8.90)$$

where $\boldsymbol{H}_0 = \{[H^1(\Omega)]^7 : (\boldsymbol{v},q,\boldsymbol{\tau})$ satisfies the corresponding homogeneous boundary conditions on $\Gamma\}$. By using Green's formulae (B.1), (B.3) and (B.5), from (8.90) we have

$$\begin{aligned}
&-\Big(\nabla\cdot(\nabla p+\nabla\times\boldsymbol{\omega}-\boldsymbol{f}),\ q\Big)+\langle\boldsymbol{n}\cdot(\nabla p+\nabla\times\boldsymbol{\omega}-\boldsymbol{f}),\ q\rangle\\
&+\Big(\nabla\times(\nabla p+\nabla\times\boldsymbol{\omega}-\boldsymbol{f}),\ \boldsymbol{\tau}\Big)-\langle\boldsymbol{n}\times(\nabla p+\nabla\times\boldsymbol{\omega}-\boldsymbol{f}),\ \boldsymbol{\tau}\rangle\\
&-\Big(\nabla(\nabla\cdot\boldsymbol{\omega}),\ \boldsymbol{\tau}\Big)+\langle\nabla\cdot\boldsymbol{\omega},\ \boldsymbol{n}\cdot\boldsymbol{\tau}\rangle\\
&+(\boldsymbol{\omega}-\nabla\times\boldsymbol{u},\ \boldsymbol{\tau})\\
&-\Big(\nabla\times(\boldsymbol{\omega}-\nabla\times\boldsymbol{u}),\ \boldsymbol{v}\Big)+\langle\boldsymbol{n}\times(\boldsymbol{\omega}-\nabla\times\boldsymbol{u}),\ \boldsymbol{v}\rangle\\
&-\Big(\nabla(\nabla\cdot\boldsymbol{u}),\ \boldsymbol{v}\Big)+\langle\nabla\cdot\boldsymbol{u},\ \boldsymbol{n}\cdot\boldsymbol{v}\rangle = 0 \qquad \forall(\boldsymbol{v},q,\boldsymbol{\tau}) \in \boldsymbol{H}_0.
\end{aligned} \quad (8.91)$$

From (8.91) after simplification by using the vector identity (A.4), we obtain the following Euler–Lagrange equations and boundary conditions:

$$\Delta p = \nabla \cdot \boldsymbol{f} \qquad \text{in } \Omega, \tag{8.92a}$$

$$p = \text{given} \quad \text{or} \quad \boldsymbol{n} \cdot (\nabla p + \nabla \times \boldsymbol{\omega} - \boldsymbol{f}) = 0 \qquad \text{on } \Gamma, \tag{8.92b}$$

$$\Delta \boldsymbol{\omega} - \boldsymbol{\omega} + \nabla \times \boldsymbol{u} = -\nabla \times \boldsymbol{f} \qquad \text{in } \Omega, \tag{8.92c}$$

$$\boldsymbol{n} \times \boldsymbol{\omega} = \text{given} \quad \text{or} \quad \boldsymbol{n} \times (\nabla p + \nabla \times \boldsymbol{\omega} - \boldsymbol{f}) = \boldsymbol{0} \qquad \text{on } \Gamma, \tag{8.92d}$$

$$\boldsymbol{n} \cdot \boldsymbol{\omega} = \text{given} \quad \text{or} \quad \nabla \cdot \boldsymbol{\omega} = 0 \qquad \text{on } \Gamma, \tag{8.92e}$$

$$\Delta \boldsymbol{u} + \nabla \times \boldsymbol{\omega} = \boldsymbol{0} \qquad \text{in } \Omega, \tag{8.92f}$$

$$\boldsymbol{n} \times \boldsymbol{u} = \text{given} \quad \text{or} \quad \boldsymbol{n} \times (\boldsymbol{\omega} - \nabla \times \boldsymbol{u}) = \boldsymbol{0} \qquad \text{on } \Gamma, \tag{8.92g}$$

$$\boldsymbol{n} \cdot \boldsymbol{u} = \text{given} \quad \text{or} \quad \nabla \cdot \boldsymbol{u} = 0 \qquad \text{on } \Gamma. \tag{8.92h}$$

Equation (8.92a) is the so called pressure Poisson equation. The condition (8.92b) implies that the pressure Poisson equation may have one of two possible boundary conditions depending on whether the pressure is or not given on the boundary.

Equation (8.92c) is the vorticity equation. In this equation, $-\boldsymbol{\omega}$ and $\nabla \times \boldsymbol{u}$ can be cancelled, and then we obtain three Poisson equations for the vorticity components. However, if $-\boldsymbol{\omega}$ and $\nabla \times \boldsymbol{u}$ are deleted, then the symmetry of the whole system will be lost, that is, if the Galerkin method is employed for system (8.92), the corresponding variational statement is not (8.90), and hence the global finite element matrix is not symmetrical.

Equation (8.92f) is the Poisson equation for the velocity components.

System (8.92) reveals that the least-squares variational formulation corresponds to seven coupled second-order elliptic equations and seven sets of boundary conditions in which the original boundary conditions serve as the essential boundary conditions and the original first-order equations serve as the natural boundary conditions. In the following we list the additional boundary conditions which should be supplemented for the second-order Stokes equations (8.92a), (8.92c) and (8.92f) in different cases:

(1) $\boldsymbol{n} \cdot \boldsymbol{u}$, $\boldsymbol{n} \times \boldsymbol{\omega}$ given

$$\begin{aligned} \boldsymbol{n} \cdot (\nabla p + \nabla \times \boldsymbol{\omega} - \boldsymbol{f}) &= 0, \\ \nabla \cdot \boldsymbol{\omega} &= 0, \\ \boldsymbol{n} \times (\boldsymbol{\omega} - \nabla \times \boldsymbol{u}) &= \boldsymbol{0}. \end{aligned}$$

(2) p, $\boldsymbol{n} \cdot \boldsymbol{u}$, $\boldsymbol{n} \cdot \boldsymbol{\omega}$ given

$$\begin{aligned} \boldsymbol{n} \times (\nabla p + \nabla \times \boldsymbol{\omega} - \boldsymbol{f}) &= \boldsymbol{0}, \\ \boldsymbol{n} \times (\boldsymbol{\omega} - \nabla \times \boldsymbol{u}) &= \boldsymbol{0}. \end{aligned}$$

(3) p, $\boldsymbol{n} \times \boldsymbol{u}$, $\boldsymbol{n} \cdot \boldsymbol{\omega}$ given

$$\begin{aligned} \boldsymbol{n} \times (\nabla p + \nabla \times \boldsymbol{\omega} - \boldsymbol{f}) &= \boldsymbol{0}, \\ \nabla \cdot \boldsymbol{u} &= 0. \end{aligned}$$

(4) $\boldsymbol{n} \times \boldsymbol{u}$, $\boldsymbol{n} \times \boldsymbol{\omega}$ given

$$\begin{aligned} \boldsymbol{n} \cdot (\nabla p + \nabla \times \boldsymbol{\omega} - \boldsymbol{f}) &= 0, \\ \nabla \cdot \boldsymbol{\omega} &= 0, \\ \nabla \cdot \boldsymbol{u} &= 0. \end{aligned}$$

(5) $\boldsymbol{u}$ given

$$\begin{aligned} \nabla p + \nabla \times \boldsymbol{\omega} - \boldsymbol{f} &= \boldsymbol{0}, \\ \nabla \cdot \boldsymbol{\omega} &= 0. \end{aligned}$$

(6) p, $\boldsymbol{n} \cdot \boldsymbol{\omega}$, $\boldsymbol{n} \times \boldsymbol{\omega}$ given

$$\begin{aligned} \boldsymbol{n} \times (\boldsymbol{\omega} - \nabla \times \boldsymbol{u}) &= \boldsymbol{0}, \\ \nabla \cdot \boldsymbol{u} &= 0. \end{aligned}$$

We emphasize again that the least-squares finite element method based on the first-order velocity–pressure–vorticity formulation (8.7) does not need any additional boundary conditions. However, if another discretization scheme such as the finite difference method is used to solve the second-order velocity–pressure–vorticity formulation (8.92a), (8.92c) and (8.92f), the additional natural boundary conditions should be included.

We notice that in the second-order velocity–pressure–vorticity formulation, the solution of the pressure Poisson equation (8.92a) is coupled with the solution of the velocity and the vorticity through the boundary conditions. The significant advantages of the present second-order formulation are that (1) it guarantees the satisfaction of the continuity of velocity and the solenoidality constraint on the vorticity without explicitly including these two divergence-free conditions; (2) it is suitable not only for the standard boundary conditions but also for non-standard boundary conditions; (3) the differential operator is self-adjoint (symmetrical).

8.8.2 The Navier–Stokes Equations

Let us consider the following first-order system of the time-dependent Navier–Stokes equations in the velocity–total pressure–vorticity formulation with the standard velocity boundary condition (see Sect. 8.4):

$$\frac{\partial \boldsymbol{u}}{\partial t} - \boldsymbol{u} \times \boldsymbol{\omega} + \nabla b + \frac{1}{\mathrm{Re}} \nabla \times \boldsymbol{\omega} = \boldsymbol{f} \qquad \text{in } \Omega \times (0, T], \tag{8.93a}$$

$$\nabla \cdot \boldsymbol{\omega} = 0 \qquad \text{in } \Omega \times (0, T], \tag{8.93b}$$

$$\boldsymbol{\omega} - \nabla \times \boldsymbol{u} = 0 \qquad \text{in } \Omega \times (0, T], \tag{8.93c}$$

$$\nabla \cdot \boldsymbol{u} = 0 \quad \text{in } \Omega \times (0,T], \tag{8.93d}$$

$$\boldsymbol{u} = \boldsymbol{u}_\Gamma \quad \text{on } \Gamma \times (0,T]. \tag{8.93e}$$

The given boundary data $\boldsymbol{u}_\Gamma$ should satisfy the global mass conservation:

$$\int_\Gamma \boldsymbol{n} \cdot \boldsymbol{u}_\Gamma ds = 0. \tag{8.93f}$$

We also assume that $\boldsymbol{u}_\Gamma$ is not time-dependent, and the divergence equations (8.93b) and (8.93d) are satisfied at the initial time. Here we use the total pressure b instead of the pressure p to make the derivation much more simple.

By virtue of the div–curl theorem 5.2, system (8.93) is equivalent to the following system:

$$\nabla \times \left(\frac{\partial \boldsymbol{u}}{\partial t} - \boldsymbol{u} \times \boldsymbol{\omega} + \nabla b + \frac{1}{\mathrm{Re}} \nabla \times \boldsymbol{\omega} - \boldsymbol{f}\right) = \boldsymbol{0} \quad \text{in } \Omega \times (0,T], \tag{8.94a}$$

$$\nabla \cdot \left(\frac{\partial \boldsymbol{u}}{\partial t} - \boldsymbol{u} \times \boldsymbol{\omega} + \nabla b + \frac{1}{\mathrm{Re}} \nabla \times \boldsymbol{\omega} - \boldsymbol{f}\right) = 0 \quad \text{in } \Omega \times (0,T], \tag{8.94b}$$

$$\boldsymbol{n} \cdot \left(\frac{\partial \boldsymbol{u}}{\partial t} - \boldsymbol{u} \times \boldsymbol{\omega} + \nabla b + \frac{1}{\mathrm{Re}} \nabla \times \boldsymbol{\omega} - \boldsymbol{f}\right) = 0 \quad \text{on } \Gamma \times (0,T], \tag{8.94c}$$

$$\nabla \times (\boldsymbol{\omega} - \nabla \times \boldsymbol{u}) = \boldsymbol{0} \quad \text{in } \Omega \times (0,T], \tag{8.94d}$$

$$\nabla \cdot (\boldsymbol{\omega} - \nabla \times \boldsymbol{u}) = 0 \quad \text{in } \Omega \times (0,T], \tag{8.94e}$$

$$\boldsymbol{n} \times (\boldsymbol{\omega} - \nabla \times \boldsymbol{u}) = \boldsymbol{0} \quad \text{on } \Gamma \times (0,T], \tag{8.94f}$$

$$\nabla \cdot \boldsymbol{\omega} = 0 \quad \text{in } \Omega \times (0,T], \tag{8.94g}$$

$$\nabla \cdot \boldsymbol{u} = 0 \quad \text{in } \Omega \times (0,T], \tag{8.94h}$$

$$\boldsymbol{u} = \boldsymbol{u}_\Gamma \quad \text{on } \Gamma \times (0,T]. \tag{8.94i}$$

Taking into account $\nabla \times \nabla p = \boldsymbol{0}$, $\nabla \cdot \nabla \times \boldsymbol{\omega} = 0$, $\nabla \cdot \nabla \times \boldsymbol{u} = 0$ and the vector equality (A.4), system (8.94) can be simplified as:

$$\frac{\partial \boldsymbol{\omega}}{\partial t} - \nabla \times (\boldsymbol{u} \times \boldsymbol{\omega}) - \frac{1}{\mathrm{Re}} \Delta \boldsymbol{\omega} = \nabla \times \boldsymbol{f} \quad \text{in } \Omega \times (0,T], \tag{8.95a}$$

$$\Delta b - \nabla \cdot (\boldsymbol{u} \times \boldsymbol{\omega}) = \nabla \cdot \boldsymbol{f} \quad \text{in } \Omega \times (0,T], \tag{8.95b}$$

$$\boldsymbol{n} \cdot \left(\nabla b - \boldsymbol{u} \times \boldsymbol{\omega} + \frac{1}{\mathrm{Re}} \nabla \times \boldsymbol{\omega} - \boldsymbol{f}\right) = 0 \quad \text{on } \Gamma \times (0,T], \tag{8.95c}$$

$$\Delta \boldsymbol{u} + \nabla \times \boldsymbol{\omega} = \boldsymbol{0} \quad \text{in } \Omega \times (0,T], \tag{8.95d}$$

$$\boldsymbol{n} \times (\boldsymbol{\omega} - \nabla \times \boldsymbol{u}) = \boldsymbol{0} \quad \text{on } \Gamma \times (0,T], \tag{8.95e}$$

$$\nabla \cdot \boldsymbol{\omega} = 0 \quad \text{in } \Omega \times (0,T], \tag{8.95f}$$

$$\nabla \cdot \boldsymbol{u} = 0 \quad \text{in } \Omega \times (0,T], \tag{8.95g}$$

$$\boldsymbol{u} = \boldsymbol{u}_\Gamma \qquad \text{on } \Gamma \times (0, T]. \tag{8.95h}$$

Since (8.95a) implies that

$$\frac{\partial(\nabla \cdot \boldsymbol{\omega})}{\partial t} - \frac{1}{\mathrm{Re}} \Delta(\nabla \cdot \boldsymbol{\omega}) = 0 \qquad \text{in } \Omega \times (0, T], \tag{8.96a}$$

if we specify that

$$\nabla \cdot \boldsymbol{\omega} = 0 \qquad \text{on } \Gamma \times (0, T], \tag{8.96b}$$

then $\nabla \cdot \boldsymbol{\omega} \equiv 0$ in Ω, that is, the divergence-free property of the vorticity vector is guaranteed. Therefore we can replace the divergence equation (8.95f) by the boundary condition (8.96b).

As explained in Sect. 5.5, the divergence equation (8.95g) can be eliminated, since (8.95d) and (8.95f) and the boundary conditions (8.95e) and (8.95h) guarantee the divergence-free of the velocity.

Finally we obtain the second-order velocity–vorticity equations and the pressure Poisson equation for the Navier–Stokes equations:

$$\frac{\partial \boldsymbol{\omega}}{\partial t} - \nabla \times (\boldsymbol{u} \times \boldsymbol{\omega}) - \frac{1}{\mathrm{Re}} \Delta \boldsymbol{\omega} = \nabla \times \boldsymbol{f} \qquad \text{in } \Omega \times (0, T], \tag{8.97a}$$

$$\boldsymbol{n} \times (\boldsymbol{\omega} - \nabla \times \boldsymbol{u}) = \boldsymbol{0} \qquad \text{on } \Gamma \times (0, T], \tag{8.97b}$$

$$\nabla \cdot \boldsymbol{\omega} = 0 \qquad \text{on } \Gamma \times (0, T], \tag{8.97c}$$

$$\Delta \boldsymbol{u} + \nabla \times \boldsymbol{\omega} = \boldsymbol{0} \qquad \text{in } \Omega, \tag{8.97d}$$

$$\boldsymbol{u} = \boldsymbol{u}_\Gamma \qquad \text{on } \Gamma \times (0, T], \tag{8.97e}$$

$$\Delta b - \nabla \cdot (\boldsymbol{u} \times \boldsymbol{\omega}) = \nabla \cdot \boldsymbol{f} \qquad \text{in } \Omega \times (0, T], \tag{8.97f}$$

$$\boldsymbol{n} \cdot (\nabla b - \boldsymbol{u} \times \boldsymbol{\omega} + \frac{1}{\mathrm{Re}} \nabla \times \boldsymbol{\omega} - \boldsymbol{f}) = 0 \qquad \text{on } \Gamma \times (0, T]. \tag{8.97g}$$

From (8.97) it is clear that the calculation of the velocity and vorticity is decoupled from that of the pressure; (8.97a) and (8.97d) are termed the velocity–vorticity formulation. For simplicity, in this section we have studied only the case with the velocity boundary conditions specified on the whole boundary. For cases with other boundary conditions the derivation of these additional Neumann type of boundary conditions is similar. It is worth noting that if the pressure is prescribed on a part of boundary, the additional boundary condition on that part of the boundary for the vorticity equation (8.97a) is more complicated than (8.97b).

In the finite difference framework one may solve the vorticity and the velocity in a sequential manner: using the velocity field known from the previous time step one may solve (8.97a) with boundary conditions (8.97b) and (8.97c) to obtain the vorticity, then by solving the Poisson equations (8.97d) update the velocity. When convergent solutions for velocity and vorticity are obtained, one can solve the pressure Poisson equation (8.97f) with the Neumann boundary condition (8.97g) to obtain pressure.

Obviously, the LSFEM based on the first-order velocity–pressure–vorticity formulation is much more simple, since (1) no Neumann-type boundary conditions (8.97b), (8.97c), and (8.97g) are involved; and (2) neither upwinding nor staggered grids are needed; (3) the pressure boundary condition can be handled easily.

8.9 Concluding Remarks

For the solution of incompressible Navier–Stokes equations, the LSFEM based on the first-order velocity–pressure–vorticity formulation is a powerful technique: together with, for example, Newton's linearization, using C^0 finite elements to discretize the equations and minimizing the L_2 norm of the equation residuals leads to a symmetric and positive-definite algebraic system which can be effectively solved by simple matrix-free iterative methods; this is a minimization problem rather than a saddle-point problem, the choice of approximating spaces is thus not subject to the restriction of the LBB condition; in other words, simple equal-order interpolation with respect to a single grid for all dependent variables and test functions may be used; only easy algebraic boundary conditions are imposed, no artificial numerical boundary conditions for the vorticity need to be devised at boundaries at which the velocity is specified; accurate vorticity approximations are obtained; neither upwinding nor adjustable parameters are needed; all variables are solved in a fully-coupled manner, neither operator splitting nor projection, which may lead to difficulty of convergence and appearance of spurious vortexes (Gresho and Chan 1995), are involved.

A possible objection to LSFEM is that LSFEM involves more degrees of freedom than the Galerkin-type methods based on the velocity–pressure formulation. However, a fair comparison should take into account the size of a problem that can be solved in the available storage of a computer. As shown in this chapter, the diagonal preconditioned conjugate gradient method gives satisfactory results for the symmetric positive-definite systems obtained by LSFEM. Here only five global vectors need to be stored. Conversely, the Galerkin-type methods lead to non-symmetric discrete systems. Generally, GMRES method (see e.g. Saad 1986) is the best choice for solving large non-symmetric algebraic systems. GMRES typically needs to store $10 \sim 30$ global vectors (Brussino and Sonnad 1989). Moreover, the success of GMRES depends on using sophisticated preconditioning techniques such as the incomplete LU factorization which implies that the assemblage of the global matrix (or at least a block of global matrix) is necessary. Obviously, the storage requirements can be prohibitive for very large problems.

The least-squares method based on the first-order velocity–pressure–vorticity formulation also provides a new mathematical tool for theoretical analysis of the Navier–Stokes equations. The div–curl structure of this formulation allows us to derive all possible boundary conditions and to prove their permissibility. The least-squares method and the div–curl method are systematic and consistent methods for deriving second-order versions of the Navier–Stokes equations and their additional boundary conditions without generating spurious solutions.

The papers by Jiang and Chang (1990), Jiang (1992), and Jiang and Povinelli (1990) are the first to introduce the LSFEM based on the velocity–pressure–vorticity formulation. Subsequently, Jiang et al. (1994a) and Jiang et al. (1995) applied it to steady-state three-dimensional problems. The merits of LSFEM have been further demonstrated by Lefebvre et al. (1992) for two-dimensional unstructured triangular meshes; by Tang and Tsang (1993, 1994, 1997) for two-dimensional flows with thermal convection, doubly-diffusive flows and three-dimensional Rayleigh–Bénard convection; by Tang et al. (1995) for three-dimensional lid-driven cavity flows; and by Yu et al. (1996c) for three-dimensional Marangoni–Bénard convection. Jiang and Sonnad (1994) demonstrated that the LSFEM can be easily implemented with p-version finite elements. Chan et el. (1995) presented impressive results obtained by using the spectral element version of LSFEM.

The theoretical analysis of the least-squares method in the first-order velocity–pressure–vorticity formulation based on the div–curl theory, the rigorous derivation of permissible non-standard boundary conditions, and the derivation of second-order velocity–pressure–vorticity equations appeared first in Jiang et al. (1994b). The proof of the sub-optimality of the least-squares method for the velocity (wall) boundary condition case in Sect. 8.2.2 is originally presented in Jiang (1998).

Chang and Jiang (1990) tried to analyze the least-squares method for the Stokes equations in the first-order velocity–pressure–vorticity formulation by using the method developed by Aziz et al. (1985) based on the ADN theory. Bochev and Gunzburger (1994) pointed out that the analysis in Chang and Jiang (1990) is incorrect for the the velocity boundary condition case. According to the ADN theory, the functional being minimized in that case should be

$$\mathcal{K}(\boldsymbol{v},q,\boldsymbol{\tau}) = \|\nabla q + \nabla\times\boldsymbol{\tau}\|_0^2 + \|\nabla\cdot\boldsymbol{\tau}\|_0^2 + \|\boldsymbol{\tau} - \nabla\times\boldsymbol{v}\|_1^2 + \|\nabla\cdot\boldsymbol{v}\|_1^2. \quad (8.98)$$

Note that this functional involves products of second derivatives of the velocity, and thus the corresponding method is not practical due to the need of C^1 elements for the approximation of the velocity. In order to overcome this difficulty, by considering the inverse inequality (2.56) for finite element shape functions they suggest that the norm $\|\cdot\|_1$ in (8.98) might be simulated by $h^{-1}\|\cdot\|_0$ where h is an appropriate measure of the grid size. Therefore, the following simple weighted functional is recommended by them:

$$\mathcal{K}_h(\boldsymbol{v},q,\boldsymbol{\tau}) = \|\nabla q+\nabla\times\boldsymbol{\tau}\|_0^2+\|\nabla\cdot\boldsymbol{\tau}\|_0^2+h^{-2}\|\boldsymbol{\tau}-\nabla\times\boldsymbol{v}\|_0^2+h^{-2}\|\nabla\cdot\boldsymbol{v}\|_0^2. \quad (8.99)$$

Their numerical experiments support this idea. Their formulation was applied to boundary control problems (Bochev and Bedivan 1997). The same weighted functional (without the $\nabla \cdot \boldsymbol{\tau}$ term) was also proposed by Harbord and Gellert (1991). However, the weighted functional (8.99) perhaps has more theoretical value than practical value, because (1) for real world problems, the mesh is often nonuniform, then the question is how to choose the mesh size h; (2) for practical problems, the boundary conditions are not always the pure standard velocity condition, they can be mixed with any one listed in Table 8.1; (3) more important, in Sect. 4.11 we point out that the least-squares finite element method with Gaussian quadrature is equivalent to the least-squares finite element collocation method. In order to obtain a good approximate solution, the number of Gaussian points must be selected such that one solves a determined or a slightly overdetermined collocated residual equations. As discussed in Sect. 4.11, for a determined set of linear algebraic equations, weighting has no effect on the least-squares solution. That is, in practical computations we must do our best to make the weighting have minimal effect.

Chang (1990) and Chang et al. (1995) also proposed a least-squares method for the Stokes equations based on the velocity–velocity derivatives–pressure formulation. This formulation achieves optimal rate of convergence in the H^1 norm for the velocity boundary condition case. However, it has too many variables.

As pointed out originally by Lighthill (1986), velocity and vorticity are the most suitable variables from the viewpoint of a fluid dynamic description of incompressible viscous flows. Therefore, the first-order velocity–pressure–vorticity formulation is preferred for the least-squares method.

9. Convective Transport

Convective transport phenomena are governed by hyperbolic equations. Convection processes arise in such diverse problems as fluid flow in water resources, oil, gas, and geothermal reservoir simulation, in heat and mass transfer in chemical and nuclear engineering, and air pollution in atmospheric environment. Such effects are also present in the convection–diffusion equations, the Navier–Stokes equations or the vorticity–transport equations of classic fluid dynamics.

In Chap. 2 we have studied convective transport problems through a simple example in one dimension. There, it is demonstrated that the central difference method as well as the intimately related Galerkin finite element method lead to oscillatory results, and the use of upwinding strategies in these methods is necessary to represent the uni-directional character of the propagation phenomenon and to suppress numerical oscillations; it is also shown that special treatment is not needed in LSFEM, and the upwinding mechanism is automaticly and naturally generated in LSFEM.

The conclusions drawn from one-dimensional problems are also correct for multi-dimensional problems studied in this chapter. Here we shall first study the LSFEM for steady-state pure convection problems with smooth solutions, and compare the LSFEM with the Galerkin finite element method and the streamline upwinding Petrov–Galerkin (SUPG) method. We shall prove that the LSFEM has the same stability estimate as the original equation, namely, in LSFEM the streamline derivative is well controlled. Numerical convergence rates are given to show that the LSFEM is almost optimal.

Then we use this LSFEM as a framework to develop the L_1 method and the iteratively reweighted least-squares finite element method (IRLSFEM) to obtain non-oscillatory and non-diffusive solutions for problems with contact discontinuities. These new methods produce a highly accurate numerical solution that has a sharp discontinuity in one element. A number of examples solved by using triangular and bilinear elements are presented to demonstrate that the method can convect contact discontinuities without error.

Finally we deal with transient problems and compare the LSFEM with the Taylor–Galerkin method.

9.1 Steady-State Problems

We consider the following steady-state convection equation:

$$u_\beta + u = f \qquad \text{in } \Omega, \tag{9.1a}$$

where Ω is a bounded convex domain in $\mathbb{R}^3$ with a piecewise smooth boundary Γ, $u = u(x, y, z)$ is the dependent variable (e.g., the concentration), $\boldsymbol{\beta} = (\beta_1, \beta_2, \beta_3)$ is a specified convection vector which may be constant or spatially-varying, $u_\beta = \boldsymbol{\beta} \cdot \nabla u$ denotes the derivative in the $\boldsymbol{\beta}$ direction, and $f \in L_2(\Omega)$ is a given source function. The inflow boundary Γ_{in} is defined by

$$\Gamma_{\text{in}} = \{(x, y, z) \in \Gamma : \boldsymbol{n}(x, y, z) \cdot \boldsymbol{\beta} < 0\},$$

where $\boldsymbol{n}$ is the outward unit normal to Γ at point $(x, y, z) \in \Gamma$. A boundary value problem for (9.1a) is given by specifying the dependent variable u on Γ_{in}. Assume that, for simplicity without loss of generality, we require

$$u = 0 \qquad \text{on } \Gamma_{\text{in}}. \tag{9.1b}$$

For the first-order equation (9.1a) the characteristic equation (4.47) takes the form:

$$\beta_1\alpha_1 + \beta_2\alpha_2 + \beta_3\alpha_3 = 0. \tag{9.2}$$

If α_1 and α_2 are arbitrarily given real numbers, we always can solve (9.2) to obtain a real root α_3. Therefore, (9.1a) is hyperbolic. The streamlines corresponding to the given velocity field $\boldsymbol{\beta} = (\beta_1, \beta_2, \beta_3)$ are determined by the curves $\boldsymbol{x}(s)$, $\boldsymbol{x} = (x, y, z)$, where $\boldsymbol{x}(s)$ is the solution of the following system of ordinary differential equations:

$$\frac{dx}{ds} = \beta_1, \tag{9.3a}$$

$$\frac{dy}{ds} = \beta_2, \tag{9.3b}$$

$$\frac{dz}{ds} = \beta_3. \tag{9.3c}$$

These curves are the characteristics, since their normal direction is $(\alpha_1, \alpha_2, \alpha_3)$. By virtue of the chain-rule

$$\begin{aligned} \frac{d}{ds}(u(\boldsymbol{x}(s))) &= \frac{\partial u}{\partial x}\frac{dx}{ds} + \frac{\partial u}{\partial y}\frac{dy}{ds} + \frac{\partial u}{\partial z}\frac{dz}{ds} \\ &= \beta_1\frac{\partial u}{\partial x} + \beta_2\frac{\partial u}{\partial y} + \beta_3\frac{\partial u}{\partial z} = \boldsymbol{\beta} \cdot \nabla u, \end{aligned}$$

i.e., along each characteristic the partial differential equation (9.1) is reduced to an ordinary differential equation:

$$\frac{d}{ds}u(\boldsymbol{x}(s)) + u(\boldsymbol{x}(s)) = 0. \tag{9.4}$$

The concentration u at any point $(\bar{x}, \bar{y}, \bar{z})$ can then be determined by integration along the characteristic passing through $(\bar{x}, \bar{y}, \bar{z})$ starting on Γ_{in}.

Throughout this chapter we also use the following notations:

$$\langle u, v\rangle = \int_\Gamma uv\boldsymbol{n}\cdot\boldsymbol{\beta}ds, \qquad |u|_\Gamma = \left(\int_\Gamma u^2|\boldsymbol{n}\cdot\boldsymbol{\beta}|ds\right)^{\frac{1}{2}},$$

$$\langle u, v\rangle_{\text{out}} = \int_{\Gamma_{\text{out}}} uv\boldsymbol{n}\cdot\boldsymbol{\beta}ds, \qquad |u|_{\text{out}} = \left(\int_{\Gamma_{\text{out}}} u^2|\boldsymbol{n}\cdot\boldsymbol{\beta}|ds\right)^{\frac{1}{2}},$$

where

$$\Gamma_{\text{out}} = \{(x, y) \in \Gamma : \boldsymbol{n}(x, y)\cdot\boldsymbol{\beta} \geq 0\}.$$

We recall Green's formula

$$(u_\beta, v) = \langle u, v\rangle - (u, v_\beta), \tag{9.5}$$

and thus

$$(u_\beta, u) = \frac{1}{2}|u|_\Gamma^2. \tag{9.6}$$

Further we also define the function space

$$S = \{u \in H^1(\Omega) : u = 0 \text{ on } \Gamma_{\text{in}}\},$$

and the corresponding finite element subspace S_h, i.e. S_h is the space of continuous piecewise polynomial functions of order r. From the finite element interpolation theory (Theorem 4.5) we have: Given a function $u \in H^{r+1}(\Omega)$, there exists an interpolant $\Pi_h u \in S_h$ such that

$$\|u - \Pi_h u\|_0 \leq Ch^{r+1}\|u\|_{r+1}, \tag{9.7a}$$

$$\|\nabla u - \nabla\Pi_h u\|_0 \leq Ch^r\|u\|_{r+1}. \tag{9.7b}$$

Moreover, if u is smooth enough in $\bar{\Omega}$, then

$$|u - \Pi_h u|_\Gamma \leq Ch^{r+1}\|u\|_{r+1}. \tag{9.7c}$$

The inequality (9.7c) will be used only for the error estimate of the Galerkin method, and has nothing to do with LSFEM. We refer to Cialet (1991) for its proof.

In the error analysis below we will often use the inequality

$$2ab \leq \varepsilon a^2 + \frac{1}{\varepsilon}b^2, \tag{9.8}$$

for a, b real numbers and $\varepsilon > 0$.

9.1.1 The Classic Galerkin Method

Now let us analyze the following classic Galerkin method for problem (9.1): Find $u_h \in S_h$ such that

$$(u_{h,\beta} + u_h, v_h) = (f, v_h) \qquad \forall v_h \in S_h. \tag{9.9}$$

Since the exact solution satisfies (9.1), we also have

$$(u_\beta + u, v_h) = (f, v_h) \qquad \forall v_h \in S_h. \tag{9.10}$$

By subtracting (9.9) from (9.10) we obtain the following orthogonality for the error $e = u - u_h$:

$$(e_\beta + e, v_h) = 0 \qquad \forall v_h \in S_h. \tag{9.11}$$

Let $\Pi_h u \in S_h$ be the interpolant of u satisfying (9.7) and write $\rho = u - \Pi_h u$ and $\eta = \Pi_h u - u_h$ so that $e = \rho + \eta$. Then, by using (9.5), (9.6), (9.11) and (9.8) we have

$$\begin{aligned} \frac{1}{2}|e|_\Gamma^2 + \|e\|^2 &= (e_\beta + e, e) = (e_\beta + e, \rho) + (e_\beta + e, \eta) \\ &= (e_\beta, \rho) + (e, \rho) = -(e, \rho_\beta) + \langle e, \rho \rangle + (e, \rho) \\ &\le \|\rho_\beta\|_0^2 + \frac{1}{4}\|e\|_0^2 + |\rho|_\Gamma^2 + \frac{1}{4}|e|_\Gamma^2 + \|\rho\|_0^2 + \frac{1}{4}\|e\|_0^2. \end{aligned} \tag{9.12}$$

By moving all terms related to e into the left-hand side, the inequality (9.12) becomes

$$\frac{1}{4}|e|_\Gamma^2 + \frac{1}{2}\|e\|_0^2 \le \|\rho_\beta\|_0^2 + |\rho|_\Gamma^2 + \|\rho\|_0^2. \tag{9.13}$$

Recalling (9.7) we have

$$\|\rho_\beta\|_0 + |\rho|_\Gamma + \|\rho\|_0 \le Ch^r \|u\|_{r+1}. \tag{9.14}$$

From (9.13) and (9.14) it is easy to obtain the error estimate for the classic Galerkin method:

$$\|e\|_0 + |e|_\Gamma \le Ch^r \|u\|_{r+1}, \tag{9.15}$$

which is one order lower than optimal. Furthermore, for the continuous problem (9.1), we have the following stability estimate:

$$\|u\|_0 + \|u_\beta\|_0 + |u|_\Gamma \le C\|f\|_0. \tag{9.16}$$

This stability estimate follows by applying the least-squares method to (9.1a), see Sect. 9.1.3. By letting $v_h = u_h$ in (9.9) it is easy to obtain the stability estimate for the classic Galerkin method:

$$\|u_h\|_0 + |u_h|_\Gamma \le C\|f\|_0, \tag{9.17}$$

which implies that the classic Galerkin method has no control of $\|u_{h,\beta}\|_0$. This explains why the Galerkin solution is oscillatory.

9.1.2 The SUPG Method

In order to get better accuracy and stability, methods incorporating some form of upwinding have appeared, see e.g., papers by Dendy (1974), Wahlbin (1974), Christie et al. (1976), Brooks and Hughes (1982), Johnson, et al. (1984), Morton and Parrot (1980). Below we shall follow Johnson (1987) to analyze the streamline upwinding Petrov–Galerkin (SUPG) method (Hughes 1987) or the streamline diffusion method (Johnson 1987). The term "Petrov–Galerkin" is used to indicate that the test functions are no longer the same as the trial functions for the approximation. The term "streamline upwind" comes form the term $\boldsymbol{\beta}\cdot\nabla v$ in the test functions. The SUPG variational statement is: Find $u_h \in S_h$ such that

$$(u_{h,\beta}+u_h, v_h+hv_{h,\beta}) = (f, v_h+hv_{h,\beta}) \qquad \forall v_h \in S_h. \tag{9.18}$$

The orthogonality condition for the error is

$$(e_\beta+e, v_h+hv_{h,\beta}) = 0 \qquad \forall v_h \in S_h. \tag{9.19}$$

By using (9.6) it is easy to show that

$$\|e\|_0^2 + h\|e_\beta\|_0^2 + \frac{1+h}{2}|e|_\Gamma^2 = (e_\beta+e, e+he_\beta). \tag{9.20}$$

On the other hand, by using (9.19) and (9.8) we have

$$\begin{aligned}(e_\beta+e, e+he_\beta) &= (e_\beta,\rho)+(e,\rho)+h(e_\beta,\rho_\beta)+h(e,\rho_\beta)\\ &\le \frac{h}{4}\|e_\beta\|_0^2 + h^{-1}\|\rho\|_0^2 + \frac{1}{4}\|e\|_0^2 + \|\rho\|_0^2\\ &+ \frac{h}{4}\|e_\beta\|_0^2 + h\|\rho_\beta\|^2 + \frac{1}{4}\|e\|_0^2 + h^2\|\rho_\beta\|_0^2. \end{aligned}\tag{9.21}$$

Gathering (9.20) and (9.21), moving the terms related to e into the left-hand side, we obtain the error estimate for the SUPG method:

$$\left(\|e\|_0^2 + h\|e_\beta\|_0^2 + \frac{(1+h)}{2}|e|_\Gamma^2\right)^{\frac{1}{2}} \le Ch^{r+\frac{1}{2}}\|u\|_{r+1}. \tag{9.22}$$

The error estimate (9.22) implies that

$$\|e\| \le Ch^{r+\frac{1}{2}}\|u\|_{r+1}, \tag{9.23a}$$

$$\|e_\beta\| \le Ch^{r}\|u\|_{r+1}. \tag{9.23b}$$

Thus the L_2-error is nearly optimal, while the L_2-error of the derivative in the streamline direction is optimal.

By taking $v_h = u_h$ in (9.18) we obtain the corresponding stability estimate for the SUPG method:

$$\|u_h\|_0 + \sqrt{h}\|u_{h,\beta}\|_0 + |u_h|_\Gamma \le C\|f\|_0, \tag{9.24}$$

which is better than that for the Galerkin method. However, the streamline derivative in the SUPG method is still less controlled than that in the original problem (cf. (9.16)). The major weakness of the SUPG method in practical calculation is the non-symmetry of the resulting algebraic equations that makes the solution of large-scale problems difficult.

In recent years, Hughes and coworkers have been promoting the Galerkin/ Least-Squares method (Hughes et al. 1989). In this method, which is more general than the SUPG method and applicable to a wide variety of other problem types, squares of residuals are added to the classic Galerkin method. These terms yield improved stability properties with respect to the classic Galerkin method without degrading the accuracy. For problem (9.1) the formulation of the Galerkin/Least-Squares method is: Find $u_h \in S_h$ such that

$$\begin{aligned}(u_{h,\beta} + u_h, v_h) &+ h(u_{h,\beta} + u_h, v_{h,\beta} + v_h) \\ &= (f, v_h) + h(f, v_{h,\beta} + v_h) \qquad \forall v_h \in S_h. \end{aligned} \tag{9.25}$$

It is easy to show that the mathematical properties (9.23) and (9.24) are still valid for the Galerkin/Least-Squares method.

9.1.3 The Least-Squares Finite Element Method

Now let us apply LSFEM to (9.1). For an arbitrary trial function $v \in S$, we define the residual function $R = v_\beta + v - f$. According to the general formulation given in Sect. 4.8, for problem (9.1) the least-squares variational formulation is: Find $u \in S$ such that

$$(u_\beta + u, v_\beta + v) = (f, v_\beta + v) \qquad \forall v \in S. \tag{9.26}$$

The corresponding LSFEM has the following form: Find $u_h \in S_h$ such that

$$(u_{h,\beta} + u_h, v_{h,\beta} + v_h) = (f, v_{h,\beta} + v_h) \qquad \forall v_h \in S_h. \tag{9.27}$$

Since the residual of the approximate solution $R(u_h) = u_{h,\beta} + u_h - f = e_\beta + e$, we can obtain the error estimates from Theorem 4.7 :

$$\|R(u_h)\|_0 = \|e_\beta + e\|_0 \le Ch^r \|u\|_{r+1}, \tag{9.28a}$$

$$\|e\|_0 \le Ch^r \|u\|_{r+1}, \tag{9.28b}$$

which show that the residual estimate is optimal, but the error estimate for e is one order lower than optimal. In numerical tests (see below) we have observed that the accuracy of the least-squares method is higher than the rth order. It is still an open question whether a better theoretical error estimate for general two- or three-dimensional problems can be obtained.

By using Green's formula (9.5) and the boundary condition (9.1b), estimate (9.28a) can be rewritten as

$$(\|e\|_0^2 + \|e_\beta\|_0^2 + \langle e, e\rangle_{\text{out}})^{\frac{1}{2}} \le Ch^r \|u\|_{r+1}, \tag{9.29}$$

which shows that the error estimate for e_β is optimal.

By taking $v = u$ in (9.26), we can obtain the stability estimate for the continuous problems:

$$\|u\|_0 + \|u_\beta\|_0 + |u|_\Gamma \leq C\|f\|_0. \tag{9.30}$$

By taking $v_h = u_h$ in (9.27), we can obtain the stability estimate for LSFEM solution:

$$\|u_h\|_0 + \|u_{h,\beta}\|_0 + |u_h|_\Gamma \leq C\|f\|_0. \tag{9.31}$$

The estimate (9.31) is of the same structure as estimate (9.30) for the original problem (9.1). It means that the least-squares method has better control of the streamline derivative than the SUPG. We also note that the bilinear form in (9.27) is symmetric and coercive. Therefore, the matrix of the resulting algebraic system is symmetric and positive definite. This is a very important advantage of the least-squares method over other methods in practical computation.

Numerical Results. We choose the following model problem:

$$\frac{\partial u}{\partial x} + \frac{\partial u}{\partial y} = \sin(x+y) \qquad \text{in } \Omega, \tag{9.32a}$$

$$u = 0 \qquad \text{on } \Gamma_{\text{in}}, \tag{9.32b}$$

where $\Omega = \{(x,y) \in \mathbb{R}^2 : 0 < x < 1,\ 0 < y < 1\}$ is the unit square, and $\Gamma_{\text{in}} = \{(x,y) \in \Gamma : x = 0 \text{ or } y = 0\}$, in which Γ is the boundary. This problem has a smooth exact solution $u = \sin(x)\sin(y)$.

A uniform meshes with $n \times n$ bilinear elements is employed. At first, one-point Gaussian quadrature was used for calculating element matrices. As pointed out in Sect. 4.11, in this case, the LSFEM is equivalent to the least-squares finite element collocation method with one collocation point at the center of each element. It is easy to check that in such a collocation method, the number of discretized algebraic equations is equal to n^2; while the number of unknowns is also equal to $N_{\text{node}} - N_{\text{bc}} = (n+1)^2 - 2 \times n - 1 = n^2$. In other words, in such a case we solve a determined system. The numerical result for the convergence rate is shown in Fig. 9.1. The optimal rate, i.e. $\|e\|_0 \sim O(h^2)$, is observed. The numerical rate of convergence with the 2×2 Gaussian rule is also included in Fig. 9.1. In this case, the LSFEM solves an overdetermined system. The convergence rate is around $O(h^{1.75})$, which is suboptimal.

We also repeated the numerical tests with specified extra boundary conditions on the outflow boundary $\Gamma_{\text{out}} = \{(x,y) \in \Gamma : x = 1 \text{ or } y = 1\}$. In this case, the least-squares method with 2×2 Gaussian rule gives the optimal rate of convergence $\|e\|_0 \sim O(h^2)$ (Fig. 9.1).

From these numerical results we may conclude that the LSFEM for the hyperbolic equation has an optimal or near optimal rate of convergence depending on the number of Gaussian points in the calculation of element matrices. More Gaussian points yield slightly less optimal results, because the least-squares method is not able to make the residual of each equation in the overdetermined system equal to zero.

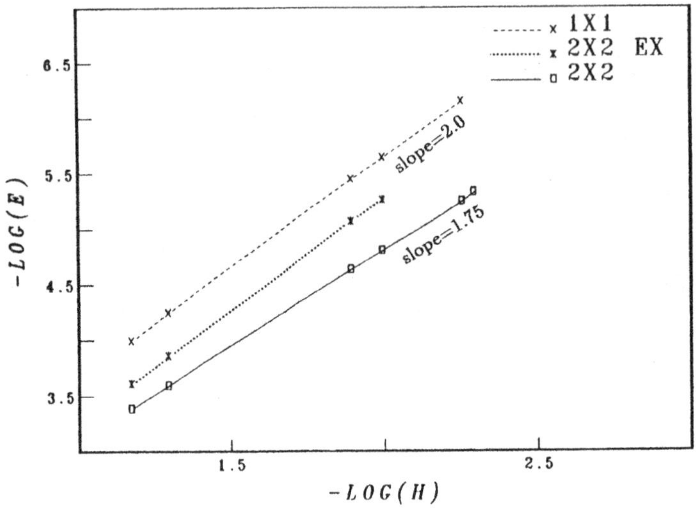

Fig. 9.1. Computed convergence rate for the pure convection problem (from Jiang 1993)

9.2 Contact Discontinuity

9.2.1 Introduction

We shall consider the following steady-state boundary value problem:

$$u_\beta = 0 \qquad \text{in } \Omega, \tag{9.33a}$$

$$u = g \qquad \text{on } \Gamma_{\text{in}}, \tag{9.33b}$$

where g is the specified data on the inflow boundary Γ_{in}. The characteristics of problem (9.33) are the streamlines defined by (9.3), and the analytic solution of problem (9.33) is constant along a characteristic. The value of this constant is equal to the given value of g at the intersection of this characteristic and the inflow boundary. Thus the solution will be discontinuous with a jump across a characteristic curve for two-dimensional cases or a characteristic surface for three-dimensional cases, if the boundary data g is discontinuous.

Usually one uses finite difference or finite element methods based on a mesh which may not be adapted to fit the characteristics of the particular problem. In such a case, if the exact solution has a jump discontinuity (contact discontinuity) across a characteristic, all conventional finite difference and finite element methods will produce approximate solutions which either oscillate or smear out a sharp front. Finding accurate approximations of the discontinuous solutions of hyperbolic equations has been a persistent difficult task in modern numerical mathematics and computational physics.

This trouble comes from the fact that all existing conventional methods are based on discretization of the following equation:

$$\beta_1 \frac{\partial u}{\partial x} + \beta_2 \frac{\partial u}{\partial y} + \beta_3 \frac{\partial u}{\partial z} = 0 \qquad \text{in } \Omega \tag{9.34}$$

in the Cartesian coordinates instead of (9.33a) in the streamline coordinates. If the boundary data g is discontinuous, then (9.33a) and (9.34) are not equivalent mathematically. Equation (9.33a) admits a solution that is discontinuous in the direction $\boldsymbol{\alpha} = (\alpha_1, \alpha_2, \alpha_3)$ which is orthogonal to the streamline direction $\boldsymbol{\beta}$, because in (9.33a) there is no derivative term with respect to the $\boldsymbol{\alpha}$ direction. However, if the solution is discontinuous, (9.34) and its discretized version do not hold, since at least one of the derivatives with respect to the coordinate direction approaches infinity across the jump. Common methods do not account about this trouble, and force (9.34), which looses meaning in the "shock" cells or elements where the front is located, to be satisfied everywhere, and this leads to an oscillatory and diffused solution around the contact front.

If we can identify the "shock" elements, and permit the equations in the "shock" elements not to be satisfied while requiring that the equations in remaining elements be satisfied exactly, the solution will then not have oscillations. This idea can be realized by minimizing the L_1 norm of the residuals of the overdetermined linear algebraic equations.

The L_1 idea can be explained as followings. In the usual L_2 (least-squares) curve fitting approach, the L_2 procedure does its best in the sense of least-squares of the residual to make the curve pass by all of the data. If the data are smooth, the L_2 fitting leads to a very good approximation. However, if the data contain abrupt changes, the L_2 procedure will produce an oscillatory and diffuse curve around sharp changes. In such a case, the L_2 fitting makes the use of individual datum equally important. The tendency of L_1 fitting is to give up the outliers in the data and to require the remaining data be satisfied exactly. Therefore, L_1 fitting is the choice for discontinuous functions. The same thing happens in the L_2 and L_1 solutions of discretized hyperbolic equations. The L_1 approach translates into a capacity to totally give up the equations in the "shock" elements, in which the discretized scheme is not valid, while making the residuals in the smooth elements be zero exactly.

9.2.2 The L_1 Solution to Linear Algebraic Equations

To reveal some important properties of the L_1 method, we considered the L_1 solution of the following overdetermined linear algebraic equations:

$$\begin{pmatrix} k_{11} & k_{12} & \dots & k_{1m} \\ k_{21} & k_{22} & \dots & k_{2m} \\ \vdots & \vdots & \ddots & \vdots \\ k_{n1} & k_{n2} & \dots & k_{nm} \end{pmatrix} \begin{pmatrix} u_1 \\ u_2 \\ \vdots \\ u_m \end{pmatrix} = \begin{pmatrix} f_1 \\ f_2 \\ \vdots \\ f_n \end{pmatrix}, \tag{9.35}$$

or simply

$$\boldsymbol{k}_i \boldsymbol{u} = f_i, \qquad i = 1, ..., n, \tag{9.36}$$

where the number of equations n is greater than the number of unknowns m, and the rank of the augmented matrix (see (4.81)) equals to $m+1$. The L_1 method is a special case of the general L_p $(1 \leq p < \infty)$ methods that minimize the following summation of the pth powered absolute value of the residuals:

$$\begin{aligned} I_p(\boldsymbol{u}) = & \quad W_1 |k_{11}u_1 + k_{12}u_2 + \ldots + k_{1m}u_m - f_1|^p \\ & + \; W_2 |k_{21}u_1 + k_{22}u_2 + \ldots + k_{2m}u_m - f_2|^p \\ & + \; \ldots \\ & + \; W_n |k_{n1}u_1 + k_{n2}u_2 + \ldots + k_{nm}u_m - f_n|^p, \end{aligned} \tag{9.37}$$

where $(W_i > 0,\ i = 1, ..., n)$ are the weighting factors. We have already studied the case $p = 2$ in Sect. 4.11. Since $I_2(\boldsymbol{u})$ is a continuously differentiable function of $\boldsymbol{u}$, the minimizer of $I_2(\boldsymbol{u})$ can be found easily by using the calculus, and the L_2 minimizer is the solution of the normal equation (4.83). However, the situation in the case $p = 1$ becomes difficult. In fact, it is not a simple task to find a L_1 minimizer. Since the residuals are linear functions of $\boldsymbol{u}$, $I_1(\boldsymbol{u})$ is a piecewise linear function of $\boldsymbol{u}$, and its derivatives with respect to $\boldsymbol{u}$ are piecewise constant. In other words, the L_1 norm is not globally differentiable. Therefore the calculus is useless for finding the L_1 minimizer, except for the problems with only one unknown. It is for this reason that people rely on linear programming to deal with L_1 problems.

In the rest of this section we review some basic knowledge about the L_1 solution to the linear algebraic equations. More details can be found in Bloomfield and Steiger (1983).

Let us first consider the simplest overdetermined system with only one unknown:

$$\begin{aligned} k_1 u &= f_1, \\ k_2 u &= f_2, \\ \ldots &= \ldots, \\ k_n u &= f_n, \end{aligned} \tag{9.38}$$

where we assume that the rank of the augmented matrix is 2, and k_i $(i = 1, ..., n)$ is not equal to zero. The L_1 solution problem is to find a minimizer of the L_1 distance function

$$I_1(u) = \sum_{i=1}^{n} |k_i u - f_i| = \sum_{i=1}^{n} |R_i(u)|. \tag{9.39}$$

Now imagine that $f_i/k_i \leq f_{i+1}/k_{i+1}$, which can always be arranged by renumbering the data. If we restrict u to the interval $(f_q/k_q,\ f_{q+1}/k_{q+1})$, $I_1(u)$ in (9.39) becomes

$$I_1(u) = \sum_{i=1}^{q} |k_i|(u - f_i/k_i) - \sum_{i=q+1}^{n} |k_i|(u - f_i/k_i). \tag{9.40}$$

Differentiation reveals that

$$\frac{dI_1(u)}{du} = \sum_{i=1}^{q} |k_i| - \sum_{i=q+1}^{n} |k_i|,$$

which is a constant that cannot decrease if q increase. Since $I_1(u) \geq 0$ is continuous, $I_1(u)$ must be piecewise linear and with a non-decreasing derivative, as shown in Fig. 9.2. As revealed by Fig. 9.2, when u increases in the given data range, $I_1(u)$ always decreases first, then increases. The minimizer always exists, even though $dI_1(u)/du$ may not be zero.

If $dI_1(u)/du = 0$ for an interval $(f_q/k_q,\ f_{q+1}/k_{q+1})$, any u in the closure of this interval minimizes $I_1(u)$. This proves the following lemma.

Lemma 9.1 $I_1(u)$ in (9.39) has a minimizer $\hat{u} = f_i/k_i$ for some $i = 1, ..., n$, say $i = q$. Thus, the L_1 solution to (9.38) satisfies exactly at least one equation, i.e., the residual $R_q(\hat{u})$ is zero.

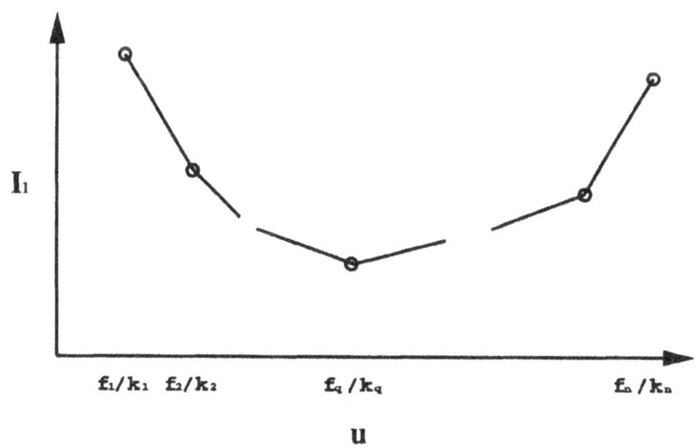

Fig. 9.2. Graph of $I_1(u)$

This motivates a crude algorithm:

Algorithm 9.1 First, compute $u_i = f_i/k_i,\ i = 1, ..., n$; Then, evaluate $I_1(u_i),\ i = 1, ..., n$ and find the minimum, say $I_1(u_q)$, and the minimizer u_q.

A distinct feature could be understood in this simplest context. As mentioned, $\hat{u}$ can be non-unique. For example, any $u \in [1, 2]$ is the L_1 solution to

$$\begin{aligned} u &= 1, \\ u &= 2. \end{aligned} \tag{9.41}$$

The next is an important theorem on the L_1 solution to system (9.35).

Theorem 9.1 (The Existence of the L_1 Minimizer) There exists a minimizer $\hat{\boldsymbol{u}} \in \mathbb{R}^m$ of $I_1(\boldsymbol{u}) = \sum_{i=1}^{n} |\boldsymbol{k}_i\boldsymbol{u} - f_i| = \sum_{i=1}^{n} |R_i(\boldsymbol{u})|$ for which $R_i(\hat{\boldsymbol{u}}) = 0$ for at least m values of i, say $i_1, ..., i_m$.

Proof: Assume that the initial $\boldsymbol{u}$ satisfies exactly some equations in (9.35). Let $\mathbf{Z_u} = \{i : f_i = \boldsymbol{k}_i\boldsymbol{u}\}$ and suppose that the number of independent equations in $\mathbf{Z_u}$ is equal to $s > 0$. Because $s < m$, we can choose $\boldsymbol{v} \neq \mathbf{0} \in \mathbb{R}^m$ such that $\boldsymbol{k}_i\boldsymbol{v} = 0$, $\forall i \in \mathbf{Z_u}$, and $\boldsymbol{k}_i\boldsymbol{v} \neq 0$ for some $i \notin \mathbf{Z_u}$. We then consider

$$\boldsymbol{u}(t) = \boldsymbol{u} + t\boldsymbol{v} \qquad t \in \mathbb{R}.$$

Since $f_i = \boldsymbol{k}_i\boldsymbol{u}$ and $\boldsymbol{k}_i\boldsymbol{v} = 0$, $\forall i \in \mathbf{Z_u}$, we have

$$\begin{aligned} I_1(\boldsymbol{u}(t)) &= \sum_{i \notin \mathbf{Z_u}} |f_i - \boldsymbol{k}_i\boldsymbol{u} - t\boldsymbol{k}_i\boldsymbol{v}| \\ &= \sum_{i \notin \mathbf{Z_u}} |w_i - tz_i|, \end{aligned} \tag{9.42}$$

where $w_i = f_i - \boldsymbol{k}_i\boldsymbol{u} = R_i(\boldsymbol{u}) \neq 0$ and $z_i = \boldsymbol{k}_i\boldsymbol{v}$. Because $z_i \neq 0$ for some $i \notin \mathbf{Z_u}$, Lemma 9.1 applies to show that the minimizer of (9.42) is $t_q = w_q/z_q$, $q \notin \mathbf{Z_u}$. Note also that $t_q \neq 0$, since $w_i \neq 0$, $\forall i \notin \mathbf{Z_u}$.

In passing from the current solution $\boldsymbol{u}$ satisfying $s < m$ equations to $\boldsymbol{u}(t_q)$, the L_1 criterion has not increased because $I_1(\boldsymbol{u}) = I_1(\boldsymbol{u}(0)) \geq I_1(\boldsymbol{u}(t_q))$. The new solution satisfies $s + 1$ equations. Since the argument applies whenever $s < m$, the theorem is proved. □

When $m = 1$ Theorem 9.1 reduces to Lemma 9.1.

Theorem 9.1 implies that the minimizer $\hat{\boldsymbol{u}}$ can be computed using the following crude algorithm:

Algorithm 9.2 For each distinct subset $J = \{j_1, ..., j_m\}$ of $\{1, ..., n\}$ of size m, when possible solve $\boldsymbol{k}_i\boldsymbol{u} = f_i$, $i \in J$ for $\boldsymbol{u}$, $\hat{\boldsymbol{u}}$ is the $\boldsymbol{u}$ that minimizes I_1.

The need to solve C_n^m linear equation systems of size m is a measure of the difficulty of the L_1 method. All current practical algorithms utilize the linear programming characterization in Theorem 9.1 in a more sensible way but are still very expensive. The details of linear programming algorithms is beyond the scope of this book, we refer the reader to Bloomfield and Steiger (1983).

At this stage we should not worry about the efficiency of the L_1 method. We appreciate its important and unique feature: For an overdetermined system of linear equations the L_1 solution exactly satisfies some equations and permits the remaining equations not satisfied.

9.2.3 The L_1 Finite Element Method

In this section we shall combine the L_1 concept with finite element interpolation to obtain highly accurate and non-oscillatory solutions for contact discontinuities. The L_1 method must be based on an overdetermined system. Fortunately, it is trivial to have an overdetermined system in the finite element context. As discussed in Sect. 4.11, the LSFEM with numerical quadrature is equivalent to the weighted least-squares finite element collocation method, in which the equation residuals are first collocated at the interior points in each element, then the algebraic system is approximately solved by the weighted least-squares method. The Gaussian points for calculating the element matrices in LSFEM correspond to the collocation points in the collocation method. If the order of the Gaussian quadrature or the number of quadrature points is appropriately chosen, the LSFEM amounts to solving an overdetermined system. For the same reason, the L_1 finite element method with the Gaussian quadrature is equivalent to the L_1 finite element collocation method.

Let us consider problem (9.33). The L_1 method is based on solving a weighted L_1 problem: Find the minimizer $\hat{u}_h$ of

$$I_1(u_h) = \sum_{j=1}^{N\text{elem}} \left(\sum_{l=1}^{N_{\text{Gauss}}} W_l w_l |R_l| |J(\xi_l, \eta_l)| \right)_j , \tag{9.43}$$

in which

$$R_l = \beta_1 \frac{\partial u_h}{\partial x}(\xi_l, \eta_l) + \beta_2 \frac{\partial u_h}{\partial y}(\xi_l, \eta_l) + \beta_3 \frac{\partial u_h}{\partial z}(\xi_l, \eta_l), \tag{9.44}$$

and R_l stands for the residual at each Gaussian point, N_{Gauss} denotes the number of Gaussian points, w_l is the weight of the Gaussian quadrature, W_l is an additional weighting factor, $|J|$ is the determinant of the Jacobian matrix, and (ξ_l, η_l) is the local coordinates of the Gaussian points. As usual, u_h can be expressed as

$$u_h(\xi, \eta) = \sum_{m=1}^{N\text{node}} \psi_m(\xi, \eta) U_m, \tag{9.45}$$

where "Nnode" is the number of nodes in an element, ψ_m denotes the shape functions, and U_m is the nodal values. In order to make problem (9.43) meaningful, the number of Gaussian points N_{Gauss} should be chosen such that $N\text{elem} \times N_{\text{Gauss}}$ is greater than the number of the total unknown nodal values.

Constant Convection Field. We consider the following two-dimensional problem with a constant convection vector:

$$\frac{\partial u}{\partial x} + \tan(\beta)\frac{\partial u}{\partial y} = 0 \qquad \text{in } \Omega, \tag{9.46a}$$

where $\Omega = \{(x, y) \in \mathbb{R}^2 : 0 < x < 1,\ 0 < y < 1\}$ is the unit square with the boundary Γ. The inflow boundary conditions are

$$u = 2 \quad \text{on } \Gamma_1 = \{(x, y) \in \Gamma : x = 0\}, \tag{9.46b}$$

$$u = 1 \quad \text{on } \Gamma_2 = \{(x, y) \in \Gamma : x > 0 \text{ and } y = 0\}. \tag{9.46c}$$

Equation (9.46) represents uniform flow along straight lines inclined at an angle of β with respect to the x-axis. The jump discontinuity occurs along the line $y = x\tan(\beta)$. In this case, the analytic solution is

$u = 2$ on and above the line $y = x\tan(\beta)$,

$u = 1$ below the line $y = x\tan(\beta)$.

Consider a uniform $n \times n$ mesh generated by linear triangular elements, such as that shown in Fig. 9.3. We use the finite element collocation method with one Gaussian point in each triangle. In this approach there are $2n^2$ elements for $n \times n$ grids, so we have $2n^2$ equations. Since there are $(n+1)^2$ nodal values and $(2n+1)$ boundary conditions, the number of unknowns is $(n+1)^2 - (2n+1) = n^2$. That is, the number of equations is two times of the number of unknowns. Therefore, the L_1 method amounts to solving an overdetermined system. It does not make sense to take more collocation points in each element, because in a linear triangular element $\partial u_h/\partial x$ and $\partial u_h/\partial y$ are constant, and thus the residuals at different points in an element are the same. Now the L_p norm of the residuals is defined as

$$I_p = \sum_{j=1}^{Nelem} W_j |R_j|^p, \tag{9.47}$$

where the Gaussian weights and the Jacobians have been suppressed, since they are constant and thus have no effect. The general L_p method is based on minimizing the total L_p norm of the equation residuals in the whole domain. Here we must emphasize the words "total" and "whole", since the correct solution or the minimizer of the total L_p norm may not minimize the L_p norm for each element or for a local subdomain.

For two typical linear triangles shown in Fig. 9.4, the residuals of (9.46a) can be expressed as

$$R_{\text{upper}} = U_{\text{ne}} - U_{\text{nw}} + \tan(\beta)(U_{\text{nw}} - U_{\text{sw}}), \tag{9.48a}$$

$$R_{\text{lower}} = U_{\text{se}} - U_{\text{sw}} + \tan(\beta)(U_{\text{ne}} - U_{\text{se}}), \tag{9.48b}$$

in which the uniform mesh size h has no effect and thus has been suppressed.

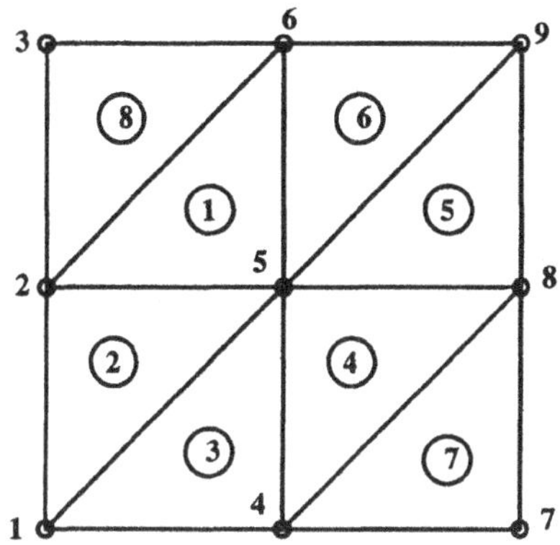

Fig. 9.3. A 2 × 2 mesh with eight triangular elements

As the simplest example, we assume that $\beta = 35°$ and the domain is divided into two triangles shown in Fig. 9.4. Due to the inflow boundary condition, we know that

$$U_{\text{nw}} = U_{\text{sw}} = U_{\text{L}}, \qquad U_{\text{se}} = U_{\text{R}}.$$

The unknown is $U_{\text{ne}} = U^*$. By using (9.48) the L_1 norm of this problem can be written as

$$I_1(U^*) = |U^* - U_{\text{L}} + \gamma(U_{\text{L}} - U_{\text{L}})| + |U_{\text{R}} - U_{\text{L}} + \gamma(U^* - U_{\text{R}})|, \tag{9.49}$$

where $\gamma = \tan(35°)$.

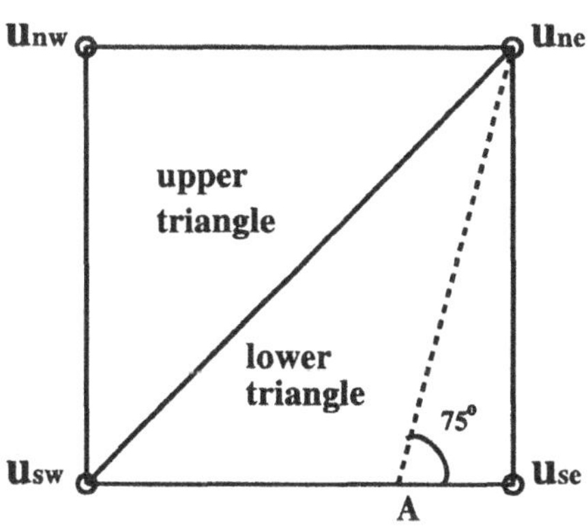

Fig. 9.4. Two typical linear triangular elements

By substituting $U_{\text{L}} = 2$ and $U_{\text{R}} = 1$ into (9.49) we have

$$I_1(U^*) = |U^* - 2| + |\gamma U^* - (1 + \gamma)|. \tag{9.50}$$

According to Algorithm 9.1 the feasible solutions of U^* are 2 and $(1+\gamma)/\gamma$. When $\beta = 35°$, then $(1+\gamma)/\gamma = 2.4281$. Since $I_1(2) = 0.2998 < I_1(2.4281) =$

0.4281, the minimizer of I_1 in (9.50) is $U^* = 2$, and the equation in the upper triangle is exactly satisfied.

In the above example, if $\beta = 75°$, then $(1+\gamma)/\gamma = 1.2679$, and $I_1(1.2679) = 0.7321$ is less than $I_1(2) = 2.7321$. Therefore, $U^* = 1.2679$ is the minimizer, and the equation residual in the lower triangle is equal to zero exactly. Although, in this case U^* equals neither 2 nor 1, it is still a correct solution. Since U^* should be equal to the given boundary value at the point A in Fig. 9.4, and according to the linear interpolation $U_A = 1.2679$.

Clearly, for these simple examples with both $\beta \leq 45°$ and $\beta > 45°$ the L_1 method produces correct solutions.

As another example, we choose a uniform 2×2 mesh with eight triangles illustrated in Fig. 9.3 to analyze. For simplicity, assume that all boundary data are given in advance, namely, the conditions on the outflow boundary are also prescribed. In this case, referring to Fig. 9.3, the boundary conditions are the following:

$$U_1 = U_2 = U_3 = U_6 = U_9 = U_L = 2,$$

$$U_4 = U_7 = U_8 = U_R = 1.$$

In this problem, the only unknown is the nodal value $U_5 = U^*$ at the middle node 5.

By using (9.47) and (9.48) the L_p norm of this problem is given by

$$\begin{aligned} I_p(U^*) = \quad & W_1|U^* - U_L + \gamma(U_L - U^*)|^p \\ + \quad & W_2|U^* - U_L + \gamma(U_L - U_L)|^p \\ + \quad & W_3|U_R - U_L + \gamma(U^* - U_R)|^p \\ + \quad & W_4|U_R - U^* + \gamma(U^* - U_R)|^p \\ + \quad & W_5|U_R - U^* + \gamma(U_L - U_R)|^p \\ + \quad & W_6|U_L - U_L + \gamma(U_L - U^*)|^p. \end{aligned} \tag{9.51}$$

At first let us examine the case $p = 1$ with $W_i = 1$. This is also the simplest L_1 minimization problem that we have studied in the previous section. Due to Lemma 9.1, the minimizer exists. As an exercise we would like to use Algorithm 9.1 to find the minimizer. According to the standard form (9.39), we may write $I_1(U^*)$ as:

$$\begin{aligned} I_1(U^*) = \quad & |(1-\gamma)U^* - (1-\gamma)U_L| \\ + \quad & |U^* - U_L| \\ + \quad & |\gamma U^* - (U_L - (1-\gamma)U_R)| \\ + \quad & |(1-\gamma)U^* - (1-\gamma)U_R)| \\ + \quad & |U^* - (\gamma U_L + (1-\gamma)U_R)| \\ + \quad & |\gamma U^* - \gamma U_L|. \end{aligned} \tag{9.52}$$

Substituting $\gamma = 0.7002, U_L = 2$ and $U_R = 1$ into (9.52), and using the first part of Algorithm 9.1 we find that the candidates for the minimizer are

2, 2, 2.428, 1, 1.700 and 2 which are the solution of the collocated equation in the corresponding element. A further computation reveals that $I_1(2) = 0.8994$ is the minimum. The correct solution $U^* = 2$ is the L_1 minimizer. In this case, three equations taken from the elements 1, 2, 6 are exactly satisfied, whereas other three equations from the elements 3, 4, 5 are given up.

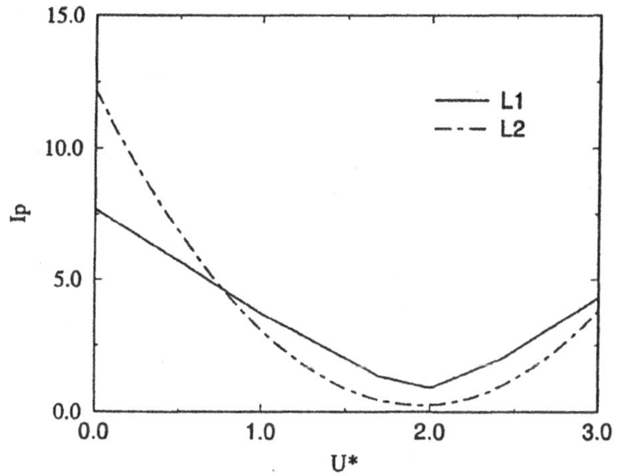

Fig. 9.5. Graph of $I_p(U^*)$

In fact, it is easy to find the minimizer of the L_p norm in (9.52) by the graphical method. Figure 9.5 shows $I_p(U^*), W_i = 1$ for $p = 1, 2$. From Fig. 9.5 we clearly see that I_1 reaches a minimum at the exact solution $U^* = 2.0$, and I_2 has a minimum at $U^* = 1.94$ which is a good approximation of the correct solution.

In the next numerical experiment for the original problem (9.46) we still choose a uniform 2×2 mesh with eight triangles shown in Fig. 9.3. This time the boundary data are given only on the inflow boundary:

$$U_1 = U_2 = U_3 = U_{\mathrm{L}} = 2,$$

$$U_4 = U_7 = U_{\mathrm{R}} = 1.$$

Now the unknowns are the nodal values U_5, U_6, U_8 and U_9 at the nodes 5, 6, 8, and 9, respectively. In this case, the equations in the elements 7 and 8 must be taken into account. The L_1 finite element method corresponds to the L_1 finite element collocation method which deals with eight equations in four unknowns. By using Algorithm 9.2 or the linear programming algorithm of Barrodale and Roberts (Bloomfield and Steiger 1983), one can find the L_1 solution: $U_5 = U_6 = U_9 = 2,\ U_8 = 1$ which are correct discontinuous solutions for the problem.

9.2.4 The Iteratively Reweighted LSFEM

The L_1 solutions are non-oscillatory, highly accurate and right up to the edge of the discontinuity. However, rigorous L_1 minimization based on linear programming is expensive; for large practical problems we need an efficient method. Since the LSFEM discussed in Sect. 9.1.3 produces a very good initial approximation to the exact discontinuos solution, we can use this initial information to find "shock" elements where the absolute value of the residuals are larger than that in the smooth elements. Then we use the least-squares method again. But this time we put a small weight for the equation residuals in the "shock" elements to suppress their interference, and repeat this procedure a few times until a convergent discontinuous solution is reached. This is the iteratively reweighted least-squares finite element method.

Reweighting must be based on an overdetermined system, because weighting has no effect for determined linear equations as explained in Sect. 4.11.1. We have emphasized many times that the LSFEM with numerical quadrature is equivalent to the weighted least-squares finite element collocation method; the LSFEM with an appropriate order of the Gaussian quadrature amounts to solving an overdetermined system.

Let us consider problem (9.33). The iteratively reweighted least-squares method is based on repeatedly solving a weighted least-squares problem: Find the minimizer $\hat{u}_h$ of

$$I(u_h) = \sum_{j=1}^{Nelem} \Big(\sum_{l=1}^{N_{Gauss}} W_l w_l (R_l)^2 |J(\xi_l, \eta_l \Big)|)_j , \tag{9.53}$$

in which

$$W_l = \frac{1}{(|R_l|^6_{\text{previous}} + \epsilon)} , \tag{9.54}$$

where W_l denotes the weight set, which depends on the information from the previous step, ϵ is a very small positive number, for example ϵ = 1.0E-20, to prevent an overflow in computation, and all other notations have been defined in (9.43) and (9.44).

The IRLSFEM would begin with the initial weight $W_l = 1$. This first step is nothing but the least-squares method introduced in Sect. 9.1.3. The result then determines a new set of weights by (9.54). In the second iteration, the residual $|R_l|$ is larger in the "shock" elements. Thus the weight W_l for the "shock" elements is smaller, and their influence becomes less important. This procedure is repeated until $\|\hat{u}_{h_{\text{current}}} - \hat{u}_{h_{\text{previous}}}\|$ is small. Our numerical experiments reveal that the difference between the residuals of "shock" elements and those of their neighboring elements in the first least-squares solution is not significant enough. We put the sixth power in (9.54) in order to additionally increase the importance of "smooth" elements and reduce the contamination of "shock" elements, and also accelerate the convergence. This is reasonable, since we want to eliminate completely equations which will have nonzero residuals from the system. This trick is applicable because the non-weighted least-squares method is good enough to locate the "shock" elements. That is, in the results of LSFEM, the absolute value of the residuals in the "shock" elements is always greater than that in other elements.

We may use another simple and reliable "shock" indicator – the variation of nodal values in each element – instead of the residual. The variation is defined as

$$V = \sum_{m=1}^{N\text{node}} |U_m - U_{m-1}| \qquad U_0 = U_{N\text{node}}. \tag{9.55}$$

Therefore, the following weight is suggested:

$$W_l = \begin{cases} 10^{15} & \text{if } |V|^6_{\text{previous}} < 10^{-7}; \\ \frac{1}{|V|^6_{\text{previous}}} & \text{otherwise.} \end{cases} \tag{9.56}$$

Here some measures have been taken to prevent overflow. The advantage of using the variation as a "shock" indicator is as follows: Once the jump in the boundary data is given, we may know the exact values of the variation in the "shock" elements in advance. There are only a few possible values, which depend on the type of finite element and are independent of the shape and size of the particular element, and have no relation with the location of quadrature points.

The implementation of this reweighted least-squares method is relatively straightforward. If a least-squares finite element code is already available, it needs only a few additional lines of FORTRAN statements.

9.2.5 Numerical Results of IRLSFEM

Constant Convection Field. We consider (9.46) with $\beta = 35^o$ as an example. Most of the computational results presented in this section are obtained in double precision on a PC-386. A direct solver with variable band-width is used to obtain the solution of the resulting linear algebraic equations. The computing time can be significantly shortened by using the preconditioned conjugate gradient method, since the least-squares solution is already close to the accurate solution and final iterations are performed often just for correcting one or two nodal values which have not yet reached 15-digit accuracy.

Many numerical experiments were carried out by using linear triangular elements on both structured and unstructured meshes and also by using bilinear elements on uniform meshes, from very coarse (1×1, 2×2 and 5×5) to fine (100×100) meshes. In all cases, we obtained perfect discontinuous solutions with nodal values of correct 15-digits. These solutions have neither oscillation nor diffusion. The transition over the discontinuity is accurately located in the vicinity of the line $y = x\tan(\beta)$, and is accomplished across just one element.

For triangular elements, we use one-point Gaussian quadrature. The IRLSFEM results for $n = 5$ (50 triangles) are listed in Table 9.1. In this table the number at the left lower corner is the nodal value at $x = 0$ and $y = 0$; the number at the right upper corner is the nodal value at $x = 1$ and $y = 1$; the rest of the numbers are the nodal values at the corresponding locations. This solution is obtained after 5 iterations and is correct to 15 digits. The IRLSFEM solution for $n = 15$ (450 triangles) is illustrated in Fig. 9.6.

Table 9.1. Nodal values of IRLSFEM solution for constant convection problem (50 triangles) (from Jiang 1993)

2.00000000000000	2.00000000000000	2.00000000000000	2.00000000000000	2.00000000000000	2.00000000000000
2.00000000000000	2.00000000000000	2.00000000000000	2.00000000000000	2.00000000000000	2.00000000000000
2.00000000000000	2.00000000000000	2.00000000000000	2.00000000000000	2.00000000000000	2.00000000000000
2.00000000000000	2.00000000000000	2.00000000000000	2.00000000000000	1.00000000000000	1.00000000000000
2.00000000000000	2.00000000000000	1.00000000000000	1.00000000000000	1.00000000000000	1.00000000000000
2.00000000000000	1.00000000000000	1.00000000000000	1.00000000000000	1.00000000000000	1.00000000000000

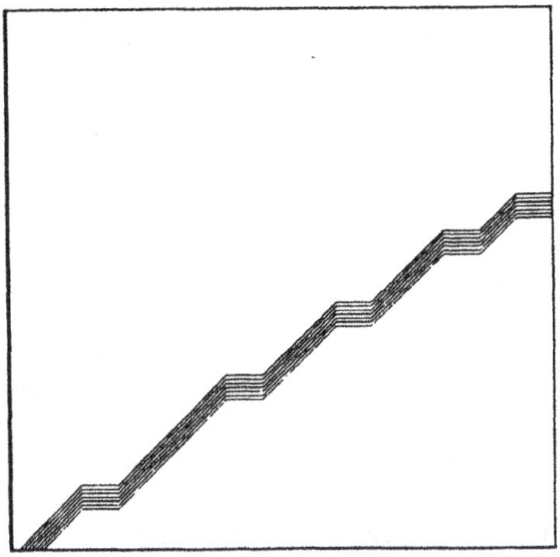

Fig. 9.6. Contours of IRLSFEM solution for constant convection problem (50 triangles) (from Jiang 1993)

For bilinear elements we use 2×2 quadrature. In each element we may write the finite element approximation of u as a bilinear function:

$$u_h(x, y) = a + bx + cy + dxy.$$

Thus the residual is

$$R_h = \frac{\partial u_h}{\partial x} + \tan(\boldsymbol{\beta})\frac{\partial u_h}{\partial y} = b + \tan(\boldsymbol{\beta})c + d(y + x\tan(\boldsymbol{\beta})),$$

which means that the four discretized equations at four Gaussian points are independent in this case. (If the flow is inclined at an angle of 45° or 135° with respect to the x-axis, we have three independent equations, because the location of Gaussian points is symmetric.) All together we have $4n^2$ equations involving n^2 unknown nodal values. Therefore, we deal with an overdetermined system.

The contours of the least-squares solution on a mesh with 40×40 bilinear elements are illustrated in Fig. 9.7. This approximate solution is reasonably good, although the discontinuity is smeared out. Taking this initial least-squares solution, and after 8 steps of processing, we obtained a clean, non-diffusive solution illustrated in Fig. 9.8.

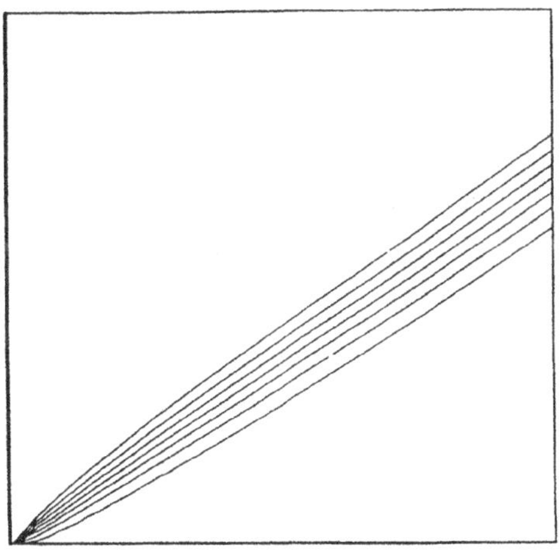

Fig. 9.7. Contours of LSFEM solution for constant convection problem (40×40 bilinear elements) (from Jiang 1993)

Spatially Varying Convection Field. Following Deconinck et al. (1991), we considered problem (9.33) in the unit square with a circular convection field:

$$\beta_1 = y, \tag{9.57a}$$

$$\beta_2 = 0.5 - x, \tag{9.57b}$$

and boundary conditions:

$$u(0, y) = 0, \tag{9.57c}$$

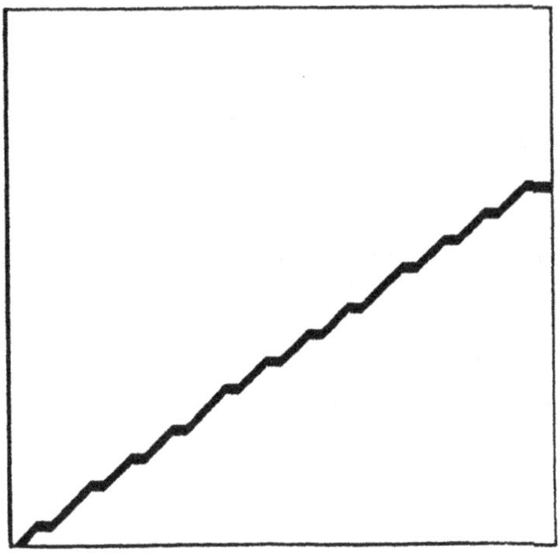

Fig. 9.8. Contours of IRLSFEM solution for constant convection problem (40×40 bilinear elements) (from Jiang 1993)

$$u(x,1) = 0, \ x \geq 0.5 \tag{9.57d}$$

$$u(x,0) = \begin{cases} 0 & x \leq 0.17; \\ 1 & 0.17 < x < 0.33; \\ 0 & 0.33 \geq x < 0.5. \end{cases} \tag{9.57e}$$

The mesh with 100×100 uniform bilinear elements is employed. The size of the elements in this mesh is 0.01 (comparing with 0.005 in the mesh used by Deconinck et al. 1991). Therefore, we have modified the data a little bit in (9.57e) in order to make them consistent with our coarser mesh. In the least-squares finite element solution with 2×2 quadrature, as seen in Fig. 9.9, the input distribution (the left half of the lower boundary) is smeared quite a bit by the time it reaches the outflow (the right half of the lower) boundary. Starting from this diffused solution, we still obtained a completely non-oscillatory and non-diffusive solution with correct jumps across just one element. However, the location of computed contact discontinuities on the outflow boundary has deviation of one or two grids from the exact solution. As indicated in Sect. 9.1.3, we may use reduced integration (one-point quadrature) to obtain a better initial solution which is oscillatory but contains correct information about the location of discontinuities. Starting from this oscillatory solution, after 8 reweighted iterations with 2×2 quadrature we obtain a perfect 14-digit correct discontinuous solution shown in Fig. 9.10. The distribution of u on the bottom boundary of the domain is shown in Fig. 9.11. To our knowledge, no other currently available methods can produce such a sharp and highly accurate resolution.

Remark. The L_1 method and the iteratively reweighted LSFEM are designed for problems with discontinuous solutions and obviously cannot be

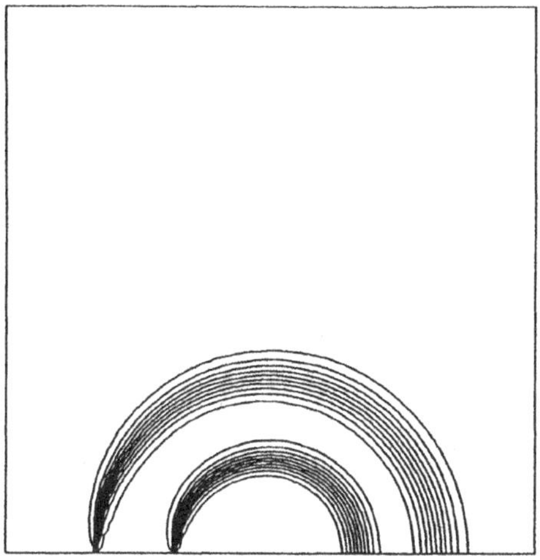

Fig. 9.9. Contours of LSFEM solution for circular convection problem (100 × 100 bilinear elements) (from Jiang 1993)

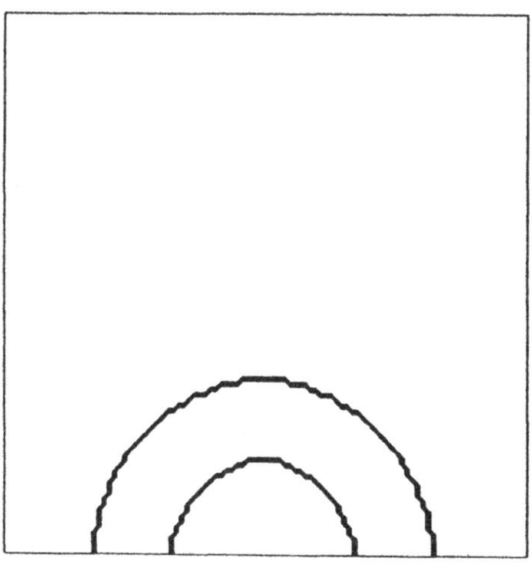

Fig. 9.10. Contours of IRLSFEM solution for circular convection problem (100 × 100 bilinear elements) (from Jiang 1993)

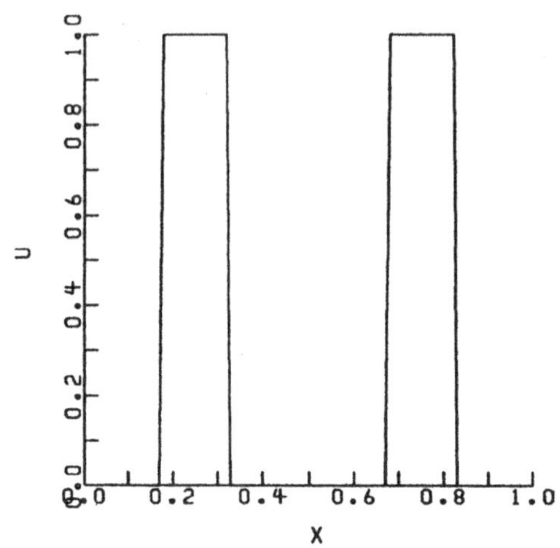

Fig. 9.11. Boundary distribution of u for circular convection problem (from Jiang 1993)

used for problems with purely smooth solutions such as problem (9.32). In the smooth case the L_1 method will take elements with larger gradients as "shock" elements and completely ignore them, and will then yield a wrong discontinuous solution. For problems with smooth data, we should use the ordinary LSFEM.

If the given data g in problem (9.33) are varying along the boundary Γ_{in} with large gradients and a very small jump, any method will have difficulty. This is related to the numerical measure of discontinuity which is different from the mathematical definition of discontinuity. That is, we must distinguish discontinuity from large gradients. The distinction can be realized easily by using fine meshes such that the difference between the nodal values at any two neighboring nodes on Γ_{in} is less than the jump and by specifying an appropriate level for a particular problem. Only if the variation indicator of an element defined in (9.55) exceeds this level, should the contribution of this element be eliminated.

The choice of the sixth power in (9.54) or (9.56) is based on our experience. In fact, the method is not sensitive to the number of the power. The 3rd, 4th, 5th, 7th...powers, all of them, work well. The only difference is that the number of iterations to convergence is different. Here the essential issue is how to completely eliminate the equations in the "shock" elements, and it doesn't matter what means are employed.

Lowrie and Roe (1994) tried to analyze the L_1 method and the iteratively reweighted least-squares method, and claimed that they had "proved" a "distressing theorem" which says that correct discontinuos solutions cannot minimize any weighted residuals and the minimizer of any weighted residuals cannot be a correct solution. Clearly, their statement is incorrect, because it is against the residual minimization theorem (Theorem 4.2) which is rigorously proved in a very general situation. In particular, their "mathematical analysis" is erroneous. They mistakenly chose a segment of solution to investigate, and found that the correct solution does not minimize the L_1 norm of a subdomain, and hence made their conclusion. As pointed out in Sect. 9.2.3, the minimizer of the global L_1 norm may not minimize the local L_1 norm of a subdomain. Moreover, the L_1 norm is the summation of the absolute values of the element residuals, and thus is a piecewise linear function; therefore as emphasized in Sect. 9.2.2, to find a L_1 minimizer is not a simple task that can be performed just by using easy calculus which they adopted. Unfortunately, in their calculation of the derivative of $|R_i|$ with respect to the unknown, the sign of the absolute value was completely ignored.

9.3 Transient Problems

In this section we consider transient convection problems:

$$\frac{\partial u}{\partial t} + \boldsymbol{\beta} \cdot \nabla u = 0 \qquad \text{in } \Omega \times (0,\ T), \tag{9.58a}$$

$$u = g \qquad \text{on } \Gamma_{\text{in}} \times (0,\ T), \tag{9.58b}$$

$$u = u_0 \qquad \text{in } \Omega \ \text{ at } \ t = 0, \tag{9.58c}$$

where the convective velocity vector $\boldsymbol{\beta}$ is assumed known and divergence-free in Ω, and is time-independent for simplicity. The smooth boundary data g is prescribed only on the inflow boundary Γ_{in}, and the initial value u_0 is sufficiently smooth.

9.3.1 The Taylor–Galerkin Method

The classic Galerkin variational problem is to find $u \in S = \{H^1(\Omega) : u = g \text{ on } \Gamma_{\text{in}}\}$ for each t, such that

$$\left(\frac{\partial u}{\partial t} + \boldsymbol{\beta} \cdot \nabla u,\ w\right) = 0 \qquad \forall v \in V = \{H^1(\Omega) : v = 0 \text{ on } \Gamma_{\text{in}}\}, \tag{9.59}$$

together with the initial condition

$$u(0) = u_0.$$

For convective transport problems, as in the steady-state case, the classic Galerkin finite element method suffers from spurious oscillations in the same way as the central finite differencing, see an example in Sect. 2.7.5. Motivated by backward differencing to reflect the 'upstream' dependence of the solution, the Petrov–Galerkin method and the streamline upwind Petrov–Galerkin (SUPG) method were developed. Their formulations are similar to those in the steady-state case (see Sect. 9.1.2).

One of appropriate approaches for transient problems is the Taylor–Galerkin Method developed by Donea (1984). It is motivated by the ideas of the Lax–Wendroff scheme in the finite difference method. Let us briefly describe the Taylor–Galerkin method. Since (9.58a) defines an evolution statement, there must exist a Taylor expansion

$$u^{n+1} = u^n + \Delta t u_t^n + \frac{1}{2}\Delta t^2 u_{tt}^n + \frac{1}{6}\Delta t^3 u_{ttt}^n + O(\Delta t^4), \tag{9.60}$$

where u^n is for $u(\boldsymbol{x}, t_n)$, $\Delta t = t_{n+1} - t_n$, and the subscript 't' denotes the order of temporal derivative at t_n. Equation (9.58a) permits restatement of the successive derivatives with respect to time in (9.60) as

$$u_t = -\boldsymbol{\beta} \cdot \nabla u = -u_\beta, \tag{9.61a}$$

$$u_{tt} = -\boldsymbol{\beta} \cdot \nabla u_t = \boldsymbol{\beta} \cdot \nabla(\boldsymbol{\beta} \cdot \nabla u) = u_{\beta\beta}, \tag{9.61b}$$

$$u_{ttt} = u_{t\beta\beta}. \tag{9.61c}$$

Substituting (9.61) into the Taylor series (9.60) and rearranging some terms yields the following semidiscrete equation:

$$\left(1 - \frac{1}{6}\Delta t^2 \partial_{\beta\beta}\right)\left(\frac{u^{n+1} - u^n}{\Delta t}\right) = -u_\beta^n + \frac{1}{2}\Delta t u_{\beta\beta}^n + O(\Delta t^3), \tag{9.62a}$$

or

$$\frac{u^{n+1} - u^n}{\Delta t} + u_\beta^n - \frac{1}{2}\Delta t u_{\beta\beta}^n - \frac{1}{6}\Delta t^2 \partial_{\beta\beta}\left(\frac{u^{n+1} - u^n}{\Delta t}\right) = 0, \tag{9.62b}$$

where ∂_β denotes the directional derivative $\boldsymbol{\beta} \cdot \nabla$.

Testing (9.62) against $v \in V$ and using Green's formula (9.5) we obtain the Taylor–Galerkin variational statement: Find $u \in S$ such that

$$\left(\frac{u^{n+1} - u^n}{\Delta t}, v\right) + \frac{1}{6}\Delta t^2 \left(\left(\frac{u^{n+1} - u^n}{\Delta t}\right)_\beta, v_\beta\right)$$

$$-\frac{1}{6}\Delta t^2 \left\langle\left(\frac{u^{n+1} - u^n}{\Delta t}\right)_\beta, v\right\rangle_{\text{out}}$$

$$= -(u_\beta^n, v) - \frac{1}{2}\Delta t (u_\beta^n, v_\beta) + \frac{1}{2}\Delta t \langle u_\beta^n, v\rangle_{\text{out}} \qquad \forall v \in V. \tag{9.63}$$

For simplicity of exploration we consider a scalar, one-dimensional version of problem. In the case of a uniform mesh of linear elements, (9.63) takes the following form at an interior node j:

$$\left[1 + \frac{1}{6}(1 - c^2)\delta^2\right](U_j^{n+1} - U_j^n)$$

$$= -\frac{c}{2}(U_{j+1}^n - U_{j-1}^n) + \frac{c^2}{2}(U_{j+1}^n - 2U_j^n + U_{j-1}^n), \tag{9.64}$$

where $c = \beta\Delta t/h$ is the CFL (Courant–Friedrichs–Lewy) number, and δ^2 is the standard notation for the central difference operators, namely,

$$(\delta^2 U)_j = U_{j+1} - 2U_j + U_{j-1}.$$

The scheme in (9.64) represents a third-order accurate time-marching method. Its stability condition is $c \leq 1$. The condition for numerical stability is $c \leq 1/(2\sqrt{2}) = 0.353$ for the two-dimensional version, and $c \leq 1/(3\sqrt{3}) = 0.192$ for the three-dimensional version. Many other forms of the Taylor–Galerkin method with different numerical properties are also available (Quartapelle 1993).

In general, the Taylor–Galerkin method is very efficient, and has been successfully applied in the solution of various problems. However, it is worth mentioning some minor distractions of the method. Since the term (u^{n+1}, v) introduces the mass matrix, the Taylor–Galerkin method requires inversion of the mass matrix. To maintain the explicit character of the method, one may use the lumped (diagonal) mass matrix. The method thus suffers from more severe stability limitations. Furthermore, the Taylor–Galerkin method is essentially a high-order scheme, and hence may promote oscillations at discontinuities. Artificial dissipation can be introduced to suppress oscillations, but the form of this added term is not unique and the associated parameters must be specified.

9.3.2 The Least-Squares Finite Element Method

To introduce the essential ideas behind LSFEM, we consider again the prototype scalar first-order hyperbolic equation (9.58). With time-differencing (9.58a) through the θ method, the resulting semidiscrete problem is

$$\frac{u^{n+1} - u^n}{\Delta t} + [(1-\theta)u^n_\beta + \theta u^{n+1}_\beta] = 0 \tag{9.65}$$

with $0 \leq \theta \leq 1$. As mentioned in Sect. 8.6, $\theta = 1$ corresponds to the backward-Euler differencing scheme and has first-order accuracy $O(\Delta t)$; $\theta = 1/2$ represents the Crank–Nicolson scheme and leads to second-order accuracy $O(\Delta t^2)$.

We are now ready to apply LSFEM. We need only to write (9.65) in the standard form of the first-order system (4.48), then use the computer program based on the general formulation given in Sect. 4.8 to find the solution for a particular problem.

It is useful to understand how the method works. For this purpose we deduce the least-squares variational formulation and its associated Euler–Lagrange equation. For any admissible $u^{n+1}(\boldsymbol{x}) \in S$, we obtain from (9.65) the residual R, and the least-squares functional becomes

$$I(u^{n+1}) = \int_\Omega R^2 d\Omega = \left\| \frac{u^{n+1} - u^n}{\Delta t} + [(1-\theta)u^n_\beta + \theta u^{n+1}_\beta] \right\|_0^2. \tag{9.66}$$

Applying the stationary condition $\delta I = 0$, and setting $\delta u^{n+1} = v$, we obtain the least-squares variational statement for (9.65): Find $u^{n+1} \in S$ such that

$$(u^{n+1} + \theta \Delta t u_\beta^{n+1},\ v + \theta \Delta t v_\beta)$$
$$= \left(u^n - (1-\theta)\Delta t u_\beta^n,\ v + \theta \Delta t v_\beta\right) \qquad \forall v \in V. \tag{9.67}$$

The least-squares method (9.67) has the property of unconditional stability for $1/2 \leq \theta \leq 1$.

Clearly (9.67) can be interpreted as a Petrov–Galerkin weighted-residual statement for (9.65) with the test function

$$\tilde{v} = (1 + \theta \Delta t \partial_\beta) v.$$

More important, we note that the convective velocity $\boldsymbol{\beta}$ enters explicitly in this test function so that this form is analogous to the SUPG method. Hence, the effect of the least-squares procedure in two or three dimensions is to add numerical dissipation preferentially in the local flow direction in a manner similar to streamline upwinding. However, by contrast with the SUPG method and the Galerkin/Least-Squares method described in Sect. 9.1.2, the least-squares method does not contain free parameters.

For transient problems, the LSFEM based on the increment form is useful. For simplicity, assuming that the boundary condition (9.58b) is independent of time, introducing the increment $\phi = \Delta u = u^{n+1} - u^n$, and considering that the increment ϕ satisfies the homogeneous boundary condition on the inflow boundary, the formulation (9.67) may be rewritten in the Δ-form: Find $\phi = \Delta u \in V$ such that

$$(\phi + \theta \Delta t \phi_\beta, v + \theta \Delta t v_\beta) = -\Delta t (u_\beta^n,\ v + \theta \Delta t v_\beta) \qquad \forall v \in V. \tag{9.68}$$

By using Green's formula (9.5) it is easy to show that

$$\|\phi\|_0^2 + \theta^2 \Delta t^2 \|\phi_\beta\|_0^2 + \theta \Delta t |\phi|_\Gamma^2 = \|\phi + \theta \Delta t \phi_\beta\|_0^2 \qquad \forall \phi \in V, \tag{9.69}$$

which implies that the bilinear form in (9.68) is bounded below in some sense, and thus the resulting matrix in LSFEM is symmetric and positive definite.

Error Estimates. Now let us analyze the LSFEM based on (9.68): Find $\phi_h \in V_h$ such that

$$(\phi_h + \theta \Delta t \phi_{h,\beta},\ v_h + \theta \Delta t v_{h,\beta}) = -\Delta t (u_\beta^n,\ v_h + \theta \Delta t v_{h,\beta}) \quad \forall v_h \in V_h. \tag{9.70}$$

From (9.70) and (9.68) we obtain the following orthogonality condition for the error $e = \phi - \phi_h$:

$$(e + \theta \Delta t e_\beta,\ v_h + \theta \Delta t v_{h,\beta}) = 0 \qquad \forall v_h \in V_h. \tag{9.71}$$

Let $\Pi_h \phi \in V_h$ be the finite element interpolant of ϕ satisfying (9.7) and write $\rho = \phi - \Pi_h \phi$ and $\eta = \Pi_h \phi - \phi_h$ so that $e = \rho + \eta$. Then, by using (9.71) we have

$$\begin{aligned} \|e + \theta \Delta t e_\beta\|_0^2 &= (e + \theta \Delta t e_\beta,\ \rho + \theta \Delta t \rho_\beta) \\ &\leq \|e + \theta \Delta t e_\beta\|_0 \|\rho + \theta \Delta t \rho_\beta)\|_0, \end{aligned}$$

or

$$\|e + \theta \Delta t e_\beta\|_0^2 \leq \|\rho + \theta \Delta t \rho_\beta)\|_0^2. \tag{9.72}$$

From (9.7), (9.72) and (9.69) we obtain

$$\begin{aligned} \|e\|_0^2 + \theta^2 \Delta t^2 \|e_\beta\|_0^2 + \theta \Delta t |e|_\Gamma^2 &\leq \|\rho + \theta \Delta t \rho_\beta)\|_0^2 \\ &\leq C(h^{2(r+1)} + \theta^2 \Delta t^2 h^{2r}) \|\phi\|_{r+1}^2, \end{aligned} \tag{9.73}$$

where C is a general constant independent of h and ϕ, and may have a different value when it appears in different places. If the time-step $\Delta t = O(h)$, we can then use (9.73) to establish the optimal error estimates for each time-step solution:

$$\|e\|_0 \leq C h^{r+1} \|\phi\|_{r+1}, \tag{9.74a}$$

$$\|e_\beta\|_0 \leq C h^r \|\phi\|_{r+1}, \tag{9.74b}$$

$$|e|_\Gamma \leq C h^{r+1/2} \|\phi\|_{r+1}. \tag{9.74c}$$

Numerical Properties. It is instructive to derive the Euler–Lagrange equation corresponding to the variational statement (9.68). Separating terms and using Green's formula (9.5), we obtain, on simplification,

$$\Big((1 - \theta \Delta t \partial_\beta)[(1 + \theta \Delta t \partial_\beta)\phi + \Delta t u_\beta^n],\ v\Big) + \text{boundary term} = 0, \tag{9.75}$$

so that the Euler–Lagrange equation is

$$(1 - \theta^2 \Delta t^2 \partial_{\beta\beta}) \Big(\frac{u^{n+1} - u^n}{\Delta t}\Big) = -(1 - \theta \Delta t \partial_\beta) u_\beta^n. \tag{9.76}$$

Note that (9.76) can also be written in the form

$$\frac{u^{n+1} - u^n}{\Delta t} + u_\beta^n - \theta \Delta t u_{\beta\beta}^n - \theta^2 \Delta t^2 \partial_{\beta\beta} \Big(\frac{u^{n+1} - u^n}{\Delta t}\Big) = 0. \tag{9.77}$$

The LSFEM can be interpreted as a Galerkin method based on (9.77); whereas the Taylor–Galerkin method is a Galerkin method based on (9.62b). The form of (9.62b) and the Euler–Lagrange equation (9.77) from LSFEM with $\theta = 1/2$ differ very slightly in the last term (factor 1/6 versus 1/4).

Some features of the qualitative behavior of the least-squares solution may be inferred from the structure of (9.77). If the first two terms are interpreted as an approximation to the operator $u_t + u_\beta$, then the next term is a numerical viscosity and the final term is a higher-order effect. The numerical viscosity varies linearly with Δt and the high-order correct depends on $\theta^2 \Delta t^2 u_{t\beta\beta}$ or equivalently $\theta^2 \Delta t^2 u_{ttt}$ or $-\theta^2 \Delta t^2 u_{\beta\beta\beta}$.

The numerical properties of LSFEM can be analyzed by considering the linear hyperbolic equation with constant coefficient in one dimension, namely,

$$u_t + \beta u_x = 0. \tag{9.78}$$

Consider a finite element discretization of the spatial domain and piecewise global basis $\{\phi\}$, see Sect. 2.5. Introducing the finite element expansion $u_h(x,t) = \sum U_j(t)\phi_j(x)$ and test functions $v_h = \phi_i(x)$ for u and v in (9.67) and simplifying yield a fully discrete system of algebraic equations for $\boldsymbol{U}^{n+1} = \{U_j(t_{n+1})\}$ at each time increment. For a representative interior node j, we have

$$\Big[\int_\Omega (\phi_i\phi_j + \beta^2\Delta t^2\theta^2\phi_{i,x}\phi_{j,x})dx\Big](U_j^{n+1} - U_j^n)$$
$$= \Big[\int_\Omega (-\beta\Delta t\phi_i\phi_{j,x} - \beta^2\Delta t^2\theta\phi_{i,x}\phi_{j,x})dx\Big]U_j^n. \tag{9.79}$$

The case for linear elements over a uniform mesh is important and readily amenable to analysis. Combining element contributions for two linear elements adjacent to node j, and arranging terms to correspond to the order in (9.76), we obtain the nodal equation

$$\Big[1 + \Big(\frac{1}{6} - \theta^2c^2\Big)\delta^2\Big](U_j^{n+1} - U_j^n)$$
$$= -\frac{c}{2}(U_{j+1}^n - U_{j-1}^n) + \theta c^2(U_{j+1}^n - 2U_j^n + U_{j-1}^n). \tag{9.80}$$

A standard von Neumann stability analysis can now be applied to the recurrence relation (9.80) for LSFEM formulation. Consider a solution to (9.78) of the form

$$u(x,t) = a_n e^{\sqrt{-1}k(x-\beta t)}, \tag{9.81}$$

where k is the wave number (wavelength $\lambda = 2\pi/k$) of mode $e^{\sqrt{-1}kx}$ so that

$$u(x, t+\Delta t) = u(x,t)e^{-\sqrt{-1}k\beta\Delta t}, \tag{9.82}$$

where $G = e^{-\sqrt{-1}k\beta\Delta t}$ is the complex amplification factor for the exact solution. The nodal values of the finite element solution can be written as

$$U_j^n = a_n e^{\sqrt{-1}kjh}, \tag{9.83}$$

$$U_j^{n+1} = a_{n+1}e^{\sqrt{-1}kjh}. \tag{9.84}$$

Substituting (9.83) and (9.84) into (9.80) and simplifying, we obtain the usual form of recurrence relation $a_{n+1} = G_{\mathrm{LS}}a_n$ with the amplification factor

$$G_{\mathrm{LS}} = \frac{Z_1}{Z_2}, \tag{9.85}$$

where

$$Z_1 = \Big[1 - 4\Big(\frac{1}{6} - \theta^2c^2\Big)\sin^2\Big(\frac{kh}{2}\Big)\Big] - 4\theta c^2\sin^2\Big(\frac{kh}{2}\Big) - \sqrt{-1}c\sin(kh),$$

$$Z_2 = \Big[1 - 4\Big(\frac{1}{6} - \theta^2c^2\Big)\sin^2\Big(\frac{kh}{2}\Big)\Big].$$

If $|Z_2| > |Z_1|$ or equivalently $|Z_2|^2 - |Z_1|^2 > 0$, then $|G_{\mathrm{LS}}| < 1$ and the scheme is stable. Using the above expressions and simplifying,

$$|Z_2|^2 - |Z_1|^2 = 4c^2 \sin^2\left(\frac{kh}{2}\right)\left\{\theta\left[\frac{2}{3} + 4\theta c^2(2\theta - 1)\right]\right.$$

$$\left.\sin^2\left(\frac{kh}{2}\right) + (2\theta - 1)\cos^2\left(\frac{kh}{2}\right)\right\} > 0 \qquad \text{for } \frac{1}{2} \le \theta \le 1.$$

Hence, the scheme is stable if $\frac{1}{2} \le \theta \le 1$ at all CFL (Courant–Friedrichs–Lewy) number.

In the limit as $kh \to 0$ we have the asymptotic expansion for G_{LS}:

$$G_{\mathrm{LS}} = 1 - \sqrt{-1}ckh - \theta c^2(kh)^2 + \sqrt{-1}\theta^2 c^3 (kh)^3 + O((kh)^4).$$

For $\theta = 1/2$, $G_{\mathrm{LS}} = e^{-\sqrt{-1}ckh} + O((kh)^3)$ so the scheme is second-order accurate in the time domain, as anticipated.

The amplification factor $|G_{\mathrm{LS}}|$ determines the growth or decay of modes in the solution. The control of oscillation is also important, and this is influenced by both the relative amplification and the phase error of the modes. The phase error can be expressed as $\arg(G_{\mathrm{LS}})/(-ckh)$ in which

$$\arg(G_{\mathrm{LS}}) = \tan^{-1}\left[\frac{Im(G_{\mathrm{LS}})}{Re(G_{\mathrm{LS}})}\right].$$

The numerical properties of the scheme are shown in Fig. 9.12, given the polar diagrams of the phase and amplitude errors versus kh for several values of the CFL number c.

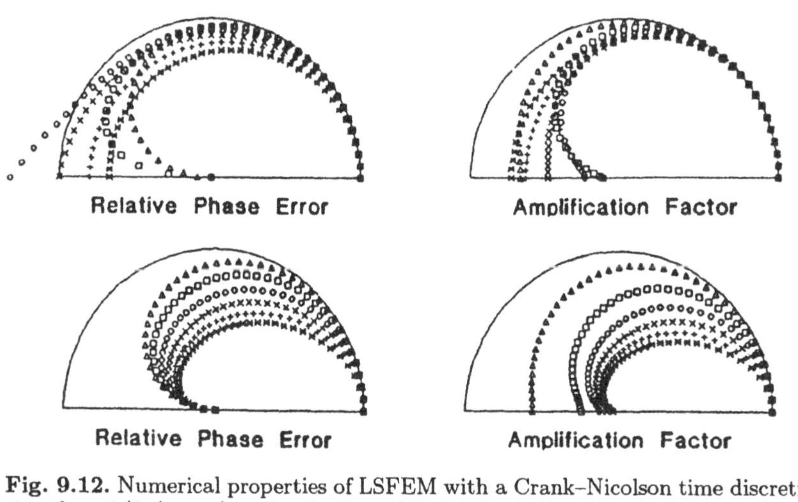

Fig. 9.12. Numerical properties of LSFEM with a Crank–Nicolson time discretization $\theta = 1/2$ (upper) and a backward-Euler time discretization $\theta = 1$ (lower): ($\triangle$) $c = 0.25$, ($\square$) $c = 0.5$, ($\circ$) $c = 0.75$, ($\times$) $c = 1.0$, ($*$) $c = 1.5$ (from Donea and Quartapelle 1992)

The Crank–Nicolson LSFEM has a rather accurate phase response provided $c \leq 1$. This approach is suitable for transient problems with continuous solutions. One may think that the method would be expensive, because of its implicit nature, for each small time-step, the inversion of matrix is required. Fortunately, as explained in Sect. 8.6, the opposite is true; a small time-step is preferred. Since the resulting algebraic equations by the least-squares method are symmetric and positive definite, simple iterative methods, such as the preconditioned conjugate gradient method (see Chap. 15) can be employed. When the time-step is small, the present solution is close to the solution at the previous step which is taken as the initial guess in the iterative method, therefore very few iterations are needed to obtain the solution in the next step.

Although the backward-Euler LSFEM has low phase accuracy, it could be useful for computing steady-state solutions and discontinuous solutions.

9.3.3 Numerical Examples of LSFEM

Propagation of a Sine Wave. In Fig. 9.13, the numerical solution for a propagating sine 'bump' is shown using LSFEM ($\theta = 0.5$) and the Taylor–Galerkin scheme described previously. The calculations employ a mesh of 50 linear elements on $(0, 1)$. The initial profile and propagated profile at $t = 0.6$ are shown in each case for calculations at a CFL number $c = \Delta t/\Delta x = 0.75$. Both methods perform well, having only a slight oscillatory 'undershoot' at the base, and exhibit little numerical diffusion. In the analysis we prove that the least-squares method is unconditionally stable. To demonstrate this point, the problem is recomputed at the CFL (Courant–Friedrichs–Lewy) number $c = 1.25 > 1.00$ to give profile in Fig. 9.13. The oscillation following the wave is more pronounced and there is some dissipation at the leading edge. However, the peak value and quality of the results are preserved, and the calculation is stable. The Taylor–Galerkin method is unstable and diverges at $c > 1$.

Propagation of a Front. The propagation of a front is a well-known problem – a Heaviside unit step function propagates from the left end into the domain. The initial discontinuity is approximated on the first element (slope=50). Results for the least-squares method ($\theta = 0.5$) and the Taylor–Galerkin method are given in Fig. 9.14, at $t = 0.5$ with $c = 0.75$. As in the previous case, both methods perform relatively well with the least-squares results again having a post-cursor oscillation of large amplitude and the Taylor–Galerkin method having both pre- and post-cursor oscillations.

Rotating Cone. The rotating cone problem by Long and Peper (1981) has been the standard problem to test numerical schemes for convective transport. A cosine hill of height 1.0 rotates in a clockwise direction with a flow field given by

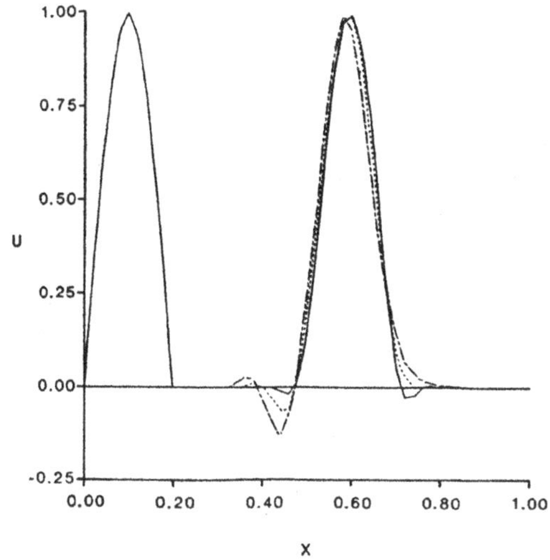

Fig. 9.13. Propagation of a sine wave, comparison of LSFEM ($\theta = 0.5$) with the Taylor–Galerkin method, initial condition at $t = 0$ (left) and solutions at $t = 0.6$: —, LS, $c = 0.75$; - - -, TG, $c = 0.75$; – . –, LS, $c = 1.25$ (from Carey and Jiang 1988)

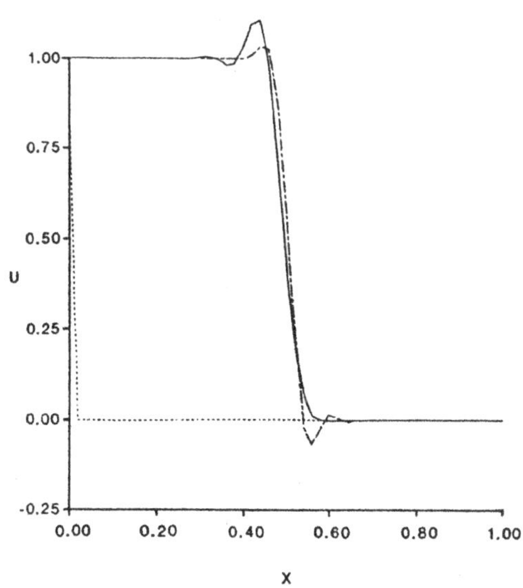

Fig. 9.14. Propagation of a front, comparison of LSFEM ($\theta = 0.5$) with the Taylor–Galerkin method, initial condition at $t = 0$ (left) and solutions at $t = 0.5$: —, LS, $c = 0.75$; - - -, TG, $c = 0.75$ (from Carey and Jiang 1988)

$$\begin{aligned}\beta_1(x,y) &= \omega y,\\ \beta_2(x,y) &= -\omega x.\end{aligned} \tag{9.86}$$

The flow domain is a square. The lower-left and the upper-right coordinates of the domain are $(-16,-16)$ and $(16,16)$, respectively. The peak of the hill is initially located at $(-8,0)$, and the initial distribution of the concentration is as follows:

$$u(x,y) = \begin{cases} \frac{1}{2}(1+\cos\frac{\pi d}{4}) & d < 4, \\ 0 & d \geq 4. \end{cases} \tag{9.87}$$

where d is the distance from the peak. The angular velocity ω is adjusted so that one revolution takes 300 time-steps. This corresponds to a CFL number of 0.16755 at the peak of the rotating cone.

The Crank–Nicolson scheme for time discretization together with LSFEM has been applied to this problem by Burrell et al. (1995). Figure 9.15 shows the computed distribution of the concentration by using 32×32 nine-node quadratical (Q_2) elements after 1, 2 and 4 and 6 revolutions. The quality of the solution can be recognized by noting that the peak maximum is maintained at about 1.0 throughout the course of six revolutions.

Rotating Puff in a Parabolic Angular Velocity Flow Field. This problem was originally proposed by Odman and Russell (1993) for further testing numerical methods, and solved by Burrell et al. (1995). In this problem, the initial puff concentration is located at $(-8,0)$ and is given by

$$u(x,y) = \exp\Big(-0.5\Big(\frac{x+8}{1.4}\Big)^2\Big)\exp\Big(-0.5\Big(\frac{y}{1.4}\Big)^2\Big). \tag{9.88}$$

Both x and y range from -16 to 16. The flow field is given as follows:

$$\beta_1(x,y) = \omega y\sqrt{x^2+y^2}\Big(1-\frac{x^2+y^2}{W^2}\Big), \tag{9.89a}$$

$$\beta_2(x,y) = -\omega x\sqrt{x^2+y^2}\Big(1-\frac{x^2+y^2}{W^2}\Big). \tag{9.89b}$$

The angular velocity is set to 0.01 so that one full revolution takes 240 time-steps. This corresponds to a CFL number of 0.2094 at the peak of the puff. The parameter W is chosen as 14.3. As compared with the results of Odman and Russell (1993), the initial puff distribution in (9.88) has much steeper gradients than that in the rotating cone problem, and the analytical solution under the conditions (9.88) and (9.89) also has steeper gradients around the puff after one revolution.

The analytical solution which can be obtained by the method of characteristics is shown in Fig. 9.16 at different stages of a revolution. Notice that after one revolution, the concentration distribution is much distorted from its initial Gaussian distribution, and parts of distribution have steep gradients.

Figure 9.17 shows the results of LSFEM using the Crank–Nicolson scheme and 32×32 nine-node quadratic elements. The results compare well with the analytic solution in Fig. 9.16.

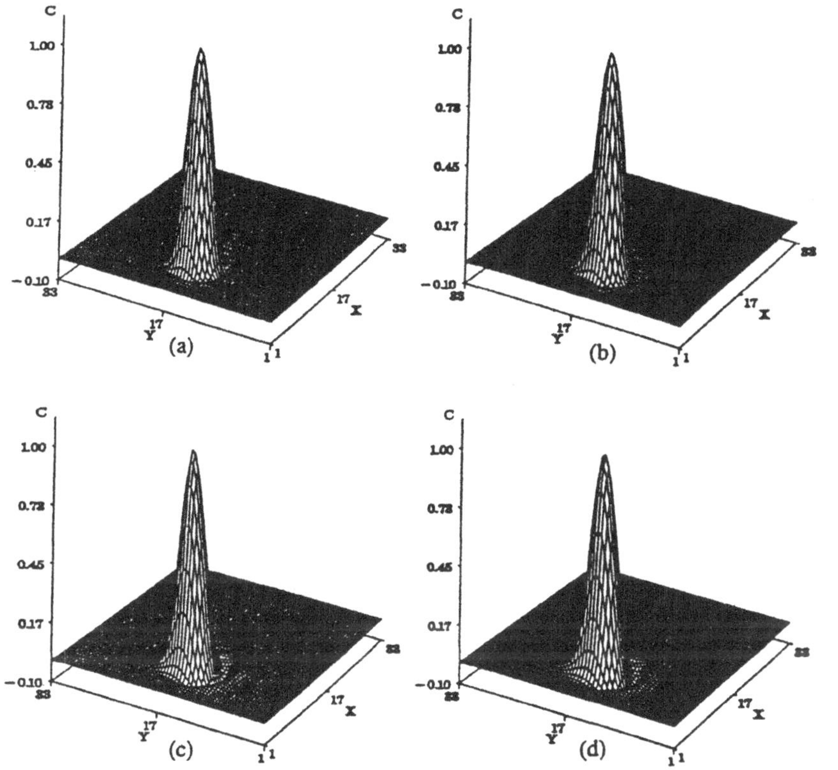

Fig. 9.15. Concentration distribution of the rotating cone after (**a**) one, (**b**) two, (**c**) four and (**d**) six revolutions by using the Crank–Nicolson LSFEM and 32×32 quadratic elements (from Burrell et al. 1995)

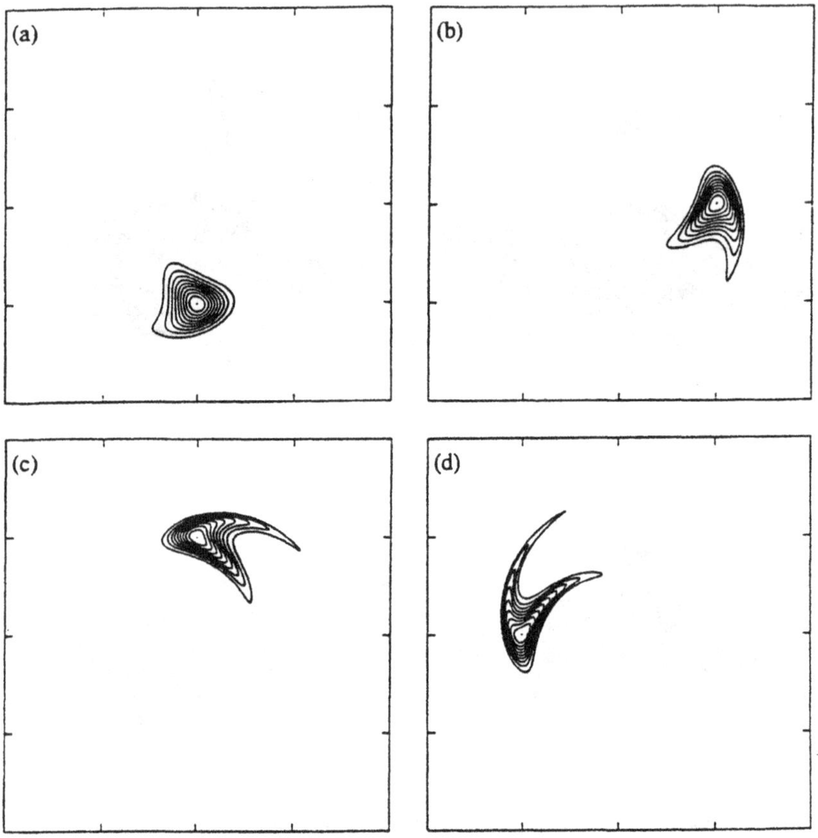

Fig. 9.16. Analytical solution for the rotating puff in a parabolic angular velocity flow after (**a**) 0.25, (**b**) 0.5, (**c**) 0.75 and (**d**) 1.0 revolution (from Burrell et al. 1995)

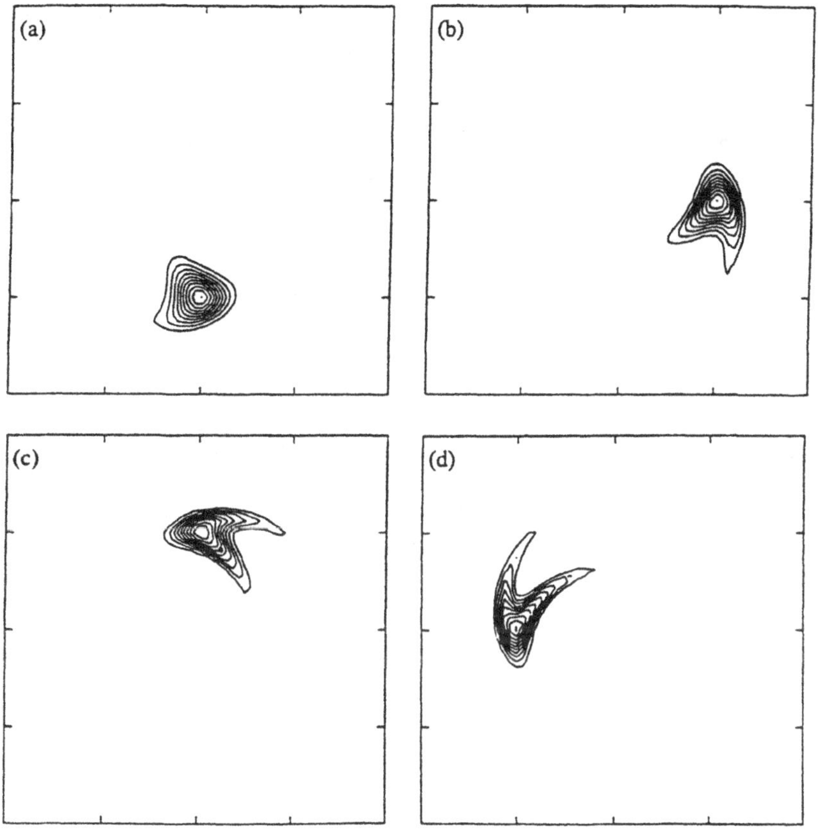

Fig. 9.17. Concentration distribution of the rotating puff computed by the Crank–Nicolson LSFEM and 32×32 quadratic elements after (**a**) 0.25, (**b**) 0.5, (**c**) 0.75 and (**d**) 1.0 revolution (from Burrell et al. 1995)

Water–Oil Displacement. As a numerical illustration of moving contact discontinuities, we consider the water–oil displacement problem in a 5-spot pattern. A displacement front propagates from the water injection well at the lower-left corner towards the oil production well at the top-right corner. The flow domain is assumed homogeneous, and the velocity field is then given by the solution of Laplace's equation for this source–sink system with symmetry (no flow) conditions on the sides to enforce periodicity:

$$\beta_1(x,y) = \frac{x}{x^2+y^2} - \frac{x-1}{(x-1)^2+(y-1)^2},$$
$$\beta_2(x,y) = \frac{y}{x^2+y^2} - \frac{y-1}{(x-1)^2+(y-1)^2}. \tag{9.90}$$

Using this specified velocity field in the least-squares finite element analysis, the finite-element system follows directly from the introduction of the finite element expansion for u into (9.67) with $\theta = 1$. The unit square domain is discretized to a uniform 40×40 grid of bilinear elements with the initial data $u(x,y,0) = 0$ and the injection value $u(0,0,t) = 1$ for $t \geq 0$. The equi-spaced solution contours for concentration $u(x,y,t)$ are shown in Fig. 9.18 at $t = 0.05$ and 0.3 and the fixed time-step $\Delta t = 0.0025$. We note that there is numerical dissipation of the propagating front and no oscillation.

Fig. 9.18. Concentration contours for the water–oil displacement problem: (**a**) at $t = 0.05$, (**b**) at $t = 0.30$ by using the backward-Euler LSFEM and 40 × 40 bilinear elements, $\Delta t = 0.005$ (from Jiang and Povinelli 1990)

9.4 Concluding Remarks

The LSFEM described in Sect. 4.8 can be used for the solution of both steady-state and time-dependent convective transport problems. The LS-FEM naturally and properly accounts for the role of the characteristics in multi-dimensional hyperbolic equations. The LSFEM automaticly acts in a manner similar to upwinding and requires no free parameters. The method has nearly optimal or optimal rate of convergence. The backward-Euler and the Crank–Nicolson LSFEM are unconditionally stable. Although the governing hyperbolic equations are non-selfadjoint, the method still produces symmetric and positive definite matrix systems, and hence iterative methods such as the conjugate gradient method can be employed. In particular, for transient problems smaller time-steps are allowed for higher accuracy without the penalty of increased computing time.

For pure convection problems with contact discontinuities, conventional finite difference and finite element solutions in general exhibit large spurious oscillations even far from the jumps and often are not close to the exact solution. New procedures based on the L_1 concept and the iteration of LSFEM for the solution of contact discontinuities have been developed. The overdetermined algebraic system is inherently obtained by choosing an appropriate number of Gaussian points in the formation of element matrices. Through L_1 minization or reweighting in the L_2 method, the contamination of "shock" elements is eliminated. These finite element methods capture discontinuities in bands of elements that are only one element wide on both coarse and fine meshes. The solution of this method has neither smearing nor oscillation, and has superior accuracy. The iterative reweighted LSFEM is simple, robust and efficient.

The LSFEM is general, very simple and easy to formulate and program. This lays down the foundation of the unified LSFEM for complex phenomena such as (1) convection–diffusion with small or vanishing diffusivity where convection dominates the fluid flow; (2) transonic flows in which the flow in subsonic regions is governed by elliptic equations and in supersonic bulbs by hyperbolic equations; and (3) two-fluid flows where the location of the interface is governed by pure convection of the so called color function. In all these cases, the LSFEM has a single unified formulation. However, other methods need different schemes in different regions or separate schemes for different equations.

The idea of LSFEM for convection problems was first appeared in Nguyen and Reynen (1984). They presented a space–time LSFEM for the advection–diffusion problems. Their numerical results show that the use of upwinding techniques or the Taylor–Galerkin approach in finite elements is not necessary when the least-squares weak formulation is extended into the time domain using standard shape functions. Donea and Quartapelle (1992) compared the space–time LSFEM with the Crank–Nicolson LSFEM described in Sect.

9.3.2, and found out that the space–time approach is less accurate and more dissipative.

The LSFEM with the backward-Euler and Crank–Nicolson time discretization was first published in Carey and Jiang (1988). This work was originally based on stimulating discussions between Srinivas Cheri and Jiang. Later, Li (1990) proposed a characteristic LSFEM. An accurate method for treating transient convection problems was introduced by Park and Liggett (1990, 1991) by combining the Taylor–Galerkin and least-squares concepts. Due to the presence of second-order spatial derivatives, elements of C^1 type, such as the Hermite cubic element must be used in this Taylor LSFEM. This method represents a valuable step towards simulations using relatively coarse meshes.

Chen (1992) provided an error analysis of the Crank–Nicolson LSFEM and her version of the method for a general linear first-order hyperbolic systems, namely, the Friedrichs system (Friedrichs 1958). The optimal error estimate given in Sect. 9.3.2 is a simplified version of her results. The analysis for steady-state problems in Sect. 9.1.3 was given in Jiang (1991, 1993).

The radical L_1 procedure for non-oscillatory discontinuous solutions was first proposed by Lavery (1988, 1989) for one-dimensional problems in the finite difference context. Because standard finite differencing leads to determined linear algebraic systems, in order to obtain an overdetermined system Lavery relies on non-traditional tricks, such as gradually adding a small viscous term to the one-dimensional Burger equation. However, it is very difficult to extend these tricks to two-dimensional problems.

The L_1 method based on overdetermined systems generated by numerical quadrature in the finite element context for multi-dimensional problems including an inexpensive solution method – the iterative reweighted LSFEM, was proposed by Jiang (1991, 1993). This L_1 approach is much more easy and general.

10. Incompressible Inviscid Rotational Flows

Incompressible inviscid rotational flows are governed by the incompressible Euler equations. This system of equations, in spite of the simplicity of the physical model from which they have been deduced, gives rise to complicated mathematical problems. Many fundamental questions concerning the existence, uniqueness and smoothness of their solutions are still open. The study of these challenging problems is a matter of current research. The discussion of these theoretical problems is beyond the scope of this book. In this chapter we concern ourselves only with the numerical solution of the incompressible Euler equations.

Interest in the numerical solution of the incompressible Euler equations can be justified at least from two aspects: (1) the capability of simulating vortex dominated inviscid flows; (2) as a first step towards the simulation of high Reynolds number flows.

Numerical solution of the incompressible Euler equations is also challenging. Firstly, the incompressible Euler equations are of first order. As explained in Chaps. 2 and 9, the classic Galerkin method and the closely related central difference method, when dealing with the first-order space derivatives, lead to non-self-adjointness and produce oscillatory solutions. Secondly, the incompressible Euler equations are neither elliptic nor hyperbolic. This makes it difficult to design appropriate finite difference formulae. Thirdly, in conventional methods, the incompressibility equation needs special treatments, such as staggered grids in the finite difference context and non-equal order elements for velocity and pressure in the finite element context, etc. Therefore, it is not surprising that the numerical solution of the incompressible Euler equations have been obtained mainly by spectral methods and vortex methods.

In this chapter we first introduce the conventional velocity–pressure formulation and the velocity–pressure–vorticity formulation of the incompressible Euler equations. Then, we use the div–curl method developed in Chap. 5 to derive the transport equation for the vorticity and the Poisson equation for the pressure. We shall describe a splitting method to construct the solution of the equations in time by solving the first-order equations of vorticity, velocity and pressure sequentially. Based on the concept of this approach we introduce an alternate velocity–pressure–vorticity formulation suitable for LSFEM. Finally, we apply LSFEM and show some numerical results.

10.1 Incompressible Euler Equations

10.1.1 The Velocity–Pressure Formulation

For simplicity, we consider fluid flow in a fixed bounded domain Ω with a piecewise smooth boundary Γ in the presence of conservative external forces $\boldsymbol{f}$ acting on the fluid. The flow of an incompressible fluid of null viscosity is governed by the incompressibility equation:

$$\nabla \cdot \boldsymbol{u} = 0 \qquad \text{in } \Omega \times (0, T], \tag{10.1a}$$

and the momentum equation:

$$\frac{\partial \boldsymbol{u}}{\partial t} + (\boldsymbol{u} \cdot \nabla)\boldsymbol{u} + \nabla p = \boldsymbol{f} \qquad \text{in } \Omega \times (0, T], \tag{10.1b}$$

where $\boldsymbol{u}(\boldsymbol{x}, t)$ is the velocity vector, $p(\boldsymbol{x}, t)$ is the pressure divided by the constant density of the fluid.

Equation (10.1b) allows the pressure to be correct up to any constant. To determine the pressure uniquely we require that the mean pressure over Ω is zero, namely,

$$\int_\Omega p dx = 0. \tag{10.1c}$$

The statement of the problem is completed by the prescription of suitable boundary and initial conditions. A typical boundary condition consists in specifying the normal component of velocity:

$$\boldsymbol{n} \cdot \boldsymbol{u} = g \qquad \text{on } \Gamma \times (0, T], \tag{10.1d}$$

where g is a given function of the location on the boundary and the time t. Since the fluid has no viscosity, the tangential components of velocity on the solid surface need not be specified. The boundary value of the normal component of the velocity g must satisfy the global conservation condition:

$$\int_\Gamma g d\Gamma = 0 \qquad \text{in } (0, T]. \tag{10.1e}$$

The initial condition consists in the specification of the velocity field at $t = 0$:

$$\boldsymbol{u}(\boldsymbol{x}, 0) = \boldsymbol{u}^0(\boldsymbol{x}) \qquad \text{in } \Omega. \tag{10.1f}$$

Moreover, the initial velocity field is assumed to be divergence-free:

$$\nabla \cdot \boldsymbol{u}^0 = 0 \qquad \text{in } \Omega. \tag{10.1g}$$

Finally, the boundary and initial data g and $\boldsymbol{u}^0$ must be compatible:

$$g|_{t=0} = \boldsymbol{n} \cdot \boldsymbol{u}^0. \tag{10.1h}$$

Equations (10.1a) and (10.1b) are the usual velocity–pressure formulation of the incompressible Euler equations.

10.1.2 The Velocity–Pressure–Vorticity Formulation

The vorticity field is fundamental in studying the behavior of rotational flows, so that it is natural to pose the governing equations of incompressible inviscid rotational flows in the velocity–pressure–vorticity formulation:

$$\frac{\partial \boldsymbol{u}}{\partial t} + (\boldsymbol{u} \cdot \nabla)\boldsymbol{u} + \nabla p = \boldsymbol{f} \qquad \text{in } \Omega \times (0, T], \tag{10.2a}$$

$$\nabla \cdot \boldsymbol{\omega} = 0 \qquad \text{in } \Omega \times (0, T], \tag{10.2b}$$

$$\boldsymbol{\omega} - \nabla \times \boldsymbol{u} = \boldsymbol{0} \qquad \text{in } \Omega \times (0, T], \tag{10.2c}$$

$$\nabla \cdot \boldsymbol{u} = 0 \qquad \text{in } \Omega \times (0, T], \tag{10.2d}$$

which can be obtained by deleting the viscous term from the transient Navier–Stokes equations (8.64). The boundary condition (10.1d) and the initial condition (10.1f) should be supplemented to (10.2). In three dimensions system (10.2) has eight equations in seven unknowns, i.e., three components of the velocity, three components of the vorticity and the pressure. As in Sect. 8.2.1, by introducing a dummy variable we can show that in three dimensions (10.2b) is not redundant, and thus system (10.2) is not overdetermined.

In order to underline some important features of the flow and develop numerically implementable algorithms, we shall derive a form of the incompressible Euler equations in terms of the velocity and vorticity. By taking the total pressure $b = p + (u^2 + v^2 + w^2)/2$ as an independent variable instead of the pressure p, (10.2a) can be rewritten as

$$\frac{\partial \boldsymbol{u}}{\partial t} - \boldsymbol{u} \times \boldsymbol{\omega} + \nabla b = \boldsymbol{f} \qquad \text{in } \Omega \times (0, T]. \tag{10.3}$$

Assuming that the velocity $\boldsymbol{u}$ is smooth enough, by virtue of the div–curl theorem (Therem 5.2), (10.3) is equivalent to the following system:

$$\nabla \times (\frac{\partial \boldsymbol{u}}{\partial t} - \boldsymbol{u} \times \boldsymbol{\omega} + \nabla b - \boldsymbol{f}) = \boldsymbol{0} \qquad \text{in } \Omega \times (0, T], \tag{10.4a}$$

$$\nabla \cdot (\frac{\partial \boldsymbol{u}}{\partial t} - \boldsymbol{u} \times \boldsymbol{\omega} + \nabla b - \boldsymbol{f}) = 0 \qquad \text{in } \Omega \times (0, T], \tag{10.4b}$$

$$\boldsymbol{n} \cdot (\frac{\partial \boldsymbol{u}}{\partial t} - \boldsymbol{u} \times \boldsymbol{\omega} + \nabla b - \boldsymbol{f}) = 0 \qquad \text{on } \Gamma \times (0, T]. \tag{10.4c}$$

By noting that the conservative force $\boldsymbol{f}$ has a potential V, namely, $\boldsymbol{f} = -\nabla V$, and using the vector identity (A.6), (10.4a) can be written as

$$\frac{\partial \boldsymbol{\omega}}{\partial t} + (\boldsymbol{u} \cdot \nabla)\boldsymbol{\omega} - (\boldsymbol{\omega} \cdot \nabla)\boldsymbol{u} = \boldsymbol{0} \qquad \text{in } \Omega \times (0,T]. \tag{10.5}$$

Equations (10.4b) and (10.4c) are the Poisson equation for the total pressure with its Neumann boundary condition:

$$\Delta b = \nabla \cdot (\boldsymbol{u} \times \boldsymbol{\omega}) + \nabla \cdot \boldsymbol{f} \qquad \text{in } \Omega \times (0,T], \tag{10.6a}$$

$$\frac{\partial b}{\partial n} = \boldsymbol{n} \cdot (\boldsymbol{u} \times \boldsymbol{\omega}) + \boldsymbol{n} \cdot \boldsymbol{f} - \frac{\partial g}{\partial t} \qquad \text{on } \Gamma \times (0,T]. \tag{10.6b}$$

It is clear from the above derivation that the computation of the pressure can be separated from that of the velocity and the vorticity.

Now we can write the governing equations of incompressible inviscid rotational flows in the following form:

$$\frac{\partial \boldsymbol{\omega}}{\partial t} + (\boldsymbol{u} \cdot \nabla)\boldsymbol{\omega} - (\boldsymbol{\omega} \cdot \nabla)\boldsymbol{u} = \boldsymbol{0} \qquad \text{in } \Omega \times (0,T], \tag{10.7a}$$

$$\nabla \cdot \boldsymbol{\omega} = 0 \qquad \text{in } \Omega \times (0,T], \tag{10.7b}$$

$$\boldsymbol{u}(\boldsymbol{x},0) = \boldsymbol{u}^0(\boldsymbol{x}) \qquad \text{in } \Omega. \tag{10.7c}$$

$$\nabla \times \boldsymbol{u} = \boldsymbol{\omega} \qquad \text{in } \Omega \times (0,T], \tag{10.7d}$$

$$\nabla \cdot \boldsymbol{u} = 0 \qquad \text{in } \Omega \times (0,T], \tag{10.7e}$$

$$\boldsymbol{n} \cdot \boldsymbol{u} = g \qquad \text{on } \Gamma \times (0,T]. \tag{10.7f}$$

We can use a splitting method to find the solution of system (10.7). At first, from the given velocity field at the previous time step, we calculate the vorticity at the boundary, and solve (10.7a) and (10.7b) to obtain the vorticity field in the domain at the current time step. Then, by using the computed vorticity we can solve the div–curl system (10.7d), (10.7e) and (10.7f) to update the velocity field. If it is necessary, we can solve the Poisson equation (10.6) to obtain the pressure from the computed velocity and vorticity.

Let us discuss more details about the above procedure. If we consider two-dimensional problems, we can shed more light on the mathematical properties of the Euler equations (10.7a) and (10.7b). In two dimensions (10.7a) and (10.7b) become much simpler: only the third component of the vorticity $\omega = \omega_z$ is present, and thus (10.7b) is automatically satisfied and the term $(\boldsymbol{\omega} \cdot \nabla)\boldsymbol{u}$ in (10.7a) vanishes. Therefore, the Euler equation for the vorticity in two dimensions becomes

$$\frac{\partial \omega}{\partial t} + (\boldsymbol{u} \cdot \nabla)\omega = \boldsymbol{0} \qquad \text{in } \Omega \times (0,T], \tag{10.8}$$

which implies that in two dimensions, the vorticity is simply transported by the velocity field. When the velocity field at the previous time step is substituted, then (10.8) becomes a pure hyperbolic convection equation studied in Sect. 9.3, and thus the time-marching LSFEM can be employed to obtain the solution for the vorticity with an optimal rate of convergence provided that the time step is properly chosen.

By contrast in three dimensions, due to the existence of the term $(\boldsymbol{\omega} \cdot \nabla)\boldsymbol{u}$ in (10.7a), vorticity is no longer conserved along the trajectories of the particles. Moreover, this term is responsible for the difficulties which arise in the construction of a global solution of the Euler equation. Since $\boldsymbol{\omega}$ is associated with the gradient of the velocity, we may conjecture that $(\boldsymbol{\omega} \cdot \nabla)\boldsymbol{u}$ is of the order ω^2 and $\partial\omega/\partial t \sim \omega^2$. Because the ordinary differential equation

$$\frac{\partial X}{\partial t} = X^2, \tag{10.9a}$$

$$X(0) = X_0 > 0 \tag{10.9b}$$

has solutions that blow up in finite time, it is not easy to exclude *a priori* situations where the vorticity becomes infinite in a finite time at some point of space.

At the second step, where we use the given current vorticity to obtain the current velocity, we solve a div–curl system that has been thoroughly investigated in Chap. 5. If the LSFEM is utilized, the rate at which the velocity solution converges is optimal.

At the third step, the pressure is uniquely determined by the Poisson equation (10.6) and condition (10.1c) from the known velocity and vorticity. In fact, when the velocity solution is available, we may directly solve (10.2a) under condition (10.1c) to obtain the pressure p by using the least-squares method. Let $\boldsymbol{a} = -(\partial\boldsymbol{u}/\partial t + \boldsymbol{u} \cdot \nabla \boldsymbol{u} - \boldsymbol{f})$ be given, the least-squares variational statement is to find $p \in S = \{H^1(\Omega) : \int_\Omega p dx = 0\}$ such that

$$(\nabla p, \nabla q) = (\boldsymbol{a}, \nabla q) \qquad \forall q \in S. \tag{10.10}$$

By virtue of the Poincaré inequality (C.1), we know that

$$\|p\|_1^2 \leq C\|\nabla p\|_0^2 \qquad \forall q \in S. \tag{10.11}$$

That is, the bilinear form in (10.10) is coercive. From Theorem 4.8, we confirm that the solution for pressure by LSFEM has an optimal rate of convergence.

Based on the above heuristic considerations, we may conclude that: (1) in two dimensions the solution of the incompressible Euler equations can be constructed uniquely; (2) for each time step the solution for the velocity, vorticity and pressure by the time-marching LSFEM is optimal; (3) the governing equations of the incompressible inviscid rotational flows can also be written in the following first-order velocity–pressure–vorticity formulation:

$$\frac{\partial \boldsymbol{u}}{\partial t} + (\boldsymbol{u} \cdot \nabla)\boldsymbol{u} + \nabla p = \boldsymbol{f} \qquad \text{in } \Omega \times (0, T], \tag{10.12a}$$

$$\frac{\partial \boldsymbol{\omega}}{\partial t} + (\boldsymbol{u} \cdot \nabla)\boldsymbol{\omega} - (\boldsymbol{\omega} \cdot \nabla)\boldsymbol{u} = \boldsymbol{0} \qquad \text{in } \Omega \times (0, T], \tag{10.12b}$$

$$\nabla \cdot \boldsymbol{\omega} = 0 \qquad \text{in } \Omega \times (0, T], \tag{10.12c}$$

$$\boldsymbol{\omega} - \nabla \times \boldsymbol{u} = \mathbf{0} \qquad \text{in } \Omega \times (0, T], \tag{10.12d}$$

$$\nabla \cdot \boldsymbol{u} = 0 \qquad \text{in } \Omega \times (0, T], \tag{10.12e}$$

with proper initial and boundary conditions on the velocity field.

10.2 Energy Conservation

Since there is no viscosity in the mathematical model of incompressible inviscid flows, that is, the fluid has neither internal friction nor friction with the boundaries, the energy defined as

$$E = \frac{1}{2} \int_{\Omega} |\boldsymbol{u}|^2 d\Omega = \frac{1}{2}(\boldsymbol{u}, \boldsymbol{u}) \tag{10.13}$$

is expected to be conserved during the motion.

For simplicity we consider the fluid flow in finite container at rest so that the boundary data $g = 0$ in (10.1d). We assume that the solution $\boldsymbol{u}$ of the Euler equation is smooth; more precisely, we suppose

$$\boldsymbol{u} \in C^1 \qquad \text{in } \Omega \times (0, T]. \tag{10.14}$$

Then by using (10.3) and $\boldsymbol{f} = -\nabla V$

$$\frac{dE}{dt} = (\boldsymbol{u}, \boldsymbol{u}_t) = (\boldsymbol{u}, \boldsymbol{u} \times \boldsymbol{\omega}) - \Big(\boldsymbol{u}, \nabla(b + V)\Big).$$

The first term in the right-hand side of the above equation equals zero, since this is a triple product of three vectors $\boldsymbol{u}$, $\boldsymbol{u}$ and $\boldsymbol{\omega}$. By using Green's formulae (B.1), the incompressibility condition $\nabla \cdot \boldsymbol{u} = 0$ and the boundary condition $\boldsymbol{n} \cdot \boldsymbol{u} = 0$, we have

$$\Big(\boldsymbol{u}, \nabla(p + V)\Big) = \langle \boldsymbol{n} \cdot \boldsymbol{u}, p + V \rangle = 0.$$

This achieves the proof of the energy conservation:

$$\frac{dE}{dt} = 0. \tag{10.15}$$

10.3 The Least-Squares Finite Element Method

Let us apply LSFEM to approximate the solution of the Euler equations in two dimensions. The time-dependent incompressible Euler equations (10.12) in two-dimensional Cartesian coordinates can be written as:

$$\frac{\partial u}{\partial x} + \frac{\partial v}{\partial y} = 0 \qquad \text{in } \Omega \times (0, T], \tag{10.16a}$$

$$\frac{\partial u}{\partial t} + u\frac{\partial u}{\partial x} + v\frac{\partial u}{\partial y} + \frac{\partial p}{\partial x} = f_x \qquad \text{in } \Omega \times (0, T], \tag{10.16b}$$

$$\frac{\partial v}{\partial t} + u\frac{\partial v}{\partial x} + v\frac{\partial v}{\partial y} + \frac{\partial p}{\partial y} = f_y \qquad \text{in } \Omega \times (0,T], \tag{10.16c}$$

$$\frac{\partial u}{\partial y} - \frac{\partial v}{\partial x} + \omega = 0 \qquad \text{in } \Omega \times (0,T], \tag{10.16d}$$

$$\frac{\partial \omega}{\partial t} + u\frac{\partial \omega}{\partial x} + v\frac{\partial \omega}{\partial y} = 0 \qquad \text{in } \Omega \times (0,T]. \tag{10.16e}$$

In Sect. 10.1 we point out that one could use the splitting method to construct the solution. Of course, one may solve all five equations in (10.16) simultaneously at each time-step. Numerical tests conducted by Wu et al. (1994), and Wu and Jiang (1995b) show that the simultaneous solution of (10.16a)–(10.16e) generally results in fast convergence and good accuracy. When the vorticity field has discontinuities such that the spatial derivative of the vorticity approaches infinity across the discontinuities, we simply do not include (10.16e) in the computation.

Before applying LSFEM, the equations should be discretized in the time domain. As usual, the general θ form can be utilized:

$$\frac{\partial u^{n+1}}{\partial x} + \frac{\partial v^{n+1}}{\partial y} = 0 \qquad \text{in } \Omega, \tag{10.17a}$$

$$\begin{aligned} &\frac{u^{n+1} - u^n}{\Delta t} + \theta\left(u^{n+1}\frac{\partial u^{n+1}}{\partial x} + v^{n+1}\frac{\partial u^{n+1}}{\partial y} + \frac{\partial p^{n+1}}{\partial x}\right) \\ &\quad +(1-\theta)\left(u^n\frac{\partial u^n}{\partial x} + v^n\frac{\partial u^n}{\partial y} + \frac{\partial p^n}{\partial x}\right) = f_x^{n+1} \qquad \text{in } \Omega, \end{aligned} \tag{10.17b}$$

$$\begin{aligned} &\frac{v^{n+1} - v^n}{\Delta t} + \theta\left(u^{n+1}\frac{\partial v^{n+1}}{\partial x} + v^{n+1}\frac{\partial v^{n+1}}{\partial y} + \frac{\partial p^{n+1}}{\partial y}\right) \\ &\quad +(1-\theta)\left(u^n\frac{\partial v^n}{\partial x} + v^n\frac{\partial v^n}{\partial y} + \frac{\partial p^n}{\partial y}\right) = f_y^{n+1} \qquad \text{in } \Omega, \end{aligned} \tag{10.17c}$$

$$\frac{\partial u^{n+1}}{\partial y} - \frac{\partial v^{n+1}}{\partial x} + \omega^{n+1} = 0 \qquad \text{in } \Omega, \tag{10.17d}$$

$$\begin{aligned} &\frac{\omega^{n+1} - \omega^n}{\Delta t} + \theta\left(u^{n+1}\frac{\partial \omega^{n+1}}{\partial x} + v^{n+1}\frac{\partial \omega^{n+1}}{\partial y}\right) \\ &\quad +(1-\theta)\left(u^n\frac{\partial \omega^n}{\partial x} + v^n\frac{\partial \omega^n}{\partial y}\right) = 0 \qquad \text{in } \Omega, \end{aligned} \tag{10.17e}$$

where the superscript 'n' denotes the previous time-step and '$n+1$' the current time-step.

Usually, two choices of θ, $\theta = 1/2$ or $\theta = 1$, can be made in the computation. $\theta = 1/2$ represents the Crank–Nicolson scheme, provides second-order accuracy in time, and is used for true transient problems. $\theta = 1$ represents the backward-Euler scheme, gives only first-order accuracy in time, and is used for a time-marching approach to obtain steady-state solutions. The numerical scheme is unconditionally stable for both choices of θ, thus there is no limitation on the size of the time-step. When steady-state solutions are of interest, we may use a large time-step to accelerate the convergence. If transient solutions are sought, however, we alway choose the time step so that the largest element CFL number is close to unity to avoid large errors introduced by the time discretization.

The convective terms in (10.17) can be linearized by Newton's method (see Sect. 8.3). To ensure time-accuracy at each time-step, Newton's linearization is performed until convergence is reached. For example, $u^{n+1}\ \partial u^{n+1}/\partial x$ is approximated as:

$$\begin{aligned}(u^{n+1})_{[k+1]}\ \frac{\partial(u^{n+1})_{[k+1]}}{\partial x} &\approx (u^{n+1})_{[k]}\ \frac{\partial(u^{n+1})_{[k+1]}}{\partial x} \\ &+(u^{n+1})_{[k+1]}\ \frac{\partial(u^{n+1})_{[k]}}{\partial x} - (u^{n+1})_{[k]}\ \frac{\partial(u^{n+1})_{[k]}}{\partial x},\end{aligned} \tag{10.18}$$

where k stands for the iteration count of the linearization and $(u^{n+1})_{[0]} = u^n$.

System (10.17), after linearization, can be written in standard matrix form as:

$$\boldsymbol{A}_1\frac{\partial \underline{\boldsymbol{u}}}{\partial x} + \boldsymbol{A}_2\frac{\partial \underline{\boldsymbol{u}}}{\partial y} + \boldsymbol{A}_0\underline{\boldsymbol{u}} = \underline{\boldsymbol{f}} \tag{10.19}$$

in which

$$\underline{\boldsymbol{u}} = \begin{pmatrix} u \\ v \\ p \\ \omega \end{pmatrix}_{[k+1]}^{n+1}, \quad \boldsymbol{A}_0 = \begin{pmatrix} 0 & 0 & 0 & 0 \\ \frac{1}{\Delta t} + \theta\frac{\partial u}{\partial x} & \theta\frac{\partial u}{\partial y} & 0 & 0 \\ \theta\frac{\partial v}{\partial x} & \frac{1}{\Delta t} + \theta\frac{\partial v}{\partial y} & 0 & 0 \\ 0 & 0 & 0 & 1 \\ \theta\frac{\partial \omega}{\partial x} & \theta\frac{\partial \omega}{\partial y} & 0 & \frac{1}{\Delta t} \end{pmatrix}_{[k]}^{n+1},$$

$$\boldsymbol{A}_1 = \begin{pmatrix} 1 & 0 & 0 & 0 \\ \theta u & 0 & 1 & 0 \\ 0 & \theta u & 0 & 0 \\ 0 & -1 & 0 & 0 \\ 0 & 0 & 0 & \theta u \end{pmatrix}_{[k]}^{n+1}, \quad \boldsymbol{A}_2 = \begin{pmatrix} 0 & 1 & 0 & 0 \\ \theta v & 0 & 0 & 0 \\ 0 & \theta v & 1 & 0 \\ 1 & 0 & 0 & 0 \\ 0 & 0 & 0 & \theta v \end{pmatrix}_{[k]}^{n+1},$$

$$\underline{\boldsymbol{f}} = \begin{pmatrix} 0 \\ f_x^{n+1} + \frac{u^n}{\Delta t} - (1-\theta)(u\ \frac{\partial u}{\partial x} + v\ \frac{\partial u}{\partial y} + \frac{\partial p}{\partial x})^n + \theta(u\ \frac{\partial u}{\partial x} + v\ \frac{\partial u}{\partial y})_{[k]}^{n+1} \\ f_y^{n+1} + \frac{v^n}{\Delta t} - (1-\theta)(u\ \frac{\partial v}{\partial x} + v\ \frac{\partial v}{\partial y} + \frac{\partial p}{\partial y})^n + \theta(u\ \frac{\partial v}{\partial x} + v\ \frac{\partial v}{\partial y})_{[k]}^{n+1} \\ 0 \\ \frac{\omega^n}{\Delta t} - (1-\theta)(u\ \frac{\partial \omega}{\partial x} + v\ \frac{\partial \omega}{\partial y})^n + \theta(u\ \frac{\partial \omega}{\partial x} + v\ \frac{\partial \omega}{\partial y})_{[k]}^{n+1} \end{pmatrix}.$$

The first-order system (10.19) is then discretized in space following the standard LSFEM procedures given in Sect. 4.8.

10.4 Numerical Results of LSFEM

Standing Vortex. This problem was originally suggested by Gresho and Chan (1988), and was chosen by Tezduyar et al. (1992c) to test their SUPG method. This standing vortex flow has been computed by Wu and Jiang (1995b) by using LSFEM. The purpose of this test is to indicate how much numerical dissipation an approximation procedure introduces. The fluid is incompressible and inviscid, and is confined in a 1×1 container. The initial condition, which is also the exact solution, consists of an axisymmetric velocity profile with zero radial velocity:

$$u_r = 0$$

and with the circumferential velocity given as

$$u_\theta = \begin{cases} 5r & \text{if } \ r \le 0.2, \\ 2 - 5r & \text{if } \ 0.2 < r \le 0.4, \\ 0 & \text{if } \ r > 0.4, \end{cases}$$

where r is the distance from the center of the container. The exact solution for the pressure p and the vorticity ω can be obtained by solving the incompressible Euler equations in polar coordinates. In Cartesian coordinates the exact solution can be expressed as if $r \le 0.2$, then

$$\begin{aligned} u &= -5y, \\ v &= 5x, \\ p &= 12.5r^2 + C1, \\ \omega &= 10; \end{aligned}$$

if $0.2 < r \le 0.4$, then

$$\begin{aligned} u &= -2y/r + 5y, \\ v &= 2x/r - 5x, \\ p &= 12.5r^2 - 20r + 4log(r) + C2, \\ \omega &= 2/r - 10; \end{aligned}$$

where

$$\begin{aligned} C2 &= 12.5 * 0.4 * 0.4 + 20.0 * 0.4^2 - 4.0 * \log(0.4), \\ C1 &= C2 - 20.0 * 0.2 + 4.0 * \log(0.2), \end{aligned}$$

if $r > 0.4$, then all variables are equal to zero.

In the LSFEM computation, the mesh consists of 20×20 uniform bilinear elements. A time-step of $\Delta t = 0.05$ is used that corresponds to a maximum element CFL number of 1.0. At each time-step three Newton's linearizations are employed. For all LSFEM computations in this chapter one-point Gaussian quadrature is used for evaluating the element matrices, and the solution at Gaussian points is redistributed by the element area to the nodes to obtain the final solutions.

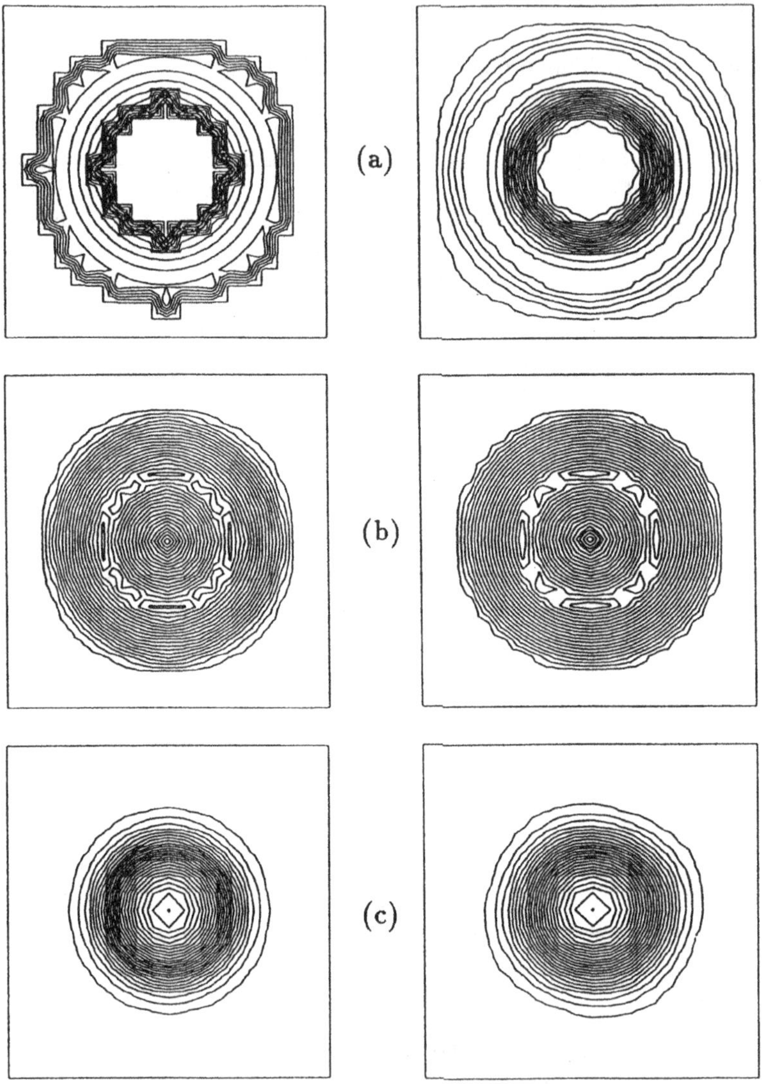

Fig. 10.1. Solution of the standing vortex problem: (**a**) Contours of vorticity, (**b**) Absolute value of velocity, (**c**) Pressure, left—analytical solutions; right—numerical solutions (from Wu and Jiang 1995b)

Solutions of the pressure, velocity and vorticity after 600 time steps that corresponds to $t = 30$ when the fluid in $r < 2$ has traveled 24 cycles are shown in Fig. 10.1. The numerical solutions and the exact solutions compare very well particularly for the velocity and pressure. Since in this case the vorticity is discontinuous, it is reasonable that the vorticity solution is smeared.

From Sect. 10.2 we know that the kinetic energy should be conserved in this flow. If significant numerical viscosity exists in a scheme, the kinetic energy of the computed fluid flow will not be conserved. To compare the numerical dissipation in the LSFEM and the SUPG method (Tezduyar et al. 1992c), Fig. 10.2 illustrates the history of the kinetic energy as the percentage of the original. Clearly, the LSFEM maintains very good accuracy for this test case all the time, and has almost no dissipation. Even after a very long period of time $t = 30$, 99.8% of the kinetic energy is still retained.

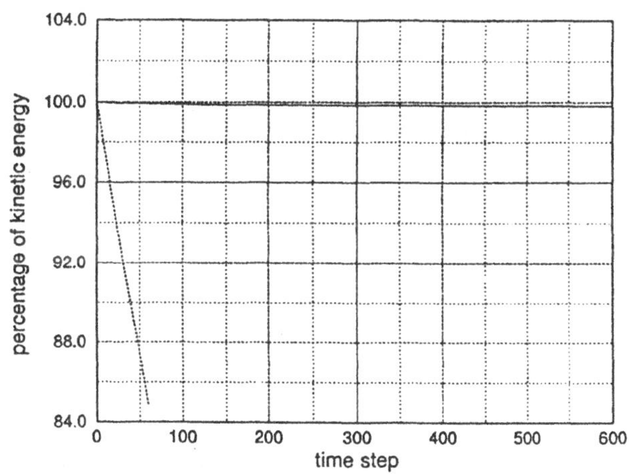

Fig. 10.2. Time history of kinetic energy, — LSFEM, ... Tezduyar et al. (1992c) (from Wu and Jiang 1995b)

Vortex Propagation. To further test the LSFEM we consider the problem of vortex propagation in uniform flow. The vortex described in the previous example is initially superimposed on the uniform flow. This problem has an exact solution which describes the vortex propagation downstream in the x-direction at the speed of the uniform flow. The computational domain for this problem is a rectangle of one unit width and four units length. A uniform mesh with 200×50 bilinear elements is chosen. The Crank–Nicolson time-marching scheme with a time-step $\Delta t = 0.02$ is used.

The following simple non-reflecting boundary condition for pressure and vorticity has been applied on the outflow boundary:

$$\frac{\partial \phi}{\partial t} + U\frac{\partial \phi}{\partial x} = 0 \qquad \phi = p \text{ or } \omega,$$

so that the vortex can accurately propagate out of the domain without being artificially reflected back into the computational domain to contaminate the rest of the flow field. This property at the outflow boundary is important in many practical applications.

The computed contours of pressure and absolute value of the velocity at different times are shown in Fig. 10.3. The distribution of the pressure and the vertical velocity at the center line at different times is illustrated in Fig. 10.4. As demonstrated in these figures, the solution of LSFEM is of high accuracy and has no visible smearing.

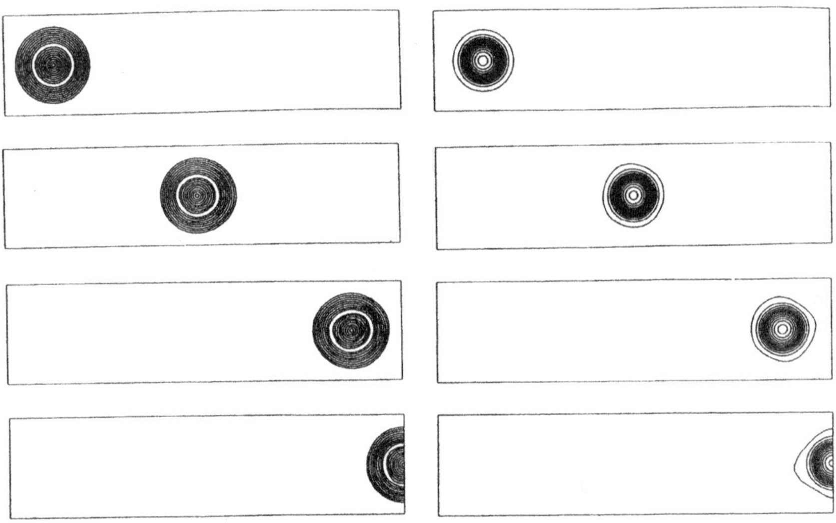

(a) (b)

Fig. 10.3. Vortex propagation: the computed solutions at different times: (**a**) Pressure, (**b**) Absolute value of the velocity (from Wu and Jiang 1995b)

Shear Flow Past Solid Bodies. We first consider steady-state shear flow past a circular cylinder. The center of this circular cylinder is placed at the origin of the rectangular coordinates. The cylinder is exposed to uniform shear flow with the velocity expressed by

$$\begin{aligned} u &= u_0 + \omega_0 y, \\ v &= 0, \end{aligned}$$

where u_0 and ω_0 are the given constants. The analytical solution to this problem (Batchelor 1970) is

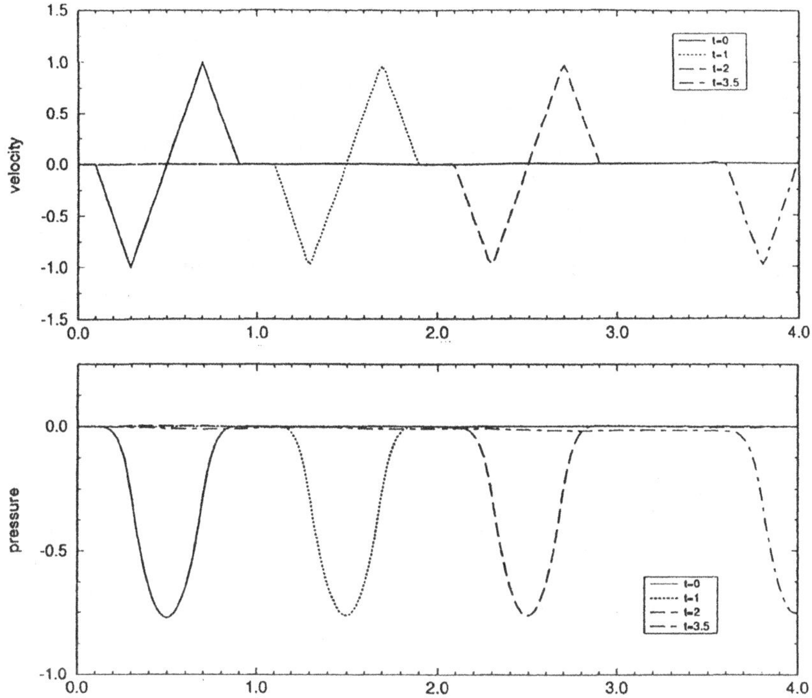

Fig. 10.4. Vortex propagation: solutions at the center line (from Wu and Jiang 1995b)

$$
\begin{aligned}
\psi &= \frac{1}{2}\omega_0 y^2 + u_0 y - u_0 a^2 \frac{y}{r^2} + \omega_0 a^4 \frac{(x^2 - y^2)}{4r^4}, \\
u &= \frac{\partial \psi}{\partial y} \\
&= u_0 + \omega_0 y - u_0 a^2 \frac{1}{r^2} + 2u_0 a^2 \frac{y^2}{r^4} - \omega_0 a^4 \frac{y}{2r^4} - \omega_0 a^4 \frac{(x^2 - y^2)y}{r^6}, \\
v &= -\frac{\partial \psi}{\partial x} = -2u_0 a^2 \frac{xy}{r^4} - \omega_0 a^4 \frac{x}{2r^4} + \omega_0 a^4 \frac{(x^2 - y^2)x}{r^6}, \\
p &= \frac{1}{2}(u^2 - u_0^2) + \omega_0 \psi \; (p_\infty = 0), \\
\omega &= \omega_0 .
\end{aligned}
$$

where ψ is the stream function, a is the radius of the circular cylinder and $r = (x^2 + y^2)^{1/2}$.

The computational domain is a circular ring. The distance between the outer boundary and the surface is 12 times the diameter of the cylinder.

The finite element mesh consists of 4302 bilinear elements with 4318 nodes. Figure 10.5a shows a part of the mesh near the cylinder. The velocity and the vorticity of uniform shear flow are prescribed on the inflow boundary; the pressure ($p_\infty = 0$) is specified on the outflow boundary. The undisturbed uniform flow is taken as an initial solution. A converged solution is obtained after six Newton's linearizations.

In Fig. 10.5b the calculated pressure distribution on the cylinder surface is compared with the analytical distribution. In Fig. 10.5c and d, the computed streamlines in the vicinity of the cylinder are compared with the analytical streamlines. The LSFEM results show excellent agreement with the analytical solutions.

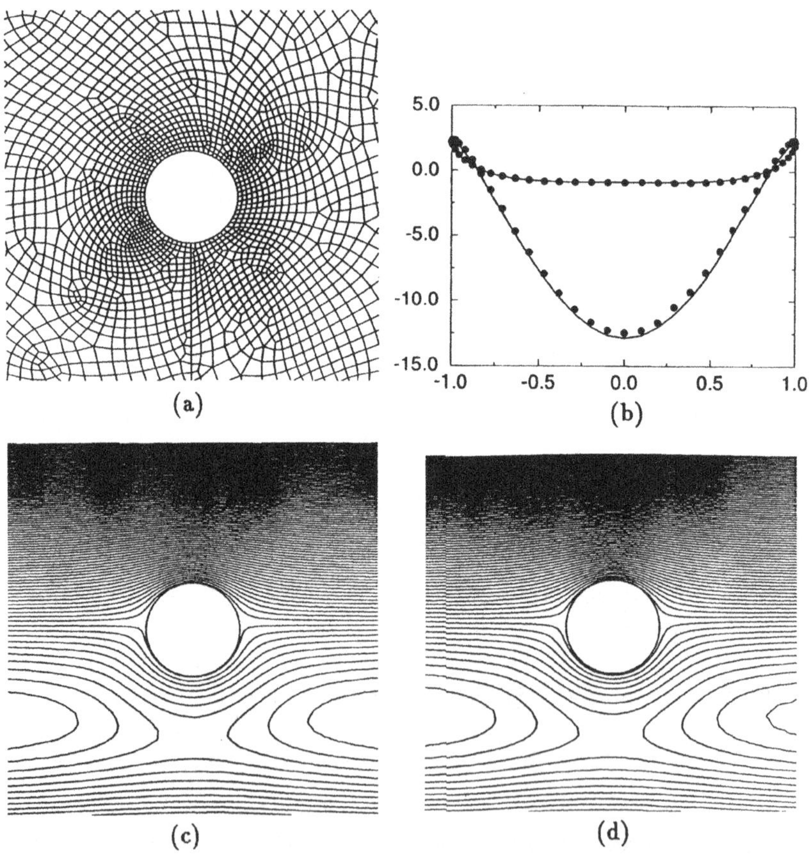

Fig. 10.5. Shear flow past a circular cylinder: (**a**) Finite element mesh, (**b**) Pressure distribution on the surface (• LSFEM, — analytical), (**c**) Analytical streamlines, (**d**) Computed streamlines (from Wu and Jiang 1995b)

We then consider the same type of flow past a NACA0012 airfoil. The outer boundary is 20 chords away from the airfoil. A mesh with 3521 bilinear elements and 3640 nodes is used. There are 170 nodes on the surface of the airfoil. The mesh near the airfoil is shown in Fig. 10.6a. The computed distribution of the pressure coefficient $C_p = (p - p_\infty)/0.5$ on the surface is shown in Fig. 10.6b.

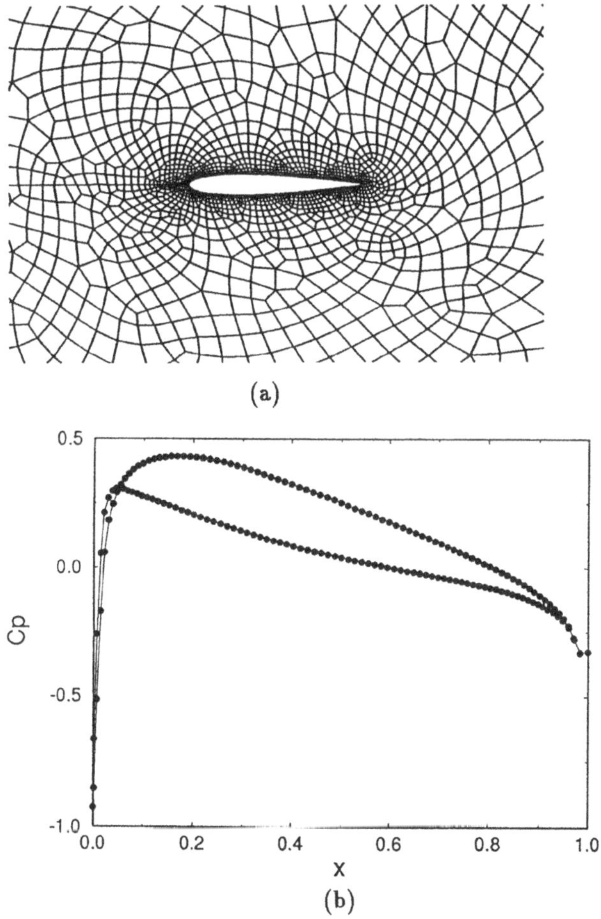

Fig. 10.6. Shear flow past a NACA0012 airfoil: (**a**) Finite element mesh, (**b**) Pressure distribution on the surface (from Wu and Jiang 1995b)

Uniform Flow Past an Ellipse. The LSFEM is further tested on the problem of uniform flow past an ellipse. The original problem was proposed by Pulliam (1989) to test Euler codes on compressible flows. Although this problem looks simple at first sight, it turns out to be quite challenging computationally. It was demonstrated by Pulliam (1989), to the surprise of many, that calculations done by almost all Euler codes available at the time for flows over cylinders and ellipses at angle of attack produced lifting solutions, and the solutions suffered from a general lack of consistency across mesh refinements, artificial viscosity parameters and algorithm types (e.g., central differences, upwind differences, implicit methods, multigrid schemes, direct solvers). Moreover, different converged solutions with lift varying from large positive to large negative values can be obtained, even within the framework of a single code and a single grid with varying algorithm parameters. Hafez and Brucker (1991) attribute the above trouble to contamination caused by artificial viscosity and vorticity due to discretization errors, and hence proposed a finite difference scheme on staggered grids with averaging of variables. Winterstein and Hafez (1993) suggested a way to avoid such difficulty by adjusting the location of the trailing edge stagnation point to produce zero lift solutions. This approach is effective, but may be cumbersome to implement.

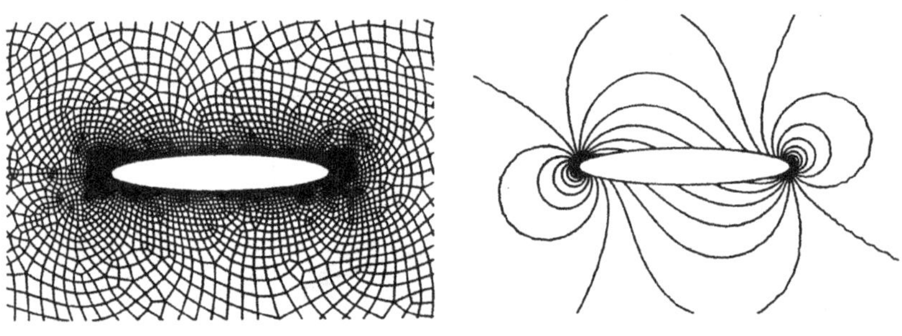

(a) (b)

Fig. 10.7. Uniform flow ($\alpha = 5°$) past an ellipse: (a) Finite element mesh, (b) Computed pressure contours (from Wu and Jiang 1995b)

Compressibility is not essential in this problem, therefore we consider a steady-state, inviscid, incompressible flow calculation. The problem definition is a $6:1$ ellipse at an angle of attack $\alpha = 5°$. Assuming that initial and boundary conditions are irrotational and contain no circulation, one would expect that the inviscid flow at any angle of attack would remain irrotational, not generate any circulation, and therefore be nonlifting.

In the LSFEM computation conducted by Wu and Jiang (1995b), the outer boundary of the computational domain is 8 chords away from the center of the ellipses; the free stream values of all variables are imposed on the outer boundary. The converged steady-state solution is obtained after six

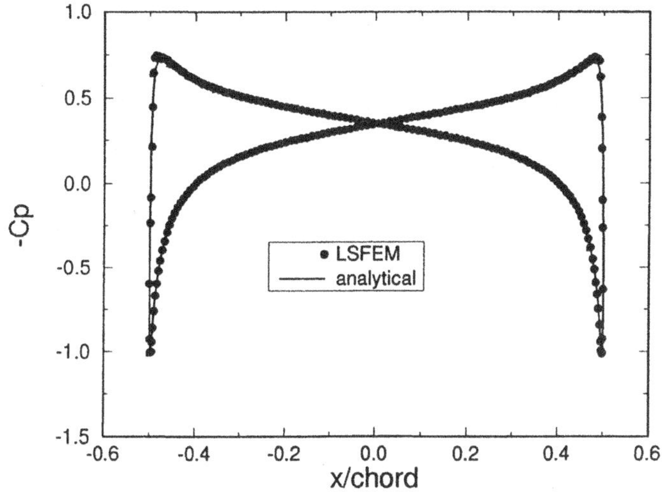

Fig. 10.8. Uniform flow ($\alpha = 5°$) past an ellipse: the pressure coefficient distribution on the surface (from Wu and Jiang 1995b)

Newton's linearizations. A part of the mesh with 238 nodes on the surface and the computed contours of the pressure are illustrated in Fig. 10.7a and b, respectively. Figure 10.8 demonstrates that the computed pressure coefficient on the surface is compared very well with the analytical solution. It is clear that the calculated flow is nonlifting and attached to the solid surface. We note that in the LSFEM no special treatment is needed.

10.5 Concluding Remarks

The solution of the incompressible Euler equations for inviscid rotational flows by using conventional finite difference and finite element methods is difficult, since the governing equations are neither elliptic nor hyperbolic, and no physical dissipation exists. Usually artificial dissipation is required for numerical stability; this contaminates the solution, leading to a deterioration of accuracy, loss of kinetic energy, and generation of artificial vorticity and thus false values for the lift.

This chapter demonstrates that the LSFEM based on the velocity–pressure–vorticity formulation is free of excessive numerical dissipation by showing that the kinetic energy for an inviscid fluid in a closed system is fully conserved over a long period of time, and excellent agreement between the LSFEM solution and the exact solution is observed. It is also shown that the LSFEM without any special treatment does not produce lift for asymmetric flow conditions.

11. Low-Speed Compressible Viscous Flows

In this chapter, low-speed compressible viscous flows are of interest. This subject has been neglected in finite element simulations, but has been intensively studied using finite difference and finite volume methods.

We shall show that by employing the velocity, pressure, vorticity, temperature, and heat flux as independent variables, the governing equations of this type of fluid flows can be written as an almost-linear first-order system. In this system, the principle part is the same div–curl–grad operator as that for the incompressible flow–heat coupling problems studied in Sect. 8.7. Therefore, the mathematical properties, such as the ellipticity for steady-state cases and the permissible boundary conditions of this system are exactly the same as that for incompressible viscous flows. Application of LSFEM based on this system to compressible buoyant flows are provided. The presentation of this chapter mainly follows the works of Yu et al. (1995a, 1996a,b).

11.1 Introduction

Low-speed flows with significant temperature variations are compressible due to the density variation induced by heat addition. For example, significant heat addition occurs in combustion related flow fields; inside chemical vapor deposition reactors, strong heat radiation also results in significant density variation. Although the flow speed is low, one must employ the compressible flow equations to simulate such flows.

It is well known that the simple explicit time-marching method, which can readily handle high-speed compressible flows for both steady-state and time-dependent problems, fails miserably when applied to low-speed or incompressible flows. This is due to the fact that for stability, the time step must be chosen inversely proportional to the largest eigenvalue of the system which is approximately the speed of sound for slow flows. Thus, this method is very time consuming. For incompressible flows, the speed of sound approaches infinity, thus simple explicit methods do not work at all. Another equivalent explanation of this difficulty is that the incompressible Navier–Stokes equations do not contain any time derivative of the pressure, hence the pressure variation propagates infinitely fast. To overcome this difficulty, Chorin (1968) added an artificial time derivative of the pressure into the continuity equation together with a multiplicative parameter β. The free parameter β is then chosen so as to reach the steady state rapidly. This idea can be extented by adding the pressure time derivative to the momentum equations and by introducing a second free parameter α. With these artificial terms the transient nature of the system is changed, and the steady state can be reached quickly by time-marching methods. Based on an analogy with the conjugate gradient method such a method is called the preconditioned method, since the object is to reduce the condition number of the matrix. Recently, noteworthy progress in simulating low-speed compressible flows has been achieved by using this preconditioning technique in the context of finite difference and finite volume methods, see e.g., Merkle and Choi (1987, 1988), Choi and Merkle (1993), Turkel (1987, 1993) and van Leer et al. (1991). However, the convergence of this preconditioned method is dependent on the choice of the free parameters. The optimal parameters depend on the dimensionalization, boundary conditions and particularly the inflow conditions of the problem. In addition, in the preconditioned equation set, one must specify the pressure boundary condition which usually is derived based on the boundary layer assumption. For recirculating flows, however, this approximate boundary condition poses a significant error.

Since for small Mach numbers the compressible equations basically have the same mathematical properties as the incompressible equations, the mixed Galerkin finite element method discussed in Sect. 8.1.1 is suitable for low-speed flows, as demonstrated by Einset and Jensen (1992) for flows inside a chemical vapor deposition furnace. It has been mentioned many times in this book that in the setting of the mixed Galerkin finite element method, only certain combinations of the approximation functions for velocity and pressure are acceptable for stable solutions. i.e., the LBB condition must be satisfied. While the LBB condition is cherished for its mathematical elegance, the condition renders no easy verification. For three-dimensional calculations, few combinations of shape functions are acceptable. In addition, the final coefficient matrix is nonsymmetric, and hence the inversion of such matrices is not trivial.

All the above mentioned difficulties associated with conventional methods have promoted the development of LSFEM for low-speed flow problems.

11.2 Two-Dimensional Case

11.2.1 The Compressible Navier–Stokes Equations

For simplicity, we assume that the gas is ideal and Newtonian, and the physical properties (viscosity μ, thermal conductivity k and the constant pressure specific heat C_p) are constant throughout the flow field. The assumptions outlined above lead to the following compressible Navier–Stokes equations:

$$\frac{\partial \rho}{\partial t}+u\frac{\partial \rho}{\partial x}+v\frac{\partial \rho}{\partial y}+\rho\left(\frac{\partial u}{\partial x}+\frac{\partial v}{\partial y}\right)=0 \qquad \text{in } \Omega\times(0,T], \tag{11.1a}$$

$$\rho\frac{\partial u}{\partial t}+\rho u\frac{\partial u}{\partial x}+\rho v\frac{\partial u}{\partial y}+\frac{\partial p}{\partial x}=\mu\frac{\partial}{\partial x}\left[2\frac{\partial u}{\partial x}-\frac{2}{3}\left(\frac{\partial u}{\partial x}+\frac{\partial v}{\partial y}\right)\right]$$
$$+\mu\frac{\partial}{\partial y}\left(\frac{\partial u}{\partial y}+\frac{\partial v}{\partial x}\right) \qquad \text{in } \Omega\times(0,T], \tag{11.1b}$$

$$\rho\frac{\partial v}{\partial t}+\rho u\frac{\partial v}{\partial x}+\rho v\frac{\partial v}{\partial y}+\frac{\partial p}{\partial y}=\mu\frac{\partial}{\partial x}\left(\frac{\partial u}{\partial y}+\frac{\partial v}{\partial x}\right)$$
$$+\mu\frac{\partial}{\partial y}\left[2\frac{\partial v}{\partial y}-\frac{2}{3}\left(\frac{\partial u}{\partial x}+\frac{\partial v}{\partial y}\right)\right]-\rho g \qquad \text{in } \Omega\times(0,T], \tag{11.1c}$$

$$\rho C_p\frac{\partial \theta}{\partial t}+\rho C_p u\frac{\partial \theta}{\partial x}+\rho C_p v\frac{\partial \theta}{\partial y}-\left(\frac{\partial p}{\partial t}+u\frac{\partial p}{\partial x}+v\frac{\partial p}{\partial y}\right)$$
$$=\Phi+k\left(\frac{\partial^2\theta}{\partial x^2}+\frac{\partial^2\theta}{\partial y^2}\right)-\rho g v \qquad \text{in } \Omega\times(0,T], \tag{11.1d}$$

$$\Phi=2\mu\left[\left(\frac{\partial u}{\partial x}\right)^2+\left(\frac{\partial v}{\partial y}\right)^2+\frac{1}{2}\left(\frac{\partial u}{\partial y}+\frac{\partial v}{\partial x}\right)^2\right]$$
$$-\frac{2}{3}\mu\left(\frac{\partial u}{\partial x}+\frac{\partial v}{\partial y}\right)^2, \tag{11.1e}$$

where ρ is the density, u and v are the velocity components in the respective directions, θ is the temperature, and Φ is the viscous dissipation. Note that the coordinate system is chosen so that gravity acts in the negative y direction. Equation (11.1a) is the continuity equation; (11.1b) and (11.1c) are the momentum equations; and (11.1d) is the energy equation.

In order to reduce (11.1) to a first-order system, the following new variables are introduced:

$$\mathcal{D}=\frac{\partial u}{\partial x}+\frac{\partial v}{\partial y}, \tag{11.2a}$$

$$\omega=\frac{\partial v}{\partial x}-\frac{\partial u}{\partial y}, \tag{11.2b}$$

$$q_x=-k\frac{\partial \theta}{\partial x}, \tag{11.2c}$$

$$q_y=-k\frac{\partial \theta}{\partial y}, \tag{11.2d}$$

where $\mathcal{D}$ is the dilatation which represents a measure of the compressibility, ω is the vorticity, and q_x and q_y are the heat conduction fluxes in the respective directions. As a result, a set of first-order equations in $\Omega \times (0,T]$ are obtained:

$$\frac{\partial \rho}{\partial t} + u\frac{\partial \rho}{\partial x} + v\frac{\partial \rho}{\partial y} + \rho\mathcal{D} = 0, \tag{11.3a}$$

$$\rho\frac{\partial u}{\partial t} + \rho u\frac{\partial u}{\partial x} + \rho v\frac{\partial u}{\partial y} + \frac{\partial p}{\partial x} = \mu\left(\frac{4}{3}\frac{\partial \mathcal{D}}{\partial x} - \frac{\partial \omega}{\partial y}\right), \tag{11.3b}$$

$$\rho\frac{\partial v}{\partial t} + \rho u\frac{\partial v}{\partial x} + \rho v\frac{\partial v}{\partial y} + \frac{\partial p}{\partial y} = \mu\left(\frac{4}{3}\frac{\partial \mathcal{D}}{\partial y} + \frac{\partial \omega}{\partial x}\right) - \rho g, \tag{11.3c}$$

$$\begin{aligned}&\rho C_p\frac{\partial \theta}{\partial t} + \rho C_p u\frac{\partial \theta}{\partial x} + \rho C_p v\frac{\partial \theta}{\partial y} - \left(\frac{\partial p}{\partial t} + u\frac{\partial p}{\partial x} + v\frac{\partial p}{\partial y}\right)\\ &= \mu\left[\frac{4}{3}\mathcal{D}^2 + \omega^2 + 4\left(\frac{\partial u}{\partial y}\frac{\partial v}{\partial x} - \frac{\partial u}{\partial x}\frac{\partial v}{\partial y}\right)\right] - \frac{\partial q_x}{\partial x} - \frac{\partial q_y}{\partial y} - \rho g v.\end{aligned} \tag{11.3d}$$

As pointed out in Sect. 6.4, the equation $\mathrm{rot}(q_x, q_y) = 0$, i.e.,

$$\frac{\partial q_y}{\partial x} - \frac{\partial q_x}{\partial y} = 0 \tag{11.4}$$

should be added to the above first-order equations to make the system have good mathematical properties.

The above governing equations (11.2)–(11.4) are closed by the equation of state

$$p = \rho R\theta, \tag{11.5}$$

where R is the gas constant.

To nondimensionalize the equations (11.2)–(11.5), the following dimensionless variables are introduced:

$$\bar{\rho} = \frac{\rho}{\rho_\infty}, \qquad \bar{u} = \frac{u}{U_\infty}, \qquad \bar{v} = \frac{v}{U_\infty},$$

$$\bar{\mathcal{D}} = \frac{\mathcal{D}L}{U_\infty}, \qquad \bar{\omega} = \frac{\omega L}{U_\infty}, \qquad \bar{p}' = \frac{p'}{\rho_\infty U_\infty^2},$$

$$\bar{\theta} = \frac{\theta}{\theta_\infty}, \qquad \bar{x} = \frac{x}{L}, \qquad \bar{y} = \frac{y}{L},$$

$$\bar{q}_x = \frac{q_x}{\rho_\infty U_\infty \theta_\infty C_p}, \qquad \bar{q}_y = \frac{q_y}{\rho_\infty U_\infty \theta_\infty C_p},$$

where ρ_∞, U_∞, θ_∞, and L are reference values of density, velocity, temperature, and length scale. Note that special care is taken to deal with the pressure. Since in low-speed flows the pressure distribution is rather uniform, pressure can be decomposed into a uniform background p^* plus a variation p', namely, $p = p^* + p'$, where $p^* = \rho_\infty R\theta_\infty$. The background pressure p^* then can be dropped out in the spatial derivatives. The pressure variation p' exists due to the flow velocity, and hence is nondimensionalized by a reference kinetic energy $\rho_\infty U_\infty^2$. Then the nondimensionalized system of equations can be expressed as:

$$\frac{\partial \rho}{\partial t} + u\frac{\partial \rho}{\partial x} + v\frac{\partial \rho}{\partial y} + \rho\mathcal{D} = 0, \tag{11.6a}$$

$$\rho\frac{\partial u}{\partial t} + \rho u\frac{\partial u}{\partial x} + \rho v\frac{\partial u}{\partial y} + \frac{\partial p'}{\partial x} = \frac{1}{\mathrm{Re}}\left(\frac{4}{3}\frac{\partial \mathcal{D}}{\partial x} - \frac{\partial \omega}{\partial y}\right), \tag{11.6b}$$

$$\rho\frac{\partial v}{\partial t} + \rho u\frac{\partial v}{\partial x} + \rho v\frac{\partial v}{\partial y} + \frac{\partial p'}{\partial y} = \frac{1}{\mathrm{Re}}\left(\frac{4}{3}\frac{\partial \mathcal{D}}{\partial y} + \frac{\partial \omega}{\partial x}\right) - \frac{\rho}{2\epsilon\mathrm{Fr}}, \tag{11.6c}$$

$$\rho\frac{\partial \theta}{\partial t} + \rho u\frac{\partial \theta}{\partial x} + \rho v\frac{\partial \theta}{\partial y} - (\gamma-1)M^2\left(\frac{\partial p'}{\partial t} + u\frac{\partial p'}{\partial x} + v\frac{\partial p'}{\partial y}\right) = \frac{(\gamma-1)M^2}{\mathrm{Re}}$$
$$\left[\frac{4}{3}\mathcal{D}^2 + \omega^2 + 4\left(\frac{\partial u}{\partial y}\frac{\partial v}{\partial x} - \frac{\partial u}{\partial x}\frac{\partial v}{\partial y}\right)\right] - \frac{\partial q_x}{\partial x} - \frac{\partial q_y}{\partial y} - \frac{(\gamma-1)M^2}{2\epsilon\mathrm{Fr}}\rho v, \tag{11.6d}$$

$$\mathcal{D} = \frac{\partial u}{\partial x} + \frac{\partial v}{\partial y}, \tag{11.6e}$$

$$\omega = \frac{\partial v}{\partial x} - \frac{\partial u}{\partial y}, \tag{11.6f}$$

$$q_x = -\frac{1}{\mathrm{Pe}}\frac{\partial \theta}{\partial x}, \tag{11.6g}$$

$$q_y = -\frac{1}{\mathrm{Pe}}\frac{\partial \theta}{\partial y}, \tag{11.6h}$$

$$\frac{\partial q_y}{\partial x} - \frac{\partial q_x}{\partial y} = 0, \tag{11.6i}$$

$$1 + \gamma M^2 p' = \rho\theta. \tag{11.6j}$$

Note that the dimensionless bar notation has been suppressed in the above equation system for convenience. The dimensionless numbers in these equations are defined as

$$M = \frac{U_\infty}{\sqrt{\gamma R\theta_\infty}}, \qquad \mathrm{Fr} = \frac{U_\infty^2}{2\epsilon g L}, \qquad \mathrm{Ra} = \frac{2\epsilon g L^3}{\nu\alpha},$$

$$\mathrm{Pr} = \frac{\nu}{\alpha}, \qquad \mathrm{Re} = \sqrt{\frac{\mathrm{RaFr}}{\mathrm{Pr}}}, \qquad \mathrm{Pe} = \mathrm{RePr},$$

$$\alpha = \frac{k}{\rho_\infty C_p}, \qquad \gamma = \frac{C_p}{C_v},$$

where M is the Mach number, Fr the Froude number, Ra the Rayleigh number, Pr the Prandtl number, Re the Reynolds number, Pe the Péclet number, α the thermal diffusivity, and γ the ratio of specific heats. The temperature difference parameter ϵ is defined as

$$\epsilon = \frac{\theta_h - \theta_c}{\theta_h + \theta_c}, \tag{11.7}$$

where θ_h and θ_c are the specified hot and cold temperatures in the thermal convection problem, respectively.

11.2.2 The First-Order System for Low-Speed Flows

In (11.6d) energy contributions from the pressure changes, viscous dissipation and gravity effects are proportional to M^2. Thus for low-Mach ($M \leq 0.3$) number flows these effects become negligible. In addition, in the nondimensionalized equation of state (11.6j) the M^2 term may also be neglected, hence the density and temperature become reciprocals of each other. With strong heat addition, the temperature and thus the speed of sound are high; the assumption of low Mach number can be widely applied to combustion phenomena and material processing procedures. Therefore, we obtain the first-order equations governing the low-speed flow:

$$\frac{\partial \rho}{\partial t} + u\frac{\partial \rho}{\partial x} + v\frac{\partial \rho}{\partial y} + \rho \mathcal{D} = 0, \tag{11.8a}$$

$$\rho\frac{\partial u}{\partial t} + \rho u\frac{\partial u}{\partial x} + \rho v\frac{\partial u}{\partial y} + \frac{\partial p'}{\partial x} = \frac{1}{\mathrm{Re}}\left(\frac{4}{3}\frac{\partial \mathcal{D}}{\partial x} - \frac{\partial \omega}{\partial y}\right), \tag{11.8b}$$

$$\rho\frac{\partial v}{\partial t} + \rho u\frac{\partial v}{\partial x} + \rho v\frac{\partial v}{\partial y} + \frac{\partial p'}{\partial y} = \frac{1}{\mathrm{Re}}\left(\frac{4}{3}\frac{\partial \mathcal{D}}{\partial y} + \frac{\partial \omega}{\partial x}\right) - \frac{\rho}{2\epsilon \mathrm{Fr}}, \tag{11.8c}$$

$$\rho\frac{\partial \theta}{\partial t} + \rho u\frac{\partial \theta}{\partial x} + \rho v\frac{\partial \theta}{\partial y} + \frac{\partial q_x}{\partial x} + \frac{\partial q_y}{\partial y} = 0, \tag{11.8d}$$

$$\mathcal{D} = \frac{\partial u}{\partial x} + \frac{\partial v}{\partial y}, \tag{11.8e}$$

$$\omega = \frac{\partial v}{\partial x} - \frac{\partial u}{\partial y}, \tag{11.8f}$$

$$q_x = -\frac{1}{\mathrm{Pe}}\frac{\partial \theta}{\partial x}, \tag{11.8g}$$

$$q_y = -\frac{1}{\mathrm{Pe}}\frac{\partial \theta}{\partial y}, \tag{11.8h}$$

$$\frac{\partial q_y}{\partial x} - \frac{\partial q_x}{\partial y} = 0, \tag{11.8i}$$

$$1 = \rho\theta. \tag{11.8j}$$

Both the original time-dependent quasi-linear first-order system (11.6) and the above simplified time-dependent quasi-linear first-order system can be directly used for the computation of low-speed or incompressible flows (Yu et al. 1995a). As usual, the time-derivatives can be discretized by the backward-Euler or Crank–Nicolson differences, the nonlinear terms can be treated by Newton's method, then the LSFEM can be employed. Note that it is trivial to include the algebraic equation (11.6j) or (11.8j) into the LSFEM formulation.

11.2.3 The Div–Curl–Grad Formulation

In order to reveal the mathematical properties of the above first-order system (11.8) as well as the permissible boundary conditions, we consider the steady state case. To proceed, the reciprocal correlation of temperature and density is used to obtain the following equation:

$$\partial \rho = -\frac{1}{\theta^2}\partial \theta. \tag{11.9}$$

Substituting (11.9) into the continuity equation (11.8a), choosing the temperature as a primitive variable, and replacing the density by the reciprocal of temperature lead to

$$u\frac{\partial \theta}{\partial x} + v\frac{\partial \theta}{\partial y} = \theta \mathcal{D}. \tag{11.10}$$

As a result, the dilatation $\mathcal{D}$ can be expressed by the following algebraic equation:

$$\mathcal{D} = -\frac{\mathrm{Pe}}{\theta}\left(uq_x + vq_y\right). \tag{11.11}$$

The left hand side of (11.10) is the material derivative of temperature, which can be substituted into the energy equation (11.8d), and we obtain

$$\frac{\partial q_x}{\partial x} + \frac{\partial q_y}{\partial y} + \mathcal{D} = 0. \tag{11.12}$$

The above equation directly correlates the compressibility effects with the heat fluxes for low-speed flows. The nonlinear convective terms of the energy equation (11.8d) in terms of the temperature now become an algebraic expression in $\mathcal{D}$.

Similarly, to transform the nonlinear convective terms of the momentum equations into algebraic expressions, the total pressure b is introduced:

$$b = p' + \frac{\rho}{2}\left(u^2+v^2\right) = p' + \frac{1}{2\theta}\left(u^2+v^2\right). \tag{11.13}$$

The convective terms of the momentum equations can be reformulated as

$$\rho u\frac{\partial u}{\partial x} + \rho v\frac{\partial u}{\partial y} + \frac{\partial p'}{\partial x} = \frac{\partial b}{\partial x} - \frac{v\omega}{\theta} + \frac{q_x \mathrm{Pe}}{2\theta^2}(u^2+v^2), \tag{11.14a}$$

$$\rho u\frac{\partial v}{\partial x} + \rho v\frac{\partial v}{\partial y} + \frac{\partial p'}{\partial y} = \frac{\partial b}{\partial y} + \frac{u\omega}{\theta} + \frac{q_y \mathrm{Pe}}{2\theta^2}(u^2+v^2). \tag{11.14b}$$

By introducing another new variable $B = \mathrm{Re}b - (4/3)\mathcal{D}$, a new set of first-order equations is obtained for $(u, v, B, \omega, \theta, q_x, q_y)$:

$$\frac{\partial u}{\partial x} + \frac{\partial v}{\partial y} = -\frac{\mathrm{Pe}}{\theta}\left(uq_x + vq_y\right), \tag{11.15a}$$

$$\frac{\partial v}{\partial x} - \frac{\partial u}{\partial y} = \omega, \tag{11.15b}$$

$$\frac{\partial \omega}{\partial y} + \frac{\partial B}{\partial x} = \mathrm{Re}\left[\frac{v\omega}{\theta} - \frac{q_x\mathrm{Pe}}{2\theta^2}(u^2+v^2)\right], \tag{11.15c}$$

$$-\frac{\partial \omega}{\partial x} + \frac{\partial B}{\partial y} = -\mathrm{Re}\left[\frac{u\omega}{\theta} + \frac{q_y\mathrm{Pe}}{2\theta^2}(u^2+v^2)\right] - \frac{\mathrm{Re}}{2\epsilon\mathrm{Fr}\theta}, \tag{11.15d}$$

$$\frac{\partial \theta}{\partial x} = -\mathrm{Pe}q_x, \tag{11.15e}$$

$$\frac{\partial \theta}{\partial y} = -\mathrm{Pe}q_y, \tag{11.15f}$$

$$\frac{\partial q_x}{\partial x} + \frac{\partial q_y}{\partial y} = +\frac{\mathrm{Pe}}{\theta}\left(uq_x + vq_y\right), \tag{11.15g}$$

$$\frac{\partial q_y}{\partial x} - \frac{\partial q_x}{\partial y} = 0. \tag{11.15h}$$

These equations can be presented concisely by the notation of vector analysis:

$$\nabla \cdot \boldsymbol{u} = -\frac{\mathrm{Pe}}{\theta}\left(uq_x + vq_y\right), \tag{11.16a}$$

$$\mathrm{rot}\ \boldsymbol{u} = \omega, \tag{11.16b}$$

$$\mathbf{curl}\,\omega + \nabla B = \boldsymbol{f}, \tag{11.16c}$$

$$\nabla \cdot \boldsymbol{q} = \frac{\mathrm{Pe}}{\theta}\left(uq_x + vq_y\right), \tag{11.16d}$$

$$\mathrm{rot}\ \boldsymbol{q} = 0, \tag{11.16e}$$

$$\nabla\theta = -\mathrm{Pe}\boldsymbol{q}, \tag{11.16f}$$

where the notations of rot and **curl** have been defined in Sect. 6.4.1; the right hand side vector $\boldsymbol{f}$ in (11.16c) is defined as

$$\boldsymbol{f} = \mathrm{Re}\begin{pmatrix} \frac{v\omega}{\theta} - \frac{q_x \mathrm{Pe}}{2\theta^2}(u^2+v^2) \\ -\frac{u\omega}{\theta} - \frac{q_y \mathrm{Pe}}{2\theta^2}(u^2+v^2) - \frac{1}{2\epsilon \mathrm{Fr}\theta} \end{pmatrix}.$$

Note that all right hand sides in (11.16) are algebraic, and thus they have nothing to do with the classification of this equation system.

The principle part of system (11.16) consists of a Stokes operator (11.16a)–(11.16c) for the velocity vector, the scalar B and the vorticity, and a div–curl–grad operator (11.16d)–(11.16f) for the temperature and heat fluxes. As such, we arrive at an almost-linear first-order system with eight equations involving seven unknowns: u, v, B, ω, θ, q_x and q_y. The inconsistency between the number of the unknowns and the number of equations results in an "overdetermined" problem. However, as emphasized many times in this book, this "overdetermined" problem is a notion borrowed from linear algebra. For partial differential equations, this interpretation leads to a misconception. By introducing a dummy variable κ into the div–curl–grad system (11.16d)–(11.16f) as explained in Sect. 6.4.1, it is easy to show that the system is elliptic and properly determined (see Yu et al. 1995a and 1996a,b for details).

In this first-order elliptic system augmented by a dummy variable there are eight unknowns governed by eight equations, thus on each boundary four boundary conditions are required. To facilitate the discussion of boundary conditions, the system of equations is divided into two groups: the flow equations (11.16a)–(11.16c), and the heat equations (11.16d)–(11.16f). Accordingly, by virtue of the knowledge given in Sects. 6.6 and 8.2.2, we can list some (not all) permissible boundary conditions for each group in Table 11.1.

Table 11.1. Boundary conditions for two-dimensional problems

Conditions	Flow Equations	Heat Equations
Wall	$u = v = 0$	$\theta =$ given, $\boldsymbol{n}\times\boldsymbol{q} =$ given $\boldsymbol{n}\cdot\boldsymbol{q} =$ given, $(\kappa = 0)$
Specified Inlet (Outlet)	$u =$ given, $v =$ given	$\theta =$ given, $\boldsymbol{n}\times\boldsymbol{q} =$ given $\boldsymbol{n}\cdot\boldsymbol{q} =$ given, $(\kappa = 0)$
Symmetry	$v = 0$, $\omega = 0$	$q_y = 0$, $(\kappa = 0)$

In Table 11.1, $\boldsymbol{n}$ denotes the outward normal vector for the boundary; $\boldsymbol{n} \cdot \boldsymbol{q}$ and $\boldsymbol{n} \times \boldsymbol{q}$ are the normal and tangential components of $\boldsymbol{q}$. Without losing generality, the symmetry condition is assumed to be with respect to the x axis. Note that for certain combinations, a null boundary condition for the dummy variable κ is invoked in order to make the system (augmented by the dummy variable) well posed. It should be emphasized however that this is only for the purpose of discussion, because the dummy variable is not used in the numerical computation. Therefore, these null boundary conditions are put in parentheses.

In the conventional second-order velocity–vorticity method used by Ern and Smooke (1993), there have been endeavors to derive wall conditions for the vorticity (see Sect. 8.8.2). For the preconditioning technique, one always needs to impose a wall condition for pressure. Here, we show that since only two boundary conditions are required from the flow equations, no wall boundary condition for vorticity or pressure is needed because the no-slip condition ($u = v = 0$) is enough to make the system well posed. Similarly, no boundary condition for vorticity is needed for specified inflow or outflow conditions.

For the heat equations, we list only two types of boundary conditions: (1) the specified temperature condition; and (2) the specified heat flux condition. A linear combination of the two is also permissible (see Remark in Sect. 6.4.2). For the first case, it is convenient to impose an additional condition for heat flux tangential to the wall. This tangential heat flux condition can be deduced from the specified wall temperatures. For the second case, we activate a null boundary condition for κ in addition to the specified normal heat flux. Note again that κ is a dummy variable, and the adoption of this pseudo boundary condition poses no theoretical difficulty.

11.2.4 The Least-Squares Finite Element Method

The almost-linear first-order system (11.16) of eight equations and seven unknowns can be written as

$$\boldsymbol{A}_1 \frac{\partial \underline{\boldsymbol{u}}}{\partial x} + \boldsymbol{A}_2 \frac{\partial \underline{\boldsymbol{u}}}{\partial y} + \boldsymbol{S} = 0, \tag{11.17}$$

where $\boldsymbol{A}_1$ and $\boldsymbol{A}_2$ are constant matrices; each entry of the vector $\boldsymbol{S}$ is a nonlinear algebraic expression in seven unknown variables to be solved, i.e.,

$$s_i = s_i(u_1, u_2, \cdots, u_7) \quad i = 1, 2, \cdots, 8$$

where $u_j, j = 1, \cdots, 7$ are the unknown variables. These nonlinear terms are linearized by Newton's method in the following fashion:

$$s_i^{n+1} = s_i^n + \sum_{j=1}^{7} \left(\frac{\partial s_i}{\partial u_j} \right)^n \delta u_j \tag{11.18}$$

or

$$S^{n+1} = S^n + A_o^n \delta \underline{u}. \tag{11.19}$$

where the superscript n denotes the previous iteration step and $n+1$ is the current step. $\delta u_j = u_j^{n+1} - u_j^n$ is the increment in the flow variables in each Newton's iteration. After manipulation, we obtain a new set of equations in matrix form:

$$A_1 \frac{\partial \delta \underline{u}}{\partial x} + A_1 \frac{\partial \underline{u}^n}{\partial x} + A_2 \frac{\partial \delta \underline{u}}{\partial y} + A_2 \frac{\partial \underline{u}^n}{\partial y} + A_o^n \delta \underline{u} + S^n = 0. \tag{11.20}$$

To proceed, the governing equations are cast into the following operator form ready for finite-element discretization:

$$L \delta \underline{u} = \underline{f}, \tag{11.21}$$

where the linear operator L is defined as

$$L = A_1 \frac{\partial}{\partial x} + A_2 \frac{\partial}{\partial y} + A_o^n, \tag{11.22}$$

and the right hand side vector is

$$\underline{f} = -A_1 \frac{\partial \underline{u}^n}{\partial x} - A_2 \frac{\partial \underline{u}^n}{\partial y} - S^n. \tag{11.23}$$

In each iteration step the standard linear first-order system (11.21) can be solved by LSFEM, see Sect. 4.8.

11.2.5 Numerical Results

Buoyancy-Driven Gas Flows in Two Dimensions. The numerical study conducted by Yu et al. (1995a, 1996a) is the simulation of buoyancy-driven gas flows in a square enclosure. The configuration consists of two insulated horizontal walls and two vertical walls at different temperatures. The gravitational force is in the negative y direction. The boundary conditions are: $u = v = q_y = 0$ everywhere; $\theta = \theta_h$ on the left wall and $\theta = \theta_c$ on the right wall. This problem has been extensively studied based on the incompressible flow equations with the Boussinesq approximation, which is appropriate only for small temperature differences (see Sect. 8.7.1). In practice, however, a large temperature difference is frequently encountered in gas flows, and the compressible formulation must be employed. Previously, Chenoweth and Paolucci (1986) used a SIMPLE type method and calculated the flow field. The heat transfer correlations are deduced and reported. By using the preconditioning technique, Choi and Merkle (1993) also successfully calculated the flow field. Their results compared favorably with Chenoweth's data.

Flow features of buoyancy-driven cavity flow depend on the Prandtl number Pr, the Rayleigh number Ra, the Froude number Fr, the aspect ratio of the cavity, and the temperature difference parameter ϵ. In the study by Yu et al. (1995a, 1996a), four Rayleigh numbers , Ra $= 10^3$, 10^4, 10^5, and 10^6, are considered with a temperature difference parameter $\epsilon = 0.6$. In all four cases, the Prandtl number Pr = 0.71 (air), the Froude number and the aspect ratio are unity. The variable B is specified to be zero at the center of the bottom. For the grid refinement study, three meshes (32×32, 64×64 and 128×128) were used. In all three meshes, grid lines were clustered near all four walls to resolve high gradients of the flow properties. In all calculations, nine-node quadratic elements are used, and the integration is performed using 2×2 Gaussian quadrature.

In the computation there are two loops of iterations: the outer loop updates the coefficient matrices and source terms (see (11.21)) by Newton's method; while the inner loop solves the variable increment $\delta \underline{u}$ using the Jacobi preconditioned conjugate gradient (JCG) method (see Chap. 15).

Figure 11.1 shows typical convergence numbers for the inner iterations by the JCG method. The case shown here is Ra $= 10^6$ with 128×128 mesh. Figure 11.1 shows that the error drops to lower than the sixth decimal digit in about 300 JCG iterations. For these calculations, the specified criteria for stopping the inner iterations is that error $\leq 10^{-7}$ or 300 iterations, whichever is satisfied first. The error in the JCG method is defined as the marching distance of the CG method divided by the absolute value of the global unknown vector.

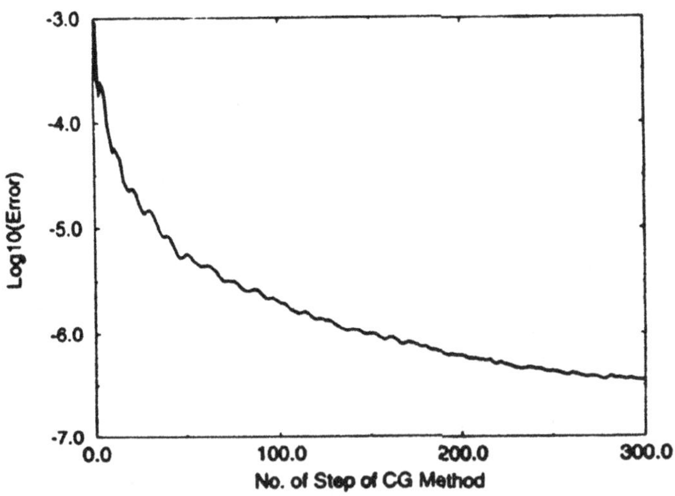

Fig. 11.1. Typical convergence history of inner iterations by the JCG method, Ra $= 10^6$, 128×128 mesh (from Yu et al. 1996a)

Figure 11.2 shows typical convergence behavior of the outer loop by Newton's method for the same case. In about six iterations, the magnitude of $\delta \underline{u}$ reduces by about seven orders of magnitude. Since the criterion for the inner iteration by the JCG method is set at 10^{-7}, the outer iterations level

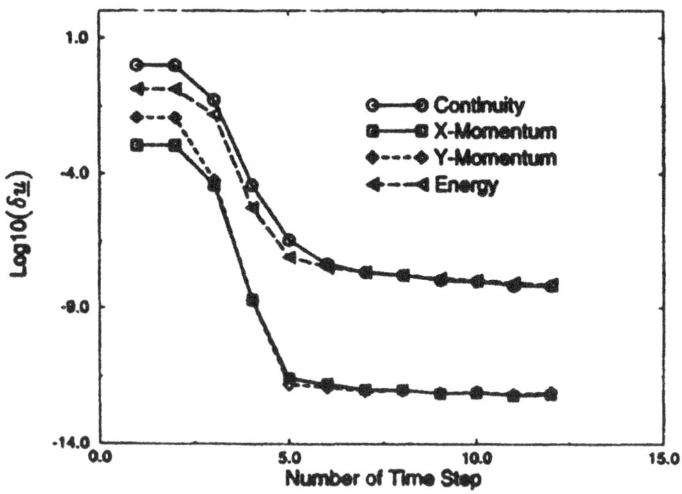

Fig. 11.2. Typical convergence history of outer iterations by Newton's method, $\mathrm{Ra} = 10^6, 128 \times 128$ mesh (from Yu et al. 1996a)

off at approximately the same order of magnitude. Figure 11.2 also shows quadratic convergence, which is a typical characteristic of Newton's method.

Figure 11.3 shows a comparison of the present result with previous data reported by Chenoweth et al. (1986). The x-axis is the Rayleigh number on a logarithmic scale, and the y-axis is the Nusselt number. Flow with large Rayleigh number results in enhanced heat transfer by natural convection, and therefore a larger Nusselt number is obtained. For all twelve calculations with four Rayleigh numbers and three meshes, favorable agreements were obtained. Only slight improvements were found for solutions using the fine mesh.

Figure 11.4 shows a comparison of the predicted heat flux on the cold wall for the case of $\mathrm{Ra} = 10^6$ by using three different meshes. A slight discrepancy can be observed for the solution between the 32×32 mesh and the other two meshes. For 64×64 and 128×128 meshes, the solution is essentially same.

In Fig. 11.5, streamlines of simulated flow fields for different Rayleigh numbers are presented. The streamlines are obtained by plotting constant contours of the stream function ψ. The distribution of ψ is obtained by solving the following equations by LSFEM:

$$\frac{\partial \psi}{\partial x} = \frac{u}{\theta},$$

$$\frac{\partial \psi}{\partial y} = -\frac{v}{\theta}.$$

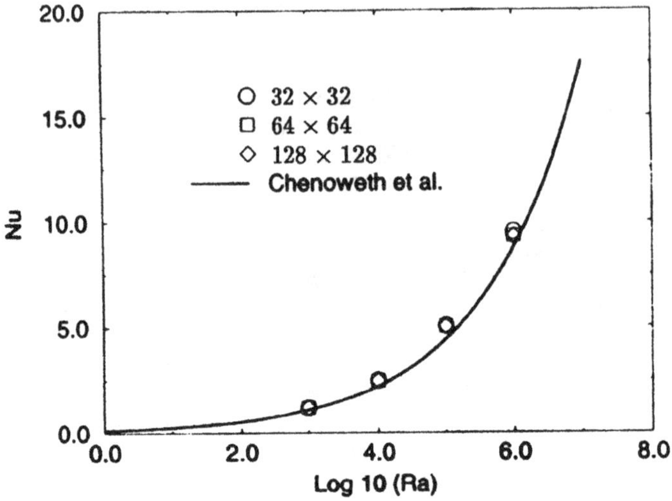

Fig. 11.3. Comparison of the calculated Nusselt numbers with Chenoweth and Paolucci's correlation (from Yu et al. 1996a)

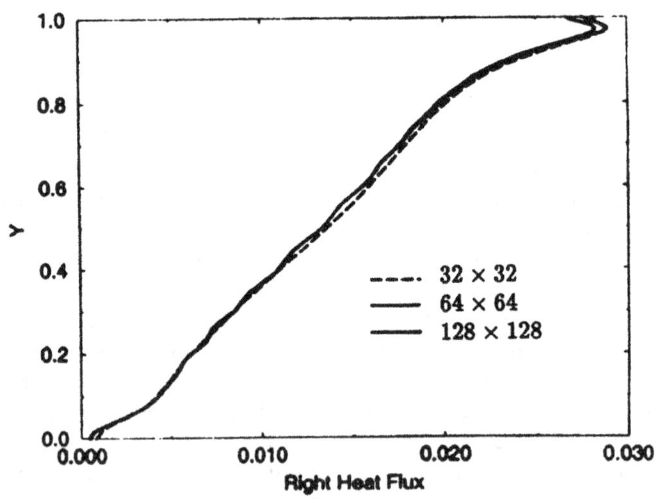

Fig. 11.4. Mesh refinement calculations at Ra $= 10^6$ (from Yu et al. 1996a)

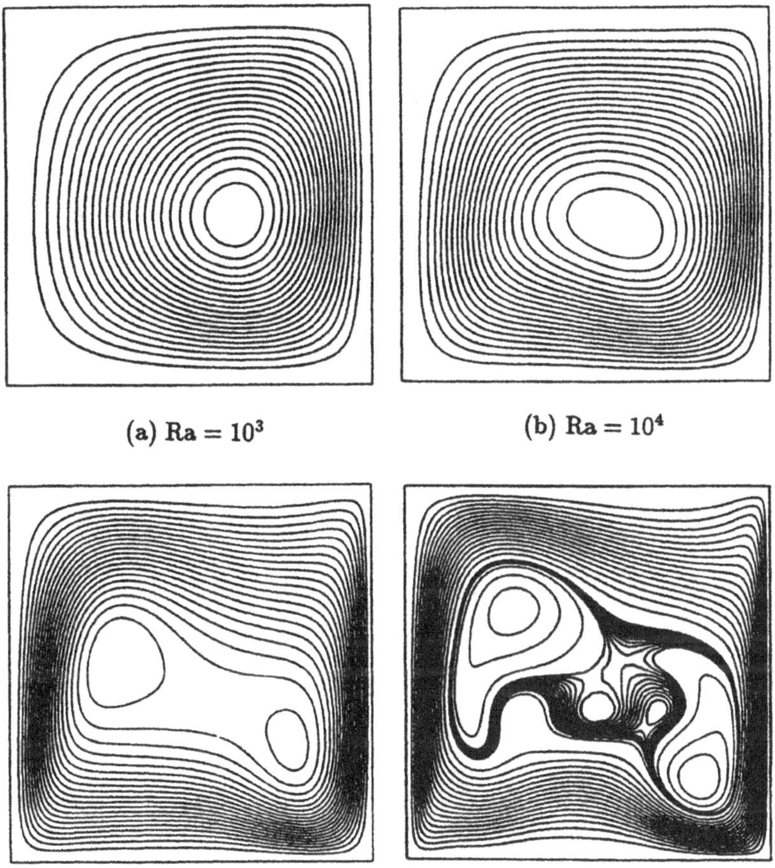

(a) Ra $= 10^3$ (b) Ra $= 10^4$

(c) Ra $= 10^5$ (d) Ra $= 10^6$

Fig. 11.5a–d. Streamlines of 2D buoyancy-driven gas flows (from Yu et al. 1996a)

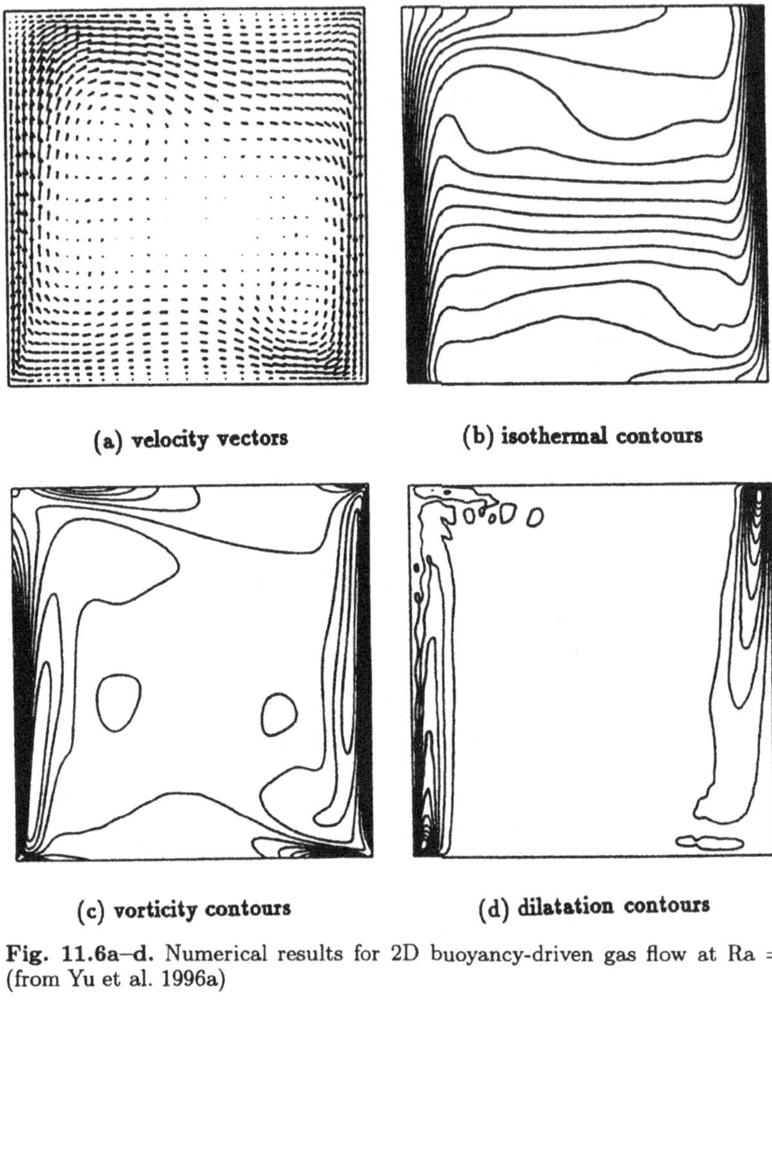

Fig. 11.6a–d. Numerical results for 2D buoyancy-driven gas flow at $\mathrm{Ra} = 10^6$ (from Yu et al. 1996a)

Since this is a post-processing procedure, u, v, and θ are known throughout the flow field. To solve the above equation system, a null boundary condition for ψ is imposed.

It is well known that the Boussinesq approximation displays a fully antisymmetric flow field with respect to the center of the cavity, see Fig. 8.22. The present calculation based on the compressible formulation shows an asymmetric flow field which has been observed experimentally. For Ra $= 10^3$ and 10^4, a shift of the vortex center towards the cold wall is observed. At Ra $= 10^5$ and 10^6, secondary rolls embedded in the primary eddy are observed. Note that for Ra $= 10^6$, four secondary vortices are observed. Although not shown, it is interesting to note that only two embedded vortices were resolved if the 32×32 mesh was used. The result of the 64×64 mesh is about the same as that of the 128×128 mesh.

Velocity vectors, isothermal contours, vorticity contours and dilatation contours of the simulated flow field at Ra $= 10^6$ are presented in Fig. 11.6. All plots presented here are obtained by using the 128×128 mesh. The velocity vectors in Fig. 11.6a show that for large Rayleigh numbers, the flow motion is most vigorous in the vicinity of the walls. In Fig. 11.6b, isothermal contours show steep temperature gradients near heat conducting walls. For the case of large Rayleigh numbers, the overall heat transfer is enhanced by fluid motion. Figure 11.6c shows the vorticity contours. It is obvious that the distribution of vorticity profiles along the walls is quite complex. In general, the vorticity and its derivatives along the walls cannot be predetermined. Figure 11.6d shows the contours of the compressibility effect $\mathcal{D}$ which is of interest for physical interpretation. The dilatation $\mathcal{D}$ can be calculated by using (11.11) in a post-processing procedure. Note that the dilatation $\mathcal{D}$ is null along the wall. It indicates that the flow is incompressible on the walls even though the flow field inside the enclosure is compressible. This can be easily verified by inspecting (11.11). For high Rayleigh number flows, the compressible region is restricted to near the lower-left and upper-right corners with steep gradients on both vertical and horizontal walls.

11.3 Three-Dimensional Case

11.3.1 The Compressible Navier–Stokes Equations

The procedure for deriving the first-order equations for three-dimensional low-speed flows is essentially the same as in two dimensions. Therefore, here we give only a brief derivation, the details can be found in Yu et al. (1996b). We consider the steady-state case. The steady-state three-dimensional compressible viscous flows are governed by the following Navier–Stokes equations:

$$(\boldsymbol{u} \cdot \nabla)\rho + \rho(\nabla \cdot \boldsymbol{u}) = 0 \qquad \text{in } \Omega, \tag{11.24a}$$

$$\rho(\boldsymbol{u}\cdot\nabla)\boldsymbol{u}+\nabla p=-\mu\nabla\times(\nabla\times\boldsymbol{u})+\frac{4}{3}\mu\nabla(\nabla\cdot\boldsymbol{u})+\boldsymbol{f}\qquad \text{in } \Omega, \tag{11.24b}$$

$$\rho C_p(\boldsymbol{u}\cdot\nabla)\theta=(\boldsymbol{u}\cdot\nabla)p+\Phi+k\Delta\theta-\rho g v\qquad \text{in } \Omega, \tag{11.24c}$$

where

$$\begin{aligned}\Phi &= 2\mu\left[\left(\frac{\partial u}{\partial x}\right)^2+\left(\frac{\partial v}{\partial y}\right)^2+\left(\frac{\partial w}{\partial z}\right)^2\right]\\ &+ \mu\left[\left(\frac{\partial u}{\partial y}+\frac{\partial v}{\partial x}\right)^2+\left(\frac{\partial v}{\partial z}+\frac{\partial w}{\partial y}\right)^2+\left(\frac{\partial w}{\partial x}+\frac{\partial u}{\partial z}\right)^2\right]\\ &- \frac{2}{3}\mu\left(\frac{\partial u}{\partial x}+\frac{\partial v}{\partial y}+\frac{\partial w}{\partial z}\right)^2.\end{aligned}$$

The coordinate system is chosen so that the gravity is in the negative y direction, and the body force term in the momentum equations is denoted by $\boldsymbol{f}=(0,-\rho g,0)^T$. As in the two-dimensional case, all physical properties are assumed to be constant throughout the flow field.

In order to cast the equations to a first-order system, the following new variables are introduced:

$$\mathcal{D}=\nabla\cdot\boldsymbol{u}, \tag{11.25a}$$

$$\boldsymbol{\omega}=\nabla\times\boldsymbol{u}, \tag{11.25b}$$

$$\boldsymbol{q}=-k\nabla\theta, \tag{11.25c}$$

where $\mathcal{D}$ is the dilatation, $\boldsymbol{\omega}=(\omega_x,\omega_y,\omega_z)$ is the vorticity vector, and $\boldsymbol{q}=(q_x,q_y,q_z)$ is the heat flux vector. As in the two-dimensional case, by introducing a pressure variation p' and a uniform background pressure p^*, we obtain the dimensionless first-order equations:

$$(\boldsymbol{u}\cdot\nabla)\rho+\rho\mathcal{D}=0, \tag{11.26a}$$

$$\rho(\boldsymbol{u}\cdot\nabla)\boldsymbol{u}+\nabla p'+\frac{1}{\text{Re}}\nabla\times\boldsymbol{\omega}-\frac{4}{3\text{Re}}\nabla\mathcal{D}-\boldsymbol{f}=\boldsymbol{0}, \tag{11.26b}$$

$$\rho(\boldsymbol{u}\cdot\nabla)\theta+\nabla\cdot\boldsymbol{q}-(\gamma-1)M^2\left[(\boldsymbol{u}\cdot\nabla)p'+\frac{\Phi}{\text{Re}}-\frac{\rho v}{2\epsilon\text{Fr}}\right]=0, \tag{11.26c}$$

$$\boldsymbol{\omega}-\nabla\times\boldsymbol{u}=\boldsymbol{0}, \tag{11.26d}$$

$$\boldsymbol{q}+\frac{1}{\text{Pe}}\nabla\theta=\boldsymbol{0}, \tag{11.26e}$$

$$\nabla\times\boldsymbol{q}=\boldsymbol{0}, \tag{11.26f}$$

$$\nabla\cdot\boldsymbol{\omega}=0, \tag{11.26g}$$

$$1+\gamma M^2 p' = \rho\theta. \tag{11.26h}$$

Note that in the above system the dimensionless bar notation has been suppressed for convenience. The dimensionless body force in the momentum equation is $\boldsymbol{f} = (0, -\rho/(2\epsilon\mathrm{Fr}), 0)$. Also note that the irrotational condition (11.26f) for the heat flux and the solenoidal condition (11.26g) for the vorticity are added to make the system have better mathematical properties, see Chaps. 6 and 8.

11.3.2 The Div–Curl–Grad Formulation

For low-speed flows ($M \leq 0.3$), the pressure derivative terms, the viscous dissipation terms, and the buoyancy term in the energy equation (11.26c) become negligible. In addition, according to the equation of state (11.26h), the density and temperature become reciprocals of each other and we have (11.9). Substituting (11.9) into the mass conservation equation (11.26a) leads to

$$(\boldsymbol{u}\cdot\nabla)\theta = \theta\mathcal{D}. \tag{11.27}$$

Here the density is replaced by the reciprocal of temperature. As a result, the dilatation $\mathcal{D}$ can be expressed by the following algebraic equation:

$$\theta\mathcal{D} = -\mathrm{Pe}(\boldsymbol{u}\cdot\boldsymbol{q}). \tag{11.28}$$

The left hand side of (11.28) is the material derivative of temperature, which can be substituted into the energy equation (11.26c) to obtain

$$\nabla\cdot\boldsymbol{q} + \mathcal{D} = 0. \tag{11.29}$$

Similarly, in order to transform the nonlinear convective terms of the momentum equations into algebraic expression, the total pressure b is introduced:

$$b = p' + \frac{1}{2\theta}(\boldsymbol{u}\cdot\boldsymbol{u}). \tag{11.30}$$

Then the convective terms in the momentum equation can be reformulated as

$$\rho(\boldsymbol{u}\cdot\nabla)\boldsymbol{u} + \nabla p' = \nabla b - \frac{\boldsymbol{u}\times\boldsymbol{\omega}}{\theta} + \frac{\mathrm{Pe}}{2\theta^2}\boldsymbol{q}(\boldsymbol{u}\cdot\boldsymbol{u}), \tag{11.31}$$

By introducing another new variable $B = \mathrm{Re}b - (4/3)\mathcal{D}$ we finally obtain an almost-linear first-order system:

$$\nabla\cdot\boldsymbol{u} = -\frac{\mathrm{Pe}}{\theta}(\boldsymbol{u}\cdot\boldsymbol{q}), \tag{11.32a}$$

$$\nabla\times\boldsymbol{u} = \boldsymbol{\omega}, \tag{11.32b}$$

$$\nabla\cdot\boldsymbol{\omega} = 0, \tag{11.32c}$$

$$\nabla \times \boldsymbol{\omega} + \nabla B = \mathrm{Re}\left[\frac{\boldsymbol{u}\times\boldsymbol{\omega}}{\theta} - \frac{\mathrm{Pe}}{2\theta^2}\boldsymbol{q}(\boldsymbol{u}\cdot\boldsymbol{u}) + \left(0, -\frac{1}{2\epsilon\theta\mathrm{Fr}}, 0\right)^T\right], \quad (11.32\mathrm{d})$$

$$\nabla\cdot\boldsymbol{q} = \frac{\mathrm{Pe}}{\theta}(\boldsymbol{u}\cdot\boldsymbol{q}), \quad (11.32\mathrm{e})$$

$$\nabla\times\boldsymbol{q} = \boldsymbol{0}, \quad (11.32\mathrm{f})$$

$$\nabla\theta = -\mathrm{Pe}\boldsymbol{q}. \quad (11.32\mathrm{g})$$

Here all left hand sides are linear first-order derivatives; and all right hand sides are algebraic, and thus they have nothing to do with the classification of this system.

This set of equations consists of a div–curl system (11.32a) and (11.32b) for the velocity, a generalized div–curl system (11.32c) and (11.32d) for the scalar B and the vorticity, and a div–curl–grad system (11.32e)–(11.32g) for the temperature and heat flux. Although this set has fifteen equations with eleven unknowns, it is elliptic and determined (see Chaps. 6 and 8).

For the purpose of discussing the boundary conditions, we divide the system of equations into two groups: the flow equations (11.32a)–(11.32d) and the heat equations (11.32e)–(11.32g). Any combination of the permissible boundary conditions for the flow equations listed in Table 8.1 and the permissible boundary conditions for the heat equations discussed in Sect. 6.4.2 is permissible for low-speed equations in the div–curl–grad formulation (11.32). Some of these combinations are listed in Table 11.2.

Table 11.2. Boundary conditions for three-dimensional problems

Conditions	Flow Equations	Heat Equations
Wall	$u=v=w=0$	$\theta=$ given, $\boldsymbol{n}\times\boldsymbol{q}=$ given $\boldsymbol{n}\cdot\boldsymbol{q}=$ given
Specified Inlet (Outlet)	$u=$ given, $v=$ given, w=given	$\theta=$ given, $\boldsymbol{n}\times\boldsymbol{q}=$ given $\boldsymbol{n}\cdot\boldsymbol{q}=$ given
Symmetry	$\boldsymbol{n}\cdot\boldsymbol{u}=0$, $\boldsymbol{n}\times\boldsymbol{\omega}=\boldsymbol{0}$	$\boldsymbol{n}\cdot\boldsymbol{q}=0$

11.3.3 Numerical Results

Buoyancy-Driven Gas Flows in Three Dimensions. A simulation of buoyancy-driven gas flows in a cubic container by using LSFEM was performed by Yu et al. (1996b). As shown in Fig. 11.7, the configuration consists of four insulated walls and two vertical walls at different temperatures θ_h and θ_c. Figure 11.7 also illustrates the specified boundary conditions. In this study, four Rayleigh numbers, Ra = 10^3, 10^4, 10^5, and 5×10^5, are considered with a temperature difference parameter $\epsilon = 0.6$. For all four cases, the Prandtl number Pr = 0.71 (air), the Froude number and the aspect ratio are unity. For the grid refinement study, three meshes ($16 \times 16 \times 8$, $32 \times 32 \times 16$ and $64 \times 64 \times 32$) were used. Fluid flows are assumed symmetric with respect to the central $x-y$ plane, and half of the grid nodes are used in the z direction. In all three meshes, grid lines were clustered near the walls to resolve high gradients of the flow properties. In all calculations, trilinear elements with one-point Gaussian quadrature are used. In addition, the calculation is carried out by using the solution of coarse mesh as initial guess for the fine mesh solution.

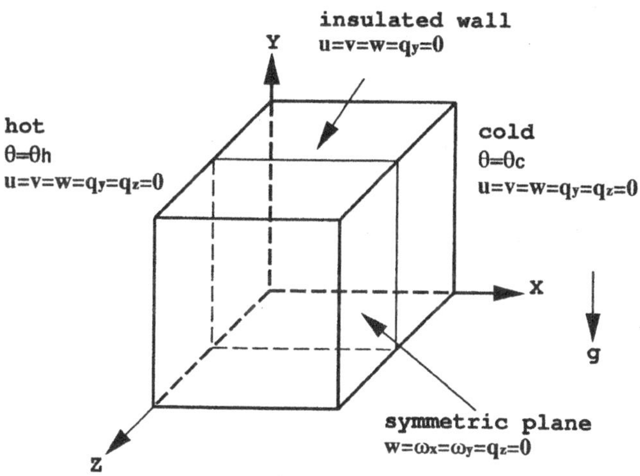

Fig. 11.7. Schematic of natural convection inside a cubic enclosure

In the computation there are two loops of iterations: the outer loop updates the coefficient matrices and source terms by Newton's method; while the inner loop solves the variable increment $\delta \underline{u}$ by using the JCG method. Typically, it takes about 10 to 15 global iterations to reduce $\delta \underline{u}$ to about eight to ten orders of magnitude. The limitation of the global convergence is due to the prescribed criterion for the inner iterations. Usually, two to three hundred inner iterations are used for each global iteration.

Figure 11.8 shows the distribution of the u velocity at the central vertical line for Ra = 10^5 using three meshes, respectively. Little difference is observed between solutions for the $32 \times 32 \times 16$ mesh and the $64 \times 64 \times 32$ mesh.

In another test, the error of overall heat transfer obtained by integrating q_x over the hot and cold conducting surfaces is evaluated. Due to the conservation of energy, the integrated heat fluxes on the cold and hot walls should

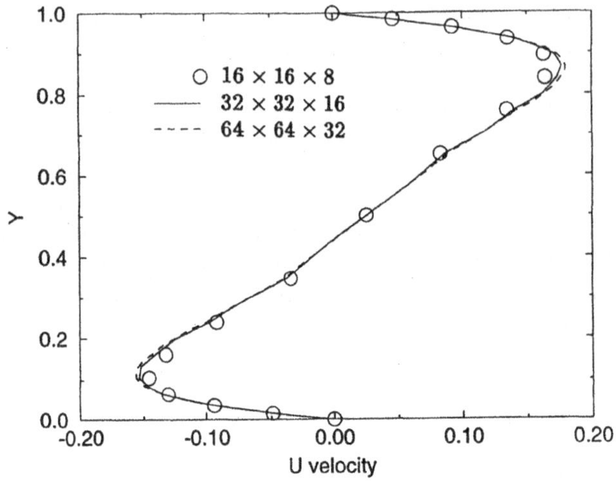

Fig. 11.8. Velocity u distribution along the vertical centerline at $\mathrm{Ra} = 10^5$ (from Yu et al. 1996b)

be equal. The error tabulated in Table 11.3 is the difference between the two integrated heat fluxes normalized by their average.

Table 11.3. Errors of overall heat transfer (%) (from Yu et al. 1996b)

Ra Mesh	17×17×9	32×32×17	64× 64 ×32
10^3	2.040	1.029	0.206
10^4	3.164	1.143	0.732
10^5	2.410	1.510	1.175
5×10^5	3.326	1.203	0.910

Figure 11.9 shows the temperature contours on several two-dimensional cross sections for the case $\mathrm{Ra} = 10^5$. For the $y-z$ planes near the conducting surfaces, several local hot and cold regions are observed.

Figure 11.10 shows the velocity vectors on the middle $x-y$ plane for flow fields at different Rayleigh numbers. In all cases, there is only one dominant vortex with its center biased to the cold wall. This situation contrasts with that in two-dimensional fluid flows, see Fig. 11.5. Figure 11.11 shows the velocity vectors on several $x-z$ planes for $\mathrm{Ra} = 10^5$. The corner vortices near the hot and cold surfaces with the stronger pair near the hot wall are observed.

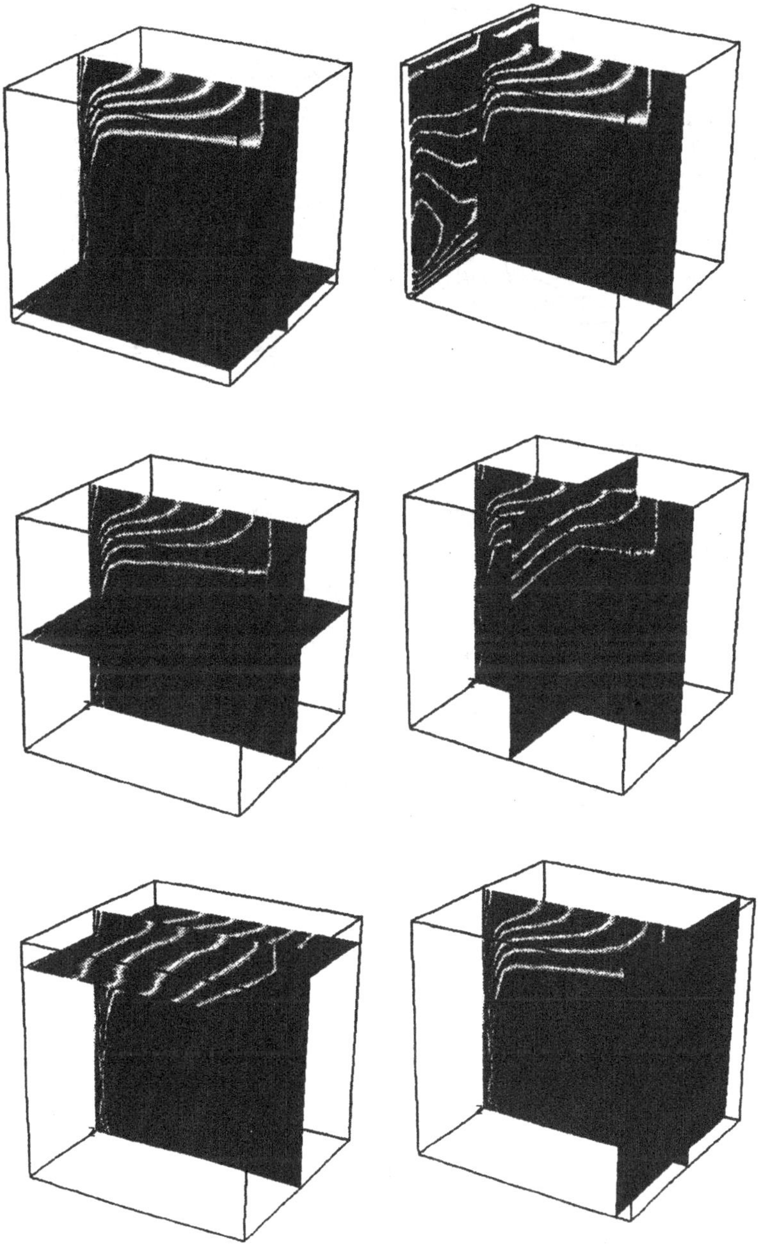

Fig. 11.9. Temperature contours of 3D buoyancy-driven gas flow on several two-dimensional cross sections at Ra = 10^5 (from Yu et al. 1996b)

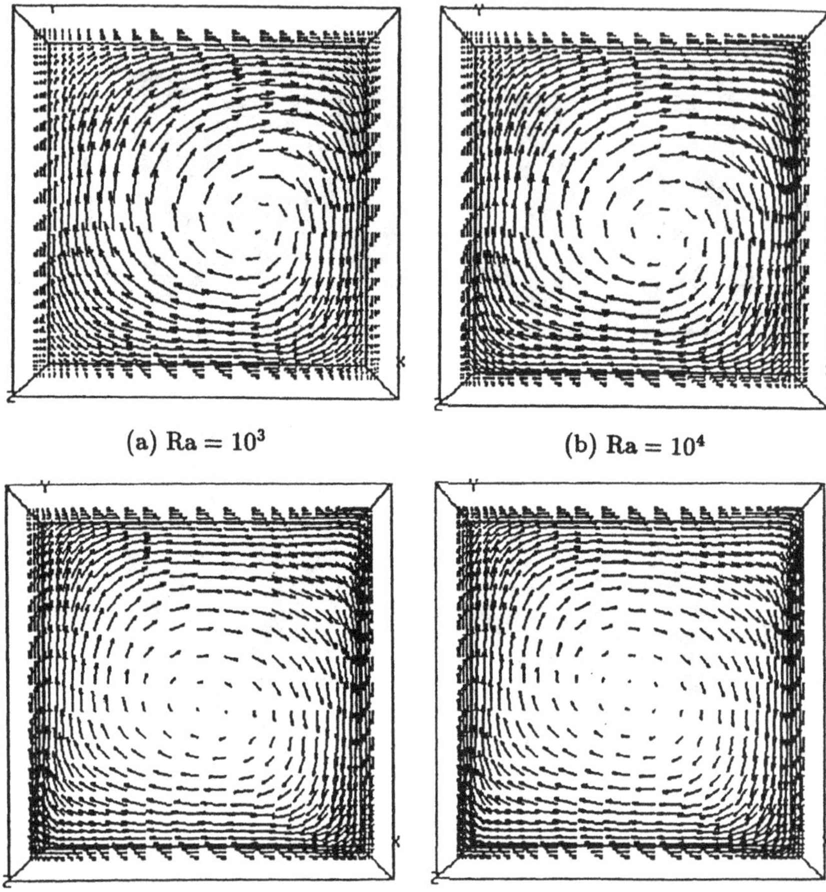

(a) Ra $= 10^3$ (b) Ra $= 10^4$

(c) Ra $= 10^5$ (d) Ra $= 10^6$

Fig. 11.10a–d. Velocity vectors of 3D buoyancy-driven gas flow on the central $x-y$ plane (from Yu et al. 1996b)

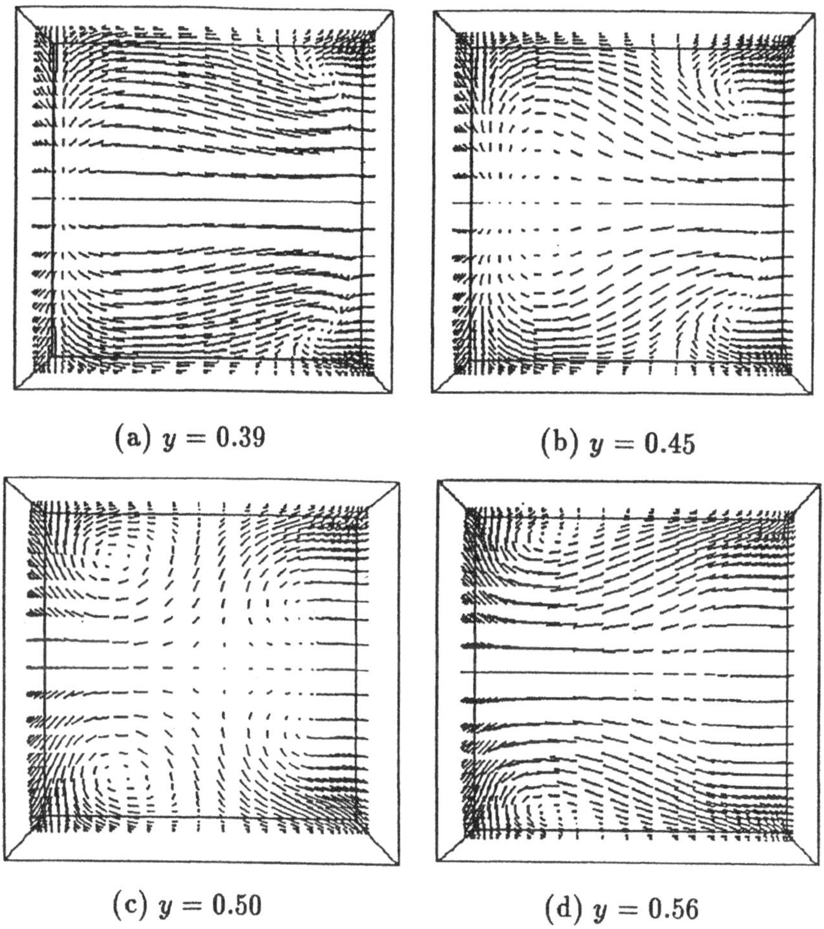

Fig. 11.11a–d. Velocity vectors of 3D buoyancy-driven gas flow on several $x-z$ planes at Ra $= 10^5$ (from Yu et al. 1996b)

The computation with Ra higher than a half million was also conducted. No steady-state solution could be obtained; this means that the flow may become unstable for Ra higher than a half million.

11.4 Concluding Remarks

By introducing the vorticity and the heat flux as additional variables, the governing equations for low-speed compressible viscous flows can be written as an almost-linear first-order system. The principle part of this system is exactly the same as that for incompressible flow–heat coupling problems. It consists of div–curl–grad operators. Consequently, the steady-state governing equation system for low-speed flows is elliptic, and the required boundary conditions become verifiable.

In contrast to conventional approaches, the LSFEM based on the quasi-linear or almost linear first-order system evades the difficulty of the lack of a time–derivative of pressure in the continuity equation, and no special treatments, such as preconditioning, staggered grids, or operator–splitting are needed. Moreover, due to the merit of LSFEM, the coefficient matrix is symmetric positive definite; its solution can be conducted by an element-by-element Jacobi preconditioned conjugate gradient method. The preconditioning matrix in LSFEM is automatically chosen by the preconditioned CG method itself after discretization, instead of the free choice of only two parameters to precondition the time derivative terms in the partial differential equations. Numerical solutions with good resolution for two and three dimensional compressible buoyant flows at various Rayleigh numbers have been obtained by LSFEM. No previously reported results showed such resolution. This illustrates the merits of LSFEM: (1) no artificial boundary condition for pressure is used; (2) no added artificial damping is employed; and (3) first-order derivative variables such as vorticity and heat fluxes are discretized to the same order of accuracy as the velocity, pressure, and temperature.

12. Two-Fluid Flows

Numerical simulation of multi-fluid flows presents a challenge, because the fluid properties are discontinuous across the fluid interface. To overcome this difficulty, fluids of different properties are identified through the use of a continuous field variable – the color function. The color function assigns a unique constant to each fluid. Interfaces between different fluids are distinct due to sharp gradients of the color function. The evolution of the interfaces is governed by the convective equation of the color function, while the fluid motion is governed by the incompressible Navier–Stokes equations. Therefore, conventional approaches are composed of two distinct methodologies, namely, a flow equation solver which is usually based on staggered grids, operator-splitting or projection methods; and a pure convection equation solver to identify the material by Lagrangian or upwind methods. Since these two totally different numerical techniques must be incorporated together, simulating two-fluid flows is regarded as a very complex problem.

In this chapter, we shall show that the LSFEM can handle the color function and the fluid variables in a unified manner without special treatment. Various benchmark tests are given to demonstrate that the LSFEM can successfully capture the interface and flow features. The material for this chapter is mainly taken from Wu et al. (1996).

12.1 Introduction

Many engineering problems involve multi-fluid/interfacial flows. Injection molding, metal casting, crystal growth and spray atomization are some significant examples. At the interface of different fluids, surface tension exists as a result of uneven molecular forces. The interface behaves in a way similar to a thinly stretched membrane. The prediction of the evolution of the interface and the treatment of the interface conditions have been a challenging task for numerical simulations.

Most existing numerical methods for multi-fluid/free-surface flows fall into two categories : (1) those which use a fixed grid; and (2) those which allow the grid to deform in time so that it remains surface-intrinsic.

In the first category, the computational grid is fixed throughout the calculation. An additional variable is used to identify the interface (front). Examples of such methods are the marker and cell (MAC) method proposed by Harlow and Welch (1965), the volume of fluid (VOF) method by Hirt and Nichols (1981), and the level set method by Zhu and Sethian (1992). The MAC method uses massless marker particles which travel with the fluid to trace the fluids and the interface. The VOF method modifies the MAC method by replacing the discrete marker particles with a continuous field variable – the color function or the level set function. This function assigns a unique constant (color) to each fluid. At fluid interfaces this color function has a sharp gradient. Numerical methods in this category are sometimes referred to as "front capturing" methods. Such methods possess great flexibility in handling large deformations and topological changes, as demonstrated by Daly (1967) and Harlow and Shannon (1967). The most difficult task with the front capturing approach is to accurately identify the interface and to impose the interface condition; these are exemplified by the elaborate work of Daly (1969). This difficulty can be alleviated by using a continuum surface force (CSF) model proposed by Brackbill et al. (1992). The CSF model interprets surface tension as a continuous, three-dimensional body force across an interface, rather than as a boundary condition on the interface. The computer implementation of the CSF model is therefore relatively simple compared with other approaches. The combination of the VOF method (or the level set method) and the CSF model has been used by a number of authors to simulate multi-phase phenomena involving surface tension and complex topological changes, see e.g., Richards et al. (1994), Lafaurie et al. (1994), Sussman et at. (1994), and Chang et al. (1996). A detailed description of the CSF model will be given in the next section.

For methods in the second category, see e.g., Fritt and Boris (1979), Fyfe et al. (1988), Jue and Ramaswamy (1992), Tezduya et al. (1992a,b), which are referred to as "front tracking" methods, imposing the interface condition is easy compared to the first, because the interface always coincides with mesh sides. However, it requires frequent updating of the computational mesh, which can be a complex and time-consuming procedure. In particular, it encounters severe difficulty when the flow experiences severe distortions and complex topological changes.

Another approach which can be regarded as a combination of the above is the front-tracking method introduced by Unverdi and Tryggvason (1992). This approach uses a fixed, structured grid to represent the flow field. A separate, unstructured grid is used to represent the interface. The interface is explicitly tracked and kept at a constant thickness of the order of the mesh size. This ensures that the interface will not be smeared by numerical diffusion. Much success has been achieved in solving a variety of two-fluid flow problems using this approach (Nobari and Tryggvason 1994a,b). The difficulty with this approach is the handling of complex topological changes.

In this chapter, numerical solutions to problems involving two immiscible fluids are sought. A large number of such problems deal with the interaction between a liquid and gas or air, which are often simply treated as free-surface problems. In the free-surface formulation the flow equations are solved only for the liquid, and zero traction is assumed on the interface. In this chapter such problems are treated as true two-fluid cases. The VOF approach and a modified version of the CSF model proposed by Jacqmin (1995) are used in the simulation. The LSFEM is used to discretize the governing system of equations.

12.2 Continuum Surface Force Model

Consider the effect of surface tension on a fluid interface. The pressure jump across the interface between two fluids (labeled 1 and 2) due to surface tension can be expressed by the Laplace formulas (Landau and Lifshitz 1959):

$$p_2 - p_1 = \sigma\kappa, \tag{12.1}$$

in which σ is the surface tension coefficient (only constant σ is considered), $\kappa = R_1^{-1} + R_2^{-1}$ is the Gaussian curvature of the interface, where R_1 and R_2 are the principal radii of curvature of the surface. The higher pressure is in the fluid on the concave side of the interface, since surface tention results in a net normal force directed toward the center of curvature of the interface.

Consider two fluids, fluid 1 and fluid 2, separated by an interface at time t. The two fluids are distinguished by some characteristic function:

$$\bar{C}(x) = \begin{cases} \bar{C}_1, & \text{in fluid 1;} \\ \bar{C}_2, & \text{in fluid 2;} \\ (\bar{C}_1 + \bar{C}_2)/2, & \text{at the interface.} \end{cases} \tag{12.2}$$

The CSF model is a means of treating surface tension effects in the VOF method. By using this model, the difficulty in imposing the interface condition (12.1) was alleviated. The basic idea of the CSF model is to regard the interface between two fluids as a transition region with a finite thickness, instead of a zero-thickness membrane. By using the CSF model, the interface condition (12.1) no longer needs to be explicitly imposed, as it is already implied in the momentum equations.

Consider replacing the discontinuous characteristic function by a smooth variation of fluid color $C(x)$ from $\bar{C}_1$ to $\bar{C}_2$ over a distance of $O(h)$, where h is a length comparable to the average finite element mesh size. This replaces the interface by a thin continuous transition layer, where $\bar{C}_1 \le C \le \bar{C}_2$. Accordingly, the surface tension effect is interpreted as a continuous body force spread across the transition region:

$$\boldsymbol{f} = \frac{\sigma}{[C]}\,\kappa\,\nabla C, \tag{12.3}$$

which acts as a source term in the momentum equations. In the above formulation, $[C]$ denotes the jump of C across the interface; the curvature κ is calculated from

$$\kappa = -(\nabla \cdot \boldsymbol{n}), \tag{12.4}$$

where $\boldsymbol{n}$ is the unit normal to the surface, which is obtained from

$$\boldsymbol{n} = \frac{\nabla C}{|\nabla C|}. \tag{12.5}$$

By using the vector equality (A.1), the x-component of the body force can be written as

$$\begin{aligned} f_x &= -\frac{\sigma}{[C]}\left[\nabla \cdot \left(\frac{\nabla C}{|\nabla C|}\right)\right]\frac{\partial C}{\partial x} \\ &= -\frac{\sigma}{[C]}\left[\nabla \cdot \left(\frac{\partial C}{\partial x}\frac{\nabla C}{|\nabla C|}\right) - \nabla\left(\frac{\partial C}{\partial x}\right)\cdot\frac{\nabla C}{|\nabla C|}\right] \\ &= -\frac{\sigma}{[C]}\left[\nabla \cdot \left(\frac{\partial C}{\partial x}\frac{\nabla C}{|\nabla C|}\right) - \frac{\partial}{\partial x}|\nabla C|\right] \\ &= -\frac{\sigma}{[C]}\left[\frac{\partial}{\partial x}\left(\frac{1}{|\nabla C|}\frac{\partial C}{\partial x}\frac{\partial C}{\partial x} - |\nabla C|\right) + \frac{\partial}{\partial y}\left(\frac{1}{|\nabla C|}\frac{\partial C}{\partial x}\frac{\partial C}{\partial y}\right)\right. \\ &\quad \left. + \frac{\partial}{\partial z}\left(\frac{1}{|\nabla C|}\frac{\partial C}{\partial x}\frac{\partial C}{\partial z}\right)\right]. \end{aligned} \tag{12.6}$$

In general, the volumetric body force caused by the surface tension is expressed as:

$$f_i = \frac{\partial \tau_{ij}}{\partial x_j}, \tag{12.7}$$

in which the stress tensor τ_{ij} is related to the color function C by:

$$\begin{aligned} \tau_{ij} &= -\frac{\sigma}{[C]}\frac{1}{|\nabla C|}\left(\frac{\partial C}{\partial x_i}\frac{\partial C}{\partial x_j} - \delta_{ij}\frac{\partial C}{\partial x_k}\frac{\partial C}{\partial x_k}\right) \\ &= -\frac{\sigma}{[C]}\left(\frac{1}{|\nabla C|}\frac{\partial C}{\partial x_i}\frac{\partial C}{\partial x_j} - \delta_{ij}|\nabla C|\right), \end{aligned} \tag{12.8}$$

where δ_{ij} is the Kronecker delta. This formulation for the body force was first derived by Jacqmin (1995). The advantage of using the formulation given by (12.7) and (12.8) instead of the original CSF model (12.3) is that (12.7) and (12.8) do not require the explicit calculation of the normalized gradient term in the right hand side of (12.5), whose definition is not clear when the denominator $|\nabla C|$ approaches zero. In (12.7) and (12.8), both $\boldsymbol{\tau}$ and $\boldsymbol{f}$ are well defined in the whole domain, and naturally vanish when $|\nabla C|$ becomes zero. An additional advantage of using the stress tensor $\boldsymbol{\tau}$ is that it can be regarded as part of the momentum flux. In many numerical procedures $\boldsymbol{\tau}$ can be used directly, and there is no need to calculate $\boldsymbol{f}$.

12.3 The First-Order Governing Equations

12.3.1 Rectangular Coordinates

The motion of both fluids is governed by a Newtonian incompressible fluid flow model. Let Ω be the fluid domain with a piecewise smooth boundary Γ. The governing equations in the velocity–pressure–vorticity form in rectangular coordinates are:

$$\nabla \cdot \boldsymbol{u} = 0 \qquad \text{in } \Omega, \tag{12.9a}$$

$$\rho \frac{\partial \boldsymbol{u}}{\partial t} + \rho(\boldsymbol{u} \cdot \nabla)\boldsymbol{u} + \nabla p + \mu(\nabla \times \boldsymbol{\omega}) - \left(\nabla\mu \cdot \nabla\boldsymbol{u} + \nabla\mu \cdot (\nabla\boldsymbol{u})^T\right) = \boldsymbol{f} \qquad \text{in } \Omega, \tag{12.9b}$$

$$\boldsymbol{\omega} - \nabla \times \boldsymbol{u} = \boldsymbol{0} \qquad \text{in } \Omega, \tag{12.9c}$$

$$\nabla \cdot \boldsymbol{\omega} = 0 \qquad \text{in } \Omega, \tag{12.9d}$$

in which ρ is the density, $\boldsymbol{u}$ is the velocity vector, p is the pressure, $\boldsymbol{\omega}$ is the vorticity vector, μ is the dynamic viscosity, and $\boldsymbol{f}$ is the body force, which generally consists of the surface tension effect given by (12.7) and the gravitational force. The terms in the last pair of brackets on the left hand side represent the effect of non-uniform viscosity.

The fluids are identified by different values of the color function C, which is convected by the flow field:

$$\frac{\partial C}{\partial t} + (\boldsymbol{u} \cdot \nabla)C = 0. \tag{12.10}$$

Fluid properties such as the density and the viscosity are assumed to be distributed in the same manner as C, i.e.,

$$\rho = \rho_1 + \frac{\rho_2 - \rho_1}{C_2 - C_1}(C - C_1), \tag{12.11a}$$

$$\mu = \mu_1 + \frac{\mu_2 - \mu_1}{C_2 - C_1}(C - C_1). \tag{12.11b}$$

The governing equations (12.9) and (12.10) are first discretized in time. The backward-Euler difference scheme is used:

$$\nabla \cdot \boldsymbol{u}^{n+1} = 0 \qquad \text{in } \Omega, \tag{12.12a}$$

$$\rho^{n+1}\frac{\boldsymbol{u}^{n+1} - \boldsymbol{u}^n}{\Delta t} + \rho^{n+1}\left(\boldsymbol{u}^{n+1} \cdot \nabla\right)\boldsymbol{u}^{n+1} + \nabla p^{n+1} + \mu^{n+1}\left(\nabla \times \boldsymbol{\omega}^{n+1}\right) - \left(\nabla\mu^{n+1} \cdot \nabla\boldsymbol{u}^{n+1} + \nabla\mu^{n+1} \cdot \left(\nabla\boldsymbol{u}^{n+1}\right)^T\right) = \boldsymbol{f}^{n+1} \qquad \text{in } \Omega, \tag{12.12b}$$

$$\boldsymbol{\omega}^{n+1} - \nabla \times \boldsymbol{u}^{n+1} = \boldsymbol{0} \qquad \text{in } \Omega, \tag{12.12c}$$

$$\nabla \cdot \boldsymbol{\omega}^{n+1} = 0 \qquad \text{in } \Omega, \tag{12.12d}$$

$$\frac{C^{n+1} - C^n}{\Delta t} + (\boldsymbol{u}^{n+1} \cdot \nabla) C^{n+1} = 0 \qquad \text{in } \Omega, \tag{12.12e}$$

where the superscript 'n' denotes the previous time-step and '$n+1$' denotes the current time-step.

The above system of equations (12.12) is further linearized. To ensure time-accuracy for time-dependent problems, linearization iterations are performed within each time-step (typically three iterations are used). For steady state problems only one linearization is performed in each time-step. Let k denote the number of linerization, the linearization is carried out as follows:

$$\nabla \cdot \boldsymbol{u}_{[k]}^{n+1} = 0, \qquad \text{in } \Omega, \tag{12.13a}$$

$$\begin{aligned} &\rho_{[k-1]}^{n+1} \frac{\boldsymbol{u}_{[k]}^{n+1}}{\Delta t} + \rho_{[k-1]}^{n+1} \left(\boldsymbol{u}_{[k-1]}^{n+1} \cdot \nabla \right) \boldsymbol{u}_{[k]}^{n+1} + \nabla p_{[k]}^{n+1} + \mu_{[k-1]}^{n+1} \nabla \times \boldsymbol{\omega}_{[k]}^{n+1} \\ &\quad = \nabla \mu_{[k-1]}^{n+1} \cdot \nabla \boldsymbol{u}_{[k-1]}^{n+1} + \nabla \mu_{[k-1]}^{n+1} \cdot \left(\nabla \boldsymbol{u}_{[k-1]}^{n+1} \right)^T \\ &\qquad + \boldsymbol{f}_{[k-1]}^{n+1} + \rho_{[k-1]}^{n+1} \frac{\boldsymbol{u}^n}{\Delta t} \qquad \text{in } \Omega, \end{aligned} \tag{12.13b}$$

$$\boldsymbol{\omega}_{[k]}^{n+1} - \nabla \times \boldsymbol{u}_{[k]}^{n+1} = \boldsymbol{0} \qquad \text{in } \Omega, \tag{12.13c}$$

$$\nabla \cdot \boldsymbol{\omega}_{[k]}^{n+1} = 0 \qquad \text{in } \Omega, \tag{12.13d}$$

$$\frac{C_{[k]}^{n+1}}{\Delta t} + \left(\boldsymbol{u}_{[k]}^{n+1} \cdot \nabla \right) C_{[k]}^{n+1} = \frac{C_{[k-1]}^n}{\Delta t} \qquad \text{in } \Omega, \tag{12.13e}$$

in which $C_{[0]}^{n+1} = C^n$, $\mu_{[0]}^{n+1} = \mu^n$ and $\boldsymbol{u}_{[0]}^{n+1} = \boldsymbol{u}^n$.

At this stage the standard LSFEM procedure is employeded for spatial discretization. We note that the solution of the flow field $\boldsymbol{u}_{[k]}^{n+1}$, $p_{[k]}^{n+1}$, $\boldsymbol{\omega}_{[k]}^{n+1}$ does not require knowledge of $C_{[k]}^{n+1}$; thus for each linearization, (12.13a)–(12.13d) are solved first. Then the newly obtained $\boldsymbol{u}_{[k]}^{n+1}$ is used in (12.13e) to solve for $C_{[k]}^{n+1}$. We use reduced integration (one point Gaussian quadrature for bilinear elements) in the solution of (12.13a)–(12.13d). The reason is that, as pointed out in Sect. 4.11, the LSFEM with Gaussian quadrature is equivalent to the least-squares finite element collocation method; the use of reduced integration makes the total number of collocation points compatible with the total number of unknowns and has proved to result in better accuracy. For (12.13e), however, we have used exact numerical integration (2×2 Gauss points for bilinear elements), because using reduced integration sometimes produces oscillatory solutions for the color function. Such oscillations will further result in unacceptable errors in the tension force, which is obtained by twice differentiating the color functions. In order to avoid excessive numerical diffusion and to maintain time-accuracy, a time-step is chosen so that the corresponding CFL number is less than or close to unity for true transient problems. If only steady state results are of interest, much larger time-steps are used. A matrix-free Jacobi-preconditioned conjugate gradient method has been used to solve the algebraic equations.

The body force term $\boldsymbol{f}^{n+1}_{[k-1]}$ in (12.13b) is calculated using (12.7) and (12.8). The finite element solution of $\boldsymbol{\tau}$ obtained from (12.8) is discontinuous across element boundaries. In order to use (12.7) to obtain the body force $\boldsymbol{f}$, the $\boldsymbol{\tau}$ field needs to be spatially continuous. Such a continuous field can be obtained by treating the discontinuous field with various recovery procedures (Zienkiewicz and Zhu 1992 and the references therein). In our numerical tests, only bilinear elements are used for the discretization and a simple averaging method proposed by Zienkiewicz and Zhu (1987) is used for the recovery procedure. For higher order elements, the superconvergent patch recovery (Zienkiewicz and Zhu 1992) should be used. The recovery procedure is also used to obtain a continuous $\boldsymbol{f}$ field.

12.3.2 Cylindrical Coordinates

The Navier–Stokes equations given by (12.9a) and (12.9b) can be written in cylindrical coordinates as:

$$\frac{\partial u_r}{\partial r} + \frac{1}{r}u_r + \frac{1}{r}\frac{\partial u_\theta}{\partial \theta} + \frac{\partial u_z}{\partial z} = 0 \qquad \text{in } \Omega, \tag{12.14a}$$

$$\begin{aligned}
&\rho\frac{\partial u_r}{\partial t} - \rho\frac{u_\theta^2}{r} + \rho u_r\frac{\partial u_r}{\partial r} + \rho u_\theta\frac{1}{r}\frac{\partial u_r}{\partial \theta} + \rho u_z\frac{\partial u_r}{\partial z} + \frac{\partial p}{\partial r} \\
&\quad +\mu\left(\frac{1}{r}\frac{\partial \omega_z}{\partial \theta} - \frac{\partial \omega_\theta}{\partial z}\right) - 2\frac{\partial \mu}{\partial r}\frac{\partial u_r}{\partial r} - \frac{1}{r}\frac{\partial \mu}{\partial \theta}\left(\frac{1}{r}\frac{\partial u_r}{\partial \theta} + \frac{\partial u_\theta}{\partial r} - \frac{u_\theta}{r}\right) \\
&\quad -\frac{\partial \mu}{\partial z}\left(\frac{\partial u_r}{\partial z} + \frac{\partial u_z}{\partial r}\right) = f_r \qquad \text{in } \Omega,
\end{aligned} \tag{12.14b}$$

$$\begin{aligned}
&\rho\frac{\partial u_\theta}{\partial t} + \rho\frac{u_r u_\theta}{r} + \rho u_r\frac{\partial u_\theta}{\partial r} + \rho u_\theta\frac{1}{r}\frac{\partial u_\theta}{\partial \theta} + \rho u_z\frac{\partial u_\theta}{\partial z} + \frac{1}{r}\frac{\partial p}{\partial \theta} \\
&\quad +\mu\left(\frac{\partial \omega_r}{\partial z} - \frac{\partial \omega_z}{\partial r}\right) - \frac{\partial \mu}{\partial r}\left(\frac{1}{r}\frac{\partial u_r}{\partial \theta} + \frac{\partial u_\theta}{\partial r} - \frac{u_\theta}{r}\right) - \frac{2}{r}\frac{\partial \mu}{\partial \theta}\left(\frac{1}{r}\frac{\partial u_\theta}{\partial \theta} + \frac{u_r}{r}\right) \\
&\quad -\frac{\partial \mu}{\partial z}\left(\frac{\partial u_\theta}{\partial z} + \frac{1}{r}\frac{\partial u_z}{\partial \theta}\right) = f_\theta \qquad \text{in } \Omega,
\end{aligned} \tag{12.14c}$$

$$\rho\frac{\partial u_z}{\partial t} + \rho u_r\frac{\partial u_z}{\partial r} + \rho u_\theta\frac{1}{r}\frac{\partial u_z}{\partial \theta} + \rho u_z\frac{\partial u_z}{\partial z} + \frac{\partial p}{\partial z}$$

$$+\mu\left(\frac{\partial \omega_\theta}{\partial r} + \frac{1}{r}\omega_\theta - \frac{1}{r}\frac{\partial \omega_r}{\partial \theta}\right) - \frac{\partial \mu}{\partial r}\left(\frac{\partial u_r}{\partial z} + \frac{\partial u_z}{\partial r}\right)$$

$$-\frac{1}{r}\frac{\partial \mu}{\partial \theta}\left(\frac{\partial u_\theta}{\partial z} + \frac{1}{r}\frac{\partial u_z}{\partial \theta}\right) - 2\frac{\partial \mu}{\partial z}\frac{\partial u_z}{\partial z} = f_z \qquad \text{in } \Omega. \tag{12.14d}$$

The definition (12.9c) of $\boldsymbol{\omega}$ is given as:

$$\omega_r - \frac{1}{r}\frac{\partial u_z}{\partial \theta} + \frac{\partial u_\theta}{\partial z} = 0 \qquad \text{in } \Omega, \tag{12.15a}$$

$$\omega_\theta - \frac{\partial u_r}{\partial z} + \frac{\partial u_z}{\partial r} = 0 \qquad \text{in } \Omega, \tag{12.15b}$$

$$\omega_z - \frac{\partial u_\theta}{\partial r} - \frac{1}{r}u_\theta + \frac{1}{r}\frac{\partial u_r}{\partial \theta} = 0 \qquad \text{in } \Omega. \tag{12.15c}$$

The compatibility condition (12.9d) in this system becomes:

$$\frac{\partial \omega_r}{\partial r} + \frac{1}{r}\omega_r + \frac{1}{r}\frac{\partial \omega_\theta}{\partial \theta} + \frac{\partial \omega_z}{\partial z} = 0 \qquad \text{in } \Omega. \tag{12.16}$$

The stress components due to surface tension are:

$$\tau_{rr} = -\frac{\sigma}{[C]}\left(\frac{1}{|\nabla C|}\frac{\partial C}{\partial r}\frac{\partial C}{\partial r} - |\nabla C|\right), \tag{12.17a}$$

$$\tau_{\theta\theta} = -\frac{\sigma}{[C]}\left(\frac{1}{|\nabla C|}\frac{1}{r^2}\frac{\partial C}{\partial \theta}\frac{\partial C}{\partial \theta} - |\nabla C|\right), \tag{12.17b}$$

$$\tau_{zz} = -\frac{\sigma}{[C]}\left(\frac{1}{|\nabla C|}\frac{\partial C}{\partial z}\frac{\partial C}{\partial z} - |\nabla C|\right), \tag{12.17c}$$

$$\tau_{r\theta} = \tau_{\theta r} = -\frac{\sigma}{[C]}\left(\frac{1}{|\nabla C|}\frac{1}{r}\frac{\partial C}{\partial r}\frac{\partial C}{\partial \theta}\right), \tag{12.17d}$$

$$\tau_{\theta z} = \tau_{z\theta} = -\frac{\sigma}{[C]}\left(\frac{1}{|\nabla C|}\frac{1}{r}\frac{\partial C}{\partial \theta}\frac{\partial C}{\partial z}\right), \tag{12.17e}$$

$$\tau_{rz} = \tau_{zr} = -\frac{\sigma}{[C]}\left(\frac{1}{|\nabla C|}\frac{\partial C}{\partial r}\frac{\partial C}{\partial z}\right), \tag{12.17f}$$

in which $|\nabla C| = \sqrt{\left(\frac{\partial u_r}{\partial r}\right)^2 + \left(\frac{1}{r}\frac{\partial u_r}{\partial \theta}\right)^2 + \left(\frac{\partial u_r}{\partial z}\right)^2}$. The body force components due to the above stress tensor is:

$$f_r = \frac{\partial \tau_{rr}}{\partial r} + \frac{1}{r}(\tau_{rr} - \tau_{\theta\theta}) + \frac{1}{r}\frac{\partial \tau_{r\theta}}{\partial \theta} + \frac{\partial \tau_{rz}}{\partial z}, \tag{12.18a}$$

$$f_\theta = \frac{\partial \tau_{r\theta}}{\partial r} + \frac{1}{r}\frac{\partial \tau_{\theta\theta}}{\partial \theta} + \frac{2}{r}\tau_{r\theta} + \frac{\partial \tau_{\theta z}}{\partial z}, \tag{12.18b}$$

$$f_z = \frac{\partial \tau_{rz}}{\partial r} + \frac{1}{r}\frac{\partial \tau_{\theta z}}{\partial \theta} + \frac{1}{r}\tau_{rz} + \frac{\partial \tau_{zz}}{\partial z}. \tag{12.18c}$$

When the flow is axisymmetric, all the θ-derivatives become zero. Moreover, if the flow is non-swirling, u_θ becomes zero. For such cases the governing equations are further simplified.

12.4 Numerical Examples

Broken Dam. The broken dam problem has been used by many as a test case for simulating free-surface problems. Experimental data for this case are available (Martin and Moyce 1952). Here the problem is solved as a two-fluid problem involving both water and air. Zero surface tension and slippery walls (left and bottom sides) are assumed. On the top and right sides zero pressure is imposed. In addition, if inflow is detected on the top and right sides, the density of air is imposed as a boundary condition. The computational domain is 2 units high and 6 units long. Initially, water occupies a 1×1 area at the bottom left corner. The computational mesh consists of 120×40 uniform linear quadrilateral elements. The time-step is $\Delta t = 0.05$. The nondimensionalized gravitational acceleration, g, is taken to be unity. The viscosity is set at 3.05×10^{-5} for water which is the same choice as in Nakayama and Mori (1996) and 3.05×10^{-8} for air. The densities for water and air are 1 and 0.001, respectively.

Figure 12.1 shows free surface profiles (a contour line of density at $\rho = 0.5$) and the pressure contours at various times. The calculated water front location and water column height are compared with the experimental data in Fig. 12.2. It can be seen that the calculated results are in good agreement with the experimental data given by Martin and Moyce (1952) and with the calculation results in Nakayama and Mori (1996).

Two-Liquid Interface. This example is taken from Tezduyar et al. (1992b). In this problem, a two-dimensional 0.8×0.6 box is fully occupied by two liquids of equal volume. The density of the fluids are 1.0 and 2.0, respectively. The gravitational acceleration is 0.294. The lighter liquid is placed on top of the heavier one. Initially the interface is a straight line, with a slope of 0.25. The dynamical viscosity is 0.001 (the same for both liquids). Zero surface tension is assumed. The computational mesh consists of 21×42 uniform bilinear elements. A time-step of $\Delta t = 0.25$ is used throughout the calculation which corresponds to a CFL number close to 1 when the velocity reaches its maximum at approximately $t = 6$. Slippery wall conditions are used on all four sides. Figure 12.3 shows the interface profile and velocity field at different times. Figure 12.4 shows the time history of the interface location on the left and right hand side walls (relative to and normalized by the average height). The results agree well with those presented in Tezduyar et al. (1992b).

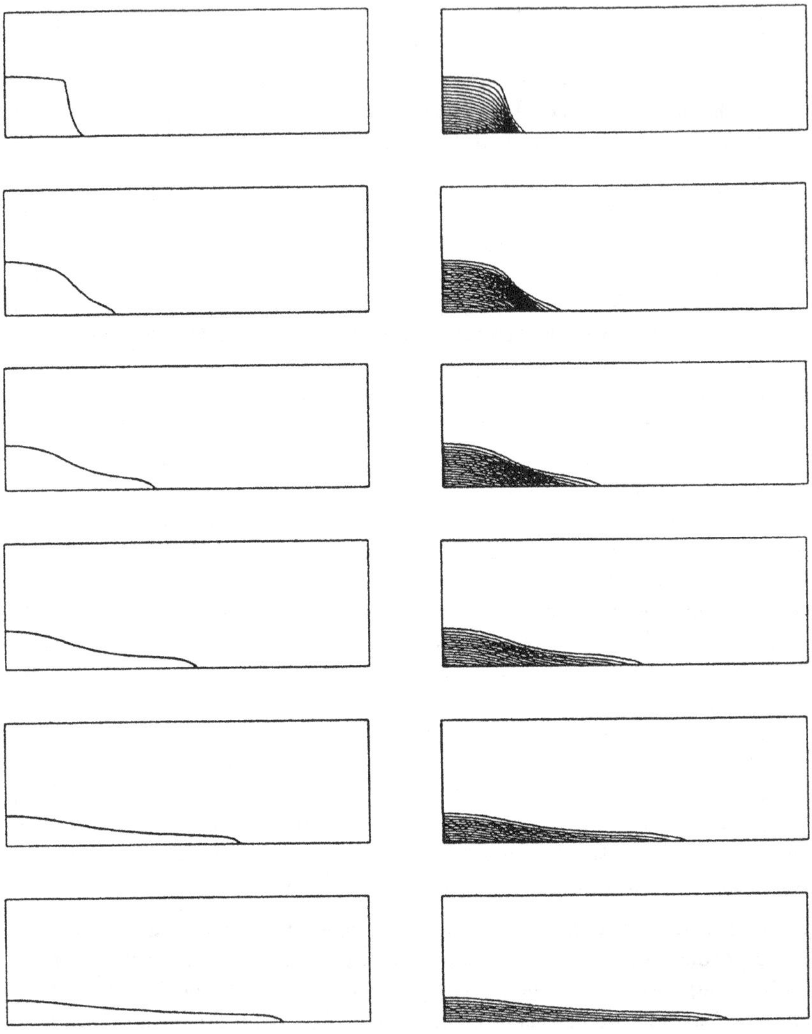

Fig. 12.1. A broken dam, free surface profile and pressure contours, from top to bottom: $t = 0.5, 1.0, 1.5, 2.0, 2.5, 3.0$ (from Wu et al. 1996)

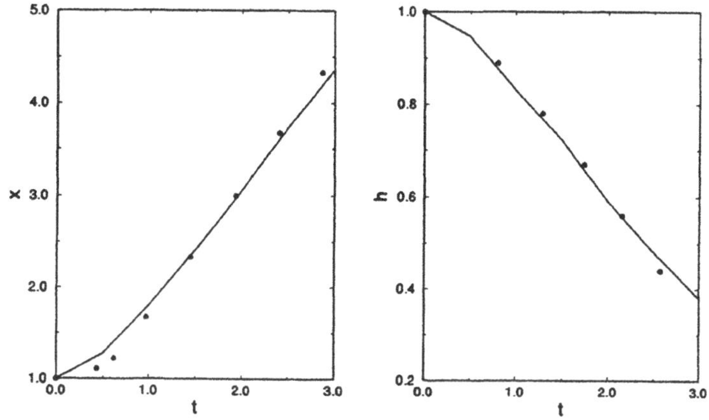

Fig. 12.2. A broken dam, front location and water column height, solid line — calculated results; dots — experimental data (from Wu et al. 1996)

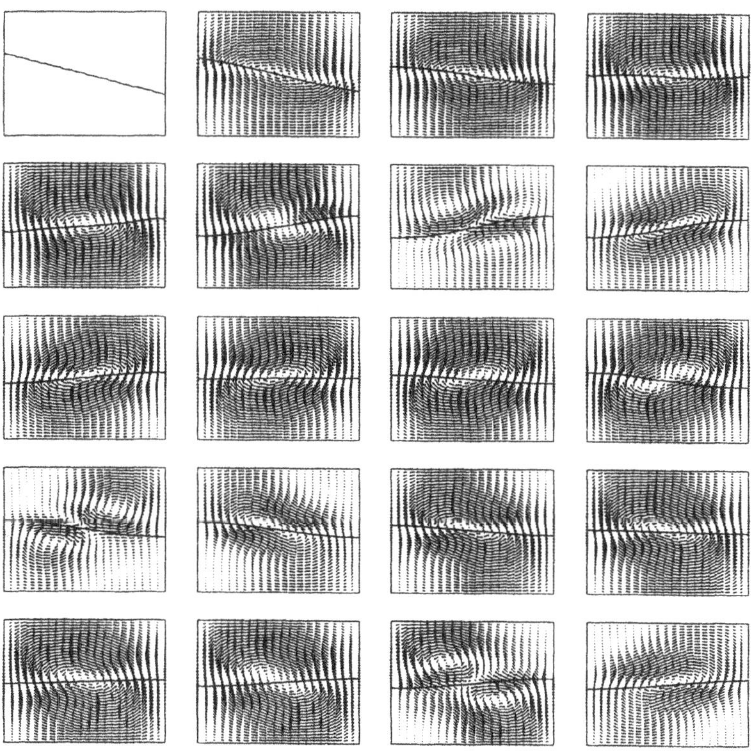

Fig. 12.3. A two-liquid interface problem, interface profile and velocity field, $t = 0, 1, 2, 3, 4, 5, 6, 7, 8, 9, 10, 11, 12, 13, 14, 15, 16, 17, 18, 19$, respectively (from Wu et al. 1996)

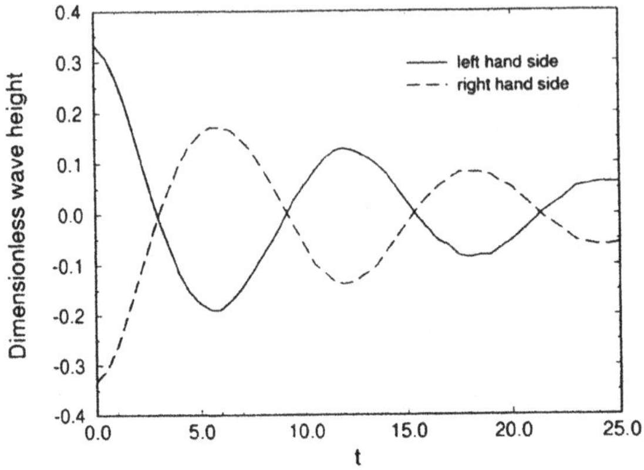

Fig. 12.4. A two-liquid interface problem, time history of the interface location on the left and right hand side walls (from Wu et al. 1996)

Oscillating Bubble. In this example the oscillation of a bubble due to surface tension is studied. Both two-dimensional planar and axisymmetric cases are considered. The two-dimensional problem was simulated by Fyfe et al. (1988) using a Lagrangian approach. The computational domain in this calculation is 4×12 units with the symmetry plane/axis along the longer side. The computational mesh consists of 40×120 uniform linear quadrilateral elements. Initially the shape of the bubble is elliptical, given by $x^2/4 + z^2 = 1$. The density of the fluid inside and outside the bubble is 1.5 and 0.5, respectively. The dynamic viscosity is 0.01 for both fluids. A time-step of $\Delta t = 0.05$ is used in both cases. After the calculation reaches $t = 15.0$, viscosity is increased to 1.0 to allow the solution to converge to steady state. When the steady state is reached the sizes of the bubbles are 1.40 and 1.58, respectively, while the theoretical values are 1.41 and 1.59. At steady state, it is verified that the Laplace formula (12.1) for the pressure jump is satisfied.

In Figs. 12.5 and 12.6 the interface shapes at different times are shown for two-dimensional and axisymmetric cases, respectively. The corresponding time history of the interface locations on the x -axis is shown in Fig. 12.7. The oscillation periods are estimated (based on the first two cycles) to be 7.6 for the two-dimensional case and 7.2 for the axisymmetric case. Theoretical solutions of the oscillation periods are available from a linear analysis for small amplitudes, see Fyfe et al. (1988) and Nobari and Tryggvason (1994) for details and references, which for this example are 6.1 for the two-dimensional case and 6.3 for the axisymmetry case. Since in the calculations both nonlinear and viscous effects are present, the theoretical solutions can be used only as references. The current numerical simulation predicts higher values of the oscillation periods than the linear theory, which is consistent with the experience of Fyfe et al. (1988).

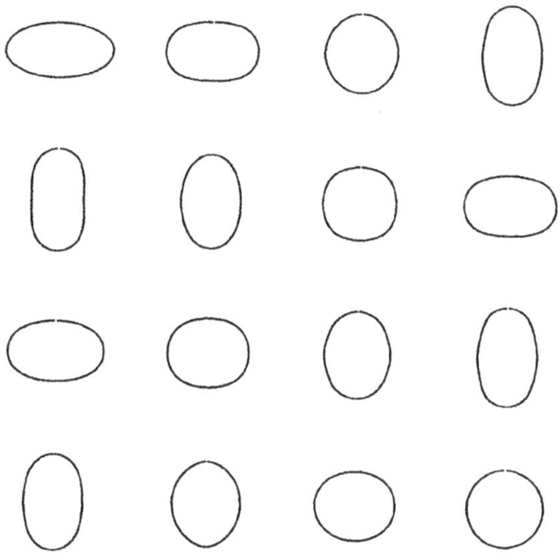

Fig. 12.5. An oscillating bubble, interface profile for the two-dimensional case, $t = 0, 1, 2, 3, 4, 5, 6, 7, 8, 9, 10, 11, 12, 13, 14, 40$, respectively (from Wu et al. 1996)

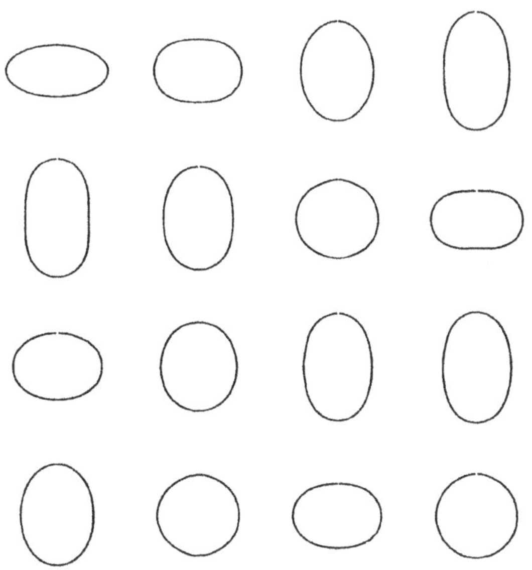

Fig. 12.6. An oscillating bubble, interface profile for the axisymmetric case, $t = 0, 1, 2, 3, 4, 5, 6, 7, 8, 9, 10, 11, 12, 13, 14, 40$, respectively (from Wu et al. 1996)

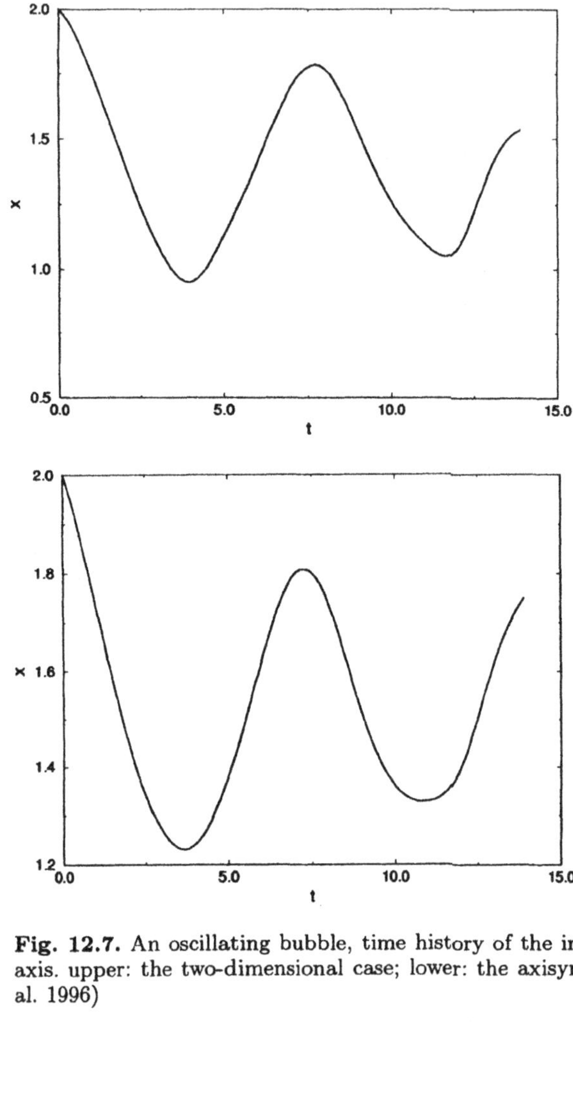

Fig. 12.7. An oscillating bubble, time history of the interface location on the x-axis. upper: the two-dimensional case; lower: the axisymmetric case (from Wu et al. 1996)

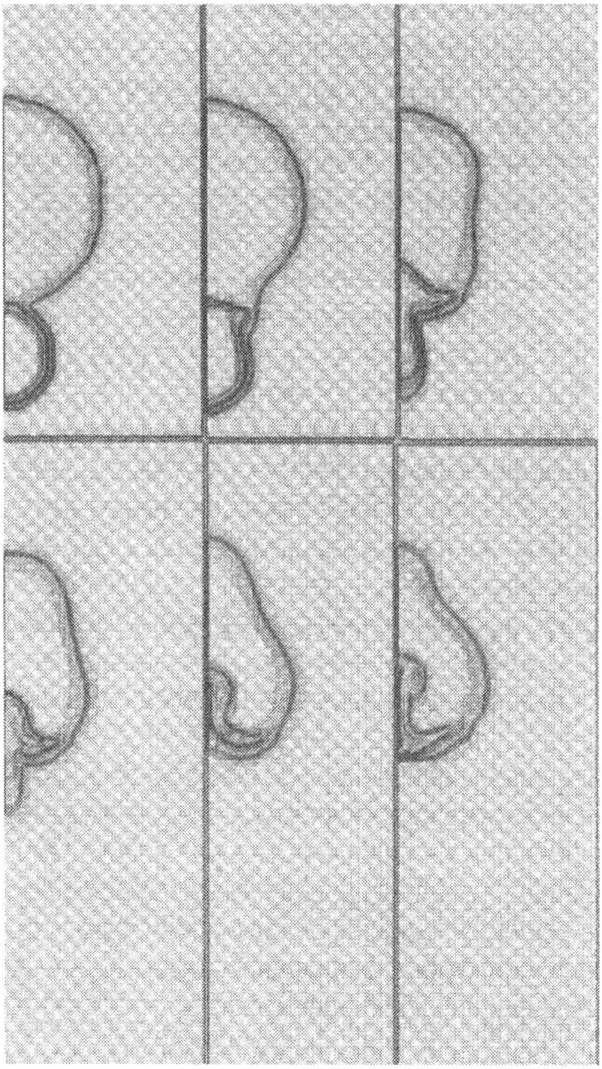

Fig. 12.8. Coalescence of two initially stationary drops, $t = 0.0, 1.0, 1.5, 2.0, 2.5, 3.0$, respectively (from Wu et al. 1996)

Coalescence of Two Initially Stationary Drops. The coalescence of two drops is a complicated process. The drops are driven by the combined effect of surface tension, inertia and viscosity. Here the axisymmetry is assumed. The computational mesh consists of 40×120 bilinear elements and is confined to $0 \le r \le 4$ and $0 \le z \le 12$. Initially the two drops, formed by the same material with radii 1 and 2 respectively, are placed next to each other with the centers of the drops two percent closer than the summation of the radii. The overlapped surfaces are assumed to be ruptured at $t = 0$. The fluid properties for this problem are chosen to be: $\rho_1 = 0.1,\ \rho_2 = 2.0,\ \mu_1 = \mu_2 = 0.01,\ \sigma = 2.5$, where the subscript 1 and 2 denote the ambient fluid and the drops, respectively. Figure 12.8 shows the calculated interface evolution. It can be observed that at first the opening resulted from the rupture is rapidly enlarged due to the high curvature at the area where the two drops contact. After this initial enlargement of the opening, the high curvature remains mainly in the small drop. As a result, the surface tension quickly pulls the small drop into the large one. Once the small drop has been completely pulled in, the inertia, built up during the "pull-in", becomes the dominating driving force. This brings the small drop further into the large drop and forms a jet with a vortex ring at its top. These results agree well with the experimental work by Anilkumar et al. (1991).

Flow in a Pressure Swirl Atomizer. This example deals with the swirling flow in a pressure swirl atomizer (simplex nozzle). Simplex nozzles are widely used as fuel atomizers in gas turbine engines. The formation of a conical liquid sheet is characteristic of such nozzles. Prediction of the thickness of the liquid sheet and the exit angle for various inlet conditions provides helpful information in the design process.

The computational domain for the current simulation is shown in Fig. 12.9. The liquid enters into the swirl chamber though a number of inlet slots. For the present calculation, axisymmetry is assumed. Gravitational effects are ignored. The computational mesh, shown in Fig. 12.10, consists of 3600 bilinear elements with 3741 nodes. Initially the whole domain is filled with motionless air. At $t > 0$ liquid begins to stream into the domain through the inlet slots. At the inlet, values for the liquid density and velocity are prescribed. The prescribed values are: $u_r = -4.8, u_\theta = 8.0, u_z = 3.6, \rho = 0.1$ The density of the air is $\rho = 0.01$. The dynamic viscosity for both fluids is $\mu = 0.005$. For nodes on the free boundary, the density of air is specified if inflow is detected, nothing is imposed if outflow is detected. Slip boundary conditions are imposed on portions of the wall; and no-slip conditions are imposed elsewhere on the wall, as shown in Fig. 12.9. The pressure is specified on the top left corner of the domain. A constant time-step $\Delta t = 0.1$ is used throughout the calculation. As only the steady state solution is of interest, a linearization iteration is performed only once in each time-step.

Figure 12.10 shows the steady state interface profile. The formation of the conical liquid sheet is successfully predicted in this simulation. This is a

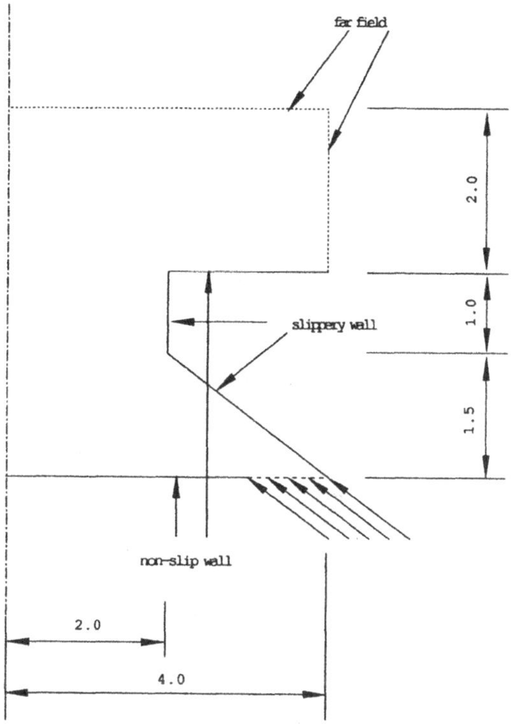

Fig. 12.9. Flow in a pressure swirl atomizer, the problem definition (from Wu et al. 1996)

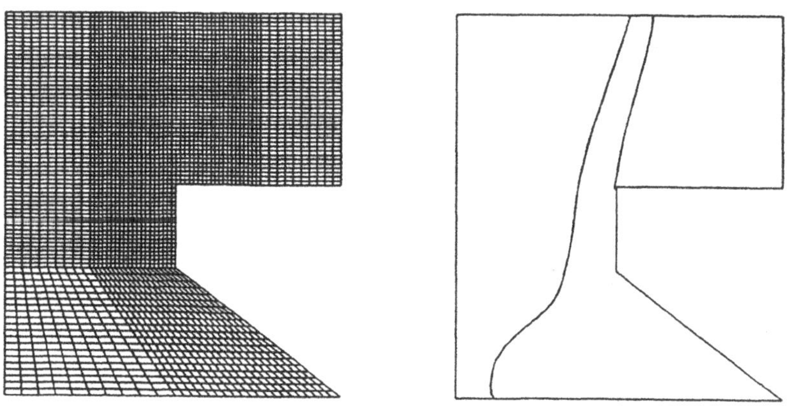

Fig. 12.10. Flow in a pressure swirl atomizer, the computational mesh and the interface profile (from Wu et al. 1996)

preliminary simulation and no attempt has been made to compare computational results with measured data. At the moment, no experimental data are available for comparison.

12.5 Concluding Remarks

In this chapter a numerical procedure based on a CSF model and the LSFEM is presented for two-fluid flow problems. Numerical tests carried out on a number of two-dimensional planar and axisymmetric flows indicate that this approach is capable of simulating such flow phenomena. The present approach has the advantage in the ability to handle complex topological changes such as the breakup of liquid jet. Another advantage is the ability to handle complex geometrical configurations in practical engineering environments, since the LSFEM is based on totally unstructured grids. In the presented examples the interface is typically spread over $3 \sim 5$ grids. A change of the interface thickness is also observed in the calculations. The analysis by Haj-Hariri and Borhan (1994) indicates that the migration velocity is strongly affected by the smearing of interface. Thus, to ensure good accuracy of the simulation a fine grid would have to be used. Improvement of the accuracy of the numerical scheme is expected by introducing higher order discretization schemes in both time and space. The tests conducted by Yu et al. (1995b) indicate that discontinuity can be modeled within 3 grids, even when the interface undergoes very large deformation due to a highly vortical flow field.

13. High-Speed Compressible Flows

This chapter is devoted to high-speed compressible inviscid flows with Mach numbers generally in excess of 0.6. Such flows, which usually involve the formation of shocks with characteristic discontinuities, are governed by the compressible Euler equations. For more than two decades, much attention in computational fluid dynamics has been focused on the efficient solution of the compressible Euler equations. This is mainly due to the fact that high-speed gas flows are of obvious practical importance and the cost of physical experiments is high. In this area, considerable success has been achieved by using finite difference and finite volume methods. There exists a large body of literature, see surveys in Hirsch (1990) and Feistauer (1993). The potential offered by finite element methods has been realized much later. The main advantage in the use of finite element approximation here is its capability for fitting complex geometry and permitting adaptive mesh refinement (Zienkiewicz and Taylor 1991).

We will show that shocks can be easily captured by the least-squares method when implemented with implicit time-difference equations. We use the one-dimensional Burger equation as a model problem to establish various least-squares schemes, then apply these schemes to the solution of one- and two-dimensional compressible Euler equations.

13.1 Various Least-Squares Schemes

13.1.1 Non-conservative L_2 Scheme

Let us consider a simple one-dimensional scalar, quasi-linear, hyperbolic equation

$$\frac{\partial u}{\partial t} + a(u)\frac{\partial u}{\partial x} = 0 \qquad \text{in } [0,1] \times (0,T], \tag{13.1a}$$

$$u(0,t) = g(t) \qquad \text{in } (0,T], \tag{13.1b}$$

$$u(x,0) = u^0(x) \qquad \text{in } [0,1], \tag{13.1c}$$

where the speed of wave propagation a is a function of u, g is specified boundary data. Depending on the problem, the boundary data may be given at $x = 1$, or specified at both $x = 0$ and $x = 1$.

The least-squares method to be defined in the following is similar to the one introduced in Sect. 9.3.2 for linear hyperbolic problems. There we pay attention to its mathematical properties including the rate of convergence, amplitude and phase errors; here we emphasize the capability of shock capturing.

For a given time step $\Delta t = t^{n+1} - t^n$, we linearize (13.1a) by setting $a^n = a(u^n)$. Backward-Euler time differencing leads to an implicit semi-discretized equation:

$$u^{n+1} - u^n + \Delta t a^n \frac{\partial u^{n+1}}{\partial x} = 0. \tag{13.2}$$

The scheme (13.2) is suitable for time-marching to stationary solutions.

For transient problems, to ensure time-accuracy, we may perform a few iterations at each time-level to improve the wave speed a, then go to the next time-level:

$$u^{n+1}_{[k+1]} - u^n + \Delta t a^{n+1}_{[k]} \frac{\partial u^{n+1}_{[k+1]}}{\partial x} = 0, \tag{13.3}$$

where k stands for the iteration count, $a^{n+1}_{[k]} = a(u^{n+1}_{[k]})$ and $u^{n+1}_{[0]} = u^n$. In the following discussion, for neatness and convenience we just use (13.2) to write the related formulations.

We can now apply the general formulation of LSFEM given in Sect. 4.8 to (13.2) to find the solution u^{n+1} at each time-level.

We proceed to discern why this method can capture shocks. For this reason we deduce the variational statement. The basic least-squares method for (13.2) amounts to minimizing the L_2 norm of the residual R for admissible u^{n+1} in (13.2), i.e., minimizing the objective functional:

$$I(u^{n+1}) = \int_0^1 R^2 dx = \int_0^1 \left(u^{n+1} - u^n + \Delta t a^n \frac{\partial u^{n+1}}{\partial x}\right)^2 dx. \tag{13.4}$$

Taking variations of I with respect to u^{n+1} and setting $v = \delta u^{n+1}$, $\delta I = 0$ leads to the least-squares variational statement: Find $u^{n+1} \in H = \{H^1(0,1) : u^{n+1} = g(t^{n+1}) \text{ at } x = 0\}$, such that

$$\int_0^1 \left(v + \Delta t a^n \frac{\partial v}{\partial x}\right)\left(u^{n+1} + \Delta t a^n \frac{\partial u^{n+1}}{\partial x}\right) dx$$

$$= \int_0^1 \left(v + \Delta t a^n \frac{\partial v}{\partial x}\right) u^n dx \qquad \forall v \in H_0, \tag{13.5}$$

where $H_0 = \{H^1(0,1) : v = 0 \text{ at } x = 0\}$. The least-squares method (13.5) corresponds to a Petrov–Galerkin scheme with test function $(v + \Delta t a^n \partial v / \partial x)$. Since a^n is the speed of wave propagation, the second term in the test function naturally provides an upwinding mechanism, as explained in Chap. 9. If the problem has a steady-state solution, then the stationary solution obtained by the least-squares method (13.5) satisfies the following variational statement:

$$\int_0^1 \left[v\left(a\frac{\partial u}{\partial x}\right) + \Delta t\left(a\frac{\partial v}{\partial x}\right)\left(a\frac{\partial u}{\partial x}\right)\right] dx = 0. \tag{13.6}$$

The first term in (13.6) corresponds to the standard Galerkin variational statement. The second term, which is proportional to Δt, is positive and acts as numerical dissipation to stabilize the solution by smoothing out discontinuities.

The scheme (13.5) is an implicit method and thus unconditionally stable, therefore the time-step can be large. However in general, the time-step Δt should be chosen such that the CFL (Courant–Friedrichs–Lewy) number $a\Delta t/h = 0.1 \sim 1$, where h is the mesh-size. If Δt is too large, the solution will diffused excessively. If Δt is too small, the solution will have oscillations. Here, small time-steps do not cause an efficiency problem, because the conjugate gradient (CG) method can always be employed due to the positive definiteness of the resulting matrix. The smaller the time step is, the less CG iterations are needed.

As an example of stationary problems, we consider the inviscid Burger equation

$$\frac{\partial u}{\partial t} + u\frac{\partial u}{\partial x} = 0 \qquad \text{in } [0,1]\times(0,T], \tag{13.7a}$$

with boundary data

$$u(0,t) = 0.8, \qquad u(1,t) = -0.8, \tag{13.7b}$$

and initial data

$$u(x,0) = u^0(x) = \begin{cases} 0.8 & 0 \le x < 0.4 \\ 0.8 - 8(x-0.4) & 0.4 \le x \le 0.6 \\ -0.8 & x > 0.6 \end{cases} \tag{13.7c}$$

corresponding to an initial slant step.

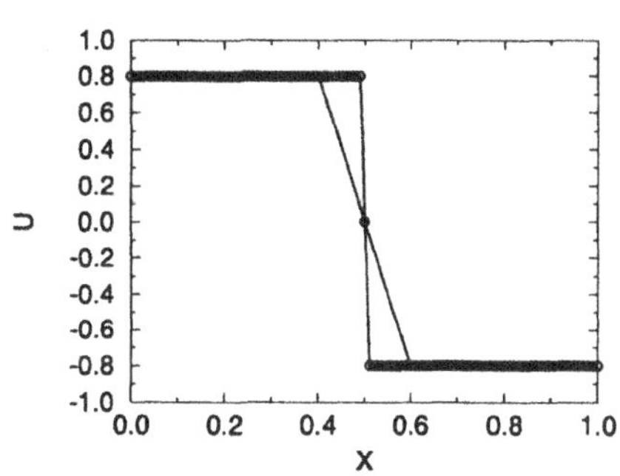

Fig. 13.1. Application of the non-conservative L_2 scheme to the Burger equation: 100 linear elements, $\Delta t = 0.0125, t = 0.20$

The slant step steepens as time increases to form a stationary 'shock' located at $x = 0.5$. In Fig. 13.1 we show the initial profile and the computed solution for scheme (13.5) at t=0.20. The computation is performed on a uniform mesh of 100 linear elements with fixed time step $\Delta t = 0.0125$. One-point Gaussian quadrature is used in the computation. It is seen that the stationary discontinuity or 'shock' can be well captured by the non-conservative L_2 scheme.

13.1.2 Non-conservative H^1 Scheme

The L_2 approach (13.5) is similar in nature to Lax–Wendroff and Taylor–Galerkin methods as demonstrated in Chap. 9. As in these methods, due to the non-linearity of the governing equations, oscillations can sometimes not be completely eliminated when discontinuities or shocks form and move, especially if the mesh is not adaptively refined, or higher-order elements are used.

Since both the solution and the equation residual become oscillatory as the solution gradient steepens during shock formation, this suggests that additional solution-derivative or residual-derivative control could provide a mechanism for stabilizing the solution. In the context of the least-squares residual formulation, this can be achieved in several ways: e.g., by adding a penalty functional to constrain the solution gradients or considering alternative forms of the functional. A natural modification is the weighted H^1-norm of R :

$$I = \int_0^1 \left[R^2 + \beta\left(\frac{\partial R}{\partial x}\right)^2\right] dx \tag{13.8}$$

for the minimization problem, where β is a positive small-scale factor. We may view (13.8) simply as a multi-objective optimization functional where the first term represents the principal objective of interest, while the second term controls the residual-derivative.

We seek to obtain a finite element solution with good shock resolution, and also suppress the numerical oscillation at the shock due to the large derivative. The additional functional acts as a control mechanism on the size of the derivative but should not be so strong that the solution is over-dissipative. Note, however, that the form in (13.8) will contain a second derivative of u which implies that C^1 elements are appropriate for a scheme based on this higher-order statement. Since in practice we want to be able to employ simpler elements with C^0 base, we consider a practical approximate form based on (13.8). Introducing the residual R of (13.2) into (13.8), omitting the second-order derivative term and the $\partial u^n/\partial x$ term, and taking variations, the resulting variational statement becomes: Find $u^{n+1} \in H = \{H^1(0,1) : u^{n+1} = g(t^{n+1}) \text{ at } x = 0\}$, such that

$$\int_0^1 \left[\left(v + \Delta t a^n \frac{\partial v}{\partial x} \right) \left(u^{n+1} + \Delta t a^n \frac{\partial u^{n+1}}{\partial x} \right) + \beta \left(1 + \Delta t \frac{\partial a^n}{\partial x} \right)^2 \frac{\partial v}{\partial x} \frac{\partial u^{n+1}}{\partial x} \right] dx$$

$$= \int_0^1 \left(v + \Delta t a^n \frac{\partial v}{\partial x} \right) u^n dx \qquad \forall v \in H_0, \tag{13.9}$$

which involves an additional artificial dissipation contribution

$$\beta \left(1 + \Delta t \frac{\partial a^n}{\partial x} \right)^2 \frac{\partial v}{\partial x} \frac{\partial u^{n+1}}{\partial x}.$$

This term corresponds to adding the stabilization functional

$$\beta \int_0^1 \left(1 + \Delta t \frac{\partial a^n}{\partial x} \right)^2 \left(\frac{\partial u^{n+1}}{\partial x} \right)^2 dx$$

to the functional in (13.4). Integration by parts reveals that the corresponding numerical viscosity term in the associated Euler–Lagrange equation is

$$-\beta \frac{\partial}{\partial x} \left[\left(1 + \Delta t \frac{\partial a^n}{\partial x} \right)^2 \frac{\partial u^{n+1}}{\partial x} \right].$$

That is, for $\beta \neq 0$ we have added an artificial dissipation with a term proportional to $(1 + \Delta t \partial a^n / \partial x)^2$, and the proposed scheme is a specific form of artificial dissipation technique.

As a numerical example of moving shock problems, we consider the inviscid Burger equation (13.7a) with boundary data

$$u(0,t) = 1.0, \qquad u(1,t) = 0.0, \tag{13.10a}$$

and initial data

$$u(x,0) = u^0(x) = \begin{cases} 1 & 0 \le x < 0.4 \\ 1 - 5(x - 0.4) & 0.4 \le x \le 0.6 \\ 0 & x > 0.6 \end{cases}. \tag{13.10b}$$

As time increases, this initial slant step steepens and forms a 'shock' located at $x = 0.6$ when $t = 0.2$. Then the shock moves with a velocity $= (1+0)/2 = 0.5$. When $t = 0.4$, the shock reaches the position $x = 0.7$.

All numerical results for this problem are obtained on a uniform mesh of 100 linear elements. As shown in Fig. 13.2a, in the absence of derivative control ($\beta = 0$), the calculations are unstable in that, as the shock forms, local oscillations develop. Figure 13.2b shows the computed solution by scheme (13.9) at $t = 0.4$ with fixed time step $\Delta t = 0.01$ and parameter $\beta = 0.00004$; in the computations, one-point Gaussian quadrature and two iterations at each time-level ($k = 1$) are used.

As mentioned many times in previous chapters, the number of Gaussian points is of importance in LSFEM. In general, the reduced integration is recommended. But for shock problems, we may use the regular intergration to smooth solutions. Figures 13.2 (**c**) and (**d**) show the effects of two-points quadrature.

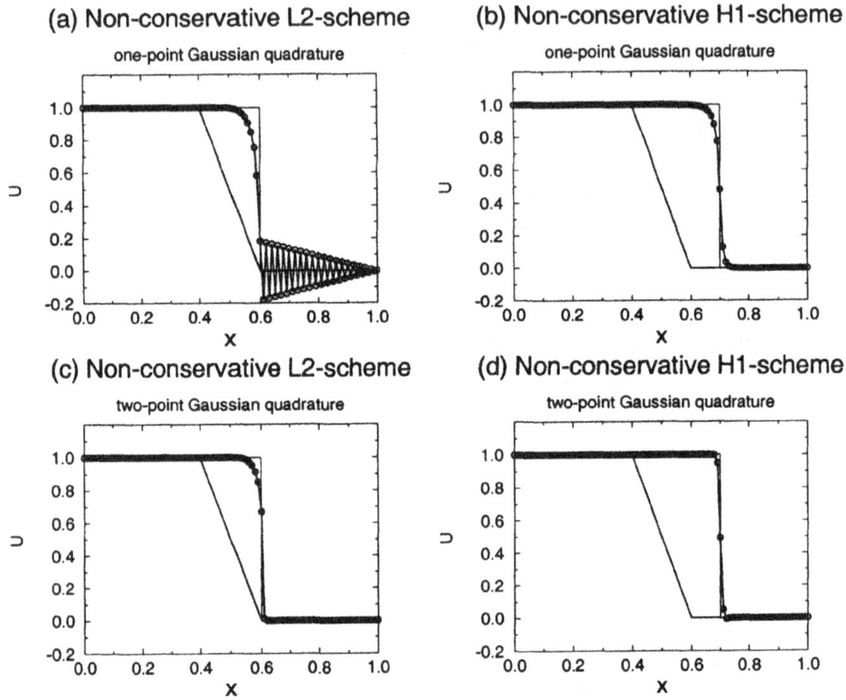

Fig. 13.2. Solution to the Burger equation from initial slant step with 100 uniform linear elements: (**a**) $\Delta t = 0.02, t = 0.2, k = 0$; (**b**) $\Delta t = 0.01, t = 0.4, \beta = 0.00004, k = 1$; (**c**) $\Delta t = 0.01, t = 0.2, k = 1$; (**d**) $\Delta t = 0.001, t = 0.4, \beta = 0.000002, k = 1$

13.1.3 Conservative Schemes

The one-dimensional scalar conservative form is

$$\frac{\partial q}{\partial t} + \frac{\partial F}{\partial x} = 0 \qquad \text{in } [0,1] \times (0,T], \tag{13.11}$$

where q is an conservative variable, and F is the conservative flux with $\partial F/\partial q = a(q)$. The boundary and initial conditions are similar to (13.1b) and (13.1c).

Introducing $\Delta q = q^{n+1} - q^n$ as an unknown, and time-differencing applied to the conservative equation (13.11) through the θ method leads to a semi-discretized equation:

$$\Delta q + \theta \Delta t \frac{\partial F^{n+1}}{\partial x} + (1-\theta)\Delta t \frac{\partial F^n}{\partial x} + \left(\theta - \frac{1}{2}\right) O(\Delta t^2) = 0. \tag{13.12}$$

The flux at time level $(n+1)$ is obtained by

$$
\begin{aligned}
F^{n+1} &= F^n + \Delta t\Big(\frac{\partial F}{\partial t}\Big)^n + O(\Delta t^2) \\
&= F^n + \Delta t\Big(a\frac{\partial q}{\partial t}\Big)^n + O(\Delta t^2) \\
&= F^n + a^n \Delta q + O(\Delta t^2).
\end{aligned} \tag{13.13}
$$

Applying the linearization procedure of (13.13) to the flux derivative $\partial F/\partial x$, (13.12) becomes

$$
\Big(1 + \theta\Delta t\frac{\partial a^n}{\partial x}\Big)\Delta q + \theta\Delta t a^n \frac{\partial \Delta q}{\partial x} = -\Delta t\frac{\partial F^n}{\partial x} + \Big(\theta - \frac{1}{2}\Big)O(\Delta t^2). \tag{13.14}
$$

When $\theta = 1$, this scheme corresponds to the backward-Euler differencing scheme and is of first-order accuracy $O(\Delta t)$ in the sense of the Taylor-series expansion; for $\theta = 1/2$ one obtains the Crank–Nicolson scheme with a second-order accuracy $O(\Delta t)^2$.

The corresponding least-squares variational statement is: Find $\Delta u \in H_0$ such that

$$
\int_0^1 \Big[\Big(1 + \theta\Delta t\frac{\partial a^n}{\partial x}\Big)v + \theta\Delta t a^n\frac{\partial v}{\partial x}\Big]\Big[\Big(1 + \theta\Delta t\frac{\partial a^n}{\partial x}\Big)\Delta q + \theta\Delta t a^n\frac{\partial \Delta q}{\partial x}\Big]dx
$$

$$
= -\int_0^1 \Big[\Big(1 + \theta\Delta t\frac{\partial a^n}{\partial x}\Big)v + \theta\Delta t a^n\frac{\partial v}{\partial x}\Big]\Big[\Delta t\frac{\partial F^n}{\partial x}\Big]dx \quad \forall v \in H_0. \tag{13.15}
$$

Applying the idea introduced in Sect. 13.1.2, we may also develop a conservative approximate H^1-residual scheme which can be obtained just by adding the following term into the right-hand side of (13.15):

$$
\int_0^1 \beta\Big(1 + 2\theta\Delta t\frac{\partial a^n}{\partial x}\Big)^2 \frac{\partial v}{\partial x}\frac{\partial \Delta q}{\partial x}dx.
$$

We note that it is not necessary to write a new LSFEM program to implement the H^1 scheme. In fact, the H^1 scheme is equivalent to solving the following system of first-order differential equations by the L_2 method:

$$
\boldsymbol{A}_1\frac{\partial \Delta q}{\partial x} + \boldsymbol{A}_0\Delta q = \boldsymbol{f} \qquad \text{in } \Omega, \tag{13.16}
$$

in which

$$
\boldsymbol{A}_0 = \begin{pmatrix} 1 + \theta\Delta t\frac{\partial a^n}{\partial x} \\ 0 \end{pmatrix}, \qquad \boldsymbol{A}_1 = \begin{pmatrix} \theta\Delta t a^n \\ \sqrt{\beta}\Big(1 + 2\theta\Delta t\frac{\partial a^n}{\partial x}\Big) \end{pmatrix},
$$

$$
\boldsymbol{f} = \begin{pmatrix} -\Delta t\frac{\partial f^n}{\partial x} \\ 0 \end{pmatrix}.
$$

The general formulation of LSFEM given in Sect. 4.8 can be directly applied to solving (13.16).

To obtain time-accurate results, at each time-level we may perform two or three iterations as discussed in Sect. 13.1.1.

Figure 13.3 shows the computed solutions to the same Burger equation tested in the previous section with the conservative H^1 scheme. As expected, the first-order scheme ($\theta = 1$) produces a diffused shock front, while the second-order scheme ($\theta = 0.5$) can provide sharp resolution of shocks but with some oscillations.

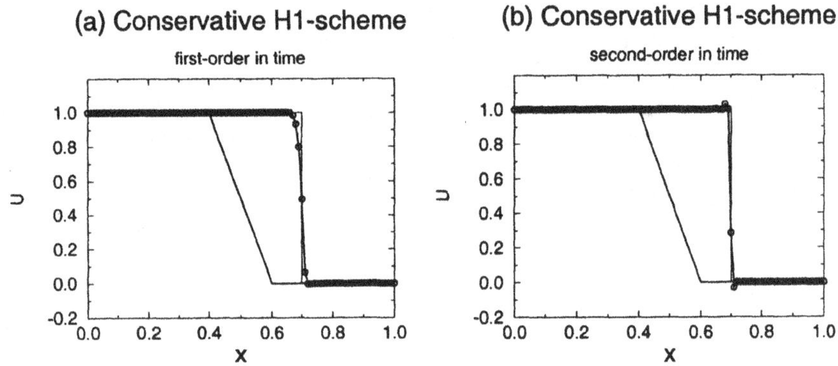

Fig. 13.3. Application of the conservative H^1 scheme to the Burger equation with 100 uniform linear elements, $\Delta t = 0.016, t = 0.4, k = 2$, one-point Gaussian quadrature: (**a**) first-order ($\theta = 1$), $\beta = 0.00003$; (**b**) second-order ($\theta = 0.5$), $\beta = 0.00003$

In summary, numerical experiments conducted for the inviscid Burger equation show that both non-conservative and conservative LSFEM can capture stationary and moving shocks with correct strength and location. The selection of time-step size and the iteration at each time-level are most important for the overall accuracy of LSFEM for moving shocks. To ensure correct location and strength of shocks, the time-step size Δt should be chosen such that the CFL number is not greater than 1; two or three iterations at each time-level are necessary.

13.2 One-Dimensional Flows

The computational schemes outlined can be extended to high-speed compressible flow problems. In this section we illustrate their performance on a few relative simple one-dimensional flow problems.

The governing equations for compressible flows can be written in standard matrix form of first-order system:

$$\frac{\partial \boldsymbol{U}}{\partial t} + \boldsymbol{A}_1 \frac{\partial \boldsymbol{U}}{\partial x} = \boldsymbol{S}, \tag{13.17}$$

where $\boldsymbol{U}$ is the primitive, vector-valued variable, $\boldsymbol{S}$ is a source term, and $\boldsymbol{A}_1$ is a matirx whose entries depend on $\boldsymbol{U}$. For a given time step $\Delta t = t^{n+1} - t^n$, we linearize the problem by setting $\boldsymbol{A}_1^n = \boldsymbol{A}_1(\boldsymbol{u}^n)$. backward-Euler differencing leads to an implicit time-differenced problem

$$\boldsymbol{U}^{n+1} + \Delta t \boldsymbol{A}_1^n \frac{\partial \boldsymbol{U}^{n+1}}{\partial x} = \boldsymbol{U}^n + \Delta t \boldsymbol{S}. \tag{13.18}$$

Now we are ready to use the non-conservative formulation of LSFEM given in Sect. 13.1 to get the solution $\boldsymbol{U}^{n+1}$ for each time-level.

Isothermal Flow in a Nozzle. The equations governing isothermal flow in a nozzle constitute a first-order system of the form (13.17) with

$$\boldsymbol{U} = \begin{pmatrix} \rho a \\ \rho a u \end{pmatrix}, \quad \boldsymbol{A}_1 = \begin{pmatrix} 0 & 1 \\ c^2 - u^2 & 2u \end{pmatrix}, \quad \boldsymbol{S} = \begin{pmatrix} 0 \\ \rho c^2 \frac{da}{dx} \end{pmatrix}, \tag{13.19}$$

where ρ is the density, u is the velocity, the speed of sound c is normalized to unit and the cross-sectional area of the nozzle is

$$a = 1.0 + \frac{(x-2.5)^2}{12.5}, \qquad 0 \le x \le 5. \tag{13.20}$$

Numerical results for subsonic inflow and outflow with interior supersonic flow regime terminating in a shock, using the non-conservative H^1 scheme, are shown in Fig. 13.4. The computations are performed on 80 uniform linear elements with a time-step $\Delta t = 0.5$ and $\beta = 0.0001$. The steady-state solution is attained after 80 time steps.

Shock Tube Problem. The one-dimensional Euler equations for compressible flow can be expressed as a system in the form (13.17) with

$$\boldsymbol{U} = \begin{pmatrix} \rho \\ u \\ p \end{pmatrix}, \quad \boldsymbol{A}_1 = \begin{pmatrix} u & \rho & 0 \\ 0 & u & \rho^{-1} \\ 0 & \gamma & u \end{pmatrix}, \quad \boldsymbol{S} = \boldsymbol{0}, \tag{13.21}$$

where ρ is the density, u is the velocity, p is the pressure and $\gamma = 1.4$ is the ratio of specific heats. For the shock tube problem the initial data are

$$\boldsymbol{U} = \begin{pmatrix} 1.0 \\ 0.0 \\ 1.0 \end{pmatrix} \quad \text{for } x \le 0.5, \qquad \boldsymbol{U} = \begin{pmatrix} 0.125 \\ 0.0 \\ 0.1 \end{pmatrix} \quad \text{for } x > 0.5. \tag{13.22}$$

Here the initial density and pressure difference between two sections of the tube is maintained by a diaphragm which is destroyed at $t = 0$. The objective is to compute accurate approximations to ρ, u and p at subsequent times as the shock and contact discontinuities propagate along the tube. The problem is solved here using the non-conservative L_2 scheme (13.5) with a uniform mesh of 100 linear elements, a fixed time step $\Delta t = 0.005$ and one-point Gaussian quadrature. The solution profiles for density, velocity and pressure at $t = 0.14$ are shown in Fig. 13.5. There is moderate smearing of the 'fronts'. The solution is neither oscillatory nor unstable.

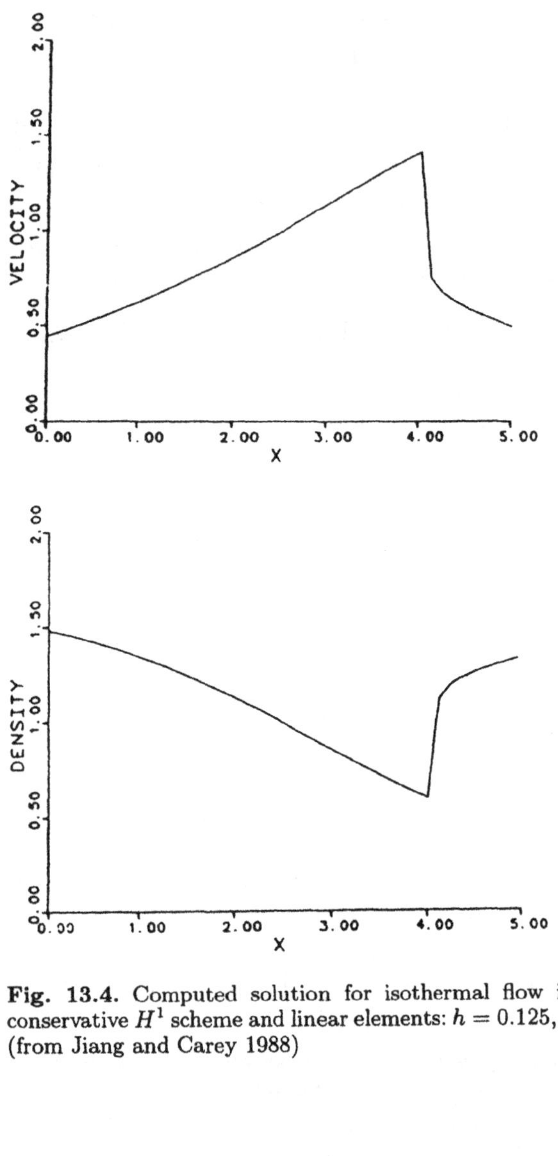

Fig. 13.4. Computed solution for isothermal flow in a nozzle using the non-conservative H^1 scheme and linear elements: $h = 0.125, \Delta t = 0.5, t = 40, \beta = 0.0001$ (from Jiang and Carey 1988)

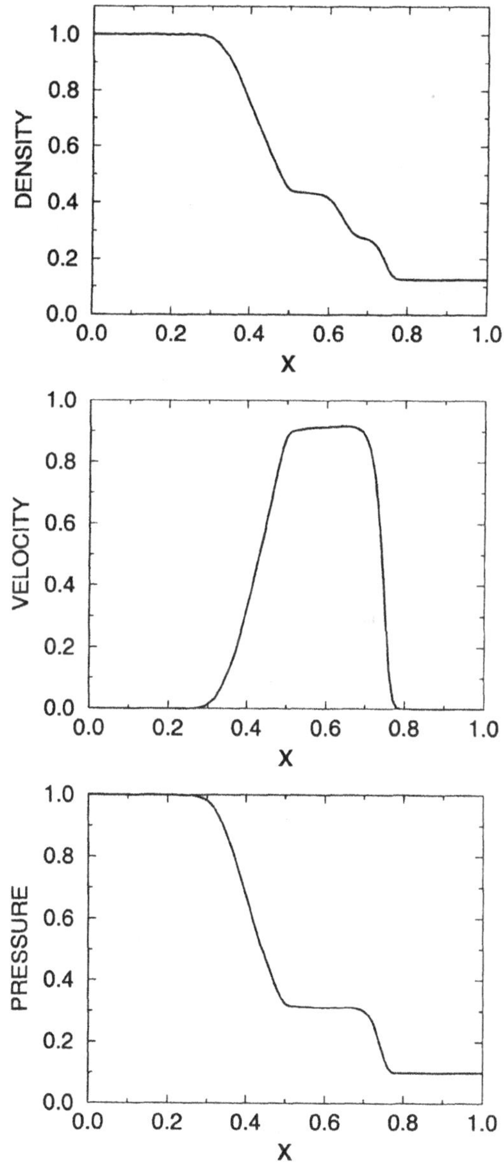

Fig. 13.5. Solutions to the shock tube problem with 100 uniform linear elements, $\Delta t = 0.005, t = 0.14, k = 1$ using a non-conservative L_2 scheme

13.3 Two-Dimensional Flows

13.3.1 Non-conservative L_2 Scheme

Least-Squares Formulation. The two-dimensional compressible Euler equations can be written as a first-order system in terms of primitive variables:

$$\frac{\partial \boldsymbol{U}}{\partial t} + \boldsymbol{A}_1 \frac{\partial \boldsymbol{U}}{\partial x} + \boldsymbol{A}_2 \frac{\partial \boldsymbol{U}}{\partial y} = \boldsymbol{0}, \tag{13.23}$$

where $\boldsymbol{U}^T = (\rho, u, v, p)$ is the vector consisting of primitive variable unknowns, and

$$\boldsymbol{A}_1 = \begin{pmatrix} u & \rho & 0 & 0 \\ 0 & u & 0 & \rho^{-1} \\ 0 & 0 & u & 0 \\ 0 & \gamma p & 0 & u \end{pmatrix}, \qquad \boldsymbol{A}_2 = \begin{pmatrix} v & 0 & \rho & 0 \\ 0 & v & 0 & 0 \\ 0 & 0 & v & \rho^{-1} \\ 0 & 0 & \gamma p & v \end{pmatrix},$$

in which ρ is the density, (u, v) are the Cartesian velocity components, p is the pressure, and γ is the specific heat ratio. Boundary conditions and an initial condition are needed to complete the description of the problem.

Introducing $\Delta\boldsymbol{U} = \boldsymbol{U}^{n+1} - \boldsymbol{U}^n$ as an unknown vector we linearize the system (13.23) by neglecting higher-order terms. After time-discretization by using the backward-Euler difference, system (13.23) becomes

$$\boldsymbol{A}\Delta\boldsymbol{U} = \boldsymbol{f} \tag{13.24}$$

with

$$\boldsymbol{A} = \boldsymbol{A}_1^n \frac{\partial}{\partial x} + \boldsymbol{A}_2^n \frac{\partial}{\partial y} + \boldsymbol{A}_0^n, \qquad \boldsymbol{f} = -\left(\boldsymbol{A}_1^n \frac{\partial \boldsymbol{U}^n}{\partial x} + \boldsymbol{A}_2^n \frac{\partial \boldsymbol{U}^n}{\partial y}\right),$$

$$\boldsymbol{A}_0 = \frac{\boldsymbol{I}}{\Delta t} + \frac{\partial \boldsymbol{A}_x^n}{\partial x} + \frac{\partial \boldsymbol{A}_y^n}{\partial y} =$$

$$\begin{pmatrix} (\frac{1}{\Delta t} + \frac{\partial u}{\partial x} + \frac{\partial v}{\partial y}) & \frac{\partial \rho}{\partial x} & \frac{\partial \rho}{\partial y} & 0 \\ 0 & (\frac{1}{\Delta t} + \frac{\partial u}{\partial x} + \frac{\partial v}{\partial y}) & 0 & \frac{-1}{\rho^2}\frac{\partial \rho}{\partial x} \\ 0 & 0 & (\frac{1}{\Delta t} + \frac{\partial u}{\partial x} + \frac{\partial v}{\partial y}) & \frac{-1}{\rho^2}\frac{\partial \rho}{\partial y} \\ 0 & \gamma\frac{\partial p}{\partial x} & \gamma\frac{\partial p}{\partial y} & (\frac{1}{\Delta t} + \frac{\partial u}{\partial x} + \frac{\partial v}{\partial y}) \end{pmatrix},$$

in which $\boldsymbol{I}$ is the identity matrix. Here we put Δt in $\boldsymbol{A}_0$. In this way, the arithmetic operation count will be reduced. For steady-state problems, the derivative terms in $\boldsymbol{A}_0$ can also be ignored.

Note that even for capturing stationary shocks, least-squares methods must be based on time-marching algorithms such as formulation (13.24). Because of a lack of the entropy contraint condition, the original compressible Euler equations (13.23) are not well posed in the sense of admitting both compression and expansion shocks. The direct application of least-squares methods to steady-state Euler equations is suitable only for subsonic flows; for supersonic flows an artificial viscosity term must be added (see examples given by Laminie 1994). As discussed in Sect. 13.1, the least-squares procedure for time-discretized equations inherently generates numerical dissipation to exclude non-physical expansion shocks.

Grid Adaptation. For the compressible Euler equations, the least-squares method is naturally diffusive, no explicit artificial viscosity is added to the formulation. In general, the inherent numerical dissipation produced by the least-squares method does not allow shocks to be sharply resolved, unless fine grids are used. However, shock resolution can be significantly improved by adapting the grid to the solution. Taghaddosi et al. (1997) have successfully obtained high quality shock solutions by combining the LSFEM with a directional-adaptive remeshing method. In the following we briefly describe the idea of their approach.

Their edge-based moving-node method (r-method) is originally developed by Ait-Ali-Yahia et al.(1996). It uses second-order derivatives of a given flow variable as an error indicator. In this approach, the number of both elements and nodes is fixed, connectivity is conserved, and the error is equally distributed over element edges. This anisotropic refinement where more resolution is introduced along those direction with rapidly changing flow variables is particularly useful for directional flow phenomena, such as shock waves.

Let us first consider a one-dimensional problem, in which the solution variable u is approximated by the linear interpolation u_h. If quadratic interpolation $\hat{u}_h$ is available and the nodal error is zero, then the local error of linear interpolation over an element can be estimated as the difference between the quadratic and linear one:

$$E_e = \hat{u}_h - u_h = \frac{\xi}{2}(h_e - \xi)\frac{d^2\hat{u}_h}{dx^2}, \tag{13.25}$$

where ξ is the local element coordinate and h_e the element length. The overall error in the element may be measured by the root-mean-square value of E_e:

$$E_e^{\text{RMS}} = \left(\int_0^{h_e} \frac{E_e^2}{h_e} d\xi\right)^{\frac{1}{2}} = \frac{1}{\sqrt{120}} h_e^2 \left|\frac{d^2\hat{u}_h}{dx^2}\right|_e. \tag{13.26}$$

An optimal mesh may be defined as one for which the error is equally distributed over elements, thus the following criteria should be satisfied by each element:

$$h_e^2 \left|\frac{d^2\hat{u}_h}{dx^2}\right|_e = C, \tag{13.27}$$

where C is a positive constant.

Extending this idea to the two-dimensional case, the second derivative in (13.26) is replaced by the Hessian tensor:

$$\sigma = \begin{pmatrix} \frac{\partial^2 \hat{u}_h}{\partial x^2} & \frac{\partial^2 \hat{u}_h}{\partial x \partial y} \\ \frac{\partial^2 \hat{u}_h}{\partial y \partial x} & \frac{\partial^2 \hat{u}_h}{\partial y^2} \end{pmatrix} \tag{13.28}$$

whose value is calculated for all nodes throughout the domain. The value of σ at the middle point of each element edge can be obtained by linear interpolation. Then the second derivative along any edge through its middle point can be calculated by a transformation of coordinates.

Since the solution is obtained by using linear elements, the quadratic solution $\hat{u}_h$ inside each element is not available. Therefore, the second derivatives have no representation inside an element. However, one may investigate the solution u_h in the elements surrounding node i. This yields an estimate of the second derivatives. For example,

$$\hat{u}_{h,xy}|_i \simeq \frac{\int_{\Omega_i} u_{h,xy}\phi_i d\Omega}{\int_{\Omega_i} \phi_i d\Omega} = -\frac{\int_{\Omega_i} u_{h,x}\phi_{i,y} d\Omega}{\int_{\Omega_i} \phi_i d\Omega}, \tag{13.29}$$

where Ω_i represents the elements sharing node i, and ϕ_i is a piecewise continuous linear function with a value unity at node i and null at other nodes.

To equally distribute the error over the entire domain, a moving-node method based on a spring analogy is used (Gnoffo 1983). Here the mesh is interpreted as a network of springs, where each edge is considered as a fictitious spring with its stiffness representing the measure of the error, see Fig. 13.6. The equilibrium of forces in this spring system will then determine the movement of each node.

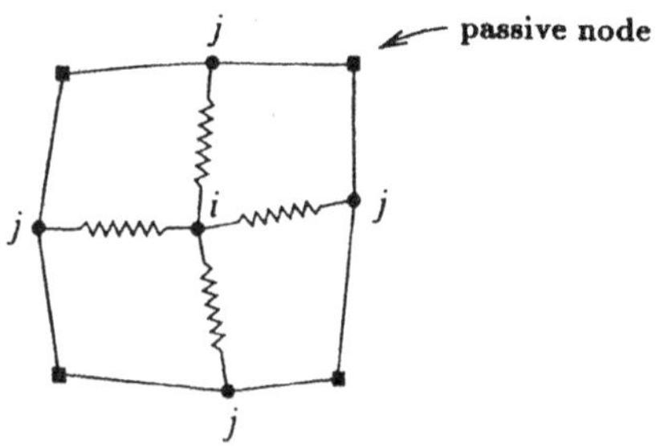

Fig. 13.6. Local spring network corresponding to node i

In the spring network, the new position of node i is determined by the potential energy of the system:

$$P = \sum_j (\boldsymbol{x}_i - \boldsymbol{x}_j)^2 k_{ij}, \tag{13.30}$$

where $\boldsymbol{x}$ is the position vector, and the stiffness k_{ij} is defined as:

$$k_{ij} = |u_h''|, \tag{13.31}$$

in which u_h'' is the second derivative along the edge $(\boldsymbol{x}_i - \boldsymbol{x}_j)$ at its middle point. Minimizing P in (13.30) with respect to $\boldsymbol{x}_i$ yields the following equation:

$$\sum_j (\boldsymbol{x}_i^{m+1} - \boldsymbol{x}_j^{m+1}) k_{ij}^{m+1} = 0, \tag{13.32}$$

which describes the equilibrium state of a spring network at the present adaptive iteration $m + 1$. By lagging $\boldsymbol{x}$ and k_{ij} at the previous iteration m, (13.32) becomes

$$\Delta \boldsymbol{x}_i = \frac{\sum (\boldsymbol{x}_i^m - \boldsymbol{x}_j^m) k_{ij}^m}{\sum_j k_{ij}^m}. \tag{13.33}$$

The new position of node i is then updated by the expression:

$$\boldsymbol{x}_i^{m+1} = \boldsymbol{x}_i^m + \alpha \Delta \boldsymbol{x}_i, \tag{13.34}$$

where α is the relaxation parameter.

The iteration process (13.34) is applied to all internal and boundary nodes. The movement of boundary nodes should be restricted to maintain the geometric integrity of the domain. The moving-node scheme is applied to nodal points in a sweeping manner. To avoid formation of elements with a negative or nearly zero Jacobian, the quality of newly oriented elements should be checked during each nodal movement. The number of iterations per adaptive cycle is chosen to be in the range of $m = 200 \sim 400$ to obtain an appropriately adapted mesh.

Shock Reflection. A standard test problem corresponding to the reflection of a shock from a solid wall is depicted in Fig. 13.7. The boundary conditions are:

$$\boldsymbol{U} = \begin{pmatrix} 1.0 \\ 2.9 \\ 0.0 \\ 0.7143 \end{pmatrix} \quad \text{for upstream,}$$

$$\boldsymbol{U} = \begin{pmatrix} 1.7 \\ 2.6193 \\ -0.5063 \\ 1.5282 \end{pmatrix} \quad \text{for upper boundary.}$$

A shock emanates from the upper left corner; this shock is reflected at the lower wall where $v = 0$, and the downstream boundary conditions remain free for outflow. The initial data were prescribed as constant at values given on the upper boundary and the specific heat ratio is $\gamma = 1.4$.

In the initial calculation, a uniform 60×20 mesh of bilinear elements was used. In the first two adaptive cycles, the solution was obtained with a time step $\Delta t = 0.1$. The numerical viscosity is reduced beginning at the third cycle by reducing the time step to 0.05. The resulting algebraic equations are solved by using the Jacobi preconditioned conjugate gradient method. Figure 13.8 shows the pressure contours and the grids after each adaptation. The improvements in the shock resolution after adaptation is quite evident in Fig. 13.9.

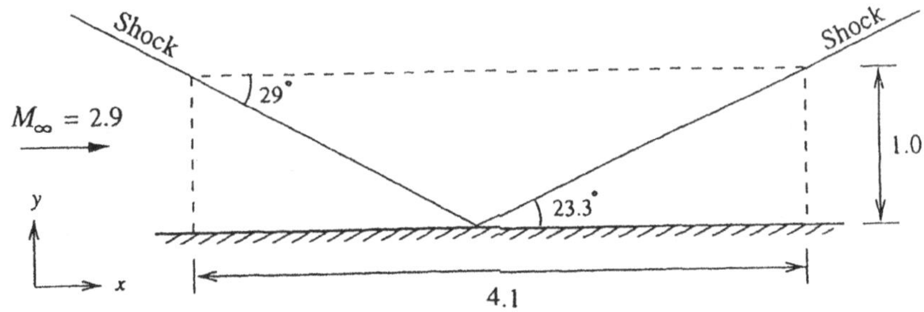

Fig. 13.7. Shock reflection from a solid wall: the computation domain (from Taghaddosi et al. 1997)

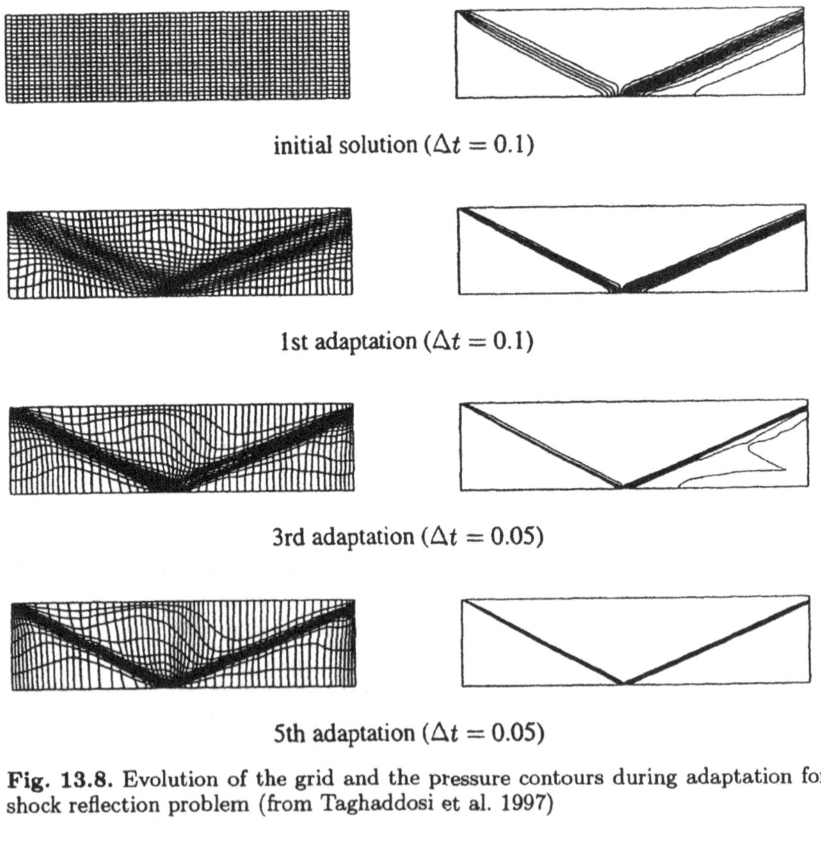

Fig. 13.8. Evolution of the grid and the pressure contours during adaptation for shock reflection problem (from Taghaddosi et al. 1997)

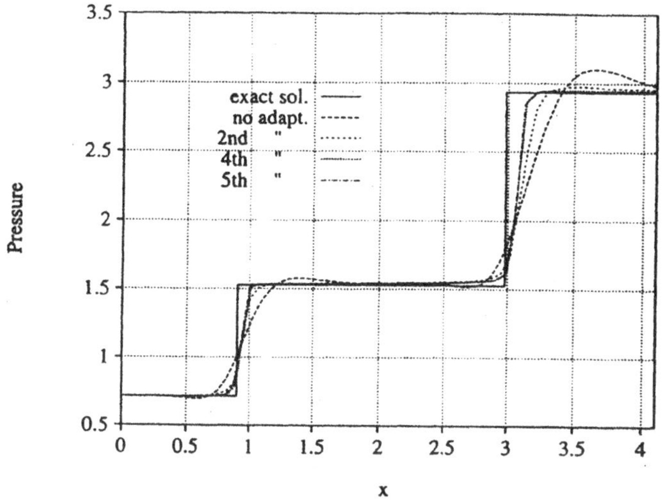

Fig. 13.9. Pressure distribution at $y = 0.5$ before and after adaptation for shock reflection problem (from Taghaddosi et al. 1997)

Transonic Channel Flow. Taghaddosi et al. (1997) also considered the transonic flow in a channel with a 10% circular arc bump. The original mesh consists of bilinear elements, with 32 elements uniformly distributed on the bump, and 16 elements on each side of it, centrally clustered with respect to the bump. The height of the channel is divided into 16 elements which are clustered toward the bottom wall, with the minimum element size on the wall equal to 0.03125. The Mach number at inlet is $M_\infty = 0.675$, which is large enough to create a supersonic bulb on the bump followed by a shock. The inlet boundary conditions are $\rho = 1.0$, $u = 0.675$, $v = 0.0$. At the outflow boundary only $p = 1.5282$ is specified. On the lower and upper walls, the no-penetration boundary condition is imposed.

Figure 13.10 shows the initial mesh and Mach number contours and adaptive results with the time step reducing from 0.3 to 0.1. In Fig. 13.11 the final adaptive solution of the Mach number distribution on the walls is compared to those of Ni (1982) and Eidelman (1984). The position of the shock is approximately at 72% of the chord, and the maximum Mach number is 1.323, both in good agreement with these results. However, there is a deviation in the Mach number distribution due to a viscous effect near the lower wall (limited to the first row of the elements).

Supersonic Channel Flow. The second test problem is the supersonic flow over a 4% circular arc bump placed in a channel. The height of the channel and its length, ahead and after the bump, are equal to the chord lenth. The boundary condition at the inlet is $U^T = (1.0, 1.65, 0.0, 0.7143)$. On the walls the normal component of the velocity vector is zero. The exit boundary is

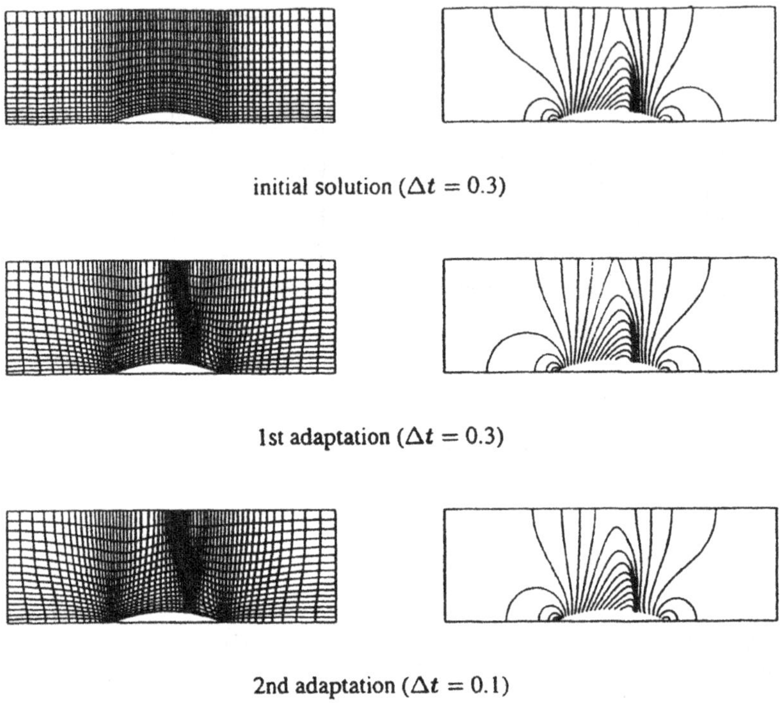

Fig. 13.10. Grids and Mach number contours before and after adaptation for transonic channel flow (from Taghaddosi et al. 1997)

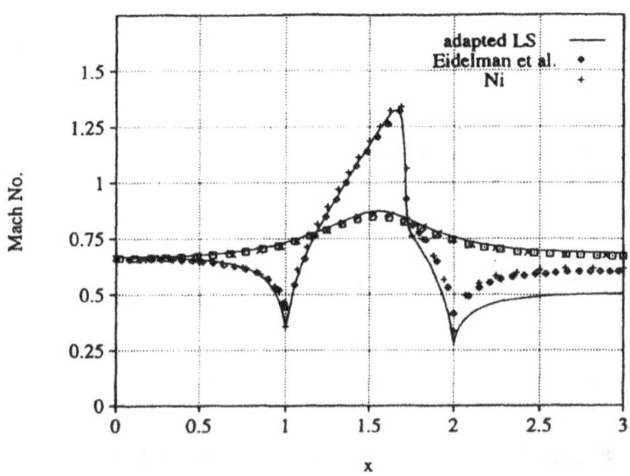

Fig. 13.11. Mach number distribution on the lower and upper walls for transonic channel flow (from Taghaddosi et al. 1997)

left free. The initial mesh consists of 64 × 16 uniformly distributed bilinear elements, with 16 elements in the y-direction, 22 elements on the bump, and 21 on each side of it.

The adaptive meshes and solutions after five cycles of adaptation are shown in Fig. 13.12. During the adaptation, the time step is also reduced. All shocks, especially the strongest leading-edge shock, are sharply captured. Figure 13.13 illustrates the improvement in shock resolution, where the leading-edge shock is quite sharp, and the trailing-edge and the reflected shock profiles are close to vertical.

13.3.2 Conservative Scheme

Least-Squares Formulation. The LSFEM described in Sect. 13.3.1 is based on a nonconservative formulation. Here we would like to construct a conservative LSFEM. The Euler equations governing two-dimensional compressible inviscid flows can be written in conservative form as

$$\frac{\partial \boldsymbol{q}}{\partial t} + \frac{\partial \boldsymbol{F}}{\partial x} + \frac{\partial \boldsymbol{G}}{\partial y} = \boldsymbol{0}, \tag{13.35a}$$

where

$$\boldsymbol{q} = \begin{pmatrix} \rho \\ \rho u \\ \rho v \\ \rho e \end{pmatrix}, \quad \boldsymbol{F} = \begin{pmatrix} \rho u \\ \rho u^2 + p \\ \rho u v \\ (\rho e + p)u \end{pmatrix}, \quad \boldsymbol{G} = \begin{pmatrix} \rho v \\ \rho u v \\ \rho v^2 + p \\ (\rho e + p)v \end{pmatrix}, \tag{13.35b}$$

in which e is the total energy, and for the case of perfect gas the equation of state is

$$p = (\gamma - 1)\Big(\rho e - \frac{1}{2}\rho(u^2 + v^2)\Big). \tag{13.35c}$$

Retaining the conservative variables as an intermediate step, we first convert (13.35) (temporarily) into the following nonconservative form:

$$\frac{\partial \boldsymbol{q}}{\partial t} + \bar{\boldsymbol{A}}_1 \frac{\partial \boldsymbol{q}}{\partial x} + \bar{\boldsymbol{A}}_2 \frac{\partial \boldsymbol{q}}{\partial y} = \boldsymbol{0}, \tag{13.36a}$$

with

$$\bar{\boldsymbol{A}}_1 = \begin{pmatrix} 0 & 1 & 0 & 0 \\ \frac{1}{2}\bar{\gamma}(u^2 + v^2) - u^2 & (3 - \gamma)u & -\bar{\gamma}v & \bar{\gamma} \\ -uv & v & u & 0 \\ (\bar{\gamma}(u^2 + v^2) - \gamma e)u & \bar{\epsilon} - \bar{\gamma}u^2 & -\bar{\gamma}uv & \gamma u \end{pmatrix}, \tag{13.36b}$$

$$\bar{\boldsymbol{A}}_2 = \begin{pmatrix} 0 & 0 & 1 & 0 \\ -uv & v & u & 0 \\ \frac{1}{2}\bar{\gamma}(u^2 + v^2) - v^2 & -\bar{\gamma}u & (3 - \gamma)v & \bar{\gamma} \\ (\bar{\gamma}(u^2 + v^2) - \gamma e)v & -\bar{\gamma}uv & \bar{\epsilon} - \bar{\gamma}v^2 & \gamma v \end{pmatrix}, \tag{13.36c}$$

in which

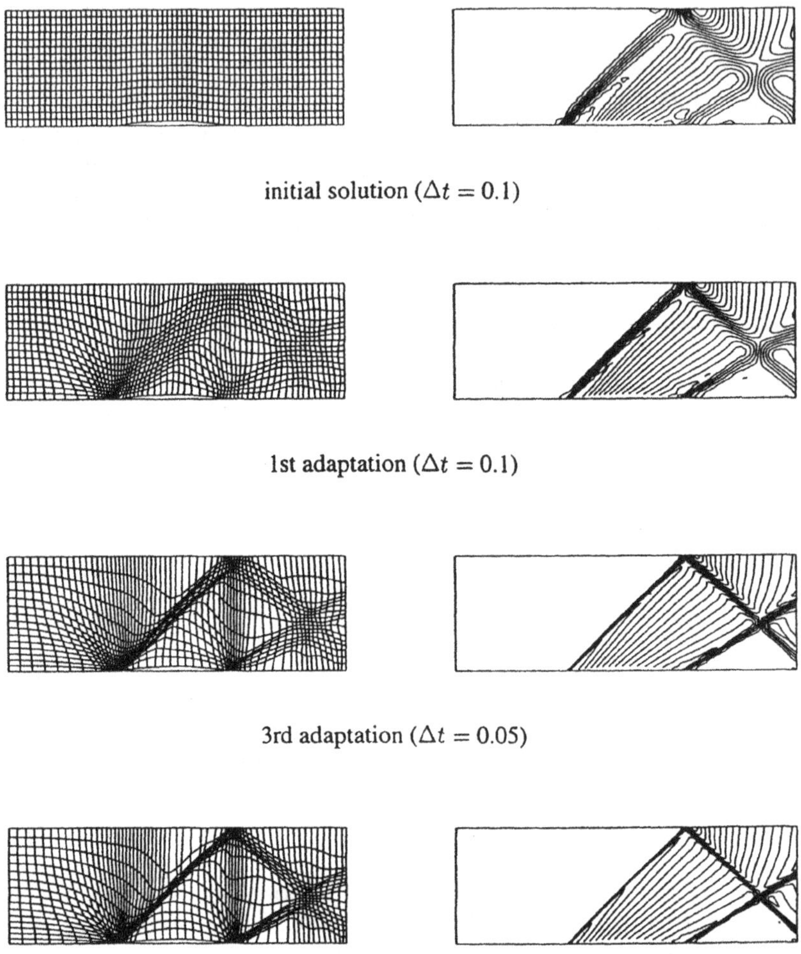

Fig. 13.12. Evolution of the grid and the pressure contours during adaptation for supersonic channel flow (from Taghaddosi et al. 1997)

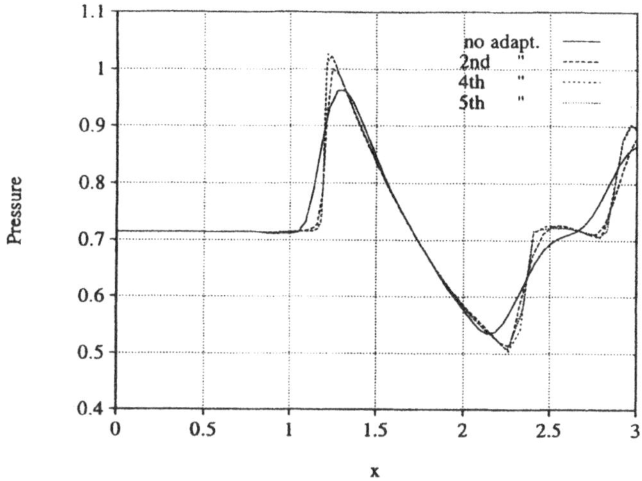

Fig. 13.13. Pressure distribution at $y = 0.2$ before and after adaptation for supersonic channel flow (from Taghaddosi et al. 1997)

$$\bar{\gamma} = \gamma - 1, \qquad \bar{\epsilon} = \gamma e - \bar{\gamma}\frac{(u^2 + v^2)}{2}. \tag{13.36d}$$

Introducing the unknown increment in conservative variables $\Delta \boldsymbol{q} = \boldsymbol{q}^{n+1} - \boldsymbol{q}^n$, and following the same procedure as in the derivation of (13.14), we obtain the time-discretized system:

$$\bar{\boldsymbol{A}}\Delta \boldsymbol{q} = \bar{\boldsymbol{f}} \tag{13.37a}$$

with

$$\bar{\boldsymbol{A}} = \bar{\boldsymbol{A}}_1^n \frac{\partial}{\partial x} + \bar{\boldsymbol{A}}_2^n \frac{\partial}{\partial y} + \bar{\boldsymbol{A}}_0^n, \qquad \bar{\boldsymbol{f}} = -\left(\frac{\partial \boldsymbol{F}^n}{\partial x} + \frac{\partial \boldsymbol{G}^n}{\partial y}\right),$$

$$\bar{\boldsymbol{A}}_0^n = \frac{\boldsymbol{I}}{\Delta t} + \frac{\partial \bar{\boldsymbol{A}}_x^n}{\partial x} + \frac{\partial \bar{\boldsymbol{A}}_y^n}{\partial y}, \tag{13.37b}$$

in which $\boldsymbol{I}$ is the unity matrix. Then the conservative least-squares variational statement is:

$$\left(\bar{\boldsymbol{A}}_0^n \boldsymbol{v} + \bar{\boldsymbol{A}}_1^n \frac{\partial \boldsymbol{v}}{\partial x} + \bar{\boldsymbol{A}}_2^n \frac{\partial \boldsymbol{v}}{\partial y},\ \bar{\boldsymbol{A}}_0^n \Delta \boldsymbol{q} + \bar{\boldsymbol{A}}_1^n \frac{\partial \Delta \boldsymbol{q}}{\partial x} + \bar{\boldsymbol{A}}_2^n \frac{\partial \Delta \boldsymbol{q}}{\partial y}\right)$$

$$= -\left(\bar{\boldsymbol{A}}_0^n \boldsymbol{v} + \bar{\boldsymbol{A}}_1^n \frac{\partial \boldsymbol{v}}{\partial x} + \bar{\boldsymbol{A}}_2^n \frac{\partial \boldsymbol{v}}{\partial y},\ \frac{\partial \boldsymbol{F}^n}{\partial x} + \frac{\partial \boldsymbol{G}^n}{\partial y}\right) \tag{13.38}$$

We may use the conservative variables at the previous time level to calculate the nodal values of components of flux $\boldsymbol{F}^n$ and $\boldsymbol{G}^n$, then use a finite element approximation to calculate $\partial \boldsymbol{F}^n/\partial x$ and $\partial \boldsymbol{G}^n/\partial y$. Once the steady conservation law

$$\frac{\partial \boldsymbol{F}}{\partial x} + \frac{\partial \boldsymbol{G}}{\partial y} = \boldsymbol{0} \tag{13.39}$$

is satisfied, the increment $\Delta \boldsymbol{q}$ becomes zero, and the calculation is terminated. Our numerical experiments on the shock wave reflection problem show that there is no essential difference between the results of nonconservative and conservative LSFEM.

Lefebvre et al. (1993) applied scheme (13.38) to problems of transonic and supersonic flows on triangular meshes. In their approach, a local time step is calculated in each element according to

$$\Delta t = \text{CFL}\frac{h}{\lambda_{\max}}, \tag{13.40}$$

where h is the element size, $\lambda_{\max}$ is the maximum eigenvalue of the present hyperbolic system (see, e.g., Hirsch 1990), and CFL (Courant–Friedrichs–Levy) number is a user specified constant. The time step is chosen so as to obtain a uniform CFL number of 5 throughout the mesh. The resulting algebraic equations is solved by the incomplete Choleski conjugate gradient method.

Transonic Flow Past an Airfoil. The first example is transonic flow past a NACA0012 airfoil with free stream Mach number of 0.95. A structured O-type mesh of quadratic (P_2) elements employing 160 nodes around the surface of the airfoil and 49 nodes in the normal direction is used. A detail of the mesh in the vicinity of the airfoil is shown in Fig 13.14. The pressure and Mach number distribution are shown in Fig. 13.15. The oblique shocks are reasonably well captured and compare favorably with highly accurate results of this test problem obtained by Pullian and Barton (1985).

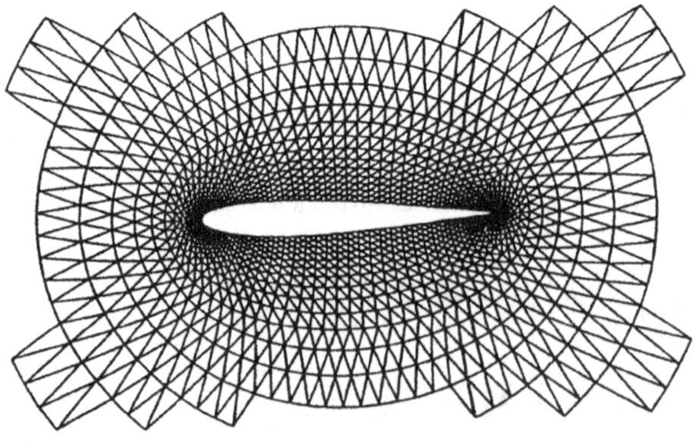

Fig. 13.14. Mesh of quadratic elements used for the solution of the transonic flow past a NACA0012 airfoil, Nnode=7840, Nelem=3840 (from Lefebvre et al. 1993)

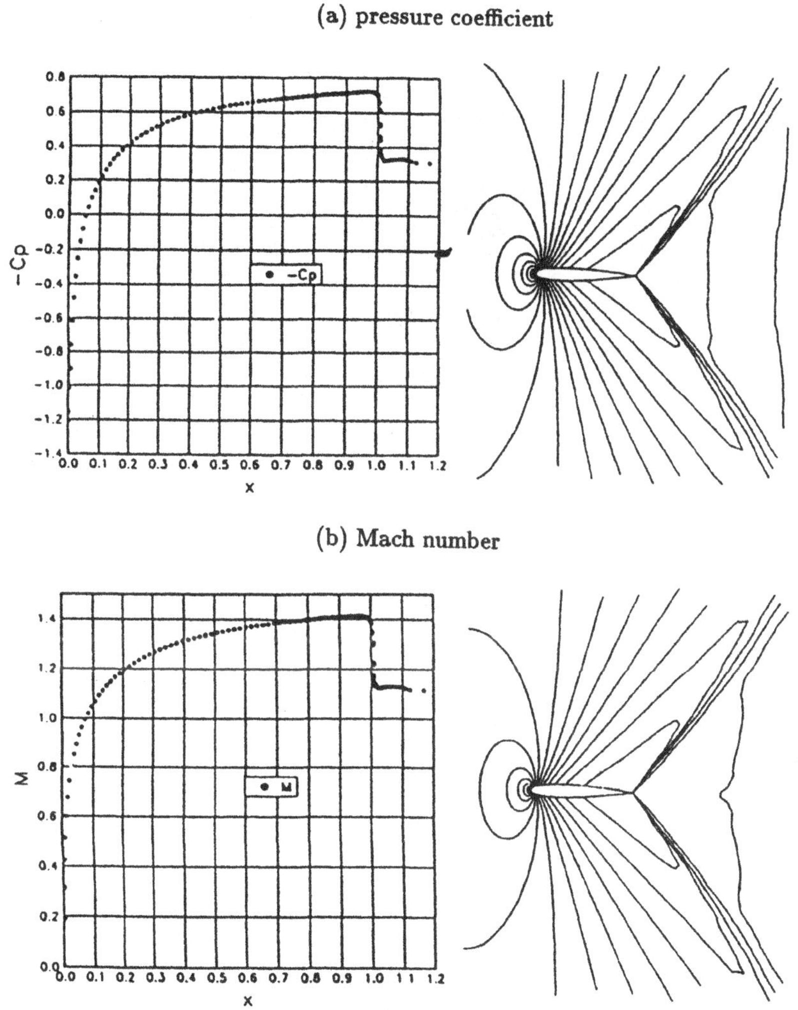

Fig. 13.15. Transonic flow past a NACA0012 airfoil, $M_\infty = 0.95$, (**a**) Pressure coefficient distribution and contours, (**b**) Mach number distribution and contours (from Lefebvre et al. 1993)

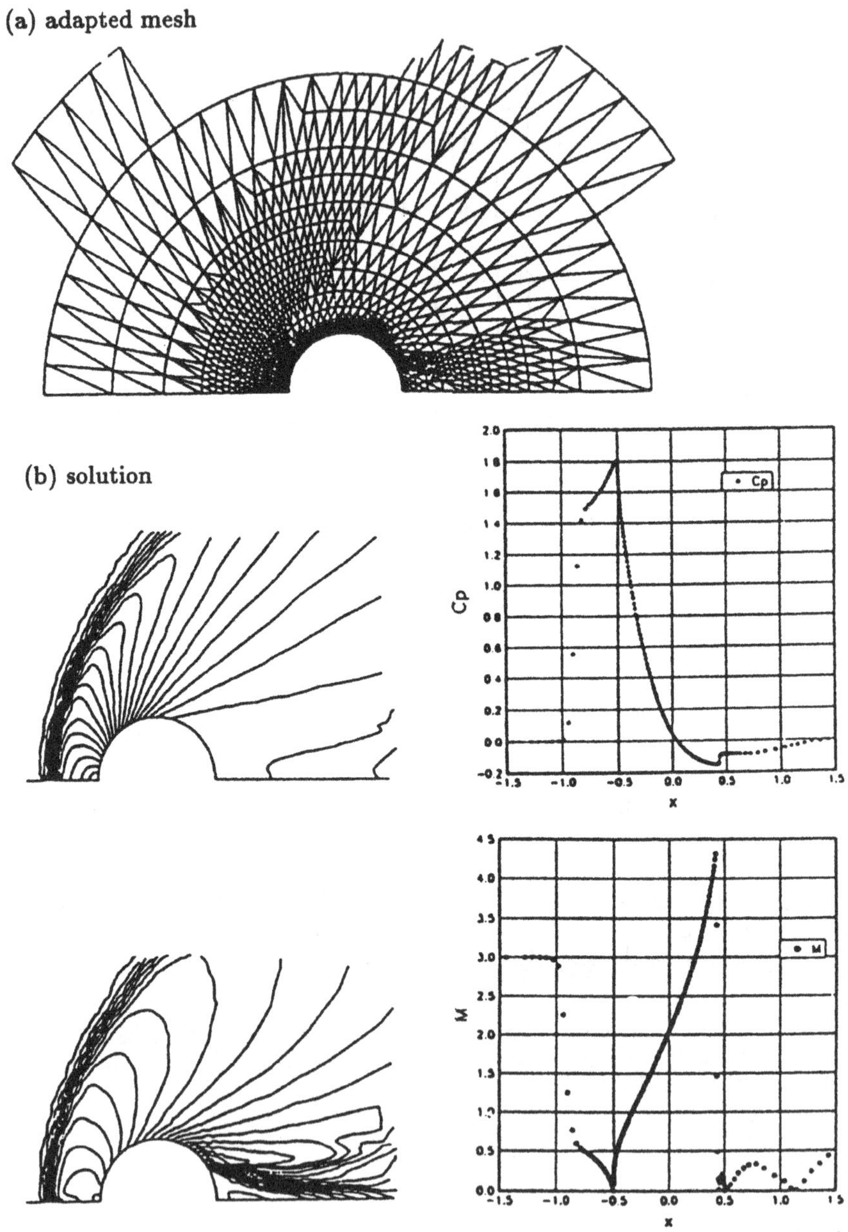

Fig. 13.16. Supersonic flow past a cylinder, $M_\infty = 3$, (**a**) Adapted mesh, (**b**) Pressure and Mach number solution (from Lefebvre et al. 1993)

Supersonic Flow Past a Cylinder. Lefebvre et al. (1993) also combined the LSFEM with adaptive mesh refinement to improve the quality of the computed solution. In their computation, the error indicator is defined as

$$b_e = \frac{E_e}{\max E_e}, \tag{13.41}$$

where E_e is calculated from the equation residual:

$$E_e = \frac{1}{\Omega_e}\int_{\Omega_e}\left(\bar{\boldsymbol{A}}_1\frac{\partial \boldsymbol{q}}{\partial x} + \bar{\boldsymbol{A}}_2\frac{\partial \boldsymbol{q}}{\partial y}\right)^T\left(\bar{\boldsymbol{A}}_1^n\frac{\partial \boldsymbol{q}}{\partial x} + \bar{\boldsymbol{A}}_2^n\frac{\partial \boldsymbol{q}}{\partial y}\right)d\Omega, \tag{13.42}$$

in which Ω_e is the element area. The refinement strategy is then to refine all these elements for which the indicator is larger than a certain specified value. Each triangle, for which refinement is needed, is subdivided into four smaller triangles, and to avoid the problem of "hanging nodes", transition elements are introduced. After the refinement, the sides of the elements are replaced by springs of unit stiffness and a few relaxation iterations are used to move the nodes until nodal equilibrium of forces is achieved, as described in Sect. 13.3.1.

This refinement technique is applied to supersonic flow past a cylinder. The free stream Mach number is 3, and due to the assumed symmetry, only one half of the cylinder is considered. The initial mesh consists of 1028 quadratic (P_2) elements and 2145 nodes. The adapted mesh and the steady-state solutions are presented in Fig. 13.16.

13.4 Concluding Remarks

We have demonstrated that the general LSFEM given in Sect. 4.8, when applied to semi-discretized compressible Euler equations obtained by various implicit time-difference schemes, can capture shocks in high-speed compressible flows. In contrast to other methods, the LSFEM does not use upwinding, Riemann solvers, flux splitting or staggered grids as building blocks. Therefore, the LSFEM is very simple. The implicit nature of LSFEM does not cause problems with computational efficiency, because the conjugate gradient method can be employed to solve the resulting positive and definite matrix. The resolution of the computed results can be significantly improved by using adaptive mesh refinement. The equation residual or the recovered second-order derivative of a given flow variable can serve as a reliable error indicator for refinement.

The time-marching LSFEM for high-speed compressible flows was originally proposed in Jiang and Carey (1988, 1990) and Jiang and Povinelli (1990). The research group led by K. Morgan (Lefebvre et al. 1992, 1993) and the group led by W.G. Habashi (Taghaddosi et al. 1997) have demonstrated that by coupling the least-squares method with adaptive mesh refinement, shocks can be captured with high resolution using relatively coarse grids.

The LSFEM proposed by Chattot et al. (1982), Bruneau (1991), Bruneau and Laminie (1993) and Laminie (1994) is based on the steady compressible Euler equations in conservation form. In this approach an artificial viscosity term is needed to eliminate expansion shocks.

Part IV

LSFEM in Electromagnetics

14. Electromagnetics

In spite of the fact that computational electromagnetics (CEM) has been applied to the prediction of many effects in electromagnetics and to the design of all kinds of devices for three decades, the success of CEM is still limited in the sense that many computational methods are inefficient and even produce erroneous solutions.

The occurrence of spurious solutions in computational electromagnetics is well known. The notable feature of these fictitious solutions has been their violation of the divergence-free conditions in cases where the physical solution is completely solenoidal. It is commonly believed that these erroneous solutions are caused by numerical inaccuracies. In this chapter we will reveal that the trouble of spurious solutions in computational electromagnetics is really caused by incorrect mathematical formulations due to misunderstanding of the first-order Maxwell equations which although were established more than one hundred years ago, and by incorrect derivation and use of the second-order curl–curl equations. Spurious solutions can only be avoided by using a correct analytical formulation of the problem to be solved.

The original full Maxwell equations in electromagnetics constitute a first-order system with two curl equations and two divergence equations. Unfortunately, many people in the CEM community consider the divergence equations in the Maxwell equations as "redundant" for time-varying problems, consequently most numerical methods in CEM solve only two first-order curl equations or the second-order curl–curl equations.

By studying the div–curl structure we will demonstrate that the full first-order Maxwell equations are not "overspecifieded", and the divergence equations must always be included to maintain the ellipticity of the system in the space domain, to guarantee the uniqueness of the solution and the accuracy of the numerical methods, and to eliminate spurious solutions.

We will show that the common derivation and use of the second-order curl–curl equations are incorrect. We will use the div–curl method introduced in Chap. 5 to derive the correct second-order Maxwell equations and their boundary conditions, and use the least-squares method to rigorously establish correct variational formulations.

In this chapter we will also show that satisfaction of divergence equations and elimination of spurious solutions can be achieved easily by application of the node-based LSFEM. Examples of numerical solutions by the LSFEM for static, time-harmonic and transient electromagnetic fields are given to demonstrate that the LSFEM is efficient and free of spurious solutions.

14.1 The First-Order Maxwell Equations

The most widely used numerical method for the solution of time-dependent electromagnetic problems has been the finite difference time-domain (FD-TD) scheme first proposed by Yee (1966) and extensively utilized and refined by Taflove and Umashankar (1989), Kunz and Luebbers (1993) and Taflove (1995) as well as other researchers (see a survey by Shlager and Schneider 1995). In the Yee scheme, only two Maxwell curl equations are solved. Some other time-domain methods are also based on two Maxwell curl equations, such as the finite volume time-domain method (FV-TD) developed by Shankar et al. (1989) and Noack and Anderson (1992), the finite difference and finite volume methods by Shang (1995) and Shang and Gaitonde (1995), and the finite element time-domain method (FE-TD) by Cangellaris et al. (1987), Madsen and Ziolkowski (1988), Lee and Madsen (1990), Ambrosiano et al. (1994), Morgan et al. (1996) and Lee et al. (1997).

As pointed out in Chap. 2, it is difficult to deal with the first-order differential operators by conventional methods. This is the reason why FD-TD needs a uniform staggered grid and FV-TD relies on upwinding schemes. FD-TD has problems near curved surfaces. Both FD-TD and FV-TD have difficulty for extension to high-order versions.

FV-TD has been based on the shock-capturing methodology used in computational fluid dynamics (CFD). Shocks will be generated in high-speed compressible fluid flows due to the nonlinearity of the governing equations (Chap. 13). However, the Maxwell equations are linear wave equations, and hence no shocks occur in electromagnetics. Therefore, one may ask why this expensive CFD method should be promoted in CEM.

In the past, very few people realized that the divergence-free conditions are really not satisfied in these time-domain solutions. Wu and Jiang (1995a, 1996) gave some evidence that clearly shows the significant violation of the divergence-free condition near the boundary of scatterers in the solutions of the curl equations. Kangro and Nicolaides (1997) also found that time marching solutions of the Maxwell curl equations are contaminated by spurious stationary components.

The first-order full Maxwell equations have a mathematical structure in which the fundamental ingredient is the div–curl system that looks "overspecified". A similar situation exists in fluid dynamics (Chap. 8). By introducing a dummy variable, however, it can be shown that the div–curl system is not "overspecified". Here we use this technique to study the full Maxwell equations and show that they are properly determined, that is, the two divergence equations should not be ignored in either the static or time-varying case.

14.1.1 Basic Equations

The Maxwell equations are a set of fundamental equations governing all macroscopic electromagnetic phenomena. For general time-varying fields, the original first-order full Maxwell equations can be written as

$$\nabla \times \boldsymbol{E} + \frac{\partial(\mu \boldsymbol{H})}{\partial t} = -\boldsymbol{K}^{\text{imp}} \quad \text{in } \Omega, \text{ (Faraday's Law)} \tag{14.1a}$$

$$\nabla\times\boldsymbol{H} - \frac{\partial(\varepsilon \boldsymbol{E})}{\partial t} - \sigma\boldsymbol{E} = \boldsymbol{J}^{\text{imp}} \quad \text{in } \Omega, \text{ (Maxwell – Ampere's Law)} \tag{14.1b}$$

$$\nabla \cdot (\varepsilon \boldsymbol{E}) = \rho^{\text{imp}} \quad \text{in } \Omega, \text{ (Gauss's Law – electric)} \tag{14.1c}$$

$$\nabla \cdot (\mu \boldsymbol{H}) = 0 \quad \text{in } \Omega, \text{ (Gauss's Law – Magnetic)} \tag{14.1d}$$

where $\Omega \subset \mathbb{R}^3$ is a bounded and simply connected domain with a piece-wise smooth boundary $\Gamma = \Gamma_1 \cup \Gamma_2$; either Γ_1 or Γ_2, not both, may be empty; $\boldsymbol{E}$ and $\boldsymbol{H}$ are the electric and magnetic field intensities respectively, ρ^{imp} is the imposed source of electric charge density, and $\boldsymbol{J}^{\text{imp}}$ and $\boldsymbol{K}^{\text{imp}}$ are imposed sources of electric and magnetic current density, respectively. All imposed sources are given functions of the space and time coordinates.

In system (14.1) we have already made use of the constitutive relations

$$\boldsymbol{D} = \varepsilon \boldsymbol{E}, \qquad \boldsymbol{B} = \mu \boldsymbol{H}, \qquad \boldsymbol{J} = \sigma \boldsymbol{E},$$

where $\boldsymbol{D}$ is the electric flux density, $\boldsymbol{B}$ is the magnetic flux density and $\boldsymbol{J}$ is the electric (eddy) current density; the constitutive parameters ε, μ and σ denote, respectively, the permittivity, permeability and conductivity of the medium. These parameters are tensors for anisotropic media, and may be functions of position and time, and may also depend on the field intensities. For simplicity, we first consider isotropic and homogeneous media, therefore they are constant scalars. We will study anisotropic media in Sect. 14.2.4.

At the interface between two media, for example, medium (+) and medium (−), the boundary conditions can be expressed as

$$\boldsymbol{n} \times (\boldsymbol{E}^+ - \boldsymbol{E}^-) = \boldsymbol{0}, \tag{14.2a}$$

$$\boldsymbol{n} \cdot (\boldsymbol{D}^+ - \boldsymbol{D}^-) = \rho_s^{\text{imp}} \tag{14.2b}$$

for electric fields, and similarly,

$$\boldsymbol{n} \times (\boldsymbol{H}^+ - \boldsymbol{H}^-) = \boldsymbol{J}_s^{\text{imp}}, \tag{14.2c}$$

$$\boldsymbol{n} \cdot (\boldsymbol{B}^+ - \boldsymbol{B}^-) = 0 \tag{14.2d}$$

for magnetic fields, where $\boldsymbol{n}$ is the unit vector normal to the interface, pointing from medium (−) to medium (+), ρ_s^{imp} is the imposed surface charge density and $\boldsymbol{J}_s^{\text{imp}}$ is the imposed surface electric current density.

The boundary conditions can be reduced to a special case when one of the media, say medium (−), becomes a perfect conductor. Since a perfect conductor cannot sustain a field within it, the corresponding boundary conditions for a perfectly conducting wall Γ_1 become

$$\boldsymbol{n} \times \boldsymbol{E} = \boldsymbol{0} \qquad \text{on } \Gamma_1, \tag{14.3a}$$

$$\boldsymbol{n} \cdot (\mu \boldsymbol{H}) = 0 \qquad \text{on } \Gamma_1, \tag{14.3b}$$

where $\boldsymbol{E}$ and $\boldsymbol{H}$ are the fields exterior to the conductor. Similarly, for a perfect magnetic wall Γ_2 we have

$$\boldsymbol{n} \times \boldsymbol{H} = \boldsymbol{0} \qquad \text{on } \Gamma_2, \tag{14.3c}$$

$$\boldsymbol{n} \cdot (\varepsilon \boldsymbol{E}) = 0 \qquad \text{on } \Gamma_2. \tag{14.3d}$$

For transient problems, the initial conditions on $\boldsymbol{E}$ and $\boldsymbol{H}$ should also be provided.

To allow system (14.1) under boundary conditions (14.3) to have a solution, the source terms must be square integrable and satisfy the following solvability conditions:

$$\nabla \cdot \boldsymbol{K}^{\text{imp}} = 0 \qquad \text{in } \Omega, \tag{14.4a}$$

$$\boldsymbol{n} \cdot \boldsymbol{K}^{\text{imp}} = 0 \qquad \text{on } \Gamma_1, \tag{14.4b}$$

$$\int_{\Gamma} \boldsymbol{n} \cdot \boldsymbol{K}^{\text{imp}} d\Gamma = 0, \tag{14.4c}$$

$$\nabla \cdot \boldsymbol{J}^{\text{imp}} + \frac{\partial \rho^{\text{imp}}}{\partial t} + \left(\frac{\sigma}{\varepsilon}\right) \rho^{\text{imp}} = 0 \qquad \text{in } \Omega, \tag{14.4d}$$

$$\boldsymbol{n} \cdot \boldsymbol{J}^{\text{imp}} = 0 \qquad \text{on } \Gamma_2. \tag{14.4e}$$

We remark that the solvability conditions (14.4a), (14.4b), (14.4d) and (14.4e) can be obtained by applying the div–curl theorem (Theorem 5.2) to the Maxwell curl equations (14.1a) and (14.1b).

14.1.2 Determinacy

The original Maxwell equations (14.1), for the three-dimensional case, consists of eight first-order equations but with only six unknown vector components; and for two-dimensional transverse electric (TE) and transverse magnetic (TM) cases, four equations involving only three unknowns. That is, the number of equations is larger than the number of unknown functions. By taking the divergence of the Faraday and Ampere laws one can see that two divergence equations will be satisfied for all time if they are satisfied initially. For these two reasons, in the CEM community it has been traditionally believed that the full first-order Maxwell equations are "overspecified", and therefore the two divergence equations are regarded as "auxiliary" or "dependent" (see e.g., Stratton 1941, p.6; Kong 1990, p.3; Kunz and Luebbers 1993, p.11; Umashankar and Taflove 1993, p.38; and Taflove 1995, p.53-54), and often neglected in numerical computation. This misconception is the true origin of spurious modes and inaccurate solutions in computational electromagnetics; they can be avoided only by using correct analytic formulations.

To show this point of view, we consider the Maxwell equations (14.1) augmented by the variables φ and χ:

$$\nabla\varphi + \nabla\times\boldsymbol{E} + \frac{\partial(\mu\boldsymbol{H})}{\partial t} = -\boldsymbol{K}^{\text{imp}} \quad \text{in } \Omega, \tag{14.5a}$$

$$\nabla\chi + \nabla\times\boldsymbol{H} - \frac{\partial(\varepsilon\boldsymbol{E})}{\partial t} - \sigma\boldsymbol{E} = \boldsymbol{J}^{\text{imp}} \quad \text{in } \Omega, \tag{14.5b}$$

$$\nabla\cdot(\varepsilon\boldsymbol{E}) = \rho^{\text{imp}} \quad \text{in } \Omega, \tag{14.5c}$$

$$\nabla\cdot(\mu\boldsymbol{H}) = 0 \quad \text{in } \Omega, \tag{14.5d}$$

$$\boldsymbol{n}\times\boldsymbol{E} = \boldsymbol{0} \quad \text{on } \Gamma_1, \tag{14.5e}$$

$$\chi = 0, \quad \boldsymbol{n}\cdot(\mu\boldsymbol{H}) = 0 \quad \text{on } \Gamma_1, \tag{14.5f}$$

$$\boldsymbol{n}\times\boldsymbol{H} = \boldsymbol{0} \quad \text{on } \Gamma_2, \tag{14.5g}$$

$$\varphi = 0, \quad \boldsymbol{n}\cdot(\varepsilon\boldsymbol{E}) = 0 \quad \text{on } \Gamma_2. \tag{14.5h}$$

The introduction of the variables φ and χ is purely for showing that the Maxwell equations are not "overspecified". They do not enter into computation in contrast to the method proposed by Assous et al. (1993).

We shall prove that φ and χ in (14.5) are identically zeros, i.e., system (14.5) is equivalent to system (14.1). In fact, by virtue of the div–curl theorem (Theorem 5.2), (14.5a) and (14.5b) are equivalent to the following equations:

$$\nabla\times\left[\nabla\varphi + \nabla\times\boldsymbol{E} + \frac{\partial(\mu\boldsymbol{H})}{\partial t} + \boldsymbol{K}^{\text{imp}}\right] = \boldsymbol{0} \quad \text{in } \Omega, \tag{14.6a}$$

$$\nabla\cdot\left[\nabla\varphi + \nabla\times\boldsymbol{E} + \frac{\partial(\mu\boldsymbol{H})}{\partial t} + \boldsymbol{K}^{\text{imp}}\right] = 0 \quad \text{in } \Omega, \tag{14.6b}$$

$$\boldsymbol{n}\cdot\left[\nabla\varphi + \nabla\times\boldsymbol{E} + \frac{\partial(\mu\boldsymbol{H})}{\partial t} + \boldsymbol{K}^{\text{imp}}\right] = 0 \quad \text{on } \Gamma_1, \tag{14.6c}$$

$$\boldsymbol{n}\times\left[\nabla\varphi + \nabla\times\boldsymbol{E} + \frac{\partial(\mu\boldsymbol{H})}{\partial t} + \boldsymbol{K}^{\text{imp}}\right] = \boldsymbol{0} \quad \text{on } \Gamma_2, \tag{14.6d}$$

$$\nabla\times\left[\nabla\chi + \nabla\times\boldsymbol{H} - \frac{\partial(\varepsilon\boldsymbol{E})}{\partial t} - \sigma\boldsymbol{E} - \boldsymbol{J}^{\text{imp}}\right] = \boldsymbol{0} \quad \text{in } \Omega, \tag{14.7a}$$

$$\nabla\cdot\left[\nabla\chi + \nabla\times\boldsymbol{H} - \frac{\partial(\varepsilon\boldsymbol{E})}{\partial t} - \sigma\boldsymbol{E} - \boldsymbol{J}^{\text{imp}}\right] = 0 \quad \text{in } \Omega, \tag{14.7b}$$

$$\boldsymbol{n} \times \left[\nabla\chi + \nabla \times \boldsymbol{H} - \frac{\partial(\varepsilon \boldsymbol{E})}{\partial t} - \sigma \boldsymbol{E} - \boldsymbol{J}^{\text{imp}}\right] = 0 \qquad \text{on } \Gamma_1, \tag{14.7c}$$

$$\boldsymbol{n} \cdot \left[\nabla\chi + \nabla \times \boldsymbol{H} - \frac{\partial(\varepsilon \boldsymbol{E})}{\partial t} - \sigma \boldsymbol{E} - \boldsymbol{J}^{\text{imp}}\right] = 0 \qquad \text{on } \Gamma_2, \tag{14.7d}$$

Taking into account the divergence-free condition (14.5d) and the solvability condition (14.4a), from (14.6b) we find that

$$\Delta\varphi = 0 \qquad \text{in } \Omega. \tag{14.8a}$$

Taking into account the boundary condition (14.5e), Theorem 5.5, the boundary condition (14.5f) and the solvability condition (14.4b), from (14.6c) we obtain

$$\boldsymbol{n} \cdot \nabla\varphi = 0 \qquad \text{on } \Gamma_1. \tag{14.8b}$$

From (14.5h) we have

$$\varphi = 0 \qquad \text{on } \Gamma_2. \tag{14.8c}$$

System (14.8) implies that $\varphi \equiv 0$ in Ω. Similarly, we can show that $\chi \equiv 0$ in Ω. Therefore, φ and χ in (14.5) are really dummy variables, and thus system (14.5) is an equivalent formulation of system (14.1).

The first-order system (14.5) has eight equations, eight unknowns, and four boundary conditions on each part of the boundary, and thus it is properly determined. This is valid for static, transient and time-harmonic cases.

In the static case, the time-derivative terms in (14.5a) and (14.5b) disappear, and $\sigma \boldsymbol{E}$ is included into the given current density. System (14.5) becomes two independent div–curl systems for the electric field and the magnetic field, respectively. It is well known that each div–curl system is elliptic (Sect. 5.2).

In the time-harmonic case, when the time factor $e^{j\omega t}$ is used and suppressed, the time-derivative terms become the zero-order terms, and system (14.5) becomes two coupled div–curl systems. The coupling is through the zero-order terms. The principle part, namely, the first-order derivative terms which classify the system, still have the div–curl structure, and thus the whole system is elliptic.

In the transient case, the whole system (14.5) is hyperbolic. However, in time-domain numerical methods, the time-derivative terms are discretized by either explicit or implicit finite differences, hence the time-derivative terms become the zero-order terms in the space domain. For each time-step, the time-discretized system is still elliptic.

In summary, in all cases, system (14.5) is properly determined, and is elliptic in the space domain. Since system (14.1) is equivalent to system (14.5), system (14.1) is indeed properly determined and also elliptic in the space domain. Therefore, the divergence equations (14.1c), (14.1d) and the boundary conditions (14.3b) and (14.3d) are not "redundant", and must always be taken into account.

14.1.3 Importance of Divergence Equations

The importance of the divergence equations (14.1c) and (14.1d) in computational electromagnetics can be further explained as the following.
(1) The original first-order full Maxwell equations (14.1a)–(14.1c) reflect general and independent laws of physics. These laws cannot be deduced from each other. They govern all electromagnetic phenomena, no matter whether the problem is static, time-harmonic or transient.

The inconsistency between the number of unknowns and the number of equations in the full first-order Maxwell equations results in the so-called "overspecified" problem. However, this "overspecified" problem is a notion borrowed from linear algebra. For partial differential equations, this interpretation leads to a misconception.
(2) By taking the divergence of (14.1a), one can claim only that $\partial[\nabla \cdot (\mu \boldsymbol{H})]/\partial t = 0$, namely, $\nabla \cdot (\mu \boldsymbol{H}) = F(\boldsymbol{x})$ which can be any function of the space coordinates. In (14.1a) there is no information about this function, that is, Gauss's law (the magnetic flux density must be divergence free) can not be induced by the Faraday's law. Of course, if the divergence of $\mu \boldsymbol{H}$ is zero at the beginning, it will be identically zero forever. The problem then is how to set $\nabla \cdot (\mu \boldsymbol{H}) = 0$ initially. Let us examine the common practice for non-homogeneous boundary condition problems: letting the initial field intensities be zero in the domain, and the boundary conditions be correctly given on the boundary. In this case, the divergence-free condition is significantly violated near the boundary at the first time-step of the computation and will be violated forever if the divergence-free condition is not enforced. If one really can set the divergence to be zero at the beginning, it is in fact equivalent to adding a divergence-free equation into the system. The discussion for the electric field runs along the same line. For similar reason, the boundary conditions (14.3b) and (14.3d) are not "dependent".

From the above discussion we know why the solution of curl equations is contaminated by stationary spurious solutions. Fortunately, erroneous near-field solutions and incorrect far-field solutions still lead to correct RCS results, since static solutions don't radiate. However, in other areas such as in plasma physics and hyperfrequency devices technology, erroneous solutions are unacceptable.
(3) From the mathematical point of view, the neglect of the divergence equations destroys both the ellipticity of the Maxwell equations in the space domain and the uniqueness of the solution. In each curl system there are only three (odd) equations and three (odd) unknowns that cannot be elliptic in the ordinary sense. The discretized system based on the curl equations without special treatment (such as the use of edge elements) cannot be positive-definite. This explains why spurious solutions occur and iterative methods do not work for the discretized curl equations.
(4) For time-harmonic problems the curl equations alone will admit infinite number of eigenfunctions corresponding to the frequency $\omega = 0$ which is

the so called infinitely degenerate eigenvalue. The inclusion of divergence equations will exclude this completely non-physical situation and guarantee that only the trivial solution $\boldsymbol{E} = \mathbf{0}$ and $\boldsymbol{H} = \mathbf{0}$ correspond to $\omega = 0$.
(5) The time marching method is often an effective approach to solve steady-state non-linear problems where the material properties depend on the electromagnetic fields. The curl equations alone are not adequate for this approach.

14.2 The Second-Order Maxwell Equations

In electromagnetics, there are mainly two reasons why the second-order curl–curl equations are often employed. First, in the curl–curl equations the electric field and the magnetic field are decoupled. Second, it is difficult to deal with non-self-adjoint first-order derivatives numerically by conventional methods, while the second-order partial differential operators can be easily treated, for example, by the classic Galerkin method.

There is a vast body of reports about spurious solutions associated with numerical solution of the curl–curl equations in the context of the finite element method, see e.g., the references in Jin (1993) and Silvester and Pelosi (1994). A majority of spurious solutions have been found in eigenvalue analyses. A spurious mode does not correspond to the physical modes which the waveguide or resonator under consideration actually supports. The spurious mode problem is severe and often renders the numerical solution useless. The spurious solutions have been also revealed in boundary-value problems, see, e.g., Crowley et al. (1988), Pinchuk et al. (1988), Wong and Cendes (1988, 1989), and Paulsen and Lynch (1991).

The phenomenon of spurious solutions for the curl–curl equations is not exclusively associated with the finite element method. This phenomenon has been also reported in the context of the finite difference method, see e.g., Corr and Davies (1972), Schwieg and Bridges (1984) and Su (1985); the boundary element method, see e.g., Ganguly and Spielman (1977) and Swaminathan et al. (1992); and the spectral method, see Farrar and Adams (1976). This fact itself undermines the common belief that the spurious solution is a result of a particular numerical process.

The popular engineering approach to removing spurious vector modes in the curl–curl equations is to modify the variational functional by penalizing the non-zero divergence. The key to success with this so-called penalty method, first used by Hara et al. (1983) and Rahman and Davies (1984), depends on the choice of a correct penalty factor – values too small or too large do not eliminate spurious solutions. Unfortunately, this is an ad hoc and problem-dependent treatment and there has been a lack of systematic study of the rationale for selecting this parameter for general problems.

Recently, the edge element method of Nedelec (1980), see e.g., Jin (1993) and the references therein, has been advocated. It is believed that many maladies in CEM could be cured by using the edge element method. The edge elements allow the normal component to jump across material interfaces and do not yield conflicting conditions at sharp corners. However, (1) such an approach can only be used in the simple divergence-free case for isotropic media; (2) the vector basis functions for the general quadrilateral elements, which are desired for modeling complex geometry with higher solution accuracy than triangles, are in fact not divergence-free (Jin 1993, p.240); (3) edge elements violate the normal field continuity between adjacent elements in the homogeneous material domain, this will cause a stability problem for numerical modeling in plasma physics (Assous et al. 1993); (4) the accuracy of edge elements is lower than that of the nodal elements for the same number of unknowns, or the computational cost of edge elements is much higher than that of nodal elements for the same accuracy (Mur 1994 and Monk 1993); (5) the edge element interpolation itself cannot solve singularity problems at sharp corners, the singularity problems can be solved only by the use of singular elements, high-order interpolations (p-version elements) or fine meshes; (6) the edge element method also needs non-conventional meshing and display programs which are not normally available.

The curl–curl equations are derived from the first-order Maxwell curl equations by applying the curl operator. As pointed out above the curl equations alone cannot determine a unique solution. Moreover, the curl–curl equations obtained by simple differentiation without additional equations and boundary conditions admit more solutions than do its progenitors.

In order to derive an equivalent higher-order system from a system of vector partial differential equations, one should use the div–curl method that is based on Theorem 5.2: if a vector is divergence-free and curl-free in a domain, and its normal component or tangential components on the boundary are zero, then this vector is identically zero. In other words, the curl and the divergence operators must act together with appropriate boundary conditions to ensure that there are no spurious solutions in the resulting higher-order equations. In this section, this div–curl method introduced in Chap. 5 is employed to derive the second-order system of time-dependent Maxwell equations and its boundary conditions; and to show that the divergence equations and additional boundary conditions must be appended to the curl–curl equations, and the Helmholtz-type equations must be solved with divergence conditions enforced on the corresponding part of boundary. We shall establish the Galerkin formulation corresponding to the Helmholtz-type equations. We shall see that this Galerkin formulation is of the same form as the popular Galerkin/penalty method with the penalty parameter $s = 1$. Or more precisely, the parameter s should be chosen such that the resulting equation becomes equivalent to the Helmholtz-type equation in homogeneous regions. We shall also give a simple least-squares look-alike method to obtain a correct variational formulation which rigorously justifies choosing $s = 1$ in the penalty method.

14.2.1 The Div–Curl Method

Assume that the field intensities and the sources in system (14.1) are sufficiently smooth. By virtue of the div–curl theorem (Theorem 5.2), system (14.1) is equivalent to

$$\nabla \times \left[\nabla \times \boldsymbol{E} + \frac{\partial(\mu \boldsymbol{H})}{\partial t} + \boldsymbol{K}^{\text{imp}}\right] = \mathbf{0} \quad \text{in } \Omega, \tag{14.9a}$$

$$\nabla \cdot \left[\nabla \times \boldsymbol{E} + \frac{\partial(\mu \boldsymbol{H})}{\partial t} + \boldsymbol{K}^{\text{imp}}\right] = 0 \quad \text{in } \Omega, \tag{14.9b}$$

$$\boldsymbol{n} \cdot \left[\nabla \times \boldsymbol{E} + \frac{\partial(\mu \boldsymbol{H})}{\partial t} + \boldsymbol{K}^{\text{imp}}\right] = 0 \quad \text{on } \Gamma_1, \tag{14.9c}$$

$$\boldsymbol{n} \times \left[\nabla \times \boldsymbol{E} + \frac{\partial(\mu \boldsymbol{H})}{\partial t} + \boldsymbol{K}^{\text{imp}}\right] = \mathbf{0} \quad \text{on } \Gamma_2, \tag{14.9d}$$

$$\nabla \times \left[\nabla \times \boldsymbol{H} - \frac{\partial(\varepsilon \boldsymbol{E})}{\partial t} - \sigma \boldsymbol{E} - \boldsymbol{J}^{\text{imp}}\right] = \mathbf{0} \quad \text{in } \Omega, \tag{14.9e}$$

$$\nabla \cdot \left[\nabla \times \boldsymbol{H} - \frac{\partial(\varepsilon \boldsymbol{E})}{\partial t} - \sigma \boldsymbol{E} - \boldsymbol{J}^{\text{imp}}\right] = 0 \quad \text{in } \Omega, \tag{14.9f}$$

$$\boldsymbol{n} \times \left[\nabla \times \boldsymbol{H} - \frac{\partial(\varepsilon \boldsymbol{E})}{\partial t} - \sigma \boldsymbol{E} - \boldsymbol{J}^{\text{imp}}\right] = \mathbf{0} \quad \text{on } \Gamma_1, \tag{14.9g}$$

$$\boldsymbol{n} \cdot \left[\nabla \times \boldsymbol{H} - \frac{\partial(\varepsilon \boldsymbol{E})}{\partial t} - \sigma \boldsymbol{E} - \boldsymbol{J}^{\text{imp}}\right] = 0 \quad \text{on } \Gamma_2, \tag{14.9h}$$

$$\nabla \cdot (\varepsilon \boldsymbol{E}) = \rho^{\text{imp}} \quad \text{in } \Omega, \tag{14.9i}$$

$$\nabla \cdot (\mu \boldsymbol{H}) = 0 \quad \text{in } \Omega, \tag{14.9j}$$

$$\boldsymbol{n} \times \boldsymbol{E} = \mathbf{0} \quad \text{on } \Gamma_1, \tag{14.9k}$$

$$\boldsymbol{n} \cdot (\mu \boldsymbol{H}) = 0 \quad \text{on } \Gamma_1, \tag{14.9l}$$

$$\boldsymbol{n} \times \boldsymbol{H} = \mathbf{0} \quad \text{on } \Gamma_2, \tag{14.9m}$$

$$\boldsymbol{n} \cdot (\varepsilon \boldsymbol{E}) = 0 \quad \text{on } \Gamma_2. \tag{14.9n}$$

Due to solvability conditions (14.4), divergence equations (14.9i), (14.9j), and boundary conditions (14.9k)–(14.9n), we may eliminate (14.9b), (14.9c), (14.9f) and (14.9h), and rewrite system (14.9) as

$$\nabla \times \left[\nabla \times \boldsymbol{E} + \frac{\partial(\mu \boldsymbol{H})}{\partial t}\right] = -\nabla \times \boldsymbol{K}^{\text{imp}} \quad \text{in } \Omega, \tag{14.10a}$$

$$\nabla \times \left[\nabla \times \boldsymbol{H} - \frac{\partial(\varepsilon \boldsymbol{E})}{\partial t} - \sigma \boldsymbol{E}\right] = \nabla \times \boldsymbol{J}^{\text{imp}} \quad \text{in } \Omega, \tag{14.10b}$$

$$\nabla \cdot (\varepsilon \boldsymbol{E}) = \rho^{\text{imp}} \quad \text{in } \Omega, \tag{14.10c}$$

$$\nabla \cdot (\mu \boldsymbol{H}) = 0 \qquad \text{in } \Omega, \tag{14.10d}$$

$$\boldsymbol{n} \times \boldsymbol{E} = \boldsymbol{0} \qquad \text{on } \Gamma_1, \tag{14.10e}$$

$$\boldsymbol{n} \cdot (\mu \boldsymbol{H}) = 0 \qquad \text{on } \Gamma_1, \tag{14.10f}$$

$$\boldsymbol{n} \times (\nabla \times \boldsymbol{H}) = \boldsymbol{n} \times \boldsymbol{J}^{\text{imp}} \qquad \text{on } \Gamma_1, \tag{14.10g}$$

$$\boldsymbol{n} \times \boldsymbol{H} = \boldsymbol{0} \qquad \text{on } \Gamma_2, \tag{14.10h}$$

$$\boldsymbol{n} \cdot (\varepsilon \boldsymbol{E}) = 0 \qquad \text{on } \Gamma_2, \tag{14.10i}$$

$$\boldsymbol{n} \times (\nabla \times \boldsymbol{E}) = -\boldsymbol{n} \times \boldsymbol{K}^{\text{imp}} \qquad \text{on } \Gamma_2. \tag{14.10j}$$

Obviously, system (14.10) is completely equivalent to system (14.1), the validation of (14.10) guarantees the validation of system (14.1). Therefore, we can use the curl equations in system (14.1) to decouple $\boldsymbol{E}$ and $\boldsymbol{H}$ in system (14.10), we then obtain

$$\nabla \times (\nabla \times \boldsymbol{E}) + \mu \frac{\partial}{\partial t}\left[\frac{\partial(\varepsilon \boldsymbol{E})}{\partial t} + \sigma \boldsymbol{E}\right] = -\nabla \times \boldsymbol{K}^{\text{imp}}$$

$$-\mu \frac{\partial \boldsymbol{J}^{\text{imp}}}{\partial t} \qquad \text{in } \Omega, \tag{14.11a}$$

$$\nabla \cdot (\varepsilon \boldsymbol{E}) = \rho^{\text{imp}} \qquad \text{in } \Omega, \tag{14.11b}$$

$$\boldsymbol{n} \times \boldsymbol{E} = \boldsymbol{0} \qquad \text{on } \Gamma_1, \tag{14.11c}$$

$$\boldsymbol{n} \cdot (\varepsilon \boldsymbol{E}) = 0 \qquad \text{on } \Gamma_2, \tag{14.11d}$$

$$\boldsymbol{n} \times (\nabla \times \boldsymbol{E}) = -\boldsymbol{n} \times \boldsymbol{K}^{\text{imp}} \qquad \text{on } \Gamma_2, \tag{14.11e}$$

and

$$\nabla \times (\nabla \times \boldsymbol{H}) + \left(\varepsilon \frac{\partial}{\partial t} + \sigma\right) \frac{\partial(\mu \boldsymbol{H})}{\partial t} = -\left(\varepsilon \frac{\partial}{\partial t} + \sigma\right) \boldsymbol{K}^{\text{imp}}$$

$$+\nabla \times \boldsymbol{J}^{\text{imp}} \qquad \text{in } \Omega, \tag{14.12a}$$

$$\nabla \cdot (\mu \boldsymbol{H}) = 0 \qquad \text{in } \Omega, \tag{14.12b}$$

$$\boldsymbol{n} \cdot (\mu \boldsymbol{H}) = 0 \qquad \text{on } \Gamma_1, \tag{14.12c}$$

$$\boldsymbol{n} \times (\nabla \times \boldsymbol{H}) = \boldsymbol{n} \times \boldsymbol{J}^{\text{imp}} \qquad \text{on } \Gamma_1, \tag{14.12d}$$

$$\boldsymbol{n} \times \boldsymbol{H} = \boldsymbol{0} \qquad \text{on } \Gamma_2. \tag{14.12e}$$

We note that the curl–curl equations in (14.11) and (14.12) do not have any information about the divergence of the field intensities, and hence they cannot stand alone; they must be supplemented by the divergence equations and the additional natural boundary conditions. In other words, the curl–curl equations admit more solutions than the first-order full system. This is the real reason that the numerical methods based on the curl–curl equations will give rise to spurious solutions.

It is difficult to solve a second-order curl–curl equation (14.11a) mixed with an explicit constraint of the first-order divergence equation (14.11b). We should look for a simple way. By virtue of Theorem 5.4 and the vector identity (A.4), system (14.11) can be reduced to

$$-\Delta \boldsymbol{E} + \mu \frac{\partial}{\partial t}\left[\frac{\partial(\varepsilon \boldsymbol{E})}{\partial t} + \sigma \boldsymbol{E}\right] = -\nabla \times \boldsymbol{K}^{\mathrm{imp}} - \mu \frac{\partial \boldsymbol{J}^{\mathrm{imp}}}{\partial t} - \left(\frac{1}{\varepsilon}\right)\nabla \rho^{\mathrm{imp}} \quad \text{in } \Omega, \tag{14.13a}$$

$$\nabla[\nabla \cdot (\varepsilon \boldsymbol{E}) - \rho^{\mathrm{imp}}] = \boldsymbol{0} \quad \text{in } \Omega, \tag{14.13b}$$

$$\boldsymbol{n} \times \boldsymbol{E} = \boldsymbol{0} \quad \text{on } \Gamma_1, \tag{14.13c}$$

$$\nabla \cdot (\varepsilon \boldsymbol{E}) = \rho^{\mathrm{imp}} \quad \text{on } \Gamma_1, \tag{14.13d}$$

$$\boldsymbol{n} \cdot (\varepsilon \boldsymbol{E}) = 0 \quad \text{on } \Gamma_2, \tag{14.13e}$$

$$\boldsymbol{n} \times (\nabla \times \boldsymbol{E}) = -\boldsymbol{n} \times \boldsymbol{K}^{\mathrm{imp}} \quad \text{on } \Gamma_2, \tag{14.13f}$$

It can be shown that (14.13b) may be eliminated (Sect. 14.2.3). That is, the divergence equation (14.13b) is implicitly satisfied by the Helmholtz-type equation (14.13a) with the boundary conditions. Therefore, system (14.13) can be further simplified as

$$-\Delta \boldsymbol{E} + \mu \frac{\partial}{\partial t}\left[\frac{\partial(\varepsilon \boldsymbol{E})}{\partial t} + \sigma \boldsymbol{E}\right] = -\nabla \times \boldsymbol{K}^{\mathrm{imp}} - \mu \frac{\partial \boldsymbol{J}^{\mathrm{imp}}}{\partial t} - \left(\frac{1}{\varepsilon}\right)\nabla \rho^{\mathrm{imp}} \quad \text{in } \Omega, \tag{14.14a}$$

$$\boldsymbol{n} \times \boldsymbol{E} = \boldsymbol{0} \quad \text{on } \Gamma_1, \tag{14.14b}$$

$$\nabla \cdot (\varepsilon \boldsymbol{E}) = \rho^{\mathrm{imp}} \quad \text{on } \Gamma_1, \tag{14.14c}$$

$$\boldsymbol{n} \cdot (\varepsilon \boldsymbol{E}) = 0 \quad \text{on } \Gamma_2, \tag{14.14d}$$

$$\boldsymbol{n} \times (\nabla \times \boldsymbol{E}) = -\boldsymbol{n} \times \boldsymbol{K}^{\mathrm{imp}} \quad \text{on } \Gamma_2. \tag{14.14e}$$

Similarly, we can have

$$-\Delta \boldsymbol{H} + \left(\varepsilon \frac{\partial}{\partial t} + \sigma\right)\frac{\partial(\mu \boldsymbol{H})}{\partial t} = -\left(\varepsilon \frac{\partial}{\partial t} + \sigma\right)\boldsymbol{K}^{\mathrm{imp}} + \nabla \times \boldsymbol{J}^{\mathrm{imp}} \quad \text{in } \Omega, \tag{14.15a}$$

$$\boldsymbol{n} \cdot (\mu \boldsymbol{H}) = 0 \quad \text{on } \Gamma_1, \tag{14.15b}$$

$$\boldsymbol{n} \times (\nabla \times \boldsymbol{H}) = \boldsymbol{n} \times \boldsymbol{J}^{\mathrm{imp}} \quad \text{on } \Gamma_1, \tag{14.15c}$$

$$\boldsymbol{n} \times \boldsymbol{H} = \boldsymbol{0} \quad \text{on } \Gamma_2, \tag{14.15d}$$

$$\nabla \cdot (\mu \boldsymbol{H}) = 0 \qquad \text{on } \Gamma_2. \tag{14.15e}$$

Since we are dealing with a second-order problem in the time-domain, in addition to initial conditions for $\boldsymbol{E}$ and $\boldsymbol{H}$, we need initial conditions for $\partial \boldsymbol{E}/\partial t$ and $\partial \boldsymbol{H}/\partial t$ which can be obtained by using (14.1a) and (14.1b).

We note that the divergence equations are required to be satisfied only on a part of boundary. This was first observed by Boyse et al. (1992) for a special case. We will rigorously prove this in Sect. 14.2.3 by using the least-squares method. The Helmholtz-type equations (14.14a) and (14.15a) can be found in most text books on electromagnetics. However, it seems that all these books (except the papers by Boyse et al. 1992, Mayergoyz and D'Angelo 1993) claim that the Helmholtz-type equation must be solved with the divergence equation satisfied in the whole domain. The rigorous derivation using the div–curl method shows that the Helmholtz-type equation can stand alone, and the divergence equation should be satisfied only on a part of boundary.

The advantages of using the Helmholtz-type equation over the curl–curl equation are obvious: one avoids the difficulty involving explicit satisfaction of the divergence equations, instead one solves three decoupled second-order equations with coupled boundary conditions. We also should remark here that the Helmholtz-type equations (14.14a) and (14.15a) are valid only for homogeneous and isotropic material regions. For inhomogeneous or anisotropic materials, we still can use the div–curl method. However, the correct second-order equations do not have $\Delta \boldsymbol{E}$ and $\Delta \boldsymbol{H}$ terms.

14.2.2 The Galerkin Method

One may elect to use, for example, the finite difference method, to solve the Helmholtz-type equations (14.14) or (4.15). Usually, the finite difference method is based on rectangular structured grids. In this case, for example, the divergence equation (14.15e) can be simplified as the Neumann boundary condition:

$$\frac{\partial}{\partial n} H_n = 0 \qquad \text{on } \Gamma_2.$$

For complex geometry it is not straightforward to implement the Neumann boundary condition in the finite difference method. By using the finite element method based on a variational principle, only the essential boundary conditions should be explicitly imposed, all natural boundary conditions including divergence conditions on the boundary will be satisfied automatically by the variational process. In the following we derive the variational formulation corresponding to (14.14).

By taking into account the vector identity (A.4), the Galerkin formulation associated with (14.14) is: Find $\boldsymbol{E} \in \boldsymbol{\mathcal{E}}$ such that

$$
\begin{aligned}
&\left(\nabla \times \left[\nabla \times \boldsymbol{E} + \boldsymbol{K}^{\text{imp}}\right], \boldsymbol{E}^*\right) + \langle \nabla \times \boldsymbol{E} + \boldsymbol{K}^{\text{imp}}, \boldsymbol{n} \times \boldsymbol{E}^* \rangle_{\Gamma_2} \\
&+\left(-\nabla\left[\nabla \cdot \boldsymbol{E} - \frac{\rho^{\text{imp}}}{\varepsilon}\right], \boldsymbol{E}^*\right) + \langle \nabla \cdot \boldsymbol{E} - \frac{\rho^{\text{imp}}}{\varepsilon}, \boldsymbol{n} \cdot \boldsymbol{E}^* \rangle_{\Gamma_1} \\
&+\left(\mu \frac{\partial}{\partial t}\left[\frac{\partial(\varepsilon \boldsymbol{E})}{\partial t} + \sigma \boldsymbol{E}\right], \boldsymbol{E}^*\right) + \left(\mu \frac{\partial \boldsymbol{J}^{\text{imp}}}{\partial t}, \boldsymbol{E}^*\right) = 0 \qquad (14.16)
\end{aligned}
$$

for all $\boldsymbol{E}^* \in \boldsymbol{\mathcal{E}}$, where $\boldsymbol{\mathcal{E}} = \{\boldsymbol{E} \in [H^1(\Omega)]^3 : \boldsymbol{n} \times \boldsymbol{E} = \boldsymbol{0} \text{ on } \Gamma_1,\ \boldsymbol{n} \cdot \boldsymbol{E} = 0 \text{ on } \Gamma_2\}$. By virtue of Green's formula (B.3) and (B.5), the statement (14.16) can be simplified to a more symmetric form: Find $\boldsymbol{E} \in \boldsymbol{\mathcal{E}}$ such that

$$
\begin{aligned}
&(\nabla \times \boldsymbol{E}, \nabla \times \boldsymbol{E}^*) + (\nabla \cdot \boldsymbol{E}, \nabla \cdot \boldsymbol{E}^*) + \left(\mu \frac{\partial}{\partial t}\left[\frac{\partial(\varepsilon \boldsymbol{E})}{\partial t} + \sigma \boldsymbol{E}\right], \boldsymbol{E}^*\right) \\
&= -(\boldsymbol{K}^{\text{imp}}, \nabla \times \boldsymbol{E}^*) + \left(\frac{\rho^{\text{imp}}}{\varepsilon}, \nabla \cdot \boldsymbol{E}^*\right) - \left(\mu \frac{\partial}{\partial t} \boldsymbol{J}^{\text{imp}}, \boldsymbol{E}^*\right) \qquad (14.17)
\end{aligned}
$$

for all $\boldsymbol{E}^* \in \boldsymbol{\mathcal{E}}$.

For time-harmonic eigenvalue problems with $\sigma = 0$, the variational formulation takes the form

$$
(\nabla \times \boldsymbol{E}, \nabla \times \boldsymbol{E}^*) + (\nabla \cdot \boldsymbol{E}, \nabla \cdot \boldsymbol{E}^*) - \omega^2 \mu \varepsilon (\boldsymbol{E}, \boldsymbol{E}^*) = 0, \qquad (14.18)
$$

where ω is the angular frequency.

The formulations for the magnetic field are analogous: Find $\boldsymbol{H} \in \boldsymbol{\mathcal{H}}$ such that

$$
\begin{aligned}
&(\nabla \times \boldsymbol{H}, \nabla \times \boldsymbol{H}^*) + (\nabla \cdot \boldsymbol{H}, \nabla \cdot \boldsymbol{H}^*) + \left(\left[\varepsilon \frac{\partial}{\partial t} + \sigma\right] \frac{\partial(\mu \boldsymbol{H})}{\partial t}, \boldsymbol{H}^*\right) \\
&= (\boldsymbol{J}^{\text{imp}}, \nabla \times \boldsymbol{H}^*) - \left(\left[\varepsilon \frac{\partial}{\partial t} + \sigma\right] \boldsymbol{K}^{\text{imp}}, \boldsymbol{H}^*\right) \qquad (14.19)
\end{aligned}
$$

for all $\boldsymbol{H}^* \in \boldsymbol{\mathcal{H}}$, where $\boldsymbol{\mathcal{H}} = \{\boldsymbol{H} \in [H^1(\Omega)]^3 : \boldsymbol{n} \times \boldsymbol{H} = \boldsymbol{0} \text{ on } \Gamma_2,\ \boldsymbol{n} \cdot \boldsymbol{H} = 0 \text{ on } \Gamma_1\}$. For time-harmonic eigenvalue problems with $\sigma = 0$, the variational formulation takes the form

$$
(\nabla \times \boldsymbol{H}, \nabla \times \boldsymbol{H}^*) + (\nabla \cdot \boldsymbol{H}, \nabla \cdot \boldsymbol{H}^*) - \omega^2 \mu \varepsilon (\boldsymbol{H}, \boldsymbol{H}^*) = 0. \qquad (14.20)
$$

The variational formulations (14.17) and (14.18) are of the same structure as the most popular Galerkin/penalty formulations in the literature. However, in contrast to the commonly used penalty formulation, there is no free parameter in the Galerkin formulations (14.17) and (14.18). In other words, the penalty parameter s should be chosen such that the Euler–Lagrange equation associated with the variational formulation becomes the Helmholtz-type equation (14.14a).

When (14.18) is discretized, its first two terms generate a positive- definite stiffness matrix by virtue of Theorem 5.3, and $(\boldsymbol{E}, \boldsymbol{E}^*)$ generates a mass matrix. The equation (14.18) then becomes a well-posed generalized eigenvalue problem.

If the second term in (14.18), which is related to the divergence-free condition, is omitted, one obtains the commonly used incorrect variational formulation that corresponds to solving only the curl–curl equation. One can see that in this case the stiffness matrix is not positive-definite in the discretized $\boldsymbol{\mathcal{E}}$ space, and thus iterative solution methods do not work; moreover, there exits an eigenvalue $\omega = 0$ of infinite multiplicity. This trouble is caused by incorrect mathematical formulation. In the correct formulation (14.18), due to Theorem 5.2, there is only a trivial solution $\boldsymbol{E} = \boldsymbol{0}$ corresponding to $\omega = 0$. In other words, the variational formulation (14.18) will not give rise to spurious modes.

We emphasize again that the spurious solutions in the computation of resonator, waveguide, scattering wave and eddy current problems are really caused by wrong analytical formulations.

14.2.3 The Least-Squares Look-Alike Method

In Sect. 14.2.1 the div–curl method is employed to derive the second-order (Helmholtz-type) Maxwell equations and their boundary conditions that guarantee no spurious solutions. But we cannot yet determine whether the divergence equations should be specified on a part of boundary or on the entire boundary. In this section we give a more powerful method to derive equivalent higher-order equations and rigorously prove the statement made in Sect. 14.2.1.

Consider the following div–curl system for the electric field:

$$\nabla \times \boldsymbol{E} = -\frac{\partial(\mu \boldsymbol{H})}{\partial t} - \boldsymbol{K}^{\text{imp}} \qquad \text{in } \Omega, \tag{14.21a}$$

$$\nabla \cdot \boldsymbol{E} = \frac{\rho^{\text{imp}}}{\varepsilon} \qquad \text{in } \Omega, \tag{14.21b}$$

$$\boldsymbol{n} \times \boldsymbol{E} = \boldsymbol{0} \qquad \text{on } \Gamma_1, \tag{14.21c}$$

$$\boldsymbol{n} \cdot (\varepsilon \boldsymbol{E}) = 0 \qquad \text{on } \Gamma_2, \tag{14.21d}$$

where $\boldsymbol{H}$ is assumed to be known and to satisfy (14.1b) and (14.1d) as well as the boundary conditions (14.3b) and (14.3c), and the source terms satisfy the solvability conditions (14.4a)–(14.4e). In other words, when the magnetic field and the sources are given, the solution of (14.21) will give the corresponding electric field.

Obviously, system (14.21) is a typical div–curl system that can be treated easily by the least-squares method, see the details in Sect. 5.4. To proceed, we define a quadratic functional:

$$I : \boldsymbol{\mathcal{E}} \longrightarrow \mathbb{R},$$

$$I(\boldsymbol{E}) = \left\| \nabla \times \boldsymbol{E} + \frac{\partial(\mu \boldsymbol{H})}{\partial t} + \boldsymbol{K}^{\text{imp}} \right\|^2 + \left\| \nabla \cdot \boldsymbol{E} - \frac{\rho^{\text{imp}}}{\varepsilon} \right\|^2. \tag{14.22}$$

The minimization of I with respect of $\boldsymbol{E}$ leads to the variational formulation:

$$\left(\nabla \times \boldsymbol{E} + \frac{\partial(\mu \boldsymbol{H})}{\partial t} + \boldsymbol{K}^{\text{imp}}, \nabla \times \boldsymbol{E}^*\right) + \left(\nabla \cdot \boldsymbol{E} - \frac{\rho^{\text{imp}}}{\varepsilon}, \nabla \cdot \boldsymbol{E}^*\right) = 0, \tag{14.23}$$

where $\boldsymbol{E}^* = \delta \boldsymbol{E} \in \boldsymbol{\mathcal{E}}$.

By using Green's formula (B.5) and considering the fact that $\boldsymbol{H}$ satisfies (14.1b) and (14.3c), the $\boldsymbol{H}$ related term in (14.23) can be written as

$$\begin{aligned}\left(\frac{\partial(\mu \boldsymbol{H})}{\partial t}, \nabla \times \boldsymbol{E}^*\right) &= \left(\mu \frac{\partial(\nabla \times \boldsymbol{H})}{\partial t}, \boldsymbol{E}^*\right) + \left\langle \frac{\partial(\mu \boldsymbol{H})}{\partial t}, \boldsymbol{n} \times \boldsymbol{E}^* \right\rangle_{\Gamma} \\ &= \left(\mu \frac{\partial}{\partial t}\left[\frac{\partial(\varepsilon \boldsymbol{E})}{\partial t} + \sigma \boldsymbol{E} + \boldsymbol{J}^{\text{imp}}\right], \boldsymbol{E}^*\right).\end{aligned} \tag{14.24}$$

Therefore, from (14.23) we obtain the variational formulation:

$$(\nabla \times \boldsymbol{E}, \nabla \times \boldsymbol{E}^*) + (\nabla \cdot \boldsymbol{E}, \nabla \cdot \boldsymbol{E}^*) + \left(\mu \frac{\partial}{\partial t}\left[\frac{\partial(\varepsilon \boldsymbol{E})}{\partial t} + \sigma \boldsymbol{E}\right], \boldsymbol{E}^*\right)$$

$$= -(\boldsymbol{K}^{\text{imp}}, \nabla \times \boldsymbol{E}^*) + \left(\frac{\rho^{\text{imp}}}{\varepsilon}, \nabla \cdot \boldsymbol{E}^*\right) - \left(\mu \frac{\partial}{\partial t} \boldsymbol{J}^{\text{imp}}, \boldsymbol{E}^*\right) \tag{14.25}$$

which is exactly the same as (14.17). By using Green's formula, from (14.25) we can obtain the Euler–Lagrange equation (14.14a) and the natural boundary condition (14.14c) and (14.14e). That is, the correctness of (14.14) or (14.15) is completely proved.

Now we understand that if the field intensities are sufficiently smooth, then the variational formulation (14.25), the Helmholtz-type equation (14.14a) with its boundary conditions, and the first-order system (14.21) are equivalent to one another. However, the finite element method based on (14.25) has advantages: (1) the sources are required only to be square integrable; (2) the test and trial functions from standard finite element spaces are required to satisfy only the essential boundary conditions (14.21c) and (14.21d); (3) the divergence equation (14.21b) is automatically satisfied,

We remark that the procedure of deriving formulation (14.25) in this section is not a true least-squares approach, because (1) we have assumed that $\boldsymbol{H}$ is given and satisfies (14.1b), and hence $\boldsymbol{H}$ is not subject to the variation; (2) the true least-squares method always leads to a symmetric bilinear form; here the σ related term is not symmetric. Even so, this procedure is mathematically justifiable. It is nothing but a rigorous method to establish the Galerkin variational formulation corresponding to the Helmholtz-type equations (14.14a) and their boundary conditions. All derivations provided in this section have rigorously proved that the penalty parameter in the Galerkin/penalty method should be chosen such that it generates the Helmholtz-type equation in homogeneous isotropic material regions.

14.2.4 Anisotropic Media

For simplicity we consider anisotropic and homogeneous nonconductors. The time-varying form of Maxwell equations are

$$\nabla \times \boldsymbol{E} + \frac{\partial(\boldsymbol{\mu}\boldsymbol{H})}{\partial t} = -\boldsymbol{K}^{\mathrm{imp}} \quad \text{in } \Omega, \tag{14.26a}$$

$$\nabla \times \boldsymbol{H} - \frac{\partial(\boldsymbol{\varepsilon}\boldsymbol{E})}{\partial t} = \boldsymbol{J}^{\mathrm{imp}} \quad \text{in } \Omega, \tag{14.26b}$$

$$\nabla \cdot (\boldsymbol{\varepsilon}\boldsymbol{E}) = 0 \quad \text{in } \Omega, \tag{14.26c}$$

$$\nabla \cdot (\boldsymbol{\mu}\boldsymbol{H}) = 0 \quad \text{in } \Omega. \tag{14.26d}$$

We also consider the homogeneous boundary conditions:

$$\boldsymbol{n} \times \boldsymbol{E} = \boldsymbol{0} \quad \text{on } \Gamma_1, \tag{14.26e}$$

$$\boldsymbol{n} \cdot (\boldsymbol{\mu}\boldsymbol{H}) = 0 \quad \text{on } \Gamma_1, \tag{14.26f}$$

$$\boldsymbol{n} \times \boldsymbol{H} = \boldsymbol{0} \quad \text{on } \Gamma_2, \tag{14.26g}$$

$$\boldsymbol{n} \cdot (\boldsymbol{\varepsilon}\boldsymbol{E}) = 0 \quad \text{on } \Gamma_2. \tag{14.26h}$$

Of course, the source terms must satisfy the following solvability conditions:

$$\nabla \cdot \boldsymbol{K}^{\mathrm{imp}} = 0 \quad \text{in } \Omega, \tag{14.27a}$$

$$\boldsymbol{n} \cdot \boldsymbol{K}^{\mathrm{imp}} = 0 \quad \text{on } \Gamma_1, \tag{14.27b}$$

$$\int_\Gamma \boldsymbol{n} \cdot \boldsymbol{K}^{\mathrm{imp}} d\Gamma = 0, \tag{14.27c}$$

$$\nabla \cdot \boldsymbol{J}^{\mathrm{imp}} = 0 \quad \text{in } \Omega, \tag{14.27d}$$

$$\boldsymbol{n} \cdot \boldsymbol{J}^{\mathrm{imp}} = 0 \quad \text{on } \Gamma_2, \tag{14.27e}$$

$$\int_\Gamma \boldsymbol{n} \cdot \boldsymbol{J}^{\mathrm{imp}} d\Gamma = 0. \tag{14.27f}$$

The permittivity $\boldsymbol{\varepsilon}$ and the permeability $\boldsymbol{\mu}$ are real symmetric positive-definite matrices, i.e., 3×3 tensors in three-dimensions. Their inverse matrices exist and are denoted by $\boldsymbol{\varepsilon}^{-1}$ and $\boldsymbol{\mu}^{-1}$, respectively. We may choose their maximum eigenvalues as their norms denoted by $|\varepsilon|$ and $|\mu|$, respectively. We always can decompose the matrix $\boldsymbol{\mu}$ such that

$$\boldsymbol{\mu} = \boldsymbol{\mu}^{\frac{1}{2}}(\boldsymbol{\mu}^{\frac{1}{2}})^T. \tag{14.28}$$

We would like to use the least-squares method to derive the second-order equations for the electric field and to establish the corresponding variational formulation. Consider the following div–curl system for the electric field:

$$\nabla \times \boldsymbol{E} = -\frac{\partial(\mu \boldsymbol{H})}{\partial t} - \boldsymbol{K}^{\mathrm{imp}} \qquad \text{in } \Omega, \tag{14.29a}$$

$$\nabla \cdot (\boldsymbol{\varepsilon E}) = 0 \qquad \text{in } \Omega, \tag{14.29b}$$

$$\boldsymbol{n} \times \boldsymbol{E} = \boldsymbol{0} \qquad \text{on } \Gamma_1, \tag{14.29c}$$

$$\boldsymbol{n} \cdot (\boldsymbol{\varepsilon E}) = 0 \qquad \text{on } \Gamma_2, \tag{14.29d}$$

where $\boldsymbol{H}$ is assumed to be known and to satisfy the equations (14.26b) and (14.26d) as well as the boundary conditions (14.26f) and (14.26g).

We define a quadratic functional:

$$I : \mathcal{E} \longrightarrow \mathbb{R},$$

$$I(\boldsymbol{E}) = |\mu| \left\| \mu^{-\frac{1}{2}} \left(\nabla \times \boldsymbol{E} + \frac{\partial(\mu \boldsymbol{H})}{\partial t} + \boldsymbol{K}^{\mathrm{imp}} \right) \right\|^2 + |\varepsilon|^{-2} \| \nabla \cdot (\boldsymbol{\varepsilon E}) \|^2, \tag{14.30}$$

where $\mathcal{E} = \{\boldsymbol{E} \in [H^1(\Omega)]^3 : \boldsymbol{n} \times \boldsymbol{E} = \boldsymbol{0} \text{ on } \Gamma_1,\ \boldsymbol{n} \cdot (\boldsymbol{\varepsilon E}) = 0 \text{ on } \Gamma_2\}$.

The minimization of I with respect to $\boldsymbol{E}$ leads to the variational formulation:

$$|\mu| \left(\mu^{-\frac{1}{2}} \left[\nabla \times \boldsymbol{E} + \frac{\partial(\mu \boldsymbol{H})}{\partial t} + \boldsymbol{K}^{\mathrm{imp}} \right], \mu^{-\frac{1}{2}} \nabla \times \boldsymbol{E}^* \right) + |\varepsilon|^{-2} \Big(\nabla \cdot (\boldsymbol{\varepsilon E}), \nabla \cdot (\boldsymbol{\varepsilon E}^*) \Big) = 0, \tag{14.31}$$

where $\boldsymbol{E}^* = \delta \boldsymbol{E} \in \mathcal{E}$.

Upon using a relation similar to (14.24), from (14.31) we finally obtain the variational formulation:

$$|\mu| \left(\mu^{-1} \nabla \times \boldsymbol{E}, \nabla \times \boldsymbol{E}^* \right) + |\varepsilon|^{-2} \Big(\nabla \cdot (\boldsymbol{\varepsilon E}), \nabla \cdot (\boldsymbol{\varepsilon E}^*) \Big) + |\mu| \left(\frac{\partial^2}{\partial t^2} (\boldsymbol{\varepsilon E}), \boldsymbol{E}^* \right) = -|\mu| \left(\mu^{-1} \boldsymbol{K}^{\mathrm{imp}}, \nabla \times \boldsymbol{E}^* \right) - |\mu| \left(\frac{\partial}{\partial t} \boldsymbol{J}^{\mathrm{imp}}, \boldsymbol{E}^* \right). \tag{14.32}$$

The correctness of (14.32) can be verified by considering two facts. First, if the medium is nonconductive and isotropic without imposed electric charge density, (14.32) reduces to (14.25). Second, upon using the technique given by Saranen (1982), one can further prove the coerciveness of the bilinear form represented by the first two terms in (14.32); in other words, when discretized, (14.32) leads to a well-posed dynamic problem (or generalized eigenvalue problem in time-harmonic cases) with symmetric positive-definite stiffness and mass matrices.

From the variational formulation (14.32) by using Green's formulae we can deduce the associated Euler–Lagrange equation and boundary conditions:

$$|\mu|\nabla\times(\mu^{-1}\nabla\times\boldsymbol{E})-|\varepsilon|^{-2}\varepsilon\nabla[\nabla\cdot(\varepsilon\boldsymbol{E})]+|\mu|\frac{\partial^2}{\partial t^2}(\varepsilon\boldsymbol{E})$$

$$=-|\mu|\nabla\times(\mu^{-1}\boldsymbol{K}^{\text{imp}})-|\mu|\frac{\partial}{\partial t}\boldsymbol{J}^{\text{imp}},\qquad \text{in } \Omega, \tag{14.33a}$$

$$\boldsymbol{n}\times\boldsymbol{E}=\boldsymbol{0}\qquad \text{on } \Gamma_1, \tag{14.33b}$$

$$\nabla\cdot(\varepsilon\boldsymbol{E})=0\qquad \text{on } \Gamma_1, \tag{14.33c}$$

$$\boldsymbol{n}\cdot(\varepsilon\boldsymbol{E})=0\qquad \text{on } \Gamma_2, \tag{14.33d}$$

$$\boldsymbol{n}\times(\mu^{-1}\nabla\times\boldsymbol{E})=-\boldsymbol{n}\times(\mu^{-1}\boldsymbol{K}^{\text{imp}})\qquad \text{on } \Gamma_2. \tag{14.33e}$$

We remark that both (14.32) and (14.33) can be generalized to spatially variable media.

Similarly, we can obtain the variational formulation for the magnetic field: Find $\boldsymbol{H}\in\mathcal{H}$ such that

$$|\varepsilon|\Big(\varepsilon^{-1}\nabla\times\boldsymbol{H},\nabla\times\boldsymbol{H}^*\Big)+|\mu|^{-2}\Big(\nabla\cdot(\mu\boldsymbol{H}),\nabla\cdot(\mu\boldsymbol{H}^*)\Big)$$

$$+|\varepsilon|\Big(\frac{\partial^2}{\partial t^2}(\mu\boldsymbol{H}),\boldsymbol{H}^*\Big)=|\varepsilon|\Big(\varepsilon^{-1}\boldsymbol{J}^{\text{imp}},\nabla\times\boldsymbol{H}^*\Big)-|\varepsilon|\Big(\frac{\partial}{\partial t}\boldsymbol{K}^{\text{imp}},\boldsymbol{H}^*\Big) \tag{14.34}$$

for all $\boldsymbol{H}^*\in\mathcal{H}$, where $\mathcal{H}=\{\boldsymbol{H}\in[H^1(\Omega)]^3:\boldsymbol{n}\times\boldsymbol{H}=\boldsymbol{0}\text{ on }\Gamma_2,\ \boldsymbol{n}\cdot(\mu\boldsymbol{H})=0\text{ on }\Gamma_1\}$.

By using the variational formulation we can show that the magnetic field satisfies the following equation and boundary conditions:

$$|\varepsilon|\nabla\times(\varepsilon^{-1}\nabla\times\boldsymbol{H})-|\mu|^{-2}\mu\nabla[\nabla\cdot(\varepsilon\boldsymbol{H})]+|\varepsilon|\frac{\partial^2}{\partial t^2}(\mu\boldsymbol{H})$$

$$=|\varepsilon|\nabla\times(\varepsilon^{-1}\boldsymbol{J}^{\text{imp}})-|\varepsilon|\frac{\partial}{\partial t}\boldsymbol{K}^{\text{imp}},\qquad \text{in } \Omega, \tag{14.35a}$$

$$\boldsymbol{n}\cdot(\varepsilon\boldsymbol{H})=0\qquad \text{on } \Gamma_1, \tag{14.35b}$$

$$\boldsymbol{n}\times(\varepsilon^{-1}\nabla\times\boldsymbol{H})=\boldsymbol{n}\times(\varepsilon^{-1}\boldsymbol{J}^{\text{imp}})\qquad \text{on } \Gamma_1, \tag{14.35c}$$

$$\boldsymbol{n}\times\boldsymbol{H}=\boldsymbol{0}\qquad \text{on } \Gamma_2, \tag{14.35d}$$

$$\nabla\cdot(\varepsilon\boldsymbol{H})=0\qquad \text{on } \Gamma_2. \tag{14.35e}$$

14.3 Electrostatic Fields

When the field quantities do not vary with time, we have a static field. In this case, there is no interaction between electric and magnetic fields, and therefore we can separate either an electrostatic case or a magnetostatic case.

In the electrostatic case, the governing equations are

$$\nabla \times \boldsymbol{E} = \boldsymbol{0} \qquad \text{in } \Omega, \tag{14.36a}$$

$$\nabla \cdot (\varepsilon \boldsymbol{E}) = \rho^{\text{imp}} \qquad \text{in } \Omega, \tag{14.36b}$$

$$\boldsymbol{n} \times \boldsymbol{E} = \boldsymbol{0} \qquad \text{on } \Gamma_c, \tag{14.36c}$$

$$\boldsymbol{n} \cdot \boldsymbol{E} = 0 \qquad \text{on } \Gamma_s, \tag{14.36d}$$

$$\boldsymbol{n} \times \boldsymbol{E}^+ = \boldsymbol{n} \times \boldsymbol{E}^- \qquad \text{on } \Gamma_{\text{int}}, \tag{14.36e}$$

$$\boldsymbol{n} \cdot (\varepsilon^+ \boldsymbol{E}^+) = \boldsymbol{n} \cdot (\varepsilon^- \boldsymbol{E}^-) \qquad \text{on } \Gamma_{\text{int}}, \tag{14.36f}$$

where Γ_c may be a perfectly conducting surface or a far-field boundary, Γ_s is the plan of symmetry, and Γ_{int} is the interface between two media.

14.3.1 Electric Potential

By introducing the electric scalar potential ϕ such that

$$\boldsymbol{E} = -\nabla \phi, \tag{14.37}$$

then the equation (14.36a) is satisfied. From the rest of equations in (14.36) we obtain

$$-\nabla \cdot (\varepsilon \nabla \phi) = \rho^{\text{imp}} \qquad \text{in } \Omega, \tag{14.38a}$$

$$\phi = g^{\text{imp}} \qquad \text{on } \Gamma_c, \tag{14.38b}$$

$$\frac{\partial \phi}{\partial n} = 0 \qquad \text{on } \Gamma_s, \tag{14.38c}$$

where g^{imp} is a given constant voltage. The equation (14.38) is the well-known Poisson equation with Dirichlet and Neumann boundary conditions The classic Galerkin finite element method can be employed to obtain the approximate solution of ϕ with an optimal accuracy. However, as we have pointed out many times, the calculated electric field intensity $\boldsymbol{E}$ by numerical differentiation of ϕ is of lower-order accuracy and discontinuous across the element boundary.

14.3.2 The Least-Squares Finite Element Method

System (14.36) is a typical div–curl system. In Sect. 5.2 we have shown by introducing a dummy variable ϑ that in the three-dimensional case this system is properly determined and elliptic. Now let us apply the least-squares method to solve the div–curl system (14.36). We construct the following quadratic functional:

$$I : \boldsymbol{\mathcal{E}} \longrightarrow \mathbb{R},$$

$$I(\boldsymbol{E}) = \|\nabla \times \boldsymbol{E}\|_0^2 + \left\|\nabla \cdot \boldsymbol{E} - \frac{\rho^{\text{imp}}}{\varepsilon}\right\|_0^2, \tag{14.39}$$

where $\boldsymbol{\mathcal{E}} = \{\boldsymbol{E} \in [H^1(\Omega)]^3 : \boldsymbol{n} \times \boldsymbol{E} = \boldsymbol{0} \text{ on } \Gamma_c,\ \boldsymbol{n} \cdot \boldsymbol{E} = 0 \text{ on } \Gamma_s\}$. Taking the variation of I with respect to $\boldsymbol{E}$, and letting $\delta\boldsymbol{E} = \boldsymbol{E}^*$ and $\delta I = 0$, we obtain the least-squares variational formulation: Find $\boldsymbol{E} \in \boldsymbol{\mathcal{E}}$ such that

$$B(\boldsymbol{E}, \boldsymbol{E}^*) = L(\boldsymbol{E}^*) \qquad \forall \boldsymbol{E}^* \in \boldsymbol{\mathcal{E}}, \tag{14.40}$$

where

$$B(\boldsymbol{E}, \boldsymbol{E}^*) = (\nabla \times \boldsymbol{E}, \nabla \times \boldsymbol{E}^*) + (\nabla \cdot \boldsymbol{E}, \nabla \cdot \boldsymbol{E}^*),$$

$$L(\boldsymbol{E}^*) = \Big(\frac{\rho^{\text{imp}}}{\varepsilon}, \nabla \cdot \boldsymbol{E}^*\Big).$$

Due to Theorem 5.7 the LSFEM based on (14.40) has an optimal rate of convergence:

$$\|\boldsymbol{E} - \boldsymbol{E}_h\|_o \le C h^{r+1} |\boldsymbol{E}|_2, \tag{14.41}$$

where r is the order of polynomials in the finite element interpolation.

It is instructive to derive the Euler–Lagrange equations associated with the least-squares weak formulation (14.40) which can be rewritten as: Find $\boldsymbol{E} \in \boldsymbol{\mathcal{E}}$ such that

$$(\nabla \times \boldsymbol{E},\ \nabla \times \boldsymbol{E}^*) + \Big(\nabla \cdot \boldsymbol{E} - \frac{\rho^{\text{imp}}}{\varepsilon},\ \nabla \cdot \boldsymbol{E}^*\Big) = 0 \qquad \forall \boldsymbol{E}^* \in \boldsymbol{\mathcal{E}}. \tag{14.42}$$

By using Green's formulae (B.3) and (B.5), the equation (14.40) can be converted to

$$\begin{aligned}&\Big(\nabla \times (\nabla \times \boldsymbol{E}), \boldsymbol{E}^*\Big) + \langle \nabla \times \boldsymbol{E},\ \boldsymbol{n} \times \boldsymbol{E}^*\rangle \\ &\quad + \Big(-\nabla\Big[\nabla \cdot \boldsymbol{E} - \frac{\rho^{\text{imp}}}{\varepsilon}\Big],\ \boldsymbol{E}^*\Big) + \Big\langle \nabla \cdot \boldsymbol{E} - \frac{\rho^{\text{imp}}}{\varepsilon},\ \boldsymbol{n} \cdot \boldsymbol{E}^*\Big\rangle = 0.\end{aligned} \tag{14.43}$$

Taking into account (A.4) and that $\boldsymbol{E}^*$ satisfies the boundary conditions (14.36c) and (14.36d), from (14.43) we obtain

$$\begin{aligned}&\Big[-\Delta \boldsymbol{E} + \nabla\Big(\frac{\rho^{\text{imp}}}{\varepsilon}\Big],\ \boldsymbol{E}^*\Big) + \langle \nabla \times \boldsymbol{E},\ \boldsymbol{n} \times \boldsymbol{E}^*\rangle_{\Gamma_s} \\ &\quad + \Big\langle \nabla \cdot \boldsymbol{E} - \frac{\rho^{\text{imp}}}{\varepsilon},\ \boldsymbol{n} \cdot \boldsymbol{E}^*\Big\rangle_{\Gamma_c} = 0\end{aligned} \tag{14.44}$$

for all admissible $\boldsymbol{E}^* \in \boldsymbol{\mathcal{E}}$, hence we have the Euler–Lagrange equation and boundary conditions:

$$\Delta \boldsymbol{E} = \nabla\Big(\frac{\rho^{\text{imp}}}{\varepsilon}\Big) \qquad \text{in } \Omega, \tag{14.45a}$$

$$\boldsymbol{n} \times \boldsymbol{E} = \boldsymbol{0}, \qquad \nabla \cdot \boldsymbol{E} - \frac{\rho^{\text{imp}}}{\varepsilon} = 0 \qquad \text{on } \Gamma_c, \tag{14.45b}$$

$$\boldsymbol{n} \cdot \boldsymbol{E} = 0, \qquad \boldsymbol{n} \times (\nabla \times \boldsymbol{E}) = \boldsymbol{0} \qquad \text{on } \Gamma_s. \tag{14.45c}$$

We may write (14.45) more explicitly as:

$$\Delta E_x = \frac{\partial}{\partial x}\left(\frac{\rho^{\text{imp}}}{\varepsilon}\right) \qquad \text{in } \Omega, \tag{14.46a}$$

$$\Delta E_y = \frac{\partial}{\partial y}\left(\frac{\rho^{\text{imp}}}{\varepsilon}\right) \qquad \text{in } \Omega, \tag{14.46b}$$

$$\Delta E_z = \frac{\partial}{\partial z}\left(\frac{\rho^{\text{imp}}}{\varepsilon}\right) \qquad \text{in } \Omega, \tag{14.46c}$$

$$E_1 = 0, \quad E_2 = 0, \quad \frac{\partial E_n}{\partial n} = \frac{\rho^{\text{imp}}}{\varepsilon} \qquad \text{on } \Gamma_c, \tag{14.46d}$$

$$E_n = 0, \quad \frac{\partial E_1}{\partial n} = 0, \quad \frac{\partial E_2}{\partial n} = 0, \qquad \text{on } \Gamma_s \tag{14.46e}$$

where 1 and 2 denote the mutual-orthogonal tangential directions on the boundary surface. Now it is clear that the least-squares method (14.40) corresponds to solving three uncoupled Poisson equations (14.46a)–(14.46c) with three original Dirichlet boundary conditions and three additional Neumann boundary conditions. In fact, the original first-order equations (14.36a) and (14.36b) serve as the additional boundary conditions here. We have already observed this in Chap. 2.

We know that it is difficult to solve the first-order div–curl system (14.36) by conventional methods. Now one may employ the finite difference method or the classic Galerkin finite element method to solve an equivalent second-order system (14.46) without much difficulty. However, the LSFEM based on (14.40) is the easiest method, because only the simple original boundary conditions (14.36c) and (14.36d) need to be dealt with.

In practical electrostatic problems, a constant electric voltage g^{imp} is often prescribed on a part of boundary. In this case we may use LSFEM to solve the following first-order div–curl–grad system:

$$\nabla \times \boldsymbol{E} = \boldsymbol{0} \qquad \text{in } \Omega, \tag{14.47a}$$

$$\nabla \cdot (\varepsilon \boldsymbol{E}) = \rho^{\text{imp}} \qquad \text{in } \Omega, \tag{14.47b}$$

$$\nabla \phi - \boldsymbol{E} = \boldsymbol{0} \qquad \text{in } \Omega, \tag{14.47c}$$

$$\phi = g^{\text{imp}} \qquad \text{on } \Gamma_c, \tag{14.47d}$$

$$\boldsymbol{n} \times \boldsymbol{E} = \boldsymbol{0} \qquad \text{on } \Gamma_c, \tag{14.47e}$$

$$\boldsymbol{n} \cdot \boldsymbol{E} = 0 \qquad \text{on } \Gamma_s, \tag{14.47f}$$

$$\phi^+ = \phi^- \qquad \text{on } \Gamma_{\text{int}}, \tag{14.47g}$$

$$\boldsymbol{n} \times \boldsymbol{E}^+ = \boldsymbol{n} \times \boldsymbol{E}^- \qquad \text{on } \Gamma_{\text{int}}, \tag{14.47h}$$

$$\boldsymbol{n} \cdot (\varepsilon^+ \boldsymbol{E}^+) = \boldsymbol{n} \cdot (\varepsilon^- \boldsymbol{E}^-) \qquad \text{on } \Gamma_{\text{int}}. \tag{14.47i}$$

As discussed in Chap. 6, the curl equation (14.47a), which seems redundant, must be included to ensure an optimal rate of convergence for the field intensity and good numerical properties for iterative solvers.

Microstrip Line. As an example, let us apply LSFEM to a two-dimensional problem of the shielded microstrip transmission line illustrated in Fig. 14.1. This problem is taken from Jin (1993, p.98). When the microstrip line operates at low frequency, its capacitance and inductance can be obtained from a static analysis, which amounts to solving (14.47) by assuming a certain potential, say 1 volt, on its strip and a different potential, say zero volts, on the shielding conductor. Due to its symmetric geometry, the domain of analysis is reduced to half by specifying $E_x = 0$ at the plane of symmetry. On the conducting surfaces, we impose $\boldsymbol{n} \times \boldsymbol{E} = \boldsymbol{0}$ as shown in Fig. 14.1. It is important not to specify the field intensity at the sharp corners of the strip. The finite element mesh consisting of 949 linear quadrilateral elements is shown in Fig. 14.2. The nodes at the material interface are doubly numbered. This is a small-scale problem, so a direct solver is employed to solve the resulting algebraic equations. For each interface node we add the following term

$$\alpha[(\phi^+ - \phi^-)^2 + (E_x^+ - E_x^-)^2 + (\varepsilon^+ E_y^+ - \varepsilon^- E_y^-)^2]$$

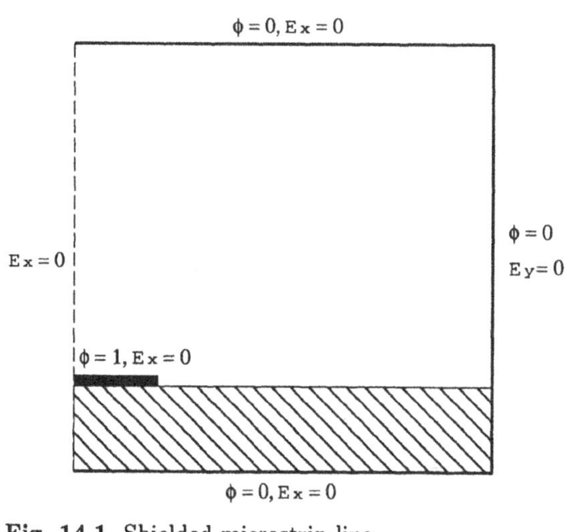

Fig. 14.1. Shielded microstrip line

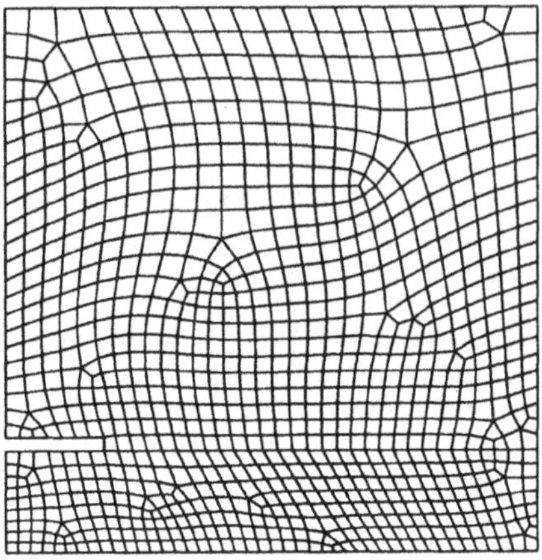

Fig. 14.2. The finite element mesh of a shielded microstrip line

into the variational functional (14.39) to enforce the interface condition, where α is a large number, say $10^5 - 10^{10}$. In this example, we take $\varepsilon^+ = 10$ for the medium (+) in the lower part, and $\varepsilon^- = 1$ for the medium (−) in the upper part. As emphasized in Sect. 4.11.3, reduced integration should be used to calculate the finite element matrix. Here we employ the one-point Gaussian quadrature. The nodal solution of the field intensity has some oscillations. However, the solution of the field intensity at Gaussian points is smooth. Therefore, we redistribute the field intensity at Gaussian points into the nodes by averaging according to element areas to obtain the final nodal solutions of the field intensity. The computed potential contours are plotted in Fig. 14.3. The nodal field intensity vectors are given in Fig. 14.4.

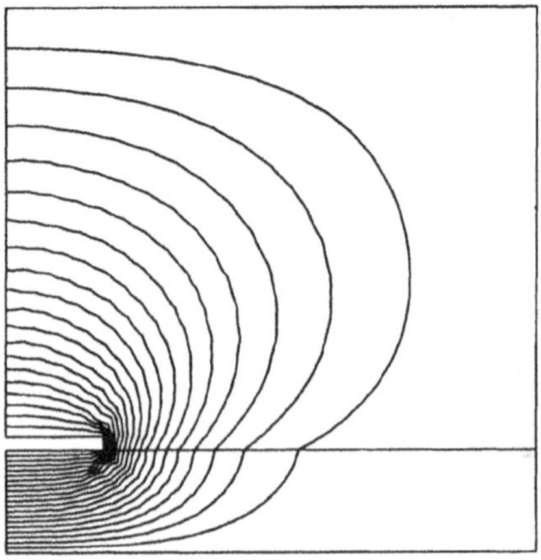

Fig. 14.3. Computed potential contours in a shielded microstrip line

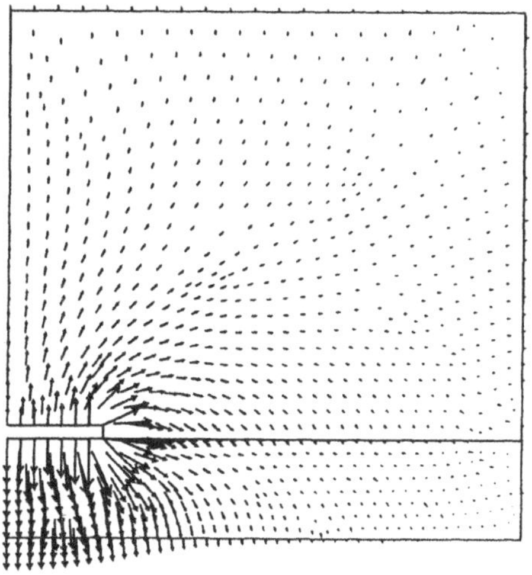

Fig. 14.4. The computed $\boldsymbol{E}$ field of a shielded microstrip line

14.4 Magnetostatic Fields

The magnetostatic fields are governed by the following div–curl system:

$$\nabla \times \boldsymbol{H} = \boldsymbol{J}^{\text{imp}} \qquad \text{in } \Omega, \tag{14.48a}$$

$$\nabla \cdot (\mu \boldsymbol{H}) = 0 \qquad \text{in } \Omega, \tag{14.48b}$$

$$\boldsymbol{n} \cdot (\mu \boldsymbol{H}) = 0 \qquad \text{on } \Gamma_1, \tag{14.48c}$$

$$\boldsymbol{n} \times \boldsymbol{H} = \boldsymbol{0} \qquad \text{on } \Gamma_2, \tag{14.48d}$$

$$\boldsymbol{n} \times \boldsymbol{H}^+ = \boldsymbol{n} \times \boldsymbol{H}^- \qquad \text{on } \Gamma_{\text{int}}, \tag{14.48e}$$

$$\boldsymbol{n} \cdot (\mu^+ \boldsymbol{H}^+) = \boldsymbol{n} \cdot (\mu^- \boldsymbol{H}^-) \qquad \text{on } \Gamma_{\text{int}}, \tag{14.48f}$$

where Γ_1 is a perfectly conducting electric wall or a symmetric plane, Γ_2 is a perfect magnetic wall, and Γ_{int} is an interface between two media without imposed surface electric current. The given electric current density should satisfy the solvability conditions:

$$\nabla \cdot \boldsymbol{J}^{\text{imp}} = 0 \qquad \text{in } \Omega, \tag{14.49a}$$

$$\int_\Gamma \boldsymbol{n} \cdot \boldsymbol{J}^{\text{imp}} ds = 0 \qquad \text{on } \Gamma. \tag{14.49b}$$

14.4.1 Magnetostatic Vector Potential

System (14.48) is of first order, and thus is difficult to solve directly by the classic Galerkin method. In order to convert (14.48) into a system of second-order equations, usually a vector potential $\boldsymbol{A}$ is introduced and is defined as

$$\mu \boldsymbol{H} = \nabla \times \boldsymbol{A}. \tag{14.50}$$

In this way, the divergence equation (14.48b) is automatically satisfied. Substituting (14.50) into (14.48a) yields

$$\nabla \times \left(\frac{1}{\mu} \nabla \times \boldsymbol{A}\right) = \boldsymbol{J}^{\text{imp}}. \tag{14.51}$$

This, however, does not determine $\boldsymbol{A}$ uniquely, because if $\boldsymbol{A}$ is a solution to (14.51), any function that can be written as $\boldsymbol{A}' = \boldsymbol{A} + \nabla f$ is also a solution regardless of the form of f. Thus, to determine $\boldsymbol{A}$ uniquely, one has to impose a condition on its divergence, that is

$$\nabla \cdot \boldsymbol{A} = 0. \tag{14.52}$$

Such a condition is called the gauge condition.

Obviously, the introduction of the magnetostatic vector potential $\boldsymbol{A}$ does not bring much benefits at all. One avoids the divergence equation (14.48b), but gets another one (14.52); the new div–curl system for $\boldsymbol{A}$, i.e., (14.50) and (14.52), has the same div–curl structure as system (14.48), and thus still needs to be converted into a second-order system which can be solved by conventional methods; moreover, the magnetic field intensity $\boldsymbol{H}$ obtained by numerical differentiation of $\boldsymbol{A}$ has lower accuracy and is discontinuous across the element boundary in the homogeneous medium region.

If one insists on using the magnetostatic vector potential $\boldsymbol{A}$, then we suggest using the least-squares method to solve the following system:

$$\nabla \times \boldsymbol{A} = \mu \boldsymbol{H} \qquad \text{in } \Omega, \tag{14.53a}$$

$$\nabla \cdot \boldsymbol{A} = 0 \qquad \text{in } \Omega, \tag{14.53b}$$

$$\boldsymbol{n} \times \boldsymbol{A} = \boldsymbol{0} \qquad \text{on } \Gamma, \tag{14.53c}$$

where $\boldsymbol{H}$ should satisfy the equation (14.48a), and for simplicity we consider the magnetostatic fields in a perfect conducting wall. The boundary condition (14.53c) is justified from the following consideration: according to Theorem 5.5, from (14.53c) we have

$$\boldsymbol{n} \cdot (\nabla \times \boldsymbol{A}) = 0 \qquad \text{on } \Gamma,$$

which is equivalent to (14.48c) due to the definition of $\boldsymbol{A}$.

We construct the following functional:

$$I : \mathcal{H} \longrightarrow \mathbb{R},$$

$$I(\boldsymbol{A}) = \|\nabla \times \boldsymbol{A} - \mu \boldsymbol{H}\|_0^2 + \|\nabla \cdot \boldsymbol{A}\|_0^2, \tag{14.54}$$

in which $\mathcal{H} = \{\boldsymbol{A} \in [H^1(\Omega)]^3 : \boldsymbol{n} \times \boldsymbol{A} = \boldsymbol{0} \text{ on } \Gamma\}$. Taking the variation of I with respect to $\boldsymbol{A}$, and letting $\delta \boldsymbol{A} = \boldsymbol{A}^*$ and $\delta I = 0$, we obtain

$$(\nabla \times \boldsymbol{A},\ \nabla \times \boldsymbol{A}^*) + (\nabla \cdot \boldsymbol{A},\ \nabla \cdot \boldsymbol{A}^*) = (\mu \boldsymbol{H},\ \nabla \times \boldsymbol{A}^*). \tag{14.55}$$

By using Green's formulae (B.5) and taking into account (14.48a) and that $\boldsymbol{A}^*$ satisfies the boundary condition (14.53c), the right-hand side term in (14.55) can be written as

$$\begin{aligned}(\mu \boldsymbol{H},\ \nabla \times \boldsymbol{A}^*) &= (\nabla \times \mu \boldsymbol{H},\ \boldsymbol{A}^*) + \langle \mu \boldsymbol{H},\ \boldsymbol{n} \times \boldsymbol{A}^* \rangle \\ &= (\mu \boldsymbol{J}^{\text{imp}},\ \boldsymbol{A}^*).\end{aligned}$$

Therefore, the least-squares variational formulation is: Find $\boldsymbol{A} \in \mathcal{H}$ such that

$$(\nabla \times \boldsymbol{A},\ \nabla \times \boldsymbol{A}^*) + (\nabla \cdot \boldsymbol{A},\ \nabla \cdot \boldsymbol{A}^*) = (\mu \boldsymbol{J}^{\text{imp}},\ \boldsymbol{A}^*) \quad \forall \boldsymbol{A}^* \in \mathcal{H}. \tag{14.56}$$

Obviously, the least-squares method amounts to minimizing the following functional:

$$I(\boldsymbol{A}) = \frac{1}{2}\left[\|\nabla \times \boldsymbol{A}\|_0^2 + \|\nabla \cdot \boldsymbol{A}\|_0^2\right] - (\mu \boldsymbol{J}^{\text{imp}},\ \boldsymbol{A}). \tag{14.57}$$

In fact, (14.57) can be obtain directly from (14.54).

14.4.2 The Least-Squares Finite Element Method

For magnetostatic problems, it is recommended that one solve the div–curl system (14.48) by LSFEM in order to obtain directly the magnetic intensity vector $\boldsymbol{H}$. The computed magnetic intensity vector has an optimal accuracy and is continuous across the element boundary. Moreover, the amount of computational work to obtain $\boldsymbol{H}$ is the same as to obtain $\boldsymbol{A}$.

Bus-Bar. As an example, let us consider the two-dimensional problem of a bus-bar illustrated in Fig. 14.5. To exploit the symmetry, we analyze a quarter of the domain ($0 \le x \le 1.0, \le y \le 1.0$). We apply the perfectly conducting electric wall condition $\boldsymbol{n} \cdot \boldsymbol{H} = 0$ to the outer boundary, and the magnetic symmetry condition $\boldsymbol{n} \times \boldsymbol{H} = \boldsymbol{0}$ on the symmetric planes. The field is excited by an imposed current density $\boldsymbol{J}^{\text{imp}} = 10$ in the region $0 \le x \le 0.5, \le y \le 0.5$ and zero elsewhere. The iron is assumed to be isotropic, having a relative permeability $\mu = 1000$, the relative permeability of the surrounding material is set to 1. The domain is divided into 400 uniform bilinear elements. As in the electrostatic case, the nodes at the interface are given two numbers, and the least-squares method is employed to enforce the interface condition. One-point Gaussian quadrature is used to evaluate the element matrices. The computed magnetic field vectors are depicted in Fig. 14.6.

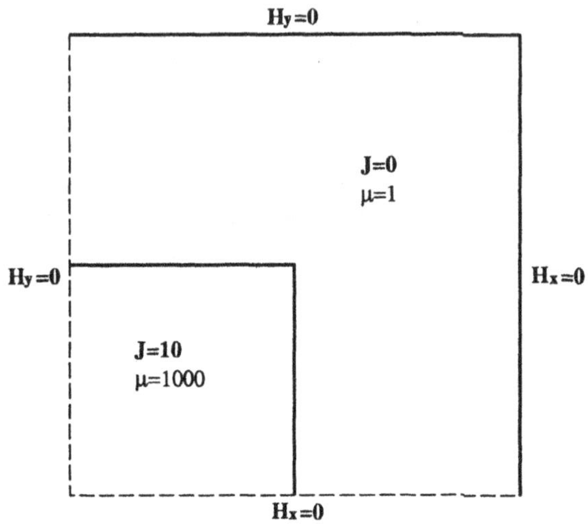

Fig. 14.5. Bus-bar

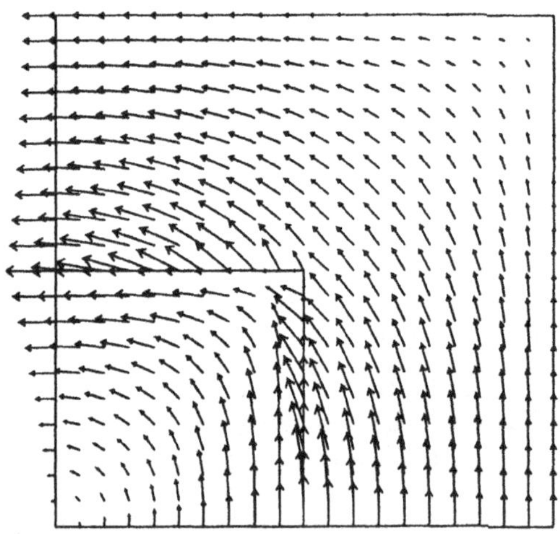

Fig. 14.6. The computed $\boldsymbol{H}$ field of a bus-bar

14.5 Time-Harmonic Fields

14.5.1 Three-Dimensional Time-Harmonic Waves

For three-dimensional time-harmonic fields, the first-order full Maxwell equations can be written as

$$\nabla \times \boldsymbol{E} + j\omega\mu\boldsymbol{H} = -\boldsymbol{K}^{\text{imp}} \qquad \text{in } \Omega, \tag{14.58a}$$

$$\nabla \times \boldsymbol{H} - j\omega\varepsilon\boldsymbol{E} = \boldsymbol{J}^{\text{imp}} \qquad \text{in } \Omega, \tag{14.58b}$$

$$\nabla \cdot \boldsymbol{E} = 0 \qquad \text{in } \Omega, \tag{14.58c}$$

$$\nabla \cdot \boldsymbol{H} = 0 \qquad \text{in } \Omega, \tag{14.58d}$$

where the time factor $e^{j\omega t}$ is used and suppressed, ω is the given angular frequency and not equal to the resonant frequencies of this problem, $\boldsymbol{E}$ and $\boldsymbol{H}$ are the complex electric and magnetic field intensities respectively, $\boldsymbol{J}^{\text{imp}}$ and $\boldsymbol{K}^{\text{imp}}$ are imposed harmonic sources of electric and magnetic current density respectively. For simplicity, ϵ and μ are constant scalars for homogeneous isotropic media. The field equations are supplemented by the homogeneous boundary conditions:

$$\boldsymbol{n} \times \boldsymbol{E} = \boldsymbol{0} \qquad \text{on } \Gamma_1, \tag{14.59a}$$

$$\boldsymbol{n} \cdot \boldsymbol{H} = 0 \qquad \text{on } \Gamma_1, \tag{14.59b}$$

$$\boldsymbol{n} \times \boldsymbol{H} = \boldsymbol{0} \qquad \text{on } \Gamma_2, \tag{14.59c}$$

$$\boldsymbol{n} \cdot \boldsymbol{E} = 0 \qquad \text{on } \Gamma_2. \tag{14.59d}$$

where Γ_1 is an electric wall, and Γ_2 is a magnetic symmetry wall.

To allow system (14.58) to have a solution, the source terms cannot be arbitrary, they must satisfy the following solvability conditions:

$$\nabla \cdot \boldsymbol{K}^{\text{imp}} = 0 \qquad \text{in } \Omega, \tag{14.60a}$$

$$\boldsymbol{n} \cdot \boldsymbol{K}^{\text{imp}} = 0 \qquad \text{on } \Gamma_1, \tag{14.60b}$$

$$\int_\Gamma \boldsymbol{n} \cdot \boldsymbol{K}^{\text{imp}} d\Gamma = 0, \tag{14.60c}$$

$$\nabla \cdot \boldsymbol{J}^{\text{imp}} = 0 \qquad \text{in } \Omega, \tag{14.60d}$$

$$\boldsymbol{n} \cdot \boldsymbol{J}^{\text{imp}} = 0 \qquad \text{on } \Gamma_2, \tag{14.60e}$$

$$\int_\Gamma \boldsymbol{n} \cdot \boldsymbol{J}^{\text{imp}} d\Gamma = 0. \tag{14.60f}$$

As in the time-varying cases, the solvability conditions (14.60a), (14.60b), (14.60d) and (14.60e) can be derived by applying the div–curl method to the curl equations (14.58a) and (14.58b), respectively. The solvability conditions (14.60c) and (14.60f) can be obtained by applying the Gauss divergence theorem to (14.60a) and (14.60d), respectively.

Separating the real and imaginary parts in system (14.58) leads to

$$\nabla \times \boldsymbol{E}_r - \omega\mu \boldsymbol{H}_i = -\boldsymbol{K}_r^{\text{imp}} \qquad \text{in } \Omega, \tag{14.61a}$$

$$\nabla \times \boldsymbol{E}_i + \omega\mu \boldsymbol{H}_r = -\boldsymbol{K}_i^{\text{imp}} \qquad \text{in } \Omega, \tag{14.61b}$$

$$\nabla \times \boldsymbol{H}_r + \omega\varepsilon \boldsymbol{E}_i = \boldsymbol{J}_r^{\text{imp}} \qquad \text{in } \Omega, \tag{14.61c}$$

$$\nabla \times \boldsymbol{H}_i - \omega\varepsilon \boldsymbol{E}_r = \boldsymbol{J}_i^{\text{imp}} \qquad \text{in } \Omega, \tag{14.61d}$$

$$\nabla \cdot \boldsymbol{E}_r = 0 \quad \text{in } \Omega, \tag{14.61e}$$

$$\nabla \cdot \boldsymbol{E}_i = 0 \quad \text{in } \Omega, \tag{14.61f}$$

$$\nabla \cdot \boldsymbol{H}_r = 0 \quad \text{in } \Omega, \tag{14.61g}$$

$$\nabla \cdot \boldsymbol{H}_i = 0 \quad \text{in } \Omega. \tag{14.61h}$$

Obviously, system (14.61) is elliptic, since its principle part consists of four div–curl systems. For the solution of (14.61) the least-squares variational formulation is: Find $\boldsymbol{u} = (\boldsymbol{E}_r, \boldsymbol{E}_i, \boldsymbol{H}_r, \boldsymbol{H}_i) \in \boldsymbol{\mathcal{H}}$ such that

$$B(\boldsymbol{u}, \boldsymbol{v}) = L(\boldsymbol{v}) \quad \forall \boldsymbol{v} = (\boldsymbol{E}_r^*, \boldsymbol{E}_i^*, \boldsymbol{H}_r^*, \boldsymbol{H}_i^*) \in \boldsymbol{\mathcal{H}}, \tag{14.62}$$

where $\boldsymbol{\mathcal{H}} = \{\boldsymbol{u} \in [H^1(\Omega)]^3 \times [H^1(\Omega)]^3 \times [H^1(\Omega)]^3 \times [H^1(\Omega)]^3 : \boldsymbol{n} \times \boldsymbol{E} = \boldsymbol{0} \text{ on } \Gamma_1, \boldsymbol{n} \cdot \boldsymbol{H} = 0 \text{ on } \Gamma_1, \boldsymbol{n} \times \boldsymbol{H} = \boldsymbol{0} \text{ on } \Gamma_2, \boldsymbol{n} \cdot \boldsymbol{E} = 0 \text{ on } \Gamma_2\}$, and $B(\cdot,\cdot)$ is the bilinear form

$$\begin{aligned}
B(\boldsymbol{u}, \boldsymbol{v}) \;=\;& (\nabla \times \boldsymbol{E}_r - \omega\mu \boldsymbol{H}_i, \nabla \times \boldsymbol{E}_r^* - \omega\mu \boldsymbol{H}_i^*) \\
+\;& (\nabla \times \boldsymbol{E}_i + \omega\mu \boldsymbol{H}_r, \nabla \times \boldsymbol{E}_i^* + \omega\mu \boldsymbol{H}_r^*) \\
+\;& (\nabla \times \boldsymbol{H}_r + \omega\varepsilon \boldsymbol{E}_i, \nabla \times \boldsymbol{H}_r^* + \omega\varepsilon \boldsymbol{E}_i^*) \\
+\;& (\nabla \times \boldsymbol{H}_i - \omega\varepsilon \boldsymbol{E}_r, \nabla \times \boldsymbol{H}_i^* - \omega\varepsilon \boldsymbol{E}_r^*) \\
+\;& (\nabla \cdot \boldsymbol{E}_r, \nabla \cdot \boldsymbol{E}_r^*) + (\nabla \cdot \boldsymbol{E}_i, \nabla \cdot \boldsymbol{E}_i^*) \\
+\;& (\nabla \cdot \boldsymbol{H}_r, \nabla \cdot \boldsymbol{H}_r^*) + (\nabla \cdot \boldsymbol{H}_i, \nabla \cdot \boldsymbol{H}_i^*),
\end{aligned} \tag{14.63a}$$

and $L(\cdot)$ is the linear form

$$\begin{aligned}
L(\boldsymbol{v}) \;=\;& (-K_r^{\text{imp}}, \nabla \times \boldsymbol{E}_r^* - \omega\mu \boldsymbol{H}_i^*) \\
+\;& (-K_i^{\text{imp}}, \nabla \times \boldsymbol{E}_i^* + \omega\mu \boldsymbol{H}_r^*) \\
+\;& (J_r^{\text{imp}}, \nabla \times \boldsymbol{H}_r^* + \omega\varepsilon \boldsymbol{E}_i^*) \\
+\;& (J_i^{\text{imp}}, \nabla \times \boldsymbol{H}_i^* - \omega\varepsilon \boldsymbol{E}_r^*).
\end{aligned} \tag{14.63b}$$

14.5.2 Time-Harmonic TE Waves

For time-harmonic TE waves the first-order Maxwell equations are

$$j\omega\mu H_z + \frac{\partial E_y}{\partial x} - \frac{\partial E_x}{\partial y} = 0 \quad \text{in } \Omega, \tag{14.64a}$$

$$j\omega\varepsilon^* E_x - \frac{\partial H_z}{\partial y} = 0 \quad \text{in } \Omega, \tag{14.64b}$$

$$j\omega\varepsilon^* E_y + \frac{\partial H_z}{\partial x} = 0 \quad \text{in } \Omega, \tag{14.64c}$$

$$\frac{\partial E_x}{\partial x} + \frac{\partial E_y}{\partial y} = 0 \quad \text{in } \Omega, \tag{14.64d}$$

in which $\varepsilon^* = \varepsilon_r + j\varepsilon_i = \varepsilon - j\sigma/\omega$ is the complex permittivity. For a complete description of TE wave problems, appropriate boundary conditions should be included. One may consider, for example,

$$H_z = \text{const} \qquad \text{on } \Gamma, \tag{14.64e}$$

$$E_x n_x + E_y n_y = 0 \qquad \text{on } \Gamma. \tag{14.64f}$$

Condition (14.64e) is an inhomogeneous version corresponding to (14.59c), and (14.64f) is a 2D version of (14.59d). We also note that the boundary conditions (14.64e) and (14.64f) satisfy the boundary compatibility condition

$$j\omega\varepsilon^*(E_x n_x + E_y n_y) = \frac{\partial H_z}{\partial y} n_x - \frac{\partial H_z}{\partial x} n_y \qquad \text{on } \Gamma, \tag{14.65}$$

which is obtained by taking the operation $\boldsymbol{n}\cdot$ to the equations (14.64b) and (14.64c).

In system (14.64) there are three unknowns and four equations, and thus the divergence-free condition (14.64d) seems "redundant". By introducing a dummy variable ϑ into system (14.64), we have

$$j\omega\mu H_z + \frac{\partial E_y}{\partial x} - \frac{\partial E_x}{\partial y} = 0 \qquad \text{in } \Omega, \tag{14.66a}$$

$$j\omega\varepsilon^* E_x + \frac{\partial \vartheta}{\partial x} - \frac{\partial H_z}{\partial y} = 0 \qquad \text{in } \Omega, \tag{14.66b}$$

$$j\omega\varepsilon^* E_y + \frac{\partial \vartheta}{\partial y} + \frac{\partial H_z}{\partial x} = 0 \qquad \text{in } \Omega, \tag{14.66c}$$

$$\frac{\partial E_x}{\partial x} + \frac{\partial E_y}{\partial y} = 0 \qquad \text{in } \Omega. \tag{14.66d}$$

By applying the operation $\partial/\partial x$ to the equation (14.66b) and the operation $\partial/\partial y$ to the equation (14.66c), and by adding the results together we obtain the Laplace equation for ϑ:

$$\frac{\partial^2 \vartheta}{\partial x^2} + \frac{\partial^2 \vartheta}{\partial y^2} = 0 \qquad \text{in } \Omega. \tag{14.67a}$$

By taking the operation $\boldsymbol{n}\cdot$ to (14.66b) and (14.66c) and using the boundary compatibility condition (14.65) we obtain

$$\frac{\partial \vartheta}{\partial n} = 0 \qquad \text{on } \Gamma. \tag{14.67b}$$

From (14.67) we know that $\vartheta = \text{const}$, that is, system (14.64) is completely equivalent to the augmented system (14.66) with four equations involving four unknowns. Since system (14.66) consists of two two-dimensional div–curl systems, and thus is elliptic. Therefore, system (14.64) is not 'overdetermined' but indeed properly determined and elliptic.

For numerical calculation, separating the real and imaginary parts in (14.64a)–(14.64d) leads to

$$-\omega\mu H_{zi} + \frac{\partial E_{yr}}{\partial x} - \frac{\partial E_{xr}}{\partial y} = 0 \qquad \text{in } \Omega, \tag{14.68a}$$

$$-\omega(\varepsilon_r E_{xi} + \varepsilon_i E_{xr}) - \frac{\partial H_{zr}}{\partial y} = 0 \qquad \text{in } \Omega, \tag{14.68b}$$

$$-\omega(\varepsilon_r E_{yi} + \varepsilon_i E_{yr}) + \frac{\partial H_{zr}}{\partial x} = 0 \qquad \text{in } \Omega, \tag{14.68c}$$

$$\frac{\partial E_{xr}}{\partial x} + \frac{\partial E_{yr}}{\partial y} = 0 \qquad \text{in } \Omega, \tag{14.68d}$$

$$\omega\mu H_{zr} + \frac{\partial E_{yi}}{\partial x} - \frac{\partial E_{xi}}{\partial y} = 0 \qquad \text{in } \Omega, \tag{14.68e}$$

$$\omega(\varepsilon_r E_{xr} - \varepsilon_i E_{xi}) - \frac{\partial H_{zi}}{\partial y} = 0 \qquad \text{in } \Omega, \tag{14.68f}$$

$$\omega(\varepsilon_r E_{yr} - \varepsilon_i E_{yi}) + \frac{\partial H_{zi}}{\partial x} = 0 \qquad \text{in } \Omega, \tag{14.68g}$$

$$\frac{\partial E_{xi}}{\partial x} + \frac{\partial E_{yi}}{\partial y} = 0 \qquad \text{in } \Omega. \tag{14.68h}$$

Of course, in (14.68) the medium properties are different for different medium regions.

We may write system (14.68) in standard matrix form:

$$\boldsymbol{A}_1 \frac{\partial \boldsymbol{u}}{\partial x} + \boldsymbol{A}_2 \frac{\partial \boldsymbol{u}}{\partial y} + \boldsymbol{A}_0 \boldsymbol{u} = \boldsymbol{f}, \tag{14.69}$$

in which

$$\boldsymbol{A}_1 = \begin{pmatrix} 0 & 0 & 1 & 0 & 0 & 0 \\ 0 & 0 & 0 & 0 & 0 & 0 \\ 1 & 0 & 0 & 0 & 0 & 0 \\ 0 & 1 & 0 & 0 & 0 & 0 \\ 0 & 0 & 0 & 0 & 0 & 1 \\ 0 & 0 & 0 & 0 & 0 & 0 \\ 0 & 0 & 0 & 1 & 0 & 0 \\ 0 & 0 & 0 & 0 & 1 & 0 \end{pmatrix}, \tag{14.70a}$$

$$\boldsymbol{A}_2 = \begin{pmatrix} 0 & -1 & 0 & 0 & 0 & 0 \\ -1 & 0 & 0 & 0 & 0 & 0 \\ 0 & 0 & 0 & 0 & 0 & 0 \\ 0 & 0 & 1 & 0 & 0 & 0 \\ 0 & 0 & 0 & 0 & -1 & 0 \\ 0 & 0 & 0 & -1 & 0 & 0 \\ 0 & 0 & 0 & 0 & 0 & 0 \\ 0 & 0 & 0 & 0 & 0 & 1 \end{pmatrix}, \tag{14.70b}$$

$$A_0 = \begin{pmatrix} 0 & 0 & 0 & -\omega\mu & 0 & 0 \\ 0 & -\omega\epsilon_i & 0 & 0 & -\omega\epsilon_r & 0 \\ 0 & 0 & -\omega\epsilon_i & 0 & 0 & -\omega\epsilon_r \\ 0 & 0 & 0 & 0 & 0 & 0 \\ \omega\mu & 0 & 0 & 0 & 0 & 0 \\ 0 & \omega\epsilon_r & 0 & 0 & -\omega\epsilon_i & 0 \\ 0 & 0 & \omega\epsilon_r & 0 & 0 & -\omega\epsilon_i \\ 0 & 0 & 0 & 0 & 0 & 0 \end{pmatrix}, \tag{14.70c}$$

$$f = \begin{pmatrix} 0 \\ 0 \\ 0 \\ 0 \\ 0 \\ 0 \end{pmatrix}, \quad u = \begin{pmatrix} H_{zr} \\ E_{xr} \\ E_{yr} \\ H_{zi} \\ E_{xi} \\ E_{yi} \end{pmatrix}. \tag{14.70d}$$

At the interface Γ_{int} between two contiguous media (+) and (−) the following general conditions should be satisfied:

$$n \times E^+ = n \times E^- \quad \text{on } \Gamma_{\text{int}}, \tag{14.71a}$$

$$n \times H^+ = n \times H^- \quad \text{on } \Gamma_{\text{int}}, \tag{14.71b}$$

$$n \cdot (\epsilon^{*+} E^+) = n \cdot (\epsilon^{*-} E^-) \quad \text{on } \Gamma_{\text{int}}, \tag{14.71c}$$

$$n \cdot (\mu^+ H^+) = n \cdot (\mu^- H^-) \quad \text{on } \Gamma_{\text{int}}. \tag{14.71d}$$

For two-dimensional TE waves, the interface conditions (14.71a) and (14.71c) become

$$n_x E_{yr}^+ - n_y E_{xr}^+ = n_x E_{yr}^- - n_y E_{xr}^-, \tag{14.72a}$$

$$n_x E_{yi}^+ - n_y E_{xi}^+ = n_x E_{yi}^- - n_y E_{xi}^-, \tag{14.72b}$$

$$n_x(\epsilon_r^+ E_{xr}^+ - \epsilon_i^+ E_{xi}^+) + n_y(\epsilon_r^+ E_{yr}^+ - \epsilon_i^+ E_{yi}^+)$$
$$= n_x(\epsilon_r^- E_{xr}^- - \epsilon_i^- E_{xi}^-) + n_y(\epsilon_r^- E_{yr}^- - \epsilon_i^- E_{yi}^-), \tag{14.72c}$$

$$n_x(\epsilon_i^+ E_{xr}^+ + \epsilon_r^+ E_{xi}^+) + n_y(\epsilon_i^+ E_{yr}^+ + \epsilon_r^+ E_{yi}^+)$$
$$= n_x(\epsilon_i^- E_{xr}^- + \epsilon_r^- E_{xi}^-) + n_y(\epsilon_i^- E_{yr}^- + \epsilon_r^- E_{yi}^-). \tag{14.72d}$$

For two-dimensional TE waves, the interface condition (14.71d) is automatically satisfied, and (14.71b) becomes

$$H_z^+ = H_z^-. \tag{14.72e}$$

In the LSFEM the treatment of the interface conditions is not difficult. As in other node-based finite element methods, the nodes on the interface should be double-numbered. If a direct solver is employed for solving the discretized system, two approaches are available: a simple way is to include the interface conditions into the least-squares functional as discussed in Sect. 14.3.2; a better way is to use the conditions (14.72a)–(14.72e) to modify the global stiffness matrix in the discretized system. If the conjugate gradient method is used, one just simply chooses the unknowns related to the medium (+) (or (−)) as the true unknowns and keeps the conditions (14.72a)–(14.72e) satisfied for each solution vector.

Since the general formulation of LSFEM has been given in Sect. 4.8, it is not necessary to write down the special one for the problem discussed in this section. One only needs to substitute the coefficients of (14.69) and the boundary conditions into a general-purpose LSFEM code.

We consider two test problems that are taken from Paulsen and Lynch (1991) (see also Jin 1993, pp. 167-170), in which the spurious solutions given by the curl–curl formulation as well as the correct solutions are illustrated.

Split Cylinder. The first example is a cylinder (R=25) which is split into two regions having different complex permittivity. For the top region, $\varepsilon_r^+ = 3.0$, $\varepsilon_i^+ = -5.0$ and $\mu^+ = 1.0$; for the bottom region, $\varepsilon_r^- = 1.0$, $\varepsilon_i^- = 0.0$ and $\mu^- = 1.0$. This cylinder is excited by a uniform $H_z|_\Gamma = (1,0)$ (with $\omega = 0.05$) imposed on the outer boundary. All variables in this problem are discretized by 932 bilinear elements with 1016 nodes shown in Fig. 14.7a. One-point Gaussian quadrature is used for evaluating the element matrices. The contours of the computed real and imaginary magnetic field intensity are shown in Fig. 14.7b and c, respectively. The vector plots of the real and imaginary electric field intensity are illustrated in Fig. 14.7d and e, respectively.

Off-center Cylinder. The second example ia a smaller off-center cylinder (R=0.1) embedded in a larger cylinder (R=0.25). The material properties for the outer region are $\varepsilon_r^+ = 0.0981$, $\varepsilon_i^+ = -0.0196$ and $\mu^+ = 1.0$; for the inner region $\varepsilon_r^- = 1.0$, $\varepsilon_i^- = 0.0$ and $\mu^- = 1.0$. A uniform $H_z|_\Gamma = (1, -0.15)$ (with $\omega = 44.7$) is imposed on the outer boundary. Figure 14.8a shows the mesh with 2027 bilinear elements and 2165 nodes. The contours of the computed real and imaginary magnetic field intensity are shown in Fig. 14.8b and c, respectively. The vector plots of the real and imaginary electric field intensity are illustrated in Fig. 14.8d and e, respectively.

As expected all computed results by LSFEM are free of spurious modes.

14.6 Transient Scattering Waves

In this section we consider the computation of scattered waves due to the existence of perfectly conducting bodies exposed to plane incident waves in free space as an example to illustrate how to use LSFEM to solve time-dependent problems and to show the difference between the solutions with and without enforcement of the divergence-free condition.

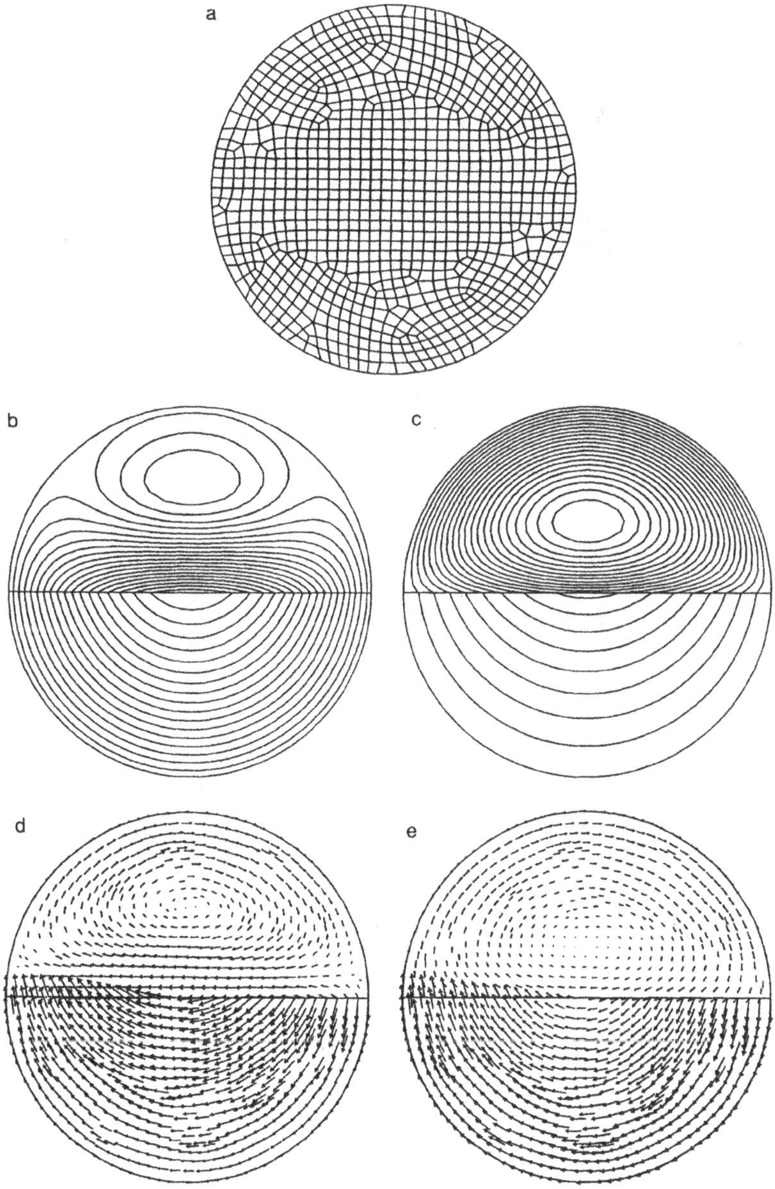

Fig. 14.7. (**a**) The split cylinder and the mesh, (**b**) Contours of constant magnetic field intensity H_r, (**c**) Contours of constant magnetic field intensity H_i, (**d**) Vectors of the computed electric field intensity $\boldsymbol{E}_r$, (**e**) Vectors of the computed electric field intensity $\boldsymbol{E}_i$ (from Jiang et al. 1996)

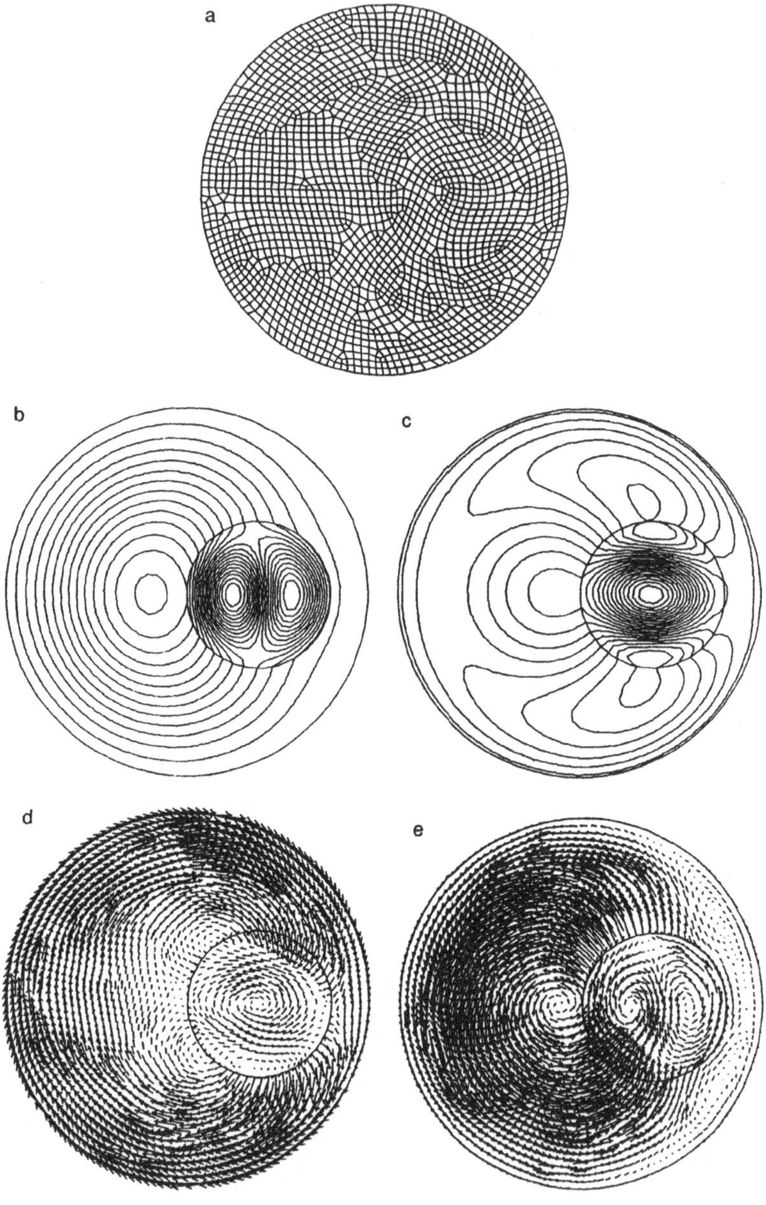

Fig. 14.8. (**a**) The off-center cylinder, (**b**) Contours of constant magnetic field intensity H_r, (**c**) Contours of constant magnetic field intensity H_i, (**d**) Vectors of the computed electric field intensity $\boldsymbol{E}_r$, (**e**) Vectors of the computed electric field intensity $\boldsymbol{E}_i$ (from Jiang et al. 1996)

14.6.1 TM and TE Waves

For two-dimensional transient transverse magnetic (TM) waves, the first-order Maxwell equations can be written as

$$\frac{\partial E_z}{\partial y} + \mu \frac{\partial H_x}{\partial t} = 0 \qquad \text{in } \Omega, \tag{14.73a}$$

$$-\frac{\partial E_z}{\partial x} + \mu \frac{\partial H_y}{\partial t} = 0 \qquad \text{in } \Omega, \tag{14.73b}$$

$$-\frac{\partial H_y}{\partial x} + \frac{\partial H_x}{\partial y} + \varepsilon \frac{\partial E_z}{\partial t} = 0 \qquad \text{in } \Omega, \tag{14.73c}$$

$$\frac{\partial H_x}{\partial x} + \frac{\partial H_y}{\partial y} = 0 \qquad \text{in } \Omega. \tag{14.73d}$$

The divergence-free condition (14.73d) is often ignored in usual time-domain computation.

The field intensity variables are conveniently split into those of the incident field and of the scattered field:

$$E_z = E_z^i + E_z^s, \tag{14.74a}$$

$$\boldsymbol{H} = \boldsymbol{H}^i + \boldsymbol{H}^s, \tag{14.74b}$$

where the superscript "i" denotes the incident field, and "s" denotes the scattered field. The incident waves are allowed to propagate in their analytical form. The scattered waves are determined by solving the Maxwell equations with appropriate boundary conditions on the surface of the scatterer and on the far-field.

On the surface of a perfectly conducting body, the tangential components of the electric intensity and the normal component of the magnetic intensity are zero, i.e., the boundary conditions (14.3a) and (14.3b) should be satisfied. When the TM scattered formulation is considered, they become:

$$E_z^s = -E_z^i \qquad \text{on } \Gamma_1, \tag{14.75a}$$

$$n_x H_x^s + n_y H_y^s = -(n_x H_x^i + n_y H_y^i) \qquad \text{on } \Gamma_1. \tag{14.75b}$$

Since the Maxwell equations are now solved in a finite domain, a proper far-field boundary condition has to be applied to ensure that waves will not artificially reflect back and contaminate the near-field solution. Here we adopt the following absorbing boundary condition:

$$\frac{1}{c}\frac{\partial \phi}{\partial t} + \frac{\phi}{2\rho} + \frac{\partial \phi}{\partial n} = 0 \qquad \text{on } \Gamma_\infty, \tag{14.76}$$

where $c = 1/\sqrt{\varepsilon\mu}$ is the wave velocity, $\boldsymbol{n}$ is the outward normal unit vector to the far-field boundary, ρ is the radius of the chosen far-field boundary, and ϕ is any component of E_z^s, H_x^s and H_y^s. After the time derivative is discretized by finite differences, this absorbing boundary condition can be easily enforced on the far-field boundary by adding the corresponding squared terms into the variational functional.

In electromagnetic scattering problems normally only the solution at the near-field is of interest. Indeed radar cross section (RCS) calculations require only the field intensities on the surface of the scatterer. Because the absorbing boundary condition (14.76) is only an approximation of the true absorbing boundary condition, the computational domain needs to be larger than those using higher-order absorbing conditions. However, at the far-field it is only required that the waves propagate out and do not reflect back to contaminate the near field solutions, the accuracy of the solution there is not of interest. For this reason it is most efficient to refine the mesh locally near the surface of the scatterer and use a coarse mesh in the far-field. The use of such non-uniform grids is easily accommodated in the LSFEM, since the calculation is carried out on totally unstructured meshes. Due to the use of non-uniform grid and large elements near the far-field boundary, the elements placed in the part of the computational domain far away from the scatterer surface contributes only to a small proportion of the total element number. Thus the present approach still provides a very good approximation of the field variables while maintaining the efficiency. We remark that this refinement strategy is not new. Ambrosiano et al. (1994) and Vinh et al. (1992) had used grids with large size variation and got satisfactory RCS. We also want to emphasis that this concept of local refinement differs from that based on uniform distribution of error, which is widely used in solid mechanics and other fields.

Similarly, the two-dimensional transient transverse electric (TE) waves should satisfy the Maxwell equations:

$$-\frac{\partial H_z}{\partial y} + \varepsilon \frac{\partial E_x}{\partial t} = 0 \qquad \text{in } \Omega, \tag{14.77a}$$

$$\frac{\partial H_z}{\partial x} + \varepsilon \frac{\partial E_y}{\partial t} = 0 \qquad \text{in } \Omega, \tag{14.77b}$$

$$\frac{\partial E_y}{\partial x} - \frac{\partial E_x}{\partial y} + \mu \frac{\partial H_z}{\partial t} = 0 \qquad \text{in } \Omega, \tag{14.77c}$$

$$\frac{\partial E_x}{\partial x} + \frac{\partial E_y}{\partial y} = 0 \qquad \text{in } \Omega. \tag{14.77d}$$

and the boundary conditions:

$$H_z^s = -H_z^i \qquad \text{on } \Gamma_1, \tag{14.78a}$$

$$n_x E_x^s + n_y E_y^s = -(n_x E_x^i + n_y E_y^i) \qquad \text{on } \Gamma_1. \tag{14.78b}$$

14.6.2 Time-Discretization

Equations (14.73) or (14.77) are first discretized in time by using the in Crank–Nicolson scheme which provides second-order accuracy in time. Although this scheme is unconditionally stable and thus poses no limitation on the time-step, the error becomes large if a very large time-step is used. Thus care should be taken when choosing the size of the time-step. To maintain the accuracy in the time-domain, we typically use a time-step corresponding to a CFL number between 1 and 2 for the smallest element.

After applying the time discretization to (14.73) and (14.77), we have, respectively,

$$\frac{1}{2}\frac{\partial E_z^{n+1}}{\partial y} + \frac{1}{2}\frac{\partial E_z^n}{\partial y} + \mu\frac{H_x^{n+1} - H_x^n}{\Delta t} = 0, \tag{14.79a}$$

$$-\frac{1}{2}\frac{\partial E_z^{n+1}}{\partial x} - \frac{1}{2}\frac{\partial E_z^n}{\partial x} + \mu\frac{H_y^{n+1} - H_y^n}{\Delta t} = 0, \tag{14.79b}$$

$$\frac{1}{2}\Big(-\frac{\partial H_y^{n+1}}{\partial x} + \frac{\partial H_x^{n+1}}{\partial y}\Big) + \frac{1}{2}\Big(-\frac{\partial H_y^n}{\partial x} + \frac{\partial H_x^n}{\partial y}\Big) + \varepsilon\frac{E_z^{n+1} - E_z^n}{\Delta t} = 0, \tag{14.79c}$$

$$\frac{\partial H_x^{n+1}}{\partial x} + \frac{\partial H_y^{n+1}}{\partial y} = 0, \tag{14.79d}$$

and

$$-\frac{1}{2}\frac{\partial H_z^{n+1}}{\partial y} - \frac{1}{2}\frac{\partial H_z^n}{\partial y} + \varepsilon\frac{E_x^{n+1} - E_x^n}{\Delta t} = 0, \tag{14.80a}$$

$$\frac{1}{2}\frac{\partial H_z^{n+1}}{\partial x} + \frac{1}{2}\frac{\partial H_z^n}{\partial x} + \varepsilon\frac{E_y^{n+1} - E_y^n}{\Delta t} = 0, \tag{14.80b}$$

$$\frac{1}{2}\Big(\frac{\partial E_y^{n+1}}{\partial x} - \frac{\partial E_x^{n+1}}{\partial y}\Big) + \frac{1}{2}\Big(\frac{\partial E_y^n}{\partial x} - \frac{\partial E_x^n}{\partial y}\Big) + \mu\frac{H_z^{n+1} - H_z^n}{\Delta t} = 0, \tag{14.80c}$$

$$\frac{\partial E_x^{n+1}}{\partial x} + \frac{\partial E_y^{n+1}}{\partial y} = 0, \tag{14.80d}$$

where the superscript n denotes the nth time-step. Equations (14.79) and (14.80) can be rewritten in a compact matrix form as:

$$\boldsymbol{A}_1\frac{\partial \boldsymbol{u}}{\partial x} + \boldsymbol{A}_2\frac{\partial \boldsymbol{u}}{\partial y} + \boldsymbol{A}_0\boldsymbol{u} = \boldsymbol{f}, \tag{14.81}$$

where for TM waves:

$$\boldsymbol{u} = \begin{pmatrix} E_z \\ H_x \\ H_y \end{pmatrix}^{n+1}, \quad \boldsymbol{f} = \begin{pmatrix} -\frac{1}{2}\frac{\partial E_z}{\partial y} + \mu\frac{H_x}{\Delta t} \\ \frac{1}{2}\frac{\partial E_z}{\partial x} + \mu\frac{H_y}{\Delta t} \\ \frac{1}{2}\left(\frac{\partial H_y}{\partial x} - \frac{\partial H_x}{\partial y}\right) + \varepsilon\frac{E_z}{\Delta t} \\ 0 \end{pmatrix}^{n},$$

$$\boldsymbol{A}_1 = \begin{pmatrix} 0 & 0 & 0 \\ -\frac{1}{2} & 0 & 0 \\ 0 & 0 & -\frac{1}{2} \\ 0 & 1 & 0 \end{pmatrix}, \quad \boldsymbol{A}_2 = \begin{pmatrix} \frac{1}{2} & 0 & 0 \\ 0 & 0 & 0 \\ 0 & \frac{1}{2} & 0 \\ 0 & 0 & 1 \end{pmatrix},$$

$$\boldsymbol{A}_0 = \begin{pmatrix} 0 & \mu\frac{1}{\Delta t} & 0 \\ 0 & 0 & \mu\frac{1}{\Delta t} \\ \varepsilon\frac{1}{\Delta t} & 0 & 0 \\ 0 & 0 & 0 \end{pmatrix};$$

and for TE waves:

$$\boldsymbol{u} = \begin{pmatrix} H_z \\ E_x \\ E_y \end{pmatrix}^{n+1}, \quad \boldsymbol{f} = \begin{pmatrix} \frac{1}{2}\frac{\partial H_z}{\partial y} + \varepsilon\frac{E_x}{\Delta t} \\ -\frac{1}{2}\frac{\partial H_z}{\partial x} + \varepsilon\frac{E_y}{\Delta t} \\ -\frac{1}{2}\left(\frac{\partial E_y}{\partial x} - \frac{\partial E_x}{\partial y}\right) + \mu\frac{H_z}{\Delta t} \\ 0 \end{pmatrix}^{n},$$

$$\boldsymbol{A}_1 = \begin{pmatrix} 0 & 0 & 0 \\ \frac{1}{2} & 0 & 0 \\ 0 & 0 & \frac{1}{2} \\ 0 & 1 & 0 \end{pmatrix}, \quad \boldsymbol{A}_2 = \begin{pmatrix} -\frac{1}{2} & 0 & 0 \\ 0 & 0 & 0 \\ 0 & -\frac{1}{2} & 0 \\ 0 & 0 & 1 \end{pmatrix},$$

$$\boldsymbol{A}_0 = \begin{pmatrix} 0 & \varepsilon\frac{1}{\Delta t} & 0 \\ 0 & 0 & \varepsilon\frac{1}{\Delta t} \\ \mu\frac{1}{\Delta t} & 0 & 0 \\ 0 & 0 & 0 \end{pmatrix}.$$

At this stage it becomes clear that once one has a general purpose LSFEM code based on the standard first-order system (14.81), the only required programming work is to write a subroutine to supply the matrices $\boldsymbol{A}$, $\boldsymbol{A}_i$ and $\boldsymbol{f}$, and the boundary conditions.

We use a Jacobi-preconditioned conjugate gradient method to solve the resulting algebraic equation system. By using this method, the iterations can be conducted in such a manner that there is no need to form the element matrices and to assemble the global matrix $\boldsymbol{K}$. This greatly reduces the requirement for computer memory. We have found that five conjugate gradient iterations are enough to guarantee the accuracy of the solution in our calculations. The use of such a small number of iterations can be attributed to the fact that when the time-step is small, the solution in the previous time-step serves as a very good initial guess for the next step, thus the conjugate gradient method needs fewer iterations than the case when a larger time-step is used. Therefore, although the method is implicit, its computational speed is competitive with an explicit method.

14.6.3 Numerical Examples

In the following we present several test cases of scattering of plane incident wave by perfect conducting bodies. The incident sinusoidal wave is given by:

$$E^{z,i} = E_0 \sin k(x\cos\alpha + y\sin\alpha - ct),$$

$$H^{x,i} = \frac{E^{z,i}\sin\alpha}{\mu c},$$

$$H^{y,i} = -\frac{E^{z,i}\cos\alpha}{\mu c}$$

for the TM case; and

$$H^{z,i} = H_0 \sin k(x\cos\alpha + y\sin\alpha - ct),$$

$$E^{x,i} = -\frac{H^{z,i}\sin\alpha}{\varepsilon c},$$

$$E^{y,i} = \frac{H^{z,i}\cos\alpha}{\varepsilon c}$$

for the TE case, where c is the wave velocity, k is the wave number, α is the angle of incidence.

Numerical results are presented in terms of the surface current (for TM case) and the radar cross section (RCS). The calculation of these requires the field variables in the frequency-domain, thus the Fourier transform is performed to transfer the field variables from the time-domain to the frequency-domain.

The surface current J_s for the TM case is defined as

$$J_s = \frac{|\boldsymbol{n} \times \boldsymbol{H}^t|}{|\boldsymbol{H}^i|},$$

where $\boldsymbol{H}^i$ and $\boldsymbol{H}^t$ are the incident and total magnetic fields in the frequency-domain, respectively.

The RCS for the TM wave, S_{TM}, is defined as:

$$S_{TM}(\theta) = \lim_{\rho\to\infty} 2\pi\rho \left|\frac{E_z^s(\rho,\theta)}{E_z^i}\right|^2,$$

where E_z^s and E_z^i are the scattered and incident electric fields, respectively, in the frequency-domain, ρ is the distance from the center of the scattering object, and θ is the angle of observation. Note that the calculation of S_{TM} from the above expression requires the knowledge of E_z^s at an infinite distance from the scatterer. This is not available, since the Maxwell equations are solved in a finite domain. However, the far-field solution can be obtained from the distribution on a near-field surface (the most convenient one being the surface of the scatterer) by a transformation using Green's function, as described by Shankar et al. (1989). This transformation is adopted to obtain the RCS.

For TE wave, the RCS S_{TE} is defined as:

$$S_{TE}(\theta) = \lim_{\rho \to \infty} 2\pi\rho \left| \frac{H_z^s(\rho, \theta)}{H_z^i} \right|^2 .$$

The calculation of S_{TE} follows the same procedure as in the TM case.

Circular Cylinder. The first example deals with a scattering field due to a perfect conducting circular cylinder. Both TM and TE cases at various wavelength are considered. The definition of the problem is given in Fig. 14.9. This problem allows an analytical solution in a series form, see, e.g., Umashankar and Taflove (1993). Due to the symmetry of this problem, only half of the domain is analyzed. The computational mesh shown in Fig. 14.10 consists of 2435 bilinear elements and 2538 nodes, and 142 divisions on the cylinder surface. Figures 14.11 and 14.12 illustrate the calculated RCS. The agreement between the LSFEM and the analytic solutions on RCS is obvious. The distributions of the scattered field variables are shown in Fig. 14.13.

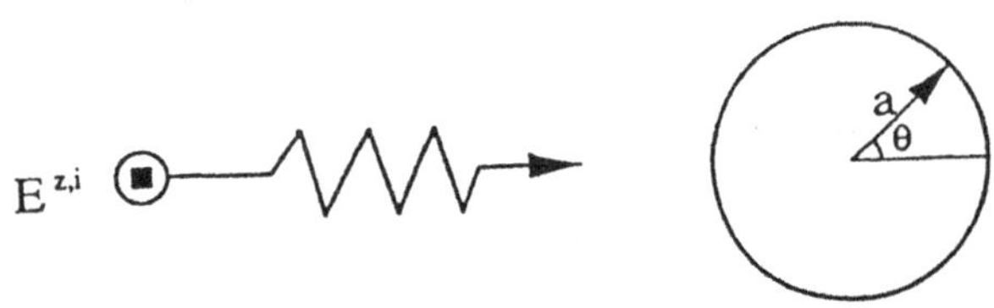

Fig. 14.9. Scattering by a circular cylinder: problem definition (from Wu and Jiang 1996)

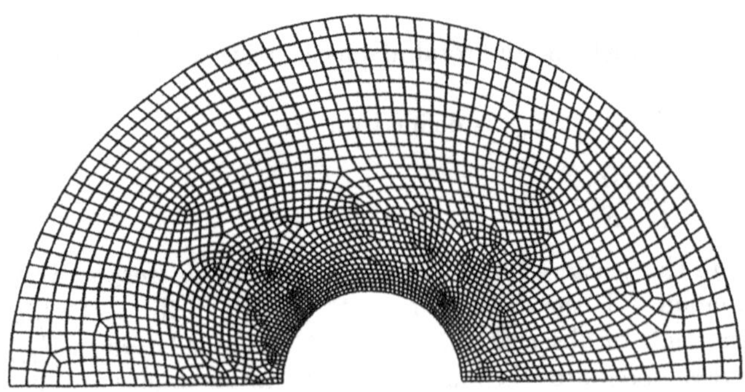

Fig. 14.10. Scattering by a circular cylinder: finite element mesh (from Wu and Jiang 1996)

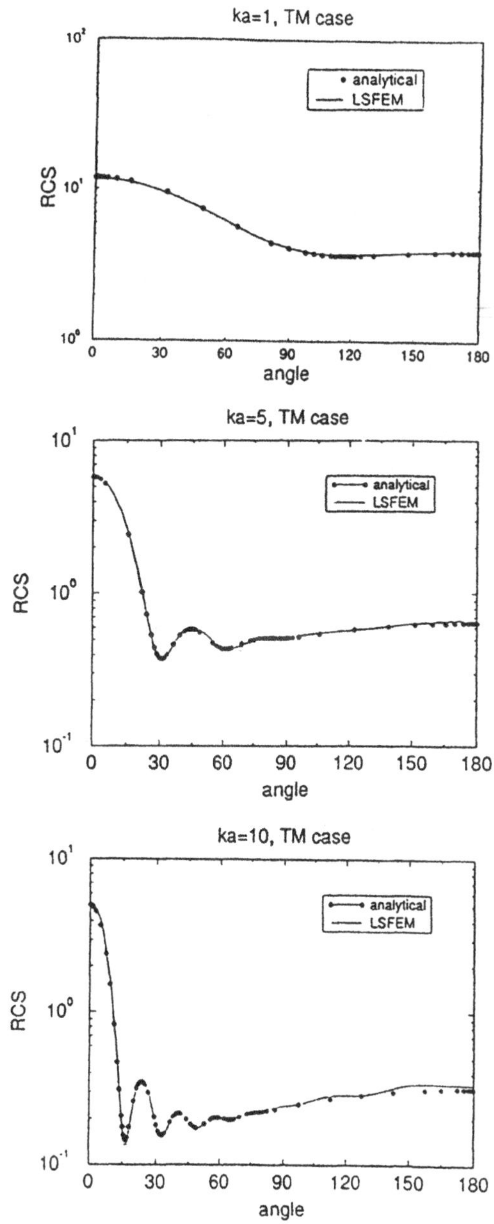

Fig. 14.11. RCS for the circular cylinder (TM mode) (from Wu and Jiang 1996)

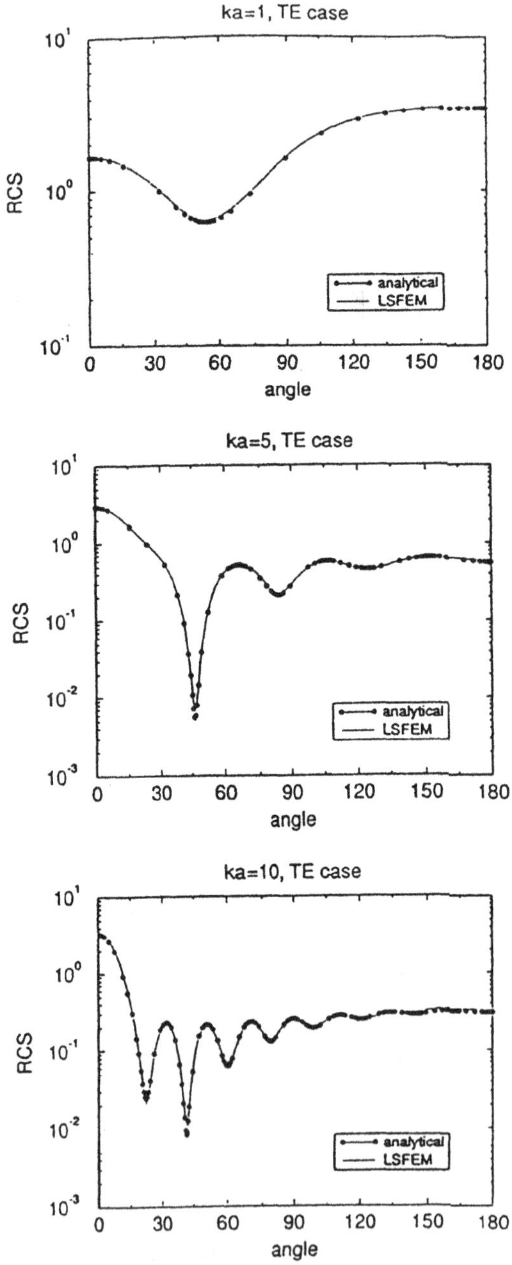

Fig. 14.12. RCS for the circular cylinder (TE mode) (from Wu and Jiang 1996)

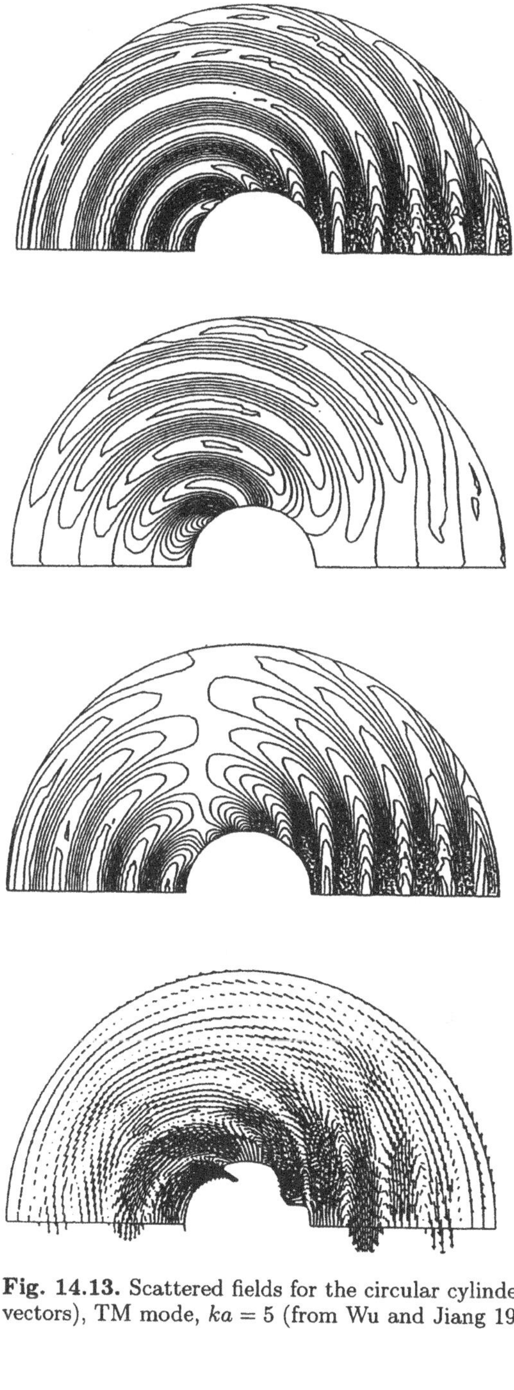

Fig. 14.13. Scattered fields for the circular cylinder (E_z, H_x, H_y contours and $\boldsymbol{H}$ vectors), TM mode, $ka = 5$ (from Wu and Jiang 1996)

Square Cylinder. The second example concerns scattering by a square cylinder defined in Fig. 14.14. The TM case with $ka = 1$ is considered. The computational domain is divided into 2749 bilinear quadrilateral elements with 2874 nodes (Fig. 14.15). Figure 14.16 shows the computed surface current distribution. Comparison is made with the results by Shankar et al. (1989) and that obtained by Umashankar and Taflove (1993) using the method of moments (MOM). The instantaneous E_z contours are shown in Fig. 14.17.

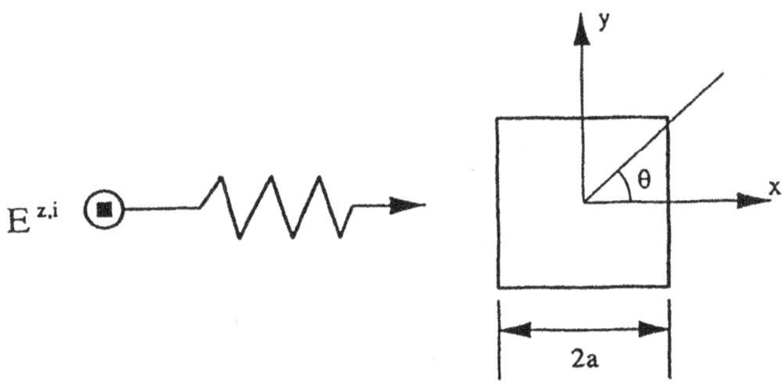

Fig. 14.14. Scattering by a square cylinder: problem definition (from Wu and Jiang 1996)

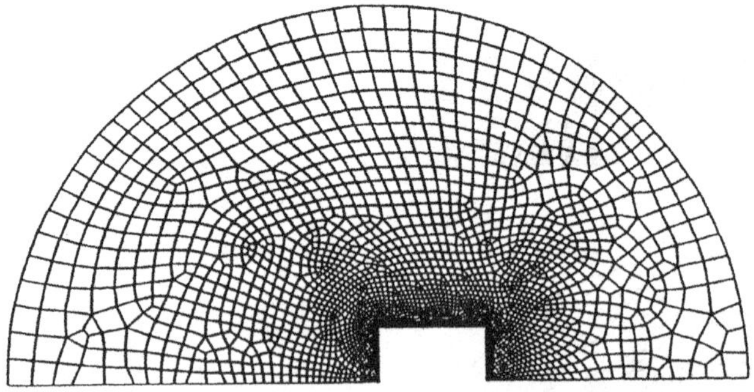

Fig. 14.15. Scattering by a square cylinder: finite element mesh (from Wu and Jiang 1996)

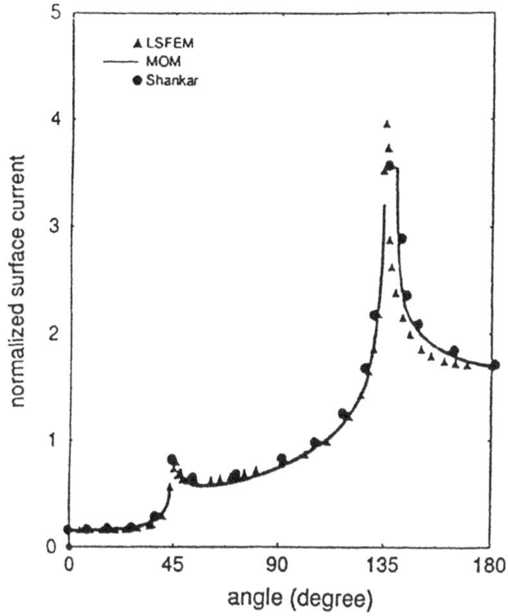

Fig. 14.16. Normalized surface current for the square cylinder (from Wu and Jiang 1996)

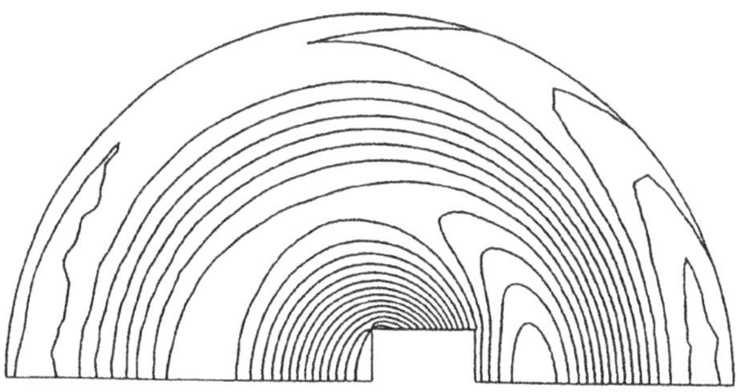

Fig. 14.17. Scattered E_z contours for the square cylinder (from Wu and Jiang 1996)

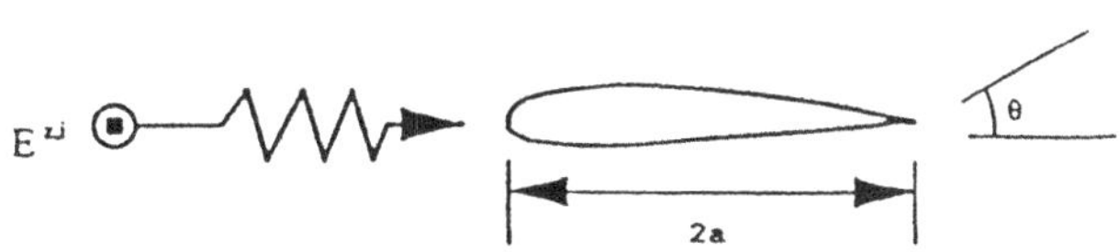

Fig. 14.18. Scattering by a NACA0012 airfoil: problem definition (from Wu and Jiang 1996)

NACA0012 Airfoil. The LSFEM is also tested for two cases of scattering TM wave by a NACA0012 airfoil defined in Fig. 14.18. The angles of incidence are 0 and 90 degrees; and the values of ka are $ka = 10$ and $ka = 10\pi$, respectively. Figure 14.19 shows the computational mesh which consists of 1782 bilinear elements and 1912 nodes. There are 156 divisions on the surface of the airfoil. Figures 14.20 and 14.21 show the computed RCS. The result in the case of 90 degree incidence is compared with those by Shankar et al. (1989) and Vinh et al. (1992). Generally, fair agreement is observed. Note that we have used a very coarse mesh and a very small domain. Finally Fig. 14.22 shows the contours of E_Z for the 0 degree incidence case.

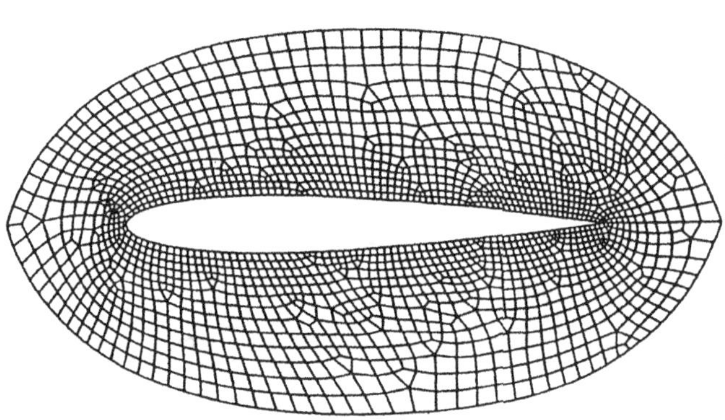

Fig. 14.19. Scattering by a NACA0012 airfoil: finite element mesh (from Wu and Jiang 1996)

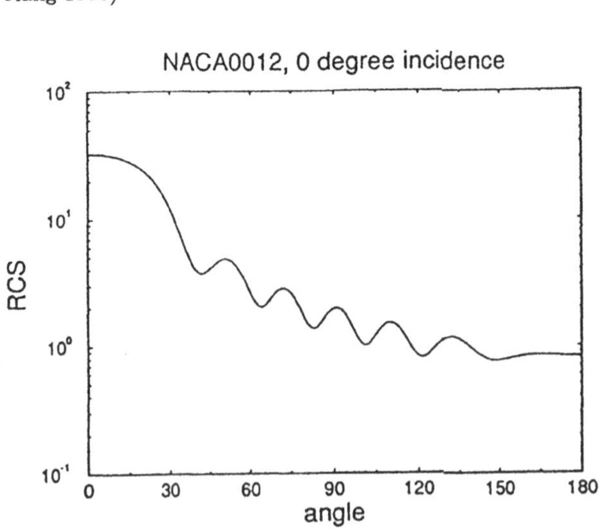

Fig. 14.20. RCS for the NACA0012 airfoil (0 degree incidence, $ka = 10$) (from Wu and Jiang 1996)

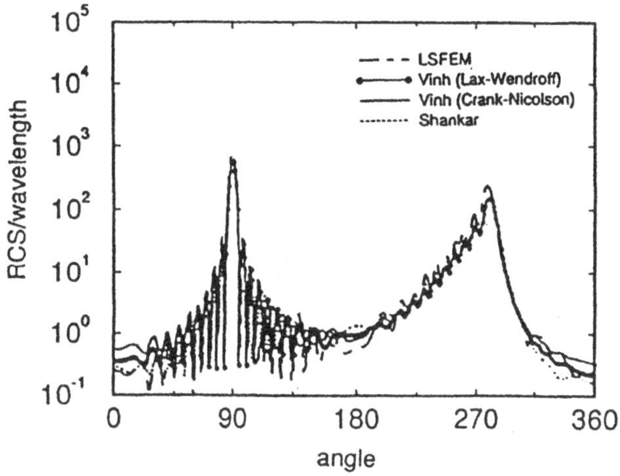

Fig. 14.21. RCS for the NACA0012 airfoil (90 degree incidence, $ka = 10\pi$) (from Wu and Jiang 1996)

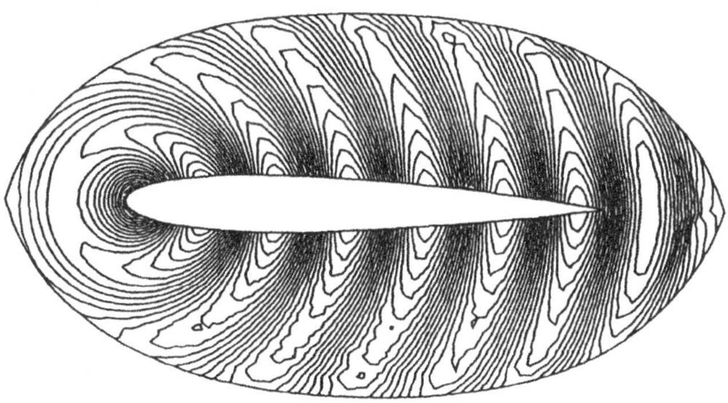

Fig. 14.22. Scattered E_z contours for the NACA0012 airfoil (0 degree incidence, $ka = 10$) (from Wu and Jiang 1996)

14.6.4 Influence of Divergence Equations

To illustrate the influence of the inclusion of divergence equation, we present here a comparison of two groups of numerical solutions, both obtained by using LSFEM: the first group is obtained by solving the full Maxwell equations, while the second group is obtained by solving the sole two curl equations. The case of TM waves scattered by a circular cylinder is considered. The values of ka used are $ka = 1$ and $ka = 10$. Figure 14.23 shows a comparison of the contours of the instantaneous field variables for $ka = 10$ between the group 1 and the group 2. It is clearly revealed that without enforcing divergence

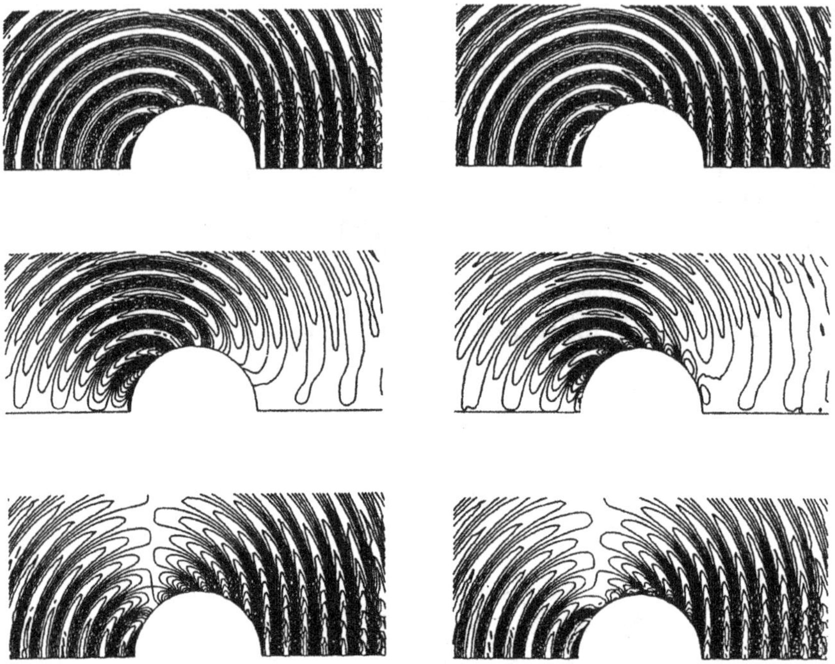

Fig. 14.23. Comparison of the solutions of the full Maxwell equations (left) and the two curl equations (right), TM case, ka=10, E_z, H_x, H_y respectively (top to bottom) (From Wu and Jiang 1996)

conditions significant spurious patterns exist for H_x and H_y near the surface. These spurious patterns are similar to those in Figs. 2–b and 2–c of Vinh et al. (1992). Further calculation shows that the group 2 solutions possess very large divergence of $\boldsymbol{H}$ in the vicinity of the scatterer surface. For the case $ka = 1$ the absolute value of $\nabla \cdot \boldsymbol{H}$ reaches its maximum of 21.28 in an element adjacent to the surface. On the other hand, the maximum value of $\nabla \cdot \boldsymbol{H}$ for the group 1 is 0.054, which occurs near the outer boundary. Since the divergence-free condition for $\boldsymbol{E}$ is automatically satisfied, almost identical E_z distributions are obtained with the two formulations.

We should remark here that though group 1 and group 2 solutions differ significantly, very little difference was observed in the RCS calculated based on the two solutions. This is due to the fact that the erroneous part of the solution is static, and the static field does not radiate. However, the erroneous solutions of the field intensities caused by neglecting the divergence conditions cannot be accepted, for example, in plasma computations and in the computation of impedance in digital circuits and electronic packaging.

14.7 Conclusion Remarks

In this chapter we have shown that:
(1) The system of the first-order full Maxwell equations seems "overspecified", because it has more equations than unknowns. By taking into account its div–curl structure and introducing the dummy variables, it proves to be properly determined and elliptic in the space domain. The information provided by the divergence equations is not completely contained in the curl equations. Therefore, the divergence equations must be explicitly included in the first-order system to assure the uniqueness of the solution in steady-state cases, to exclude the infinitely degenerate eigenvalue in time-harmonic cases, and to guarantee the accuracy of the numerical solution for time-varying cases.
(2) The least-squares method and the div–curl method are mathematically rigorous tools for derivation of correct second-order Maxwell equations and their boundary conditions. The curl–curl equations cannot stand alone, they must be supplemented by divergence equations and additional natural boundary conditions to eliminate spurious solutions.
(3) When the field intensities are smooth enough, the Helmholtz-type equations with appropriate natural boundary conditions, derived by the div–curl method or the least-squares method, can guarantee implicit satisfaction of divergence equations. For the solution of the Helmholtz-type equations divergence conditions of the electric field and the magnetic field need to be enforced only on the electric wall and the magnetic symmetry wall, respectively.
(4) The variational formulation corresponding to the Helmholtz-type equations can be derived by using the least-squares look-alike method. This formulation theoretically demonstrates that the penalty parameter in the Galerkin/penalty method should be chosen such that the Euler–Lagrange equation is the Helmholtz-type equation in homogeneous regions. The advantage of this formulation is that the trial and test functions from standard node-based finite element spaces need to satisfy only the essential boundary conditions.
(5) The node-based least-squares finite element method (LSFEM) can be used to solve both static and time-varying first-order Maxwell equations directly and efficiently with divergence equations satisfied easily. Introduction of vector potentials and the gauge condition (except for multi-layer materials), edge elements (except at interfaces), staggered grid, upwinding and non-equal-order elements, etc. are unnecessary.

The ideas presented in this chapter first appeared in Jiang et al. (1996), Wu and Jiang (1996) and Jiang (1997). Many LSFEM numerical results for scattering problems have been reported by Wu et al. (1994) and Wu and Jiang (1995a). Hillion (1997) gave a simple example to show that one has to work with divergence equations.

The application of LSFEM to three-dimensional stationary and static magnetic fields can be found in Lager and Mur (1996) and Lager (1996). In their approach, edge elements are placed at interfaces to ensure that all local continuity conditions can be met, and nodal elements are used elsewhere.

In the past, Assous et al. (1993) realized the importance of divergence equations in the time-domain Maxwell equations and correctly worked with second-order wave equations. Paulsen and Lynch (1991) and Boyse et al. (1992) first gave a correct variational formulation corresponding to the Helmholz-type operator in homogeneous regions.

Part V

Solution of Discrete Equations

15. The Element-by-Element Conjugate Gradient Method

Application of the finite element method to a continuum problem results in a linear system of algebraic equations. The system can be solved by using either direct elimination methods or iterative methods.

Direct elimination methods are most extensively used in finite element calculations, and have been highly refined (Carey and Oden 1983b). They provide the solution after a fixed number of steps and are less sensitive to the conditioning of the matrix. Nevertheless, these methods are less appealing for large problems as they require the global matrix to be assembled and factored, and thus demand prohibitively large amounts of computing time and storage.

Iterative methods have been receiving increased attention for very large systems and for use on parallel computers. In this chapter we study a well-known iterative method – the conjugate gradient (CG) method suitable for symmetric positive definite systems of linear equations arising from LSFEM. Since many excellent books about the CG method are available (see, e.g., Hageman and Young 1981, Axelsson and Barker 1984, Johnson 1987, Golub and Loan 1989 and Axelsson 1994), our attention is focused on basic concepts and matrix-free element-by-element implementation of the CG method.

15.1 Element-by-Element Technique

The main computations in a traditional finite element analysis involve calculating element contributions, assembling the global system from these contributions, and solving the resulting system. The element matrices are small and dense, and the resulting global matrix is large and sparse due to the local nature of the finite element basis. Several sparse elimination techniques have been developed for solving the linear system efficiently, but cost and hardware limitations still restrict the size of the problem.

The element-by-element sequence of calculations in a finite element program and the dense structure of the element matrices suggest that techniques which utilize these features might be advantageous. The phrase element-by-element refers to a particular data structure wherein information is stored and maintained at the element level than assembled into a global data structure. In this approach matrix–vector multiplications are carried out at the element level and assemblage is performed on the resulting vector.

To introduce the essential idea of the element-by-element approach, let us consider a symmetric positive definite system of the form:

$$\boldsymbol{K}\boldsymbol{U} = \boldsymbol{F}. \tag{15.1}$$

Most iterative methods for the solution of (15.1) can be described by the general formula:

$$\boldsymbol{U}^{n+1} = \boldsymbol{U}^n - \boldsymbol{Q}^{-1}(\boldsymbol{K}\boldsymbol{U}^n - \boldsymbol{F}), \tag{15.2}$$

where $\boldsymbol{Q}$ is a non-singular preconditioning matrix, designed to accelerate convergence of the iteration. For example, if the main diagonal of $\boldsymbol{K}$ is used to construct a diagonal matrix $\boldsymbol{Q}$, we have the well-known Jacobi iteration.

In a typical finite element analysis, the matrix $\boldsymbol{K}$ is sparse and may be assembled from element contributions as

$$\boldsymbol{K} = \sum_{e=1}^{N\text{elem}} \hat{\boldsymbol{K}}_e, \tag{15.3}$$

in which $\hat{\boldsymbol{K}}_e$ is an $N \times N$ matrix (where N is the total number of degrees-of-freedom of the problem) and consists of the element matrix $\boldsymbol{K}_e$ expanded to the full global system size, and Nelem is the number of elements. Note that the non-zero entries of $\hat{\boldsymbol{K}}_e$ are in the appropriate rows and columns as determined by the global node numbering, where $\boldsymbol{K}_e$ is dense of size $n_e \times n_e$ for an element with n_e local degrees-of-freedom. In practice, only the band or envelope of $\boldsymbol{K}$ might be stored, but this would still contain many zeros.

Next, we note that in the main calculation of the basic iterative method the residual

$$\boldsymbol{R}^n = \boldsymbol{K}\boldsymbol{U}^n - \boldsymbol{F}$$

in (15.2) requires computation of the matrix–vector multiplication $\boldsymbol{K}\boldsymbol{U}^n$. From (15.3) we can write

$$\boldsymbol{V}^n = \boldsymbol{K}\boldsymbol{U}^n = \Big(\sum_{e=1}^{N\text{elem}} \hat{\boldsymbol{K}}_e\Big)\boldsymbol{U}^n = \sum_{e=1}^{N\text{elem}} (\hat{\boldsymbol{K}}_e\hat{\boldsymbol{U}}_e^n) = \sum_{e=1}^{N\text{elem}} \hat{\boldsymbol{V}}_e^n, \tag{15.4}$$

where $\hat{\boldsymbol{U}}_e^n$ is an 'expanded' vector with non-zero values only in those components corresponding to nodes of element e. Only non-zero entries of each element matrix and element vector enter in the sum in (15.4). Hence the individual matrix products $\hat{\boldsymbol{V}}_e^n$ for the sum in (15.4) can be computed as dense element matrix–vector products at the element level rather than in the expanded form shown. That is, we may calculate

$$\boldsymbol{V}_e = \boldsymbol{K}_e\boldsymbol{U}_e \tag{15.5}$$

independently and concurrently, and accumulate the vectors $\boldsymbol{V}_e$ into appropriate positions of the global vector $\boldsymbol{V}^n$ for (15.4).

We note that the associated vectors in the operation of matrix–vector multiplication are stored globally, not element-wise.

15.2 Matrix-Free Algorithm

If several iterations are carried out, then the element matrices may be saved in a compact form. In fact, it is not even imperative that we store the element matrices in cases where storage is extremely limited, since we can perform the matrix–vector multiplication without forming the element matrices.

To illustrate the idea of matrix-free algorithm, we consider the computation of (15.5), i.e., the product $\boldsymbol{V}_e$ of an element matrix $\boldsymbol{K}_e$ by an element vector $\boldsymbol{U}_e$, where the element matrix is given by (4.59); that is, we wish to evaluate the expression

$$\boldsymbol{V}_e = \int_{\Omega_e} (\boldsymbol{A}\psi_1, \boldsymbol{A}\psi_2, ..., \boldsymbol{A}\psi_{N_n})^T (\boldsymbol{A}\psi_1, \boldsymbol{A}\psi_2, ..., \boldsymbol{A}\psi_{N_n}) \boldsymbol{U}_e d\Omega. \tag{15.6}$$

When we perform numerical integration, we replace integrals by weighted sums of the integrands evaluated at Gaussian points. We therefore need to evaluate the following expression at Gaussian points

$$(\boldsymbol{A}\psi_1, \boldsymbol{A}\psi_2, ..., \boldsymbol{A}\psi_{N_n})^T (\boldsymbol{A}\psi_1, \boldsymbol{A}\psi_2, ..., \boldsymbol{A}\psi_{N_n}) \boldsymbol{U}_e, \tag{15.7}$$

where $\boldsymbol{A}\psi_j$ is given by (4.61), and

$$\boldsymbol{U}_e = \left((u_1, u_2, ..., u_m)_1, (u_1, u_2, ..., u_m)_2, ..., (u_1, u_2, ..., u_m)_{N_n}\right)^T, \tag{15.8}$$

in which all notations have been defined in Sect. 4.8.

Now two algorithms can be chosen. Both of them avoid formation and storage of the element matrices. In the first algorithm, one forms the matrix $\boldsymbol{A}\psi_j$, and computes

$$\boldsymbol{r}_e = (\boldsymbol{A}\psi_1, \boldsymbol{A}\psi_2, ..., \boldsymbol{A}\psi_{N_n}) \boldsymbol{U}_e, \tag{15.9a}$$

and then

$$\boldsymbol{V}_e = (\boldsymbol{A}\psi_1, \boldsymbol{A}\psi_2, ..., \boldsymbol{A}\psi_{N_n})^T \boldsymbol{r}_e. \tag{15.9b}$$

This approach benefits academic researchers for quick tests of new ideas, since the operations in (15.9) can be programmed in a general setting without knowledge of $\boldsymbol{A}$. For a new system of partial differential equations, one needs very little work – only to add a subroutine to provide the entries of the matrices $\boldsymbol{A}_1$, $\boldsymbol{A}_2$, $\boldsymbol{A}_3$ and $\boldsymbol{A}_0$.

However, the matrices $\boldsymbol{A}_1$, $\boldsymbol{A}_2$, $\boldsymbol{A}_3$ and $\boldsymbol{A}_0$ may contain many zero entries, and zero-related operations are wasteful, thus the above approach may be slow. Therefore, for commercial applications, we recommend the following algorithm which needs more programming work initially but runs much faster.

By using (4.57) we can verify that

$$\boldsymbol{A}\boldsymbol{u}_h^e(\xi_l, \eta_l, \zeta_l) = (\boldsymbol{A}\psi_1, \boldsymbol{A}\psi_2, ..., \boldsymbol{A}\psi_{N_n})\boldsymbol{U}_e, \tag{15.10}$$

in which the left-hand side term $\boldsymbol{A}\boldsymbol{u}_h^e$ is a vector with N_{eq} components evaluated at the Gaussian point (ξ_l, η_l, ζ_l), where N_{eq} is the number of equations in the first-order system $\boldsymbol{A}\boldsymbol{u} = \boldsymbol{f}$. Also by using formula (15.9b) we can write an explicit expression for the vector components of $\boldsymbol{V}_e$. Therefore, we do not need to form $\boldsymbol{A}\psi_j$, we can directly compute

$$\boldsymbol{r}_e = \boldsymbol{A}\boldsymbol{u}_h^e(\xi_l, \eta_l, \zeta_l), \tag{15.11}$$

and then the components of $\boldsymbol{V}_e$. This approach is more efficient than the first one, because the multiplication of a matrix and vector is avoided and all zero-related operations are pre-eliminated.

The diagonal entries of the element stiffness matrix consist of the diagonal entries of the matrix $(\boldsymbol{A}\psi_j)^T(\boldsymbol{A}\psi_j)$ Therefore, the diagonal preconditioner $\boldsymbol{Q}$ can be calculated also in matrix-free manner in the same routine as the formation of $\boldsymbol{V}$.

In matrix-free algorithms one still needs to store the values of shape functions and their derivatives at Gaussian points for each element. Of course, even storing these values is not imperative at the expense of computer time.

15.3 The Conjugate Gradient Method

The matrix-free element-by-element notion can be exploited with any residual based iterative method. These fundamental iterative methods may also be accelerated by the conjugate gradient method which was first presented by Hestenes and Stieffel (1952). The popularity of the CG method is due to its attractive properties: (1) the method will give the exact solution after at most N steps in the absence of roundoff errors; in practice, the convergence to an acceptable accuracy can be achieved with far few steps than the number N; (2) the rate of convergence can be significantly improved with various preconditioning techniques; (3) the method is parameter-free, i.e., the user is not required to empirically choose parameters.

15.3.1 The Steepest Descent Method

Since the CG method can be regarded as a modification of the steepest descent method, we first consider the steepest descent method as applied to the linear system (15.1). Let $(\boldsymbol{U}, \boldsymbol{V})$ be an inner product in $\mathbb{R}^N$. The problem of solving (15.1) is equivalent to the problem of finding the minimizer of the quadratic form:

$$g(\boldsymbol{U}) = \frac{1}{2}(\boldsymbol{U}, \boldsymbol{K}\boldsymbol{U}) - (\boldsymbol{F}, \boldsymbol{U}). \tag{15.12}$$

This is obvious, because we obtain (15.1) from minimizing a quadratic form in the least-squares method, or we simply note that

$$\frac{1}{2}(\boldsymbol{KU}-\boldsymbol{F})^T\boldsymbol{K}^{-1}(\boldsymbol{KU}-\boldsymbol{F}) = \frac{1}{2}(\boldsymbol{U},\boldsymbol{KU}) - (\boldsymbol{F},\boldsymbol{U}) + \frac{1}{2}(\boldsymbol{F},\boldsymbol{K}^{-1}\boldsymbol{F}),$$

where the last term is constant. In fact, the steepest descent method can be viewed as a least-squares method, and the CG method can be viewed as a generalized least-squares method where the minimization takes place on a particular vector space, i.e., the Krylov space which is spanned by the vectors $(\boldsymbol{R}^0, \boldsymbol{KR}^0, ..., \boldsymbol{K}^n\boldsymbol{R}^0)$, where

$$\boldsymbol{R}^0 = \boldsymbol{KU}^0 - \boldsymbol{F}$$

is the residual for the initial vector $\boldsymbol{U}^0$ (Axelsson 1994).

The gradient of $g(\boldsymbol{U})$ is given by

$$\nabla g(\boldsymbol{U}) = \boldsymbol{KU} - \boldsymbol{F}.$$

The direction represented by the vector $\nabla g(\boldsymbol{U})$ is the direction for which the form $g(\boldsymbol{U})$ at the point $\boldsymbol{U}$ has the greatest instantaneous rate of change. If $\boldsymbol{U}^n$ is some approximation to the exact solution $\bar{\boldsymbol{U}}$, then an improved approximation $\boldsymbol{U}^{n+1}$ can be obtained by moving in the direction of $-\nabla g(\boldsymbol{U})$ to a point where $g(\boldsymbol{U}^{n+1})$ is minimal, that is,

$$\boldsymbol{U}^{n+1} = \boldsymbol{U}^n - \tau_n \nabla g(\boldsymbol{U}^n),$$

in which τ_n is chosen to minimize $g(\boldsymbol{U}^{n+1})$. Using (15.12), it is easy to see that

$$\tau_n = \frac{(\boldsymbol{R}^n, \boldsymbol{R}^n)}{(\boldsymbol{R}^n, \boldsymbol{KR}^n)},$$

where

$$\boldsymbol{R}^n = \nabla g(\boldsymbol{U}^n) = \boldsymbol{KU}^n - \boldsymbol{F}$$

denotes the nth residual. Hence the steepest descent method can be expressed in the form:

$$\boldsymbol{R}^n = \boldsymbol{KU}^n - \boldsymbol{F}, \tag{15.13a}$$

$$\tau_n = \frac{(\boldsymbol{R}^n, \boldsymbol{R}^n)}{(\boldsymbol{R}^n, \boldsymbol{KR}^n)}, \tag{15.13b}$$

$$\boldsymbol{U}^{n+1} = \boldsymbol{U}^n - \tau_n \boldsymbol{R}^n, \tag{15.13c}$$

where $n = 0, 1, ...$ and $\boldsymbol{U}^0$ is arbitrary.

It can be shown that the required number of iterations in the steepest descent method to reduce the initial error $|\boldsymbol{U}^0 - \bar{\boldsymbol{U}}|$ by a certain given factor $\epsilon \geq 0$ is proportional to the condition number $\kappa(\boldsymbol{K})$ and the number of decimals in ϵ:

$$n \geq \kappa(\boldsymbol{K}) \log\frac{1}{\epsilon}. \tag{15.14}$$

In Sect. 2.10, we have shown that for LSFEM $\kappa(\boldsymbol{K}) = O(h^{-2})$ and thus here we would have $n = O(h^{-2})$, i.e., a very large number of iterations have to be performed. However, by choosing the direction vectors differently, one obtains a more efficient method – the CG method.

15.3.2 The Conjugate Gradient Method

Let successive approximations to the exact solution be given by

$$\boldsymbol{U}^{n+1} = \boldsymbol{U}^n + \tau_n \boldsymbol{D}^n, \tag{15.15}$$

where $\boldsymbol{D}^n$ is a direction vector. For the CG method, we let $\boldsymbol{D}^0 = -\boldsymbol{R}^0$ and

$$\boldsymbol{D}^{n+1} = -\boldsymbol{R}^{n+1} + \beta_n \boldsymbol{D}^n \text{ for } n \geq 0,$$

where β_n is chosen so that $\boldsymbol{D}^{n+1}$ is $\boldsymbol{K}$-conjugate to $\boldsymbol{D}^n$, i.e., $(\boldsymbol{D}^{n+1}, \boldsymbol{K}\boldsymbol{D}^n) = 0$. Obviously,

$$\beta_n = \frac{(\boldsymbol{R}^{n+1}, \boldsymbol{K}\boldsymbol{D}^n)}{(\boldsymbol{D}^n, \boldsymbol{K}\boldsymbol{D}^n)}.$$

As before, choosing τ_n to minimize $g(\boldsymbol{U}^{n+1})$, we obtain

$$\tau_n = -\frac{(\boldsymbol{D}^n, \boldsymbol{R}^n)}{(\boldsymbol{D}^n, \boldsymbol{K}\boldsymbol{D}^n)}.$$

It can be shown that the following orthogonality or conjugate orthogonality properties hold:

$$(\boldsymbol{R}^{n+1}, \boldsymbol{R}^j) = 0 \qquad 0 \leq j \leq n, \tag{15.16a}$$

$$(\boldsymbol{R}^{n+1}, \boldsymbol{D}^j) = 0 \qquad 0 \leq j \leq n, \tag{15.16b}$$

$$(\boldsymbol{D}^{n+1}, \boldsymbol{K}\boldsymbol{D}^j) = 0 \qquad 0 \leq j \leq n. \tag{15.16c}$$

By using the $\boldsymbol{K}$-orthogonality and the linear independence of the $\boldsymbol{D}^j$ vectors, the Nth residual $\boldsymbol{R}^N$ can be shown to be orthogonal to N linearly independent vectors and hence is the null vector. Therefore, in the absence of round-off errors the method produces the solution in no more than N iterations.

From computational considerations, we can have better formulas for the CG method. Using

$$\boldsymbol{D}^n = -\boldsymbol{R}^n + \beta_{n-1}\boldsymbol{D}^{n-1},$$

$$\boldsymbol{K}\boldsymbol{D}^n = \frac{1}{\tau_n}(\boldsymbol{R}^{n+1} - \boldsymbol{R}^n),$$

and orthogonality properties (15.16), we have

$$-(\boldsymbol{R}^n, \boldsymbol{D}^n) = (\boldsymbol{R}^n, \boldsymbol{R}^n),$$

$$(\boldsymbol{R}^{n+1}, \boldsymbol{K}\boldsymbol{D}^n) = \frac{1}{\tau_n}(\boldsymbol{R}^{n+1}, \boldsymbol{R}^{n+1} - \boldsymbol{R}^n) = \frac{1}{\tau_n}(\boldsymbol{R}^{n+1}, \boldsymbol{R}^{n+1}),$$

and

$$\tau_n(\boldsymbol{D}^n, \boldsymbol{K}\boldsymbol{D}^n) = (\boldsymbol{D}^n, \boldsymbol{R}^{n+1} - \boldsymbol{R}^n) = -(\boldsymbol{D}^n, \boldsymbol{R}^n) = (\boldsymbol{R}^n, \boldsymbol{R}^n).$$

Therefore, τ_n and β_n can be given equivalently by

$$\tau_n = \frac{(\boldsymbol{R}^n, \boldsymbol{R}^n)}{(\boldsymbol{D}^n, \boldsymbol{K}\boldsymbol{D}^n)}), \tag{15.17}$$

$$\beta_n = \frac{(\boldsymbol{R}^{n+1}, \boldsymbol{R}^{n+1})}{(\boldsymbol{R}^n, \boldsymbol{R}^n)}. \tag{15.18}$$

Multiplying (15.15) by $\boldsymbol{K}$ and subtracting the vector $\boldsymbol{F}$ form both sides, we obtain

$$\boldsymbol{R}^{n+1} = \boldsymbol{R}^n + \tau_n \boldsymbol{K}\boldsymbol{D}^n.$$

Note that with the formulas (15.17) and (15.18), there is no need to compute the inner product $(\boldsymbol{R}^n, \boldsymbol{D}^n)$ and $(\boldsymbol{R}^{n+1}, \boldsymbol{K}\boldsymbol{D}^n)$. Furthermore, the inner product $(\boldsymbol{R}^{n+1}, \boldsymbol{R}^{n+1})$ can be used to test when to stop the iterations. Then the conjugate gradient method has the form:

$$\tau_n = \frac{(\boldsymbol{R}^n, \boldsymbol{R}^n)}{(\boldsymbol{D}^n, \boldsymbol{K}\boldsymbol{D}^n)}, \tag{15.19a}$$

$$\boldsymbol{U}^{n+1} = \boldsymbol{U}^n + \tau_n \boldsymbol{D}^n, \tag{15.19b}$$

$$\boldsymbol{R}^{n+1} = \boldsymbol{R}^n + \tau_n \boldsymbol{K}\boldsymbol{D}^n, \tag{15.19c}$$

$$\beta_n = \frac{(\boldsymbol{R}^{n+1}, \boldsymbol{R}^{n+1})}{(\boldsymbol{R}^n, \boldsymbol{R}^n)}, \tag{15.19d}$$

$$\boldsymbol{D}^{n+1} = -\boldsymbol{R}^{n+1} + \beta_n \boldsymbol{D}^n \tag{15.19e}$$

for $n = 0, 1, \ldots$ until $(\boldsymbol{R}^{n+1}, \boldsymbol{R}^{n+1}) \leq \epsilon$. The computer implementation of this algorithm can be described as follows:

$\boldsymbol{U} := \boldsymbol{U}^0;\ \boldsymbol{R} := \boldsymbol{F};$

$\boldsymbol{R} := \boldsymbol{K}\boldsymbol{U} - \boldsymbol{R};\ \delta 0 := (\boldsymbol{R}, \boldsymbol{R});$

IF $\delta 0 \leq \epsilon$ THEN STOP;

$\boldsymbol{D} := -\boldsymbol{R};$

$IT:$ $\boldsymbol{B} := \boldsymbol{K}\boldsymbol{D};$

$\tau := \delta 0/(\boldsymbol{D}, \boldsymbol{B});$

$\boldsymbol{U} := \boldsymbol{U} + \tau \boldsymbol{D};$

$\boldsymbol{R} = \boldsymbol{R} + \tau \boldsymbol{B};\ \delta 1 := (\boldsymbol{R}, \boldsymbol{R});$

IF $\delta 1 \leq \epsilon$ THEN STOP;

$\beta := \delta 1/\delta 0;\ \delta 0 := \delta 1;$

$\boldsymbol{D} := -\boldsymbol{R} + \beta \boldsymbol{D};$

GOTO *IT*;

It can be shown that to reduce the error $|\boldsymbol{U}^n - \bar{\boldsymbol{U}}|$ to less than ϵ times the initial error $|\boldsymbol{U}^0 - \bar{\boldsymbol{U}}|$, the required number of iteration necessary in the CG method is roughly given by

$$n \geq \frac{1}{2}\sqrt{\kappa(\boldsymbol{K})}\log\frac{1}{\epsilon}, \tag{15.20}$$

which should be compared with $\kappa(\boldsymbol{K})$ in the case of the steepest descent method. Thus, for large $\kappa(\boldsymbol{K})$, the conjugate gradient method is much more efficient than the steepest descent method. For a typical application of LS-FEM, the required number of iterations is proportional to order h^{-1} which is greatly superior to h^{-2} for the steepest descent method.

15.3.3 The Preconditioned Conjugate Gradient Method

To construct the preconditioned version of the CG method, we multiply both side of (15.1) by a matrix $\boldsymbol{Q}^{-1}$ to obtain

$$\boldsymbol{Q}^{-1}\boldsymbol{K}\boldsymbol{U} = \boldsymbol{Q}^{-1}\boldsymbol{F}, \tag{15.21}$$

where $\boldsymbol{Q}$ is symmetric positive definite and is an approximation to $\boldsymbol{K}$. There exists a nonsingular matrix $\boldsymbol{W}$ such that $\boldsymbol{Q} = \boldsymbol{W}^T\boldsymbol{W}$. That is, $\boldsymbol{W}$ is the Cholesky factorization of $\boldsymbol{Q}$. To use the above CG method it is necessary to make (15.21) symmetric. This can be done by multiplying both sides of (15.21) by $\boldsymbol{W}$. Consequently, (15.21) is replaced by

$$\tilde{\boldsymbol{K}}\tilde{\boldsymbol{U}} = \tilde{\boldsymbol{F}}, \tag{15.22}$$

where

$$\tilde{\boldsymbol{K}} = \boldsymbol{W}\boldsymbol{Q}^{-1}\boldsymbol{K}\boldsymbol{W}^{-1}, \qquad \tilde{\boldsymbol{U}} = \boldsymbol{W}\boldsymbol{U} \ \text{and}\ \tilde{\boldsymbol{F}} = \boldsymbol{W}\boldsymbol{Q}^{-1}\boldsymbol{F}.$$

System (15.22) is called a preconditioned system. If $\kappa(\boldsymbol{W}^{-T}\boldsymbol{K}\boldsymbol{W}^{-1}) << \kappa(\boldsymbol{K})$, then the CG method based on (15.22) will converge much faster than the same method applied to (15.1). If we apply the CG method to the preconditioned system (15.22), we obtain the scheme:

$$\tau_n = \frac{(\boldsymbol{B}^n, \boldsymbol{Q}\boldsymbol{R}^n)}{(\boldsymbol{D}^n, \boldsymbol{K}\boldsymbol{D}^n)}, \tag{15.23a}$$

$$\boldsymbol{U}^{n+1} = \boldsymbol{U}^n + \tau_n \boldsymbol{D}^n, \tag{15.23b}$$

$$\boldsymbol{B}^{n+1} = \boldsymbol{B}^n + \tau_n \boldsymbol{Q}^{-1}\boldsymbol{K}\boldsymbol{D}^n, \tag{15.23c}$$

$$\beta_n = \frac{(\boldsymbol{B}^{n+1}, \boldsymbol{Q}\boldsymbol{B}^{n+1})}{(\boldsymbol{B}^n, \boldsymbol{Q}\boldsymbol{B}^n)}, \tag{15.23d}$$

$$\boldsymbol{D}^{n+1} = -\boldsymbol{R}^{n+1} + \beta_n \boldsymbol{D}^n. \tag{15.23e}$$

The corresponding algorithm takes the form:

$$
\begin{array}{ll}
 & \boldsymbol{U} := \boldsymbol{U}^0;\ \boldsymbol{R} = \boldsymbol{K}\boldsymbol{U} - \boldsymbol{F};\\
 & \boldsymbol{B} := \boldsymbol{Q}^{-1}\boldsymbol{R};\\
 & \boldsymbol{D} := -\boldsymbol{B};\ \delta 0 := (\boldsymbol{R}, \boldsymbol{B});\\
 & \text{IF } \delta 0 \leq \epsilon \text{ THEN STOP;}\\
IT: & \boldsymbol{B} := \boldsymbol{K}\boldsymbol{D};\\
 & \tau := \delta 0/(\boldsymbol{D}, \boldsymbol{B});\\
 & \boldsymbol{U} := \boldsymbol{U} + \tau \boldsymbol{D};\\
 & \boldsymbol{R} = \boldsymbol{R} + \tau \boldsymbol{B};\\
 & \boldsymbol{B} := \boldsymbol{Q}^{-1}\boldsymbol{R};\ \delta 1 := (\boldsymbol{R}, \boldsymbol{B});\\
 & \text{IF } \delta 1 \leq \epsilon \text{ THEN STOP;}\\
 & \beta := \delta 1/\delta 0;\ \delta 0 := \delta 1\\
 & \boldsymbol{D} := -\boldsymbol{B} + \beta \boldsymbol{D};\\
 & \text{GOTO } IT;
\end{array}
$$

The statement $\boldsymbol{B} := \boldsymbol{Q}^{-1}\boldsymbol{R}$ should be interpreted as solving the system $\boldsymbol{Q}\boldsymbol{B} = \boldsymbol{R}$. If the main diagonal of $\boldsymbol{K}$ is chosen to construct a diagonal matrix $\boldsymbol{Q}$, we have the well-known Jacobi preconditioned CG method (JCG). The best choice would be $\boldsymbol{Q} = \boldsymbol{K} = \boldsymbol{W}^T\boldsymbol{W}$ with $\boldsymbol{W}^T\boldsymbol{W}$ the Cholesky factorization of $\boldsymbol{K}$, in which case $\kappa(\boldsymbol{W}^{-T}\boldsymbol{K}\boldsymbol{W}^{-1}) = 1$. However, this results in a method which is not different from a direct solver. Since $\boldsymbol{K}$ is a sparse matrix, we may require $\boldsymbol{W}$ to have the same sparsity as that of $\boldsymbol{K}$. This leads to an approximate Cholesky factorization of $\boldsymbol{K}$. For example, we may allow the entry w_{ij} of $\boldsymbol{W}$ to be non-zero only if the corresponding entry k_{ij} of $\boldsymbol{K}$ is non-zero. This is a common variant of the so-called incomplete factorization which takes only $O(N)$ operations with considerable reduction of the condition number, that is, $\kappa(\boldsymbol{W}^{-T}\boldsymbol{K}\boldsymbol{W}^{-1}) = O(h^{-1})$ (Axelsson 1994).

We should mention that, although various incomplete factorization preconditioning techniques can significantly reduce the number of iterations, their overall performance should be evaluated by considering the cost of additional memory requirement and total computer time.

Most LSFEM results presented in this book are obtained by using the JCG method, since the diagonal preconditioner is the simplest. Only five N-dimensional arrays are needed for $\boldsymbol{B}, \boldsymbol{U}, \boldsymbol{R}, \boldsymbol{D}$ and $\boldsymbol{Q}$ in the JCG method.

15.3.4 Numerical Results and Comparisons

The experiments conducted by Jiang (1986) and Carey and Jiang (1986) show that, even for small-scale (several hundred unknowns) LSFEM computations, the element-by-element JCG method performs better than the sparse direct solvers in storage requirement and computer time, particularly for non-linear problems.

To show that the matrix-free LSFEM-JCG is an efficient method, Tang and Tsang (1995) compared its performance with the LSFEM-LINPACK (DPBFA with BLAS level 1 using Cholesky decomposition) and the Galerkin FEM-NAG (F01LBF, F04LDF using LU decomposition) for four types of two-dimensional flows: lid-driven cavity flow (LDCF, see Sect. 8.3.1), thermally-driven cavity flow (TDCF, see Sect. 8.7.1), Rayleigh–Bénard convection (RBC, see Sect. 8.7.2) and doubly-diffusive flow (DDF, see Sect. 8.7.4). For the LSFEM computations, a mesh with 50×50 non-uniform bilinear elements was used for LDCF, TDCF and DDF, and a mesh with 50×25 non-uniform bilinear elements for RBC. For the mixed Galerkin FEM (GFEM) computations, linear velocity and constant pressure elements were used. In these tests the Crank–Nicolson time marching scheme with $\Delta t = 0.5$ and Newton linearization were used. The tolerance for the steady state solution was set as

$$\max_{1 \leq i \leq N} |U_i^{n-1} - U_i^n| < 10^{-5},$$

where U_i is the ith component of $\boldsymbol{U}$. The stopping criterion for JCG was error $< 10^{-6}$ for LDCF and RBC, and error $< 10^{-7}$ for TDCF and DDF. The error is defined as the marching distance of the CG method divided by the absolute value of the global unknown vector. The flows were all impulsively started. Table 15.1 shows that LSFEM-JCG is fast and requires much less storage than both LSFEM-LINPACK and GFEM-NAG.

Table 15.1. Comparison of total CPU time and computer storage of LSFEM and GFEM (from Tang and Tsang 1995)

Flow	Method	Solver	Degree of Freedom	Total CPU Time(sec)	Stopping Time	Memmory (Mb)
	LSFEM	JCG	10404	1795	80	1.07
LDCF		LINPACK	10404	11426	80	18.42
	GFEM	NAG	7702	5111	80	30.39
	LSFEM	JCG	18207	16404	130	1.57
TDCF		LINPACK	18207	64982	130	55.02
	GFEM	NAG	10303	18022	130	53.31
	LSFEM	JCG	9282	1336	75	0.80
RBC		LINPACK	9282	8756	100	15.06
	GFEM	NAG	5228	1791	100	14.50
	LSFEM	JCG	26010	14510	180	2.07
DDF		LINPACK	26010	98397	179	111.52
	GFEM	NAG	12904	49574	180	82.86

They also compared the matrix-free LSFEM-JCG with FIDAP's mixed formulation of the GFEM for a lid-driven cavity flow with $Re = 5000$. A non-uniform mesh system with 30×30 biquadratic elements were used. The time step was 0.5 for LSFEM, and 0.25 for the GFEM, respectively (the Newton's method in the GFEM does not converge for $\Delta t = 0.5$). The LSFEM-JCG required 1.3 Mb memory, whereas the GFEM in FIDAP required about 11 Mb. The CPU time on a HP-720 workstation were 8756 seconds and 39 257 seconds for LSFEM and GFEM, respectively.

The matrix-free LSFEM-JCG has been successfully implemented on the parallel computer CRAY-T3D by J. Wu (1994). Table 15.2 shows the speedup for a typical computation of two-dimensional incompressible viscous flow.

Table 15.2. Speedup on CRAY-T3D

No. Processors	1	4	8
Time (sec)	510	154	80
Speedup	1	3.31	6.38

15.4 Concluding Remarks

The LSFEM gives rise to symmetric, positive definite systems of linear algebraic equations. This system of equations can be efficiently solved by the matrix-free element-by-element preconditioned conjugate gradient method.

The matrix-free element-by-element CG method proves to be very attractive when solving large problems, because

(1) The need for assemblage, storage and factorization of the global matrix is circumvented, and thus the total storage and computational costs are low;

(2) For direct methods, it is advisable that the elements and nodes be numbered judiciously to reduce the bandwidth for both storage and efficiency. In some cases, this is difficult to accomplish and one must rely on a sophisticated preprocessor. Moreover, once the mesh in a local area is changed or modified, as in adaptive refinement procedures, the elements and nodes must be renumbered if a good band structure is to be preserved. However, for the element-by-element CG method, the mesh structure and the numbering of nodes can be arbitrary, the amount of storage is independent of meshing and node numbering, and depends only on the number and the type of elements in the mesh;

(3) Since element contributions are computed independently, the element-by-element CG method can be easily implemented in parallel on multi-processors without any special partitioning of the domain, topology of the mesh, or node numbering;

(4) An interpolation of the coarse-mesh solution can serve as a very good initial guess for the fine-mesh solution.

(5) The CG method is efficient for non-linear applications, such as incompressible Navier–Stokes equations, where a sequence of linear problems is to be solved, the method offers the possibility of adapting the iteration sequence to improve overall efficiency in the coupled iteration of linear subproblems within the main nonlinear iteration sequence. That is, the solution of the previous nonlinear iteration step can be taken as an initial guess for the solution of the next step to reduce the computing time. The same reasoning applies to implicit time marching procedures for transient problems; the solution for the previous time step can serve as an good initial guess for the next time step, hence an implicit time marching computation with very small time step, which is preferable due to its accuracy, becomes inexpensive.

The idea of using element-by-element data structures was first proposed by Fox and Stanton (1968) and Fried (1969a,b, 1970) stemming from the additive property of the finite element functionals.

Various applications of the element-by-element technique can be found in the works by Hughes et al. (1983a,b and 1987), Ortiz et al. (1983), Jiang and Carey (1984), Hayes and Devloo (1985, 1986), Winget and Hughes (1985), Jiang (1986), Carey and Jiang (1986) Hayes (1989), Wathen (1989), Shakib et al. (1889), Tezduyar and Liou (1989), Reddy et al. (1992) and among others. Parallel implementation of the element-by-element scheme can be found in Carey (1986), Carey et al. (1988), Barragy and Carey (1988), and Foresti et al. (1990). The rapid operation of the matrix–vector product based on the matrix-free element by element notion and p-version tensor-product shape functions is described in Sonnad et al. (1990) and Jiang and Sonnad (1994).

Appendices

A. Operations on Vectors

$$\nabla \cdot (q\boldsymbol{v}) = q\nabla \cdot \boldsymbol{v} + \nabla q \cdot \boldsymbol{v}, \tag{A.24}$$

$$\nabla \times (q\boldsymbol{v}) = q\nabla \times \boldsymbol{v} + \nabla q \times \boldsymbol{v}, \tag{A.25}$$

$$\nabla \cdot (\boldsymbol{u} \times \boldsymbol{v}) = (\nabla \times \boldsymbol{u}) \cdot \boldsymbol{v} - \boldsymbol{u} \cdot (\nabla \times \boldsymbol{v}), \tag{A.26}$$

$$\nabla \times \nabla \times \boldsymbol{v} = \nabla(\nabla \cdot \boldsymbol{v}) - \Delta \boldsymbol{v}, \tag{A.27}$$

$$\boldsymbol{v} \times (\nabla \times \boldsymbol{v}) = \frac{1}{2}\nabla(\boldsymbol{v}^2) - (\boldsymbol{v} \cdot \nabla)\boldsymbol{v}, \tag{A.28}$$

$$\nabla \times (\boldsymbol{u} \times \boldsymbol{v}) = (\boldsymbol{v} \cdot \nabla)\boldsymbol{u} - \boldsymbol{v}(\nabla \cdot \boldsymbol{u}) - (\boldsymbol{u} \cdot \nabla)\boldsymbol{v} + \boldsymbol{u}(\nabla \cdot \boldsymbol{v}). \tag{A.29}$$

B. Green's Formula

Assume that $\boldsymbol{u}$, $\boldsymbol{v}$ and q are smooth enough. Integrating (A.1) and using the Gauss divergence theorem lead to

$$(\nabla \cdot \boldsymbol{v},\ q) + (\boldsymbol{v},\ \nabla q) = \langle \boldsymbol{n} \cdot \boldsymbol{v},\ q \rangle. \tag{B.1}$$

Substituting $\boldsymbol{v} = \nabla p$ into (B.1) yields

$$(\Delta p,\ q) + (\nabla p,\ \nabla q) = \langle \boldsymbol{n} \cdot \nabla p,\ q \rangle. \tag{B.2}$$

Substituting $q = \nabla \cdot \boldsymbol{u}$ into (B.1) yields

$$(\nabla \cdot \boldsymbol{v},\ \nabla \cdot \boldsymbol{u}) + \big(\boldsymbol{v},\ \nabla(\nabla \cdot \boldsymbol{u})\big) = \langle \boldsymbol{n} \cdot \boldsymbol{v},\ \nabla \cdot \boldsymbol{u} \rangle. \tag{B.3}$$

Replacing $\boldsymbol{v}$ by $\nabla \times \boldsymbol{v}$ in (B.1) leads to

$$(\nabla \times \boldsymbol{v},\ \nabla q) = \langle \boldsymbol{n} \cdot (\nabla \times \boldsymbol{v}),\ q \rangle. \tag{B.4}$$

Integrating (A.3) and using the Gauss divergence theorem lead to

$$(\nabla \times \boldsymbol{u},\ \boldsymbol{v}) - (\boldsymbol{u},\ \nabla \times \boldsymbol{v}) = \langle \boldsymbol{n} \times \boldsymbol{u},\ \boldsymbol{v} \rangle. \tag{B.5}$$

Substituting $\boldsymbol{u} = \nabla q$ into (B.5) yields

$$(\nabla \times \boldsymbol{v},\ \nabla q) = -\langle \boldsymbol{n} \times \nabla q,\ \boldsymbol{v} \rangle = \langle \nabla q,\ \boldsymbol{n} \times \boldsymbol{v} \rangle. \tag{B.6}$$

Replacing $\boldsymbol{v}$ by $\nabla \times \boldsymbol{v}$ in (B.5) yields

$$(\nabla \times \boldsymbol{u},\ \nabla \times \boldsymbol{v}) - (\boldsymbol{u},\ \nabla \times \nabla \times \boldsymbol{v}) = \langle \boldsymbol{n} \times \boldsymbol{u},\ \nabla \times \boldsymbol{v} \rangle. \tag{B.7}$$

C. Poincaré Inequality

Let Ω be a bounded domain with a piecewise C^1 boundary Γ, then

$$\|p\|_1^2 \le C\left\{\|\nabla p\|_0^2 + \left(\int_\Omega p dx\right)^2\right\} \quad \forall p \in H^1(\Omega), \tag{C.1}$$

$$\|p\|_1^2 \le C\left\{\|\nabla p\|_0^2 + \left(\int_\Gamma p dx\right)^2\right\} \quad \forall p \in H^1(\Omega). \tag{C.2}$$

D. Lax–Milgram Theorem

Let V be a Hilbert space with corresponding norm $\|\cdot\|_V$, let $a(\cdot,\cdot)$: $V \times V \to \mathbb{R}$ be a bilinear form, and let $l : V \to \mathbb{R}$ be a linear form such that
(1) $a(\cdot,\cdot)$ is symmetric,
(2) $a(\cdot,\cdot)$ is continuous, i.e., there exists a constant $M > 0$ such that

$$|a(u,v)| \le M\|u\|_V\|v\|_V \qquad \forall u, v \in V,$$

(3) $a(\cdot,\cdot)$ is V-elliptic, i.e., there exists a constant $\alpha > 0$ such that

$$a(v,v) \ge \alpha\|v\|_V^2 \qquad \forall v \in V,$$

(4) l is continuous, i.e., there exists a constant $M > 0$ such that

$$|l(v)| \le \Lambda\|v\|_V \qquad \forall v \in V.$$

Then the abstract variational problem: find an element u such that

$$u \in V,$$

$$a(u,v) = l(v) \qquad \forall v \in V,$$

has one and only one solution.

References

Agmon, S., Douglis, A., Nirenberg, L. (1964): Estimates near the boundary for solutions of elliptic partial differential equations satisfying general boundary conditions II. Comm. Pure. Appl. Math. **17**, 35–92

Ait-Ali-Yahia, D., Habashi, W.G., Tam, A., Vallet, M.-G., Fortin, M. (1996): A directionally-adaptive finite element method for high-speed flows. Int. J. Num. Meth. Fluids **23**, 673–690

Ambrosiano, J.L., Brandon, S.T., Lohner, R. (1994): Electromagnetics via the Taylor–Galerkin finite element method on unstructured grids. J. Comp. Phys. **110**, 310–319

Anilkumar, A.V., Lee, C.P., Wang, T.G. (1991): Surface-tension-induced mixing following coalescence of initially stationary drops. Phys. Fluids A **3**, 2587–2591

Armaly, B.F., Durst, F., Pereira, J.C.F., Schonung, B. (1982): Experimental and theoretical investigation of backward-facing step flow. J. Fluid Mech. **127**, 473–496

Assous, F., Degond, P., Heintze, E., Raviart, P.A., Seger, J. (1993): On a finite-element method for solving the three-dimensional Maxwell equations. J. Comp. Phys. **109**, 222–237

Axelsson, O. (1994): *Iterative Solution Methods* (Cambridge University Press, Cambridge, England)

Axelsson, O., Barker, V.A. (1984): *Finite Element Solutions of Boundary Value Problems* (Academic Press, Orlando, Fla.)

Aziz, A.K., Kellogg, R.B., Stephens, A.B. (1985): Least squares methods for elliptic systems. Math. Comp. **44**, 53–70

Babuška, I. (1971): Error bounds for finite element method. Numer. Math. **16**, 322–333

Barragy, E., Carey, G.F. (1988): A parallel element-by-element solution scheme. Int. J. Num. Meth. Engrg. **26**, 2367–2382

Batchelor, G.K. (1970): *An Introduction to Fluid Dynamics* (Cambridge University Press, London)

Becker, E.B., Carey, G.F., Oden J.T. (1983): *Finite Elements: An Introduction*, Vol. I (Prentice-Hall, Englewood Cliffs, NJ)

Béghein, C., Haghighat, F., Allard, F. (1992): Numerical study of double-diffusive natural convection in a square cavity. Int. J. Heat Mass Trans. **35**, 833–846

Bentley, L.R., Pinder, G.F. (1992): Least-squares method for solving the mixed form of the groundwater flow equations. Int. J. Num. Meth. Fluids **14**, 729–751

Bestehorn, M. (1993): Phase and amplitude instabilities for Bénard–Marangoni convection in fluid layers with large aspect ratio. Physical Review E **48**, 3622–3634

Bloomfield, P., Steiger, W.L. (1983): *Least Absolute Deviations, Theory, Applications, and Algorithms* (Birkhäuser, Boston)

Bochev, P.B., Gunzburger, M.D. (1994): Analysis of least-squares finite element methods for the Stokes equations. Math. Comp. **63**, 479–506

Bochev, P.B., Bedivan, D.M. (1997): Least-squares methods for Navier–Stokes boundary control problems. Inter. J. Comput. Fluid Dynamics (in printing)

Boyse, W.E., Lynch, D.R., Paulsen, K.D., Minerbo, G.N. (1992): Nodal-based finite element modeling of Maxwell's equations. IEEE Trans. Antennas and Popagat. **40**, 642–651

Brackbill, J.U., Kothe, D.B., Zemach, C. (1992): A continuum method for modeling surface tension. J. Comput. Phys. **100**, 335–354

Bramble, J.H., Shatz, A.H. (1970a): "On the numerical solution of elliptic boundary-value problems by least-squares approximation of the data", in *Numerical Solution of PDE*, ed. by B. Hubberd (Academic Press, New York) Vol. 2, 107–133

Bramble, J.H., Shatz, A.H. (1970b): Rayleigh–Ritz–Galerkin methods for Dirichlet's problem using subspaces without boundary conditions. Comm. Pure Appl. Math. **23**, 653–675

Bramble, J.H., Lazarov, R., Pasciak, J.E. (1997): Least-squares for second order elliptic problems. Presented in the Symposium on Advances in Computational Mechanics, a conference in honor of Prof. J. Tinsley Oden on the occasion of his 60th birthday, Austin, Texas, January 13–15

Brezzi, F. (1974): On the existence, uniqueness and approximation of saddle-point problems arising from Lagrange multipliers. Rech. Oper., Ser. Rouge Anal. Numer. **8, R-2**, 129–151

Brezzi, F., Rappaz, J., Raviart, P.-A. (1980): Finite-dimensional approximation of nonlinear problems, Part I: Branches of nonsingular solutions. Numer. Math. **36**, 1–25

Brezzi, F., Fortin, M. (1991): *Mixed and Hybrid Finite Element Methods* (Springer-Verlag, New York)

Brooks, A., Hughes, T.J.R. (1982): Streamline upwind/Petrov–Galerkin formulations for convection dominated flows with particular emphasis on the incompressible Navier–Stokes equations. Comp. Meth. Appl. Mech. Engrg. **59**, 199–259

Bruneau, Ch.H. (1991): Computation of hypersonic flows round a blunt body. Computers & Fluids **19**, 231–242

Bruneau, Ch.H., Laminie, J. (1993): A method to compute 3D hypersonic flows with Euler model. Inter. J. Comput. Fluid Dynamics J. **1**, 347–360

Brussino, G., Sonnad, V. (1989): A comparison of direct and preconditioned iterative techniques for sparse unsymmetric systems of linear equations. Inter. J. Numer. Meth. Engrg. **28**, 801–815

Burrell, L.L., Tang, L.Q., Tsang, T.T.H. (1995): On a least-squares finite element method for advective transport in air pollution modeling. Atmospheric Environment **29**, 1352–2310

Cai, Z., Lazarov, R., Manteuffel, T.A., McCormick, S.F. (1994): First-order system least-squares for second-order partial differential equations: Part I. SIAM J. Numer. Anal. **31**, 1785–1799

Cai, Z., Manteuffel, T.A., McCormick, S.F. (1997): First-order system least-squares for second-order partial differential equations: Part II. SIAM J. Numer. Anal. **34**, 425–454

Cangellaris, A.C., Lin, C.C., Mei, K.K. (1987): Point-matched time domain finite element methods for electromagnetic radiation and scattering. IEEE Trans. Antennas and Popagat. **35**, 1160–1173.

Carey, G.F., Oden, J.T. (1983a): *Finite Elements: A Second Course*, Vol. II (Prentice-Hall, Englewood Cliffs, NJ)

Carey, G.F., Oden, J.T. (1983b): *Finite Elements: Computational Aspects*, Vol. III (Prentice-Hall, Englewood Cliffs, NJ)

Carey, G.F. (1986): Parallelism in finite element modeling. Commun. Appl. Num. Methods **2**, 281–287

Carey, G.F., Jiang, B.N. (1986): Element-by-element linear and nonlinear solution schemes. Commun. Appl. Num. Methods **2**, 145–153

Carey, G.F., Oden, J.T. (1986): *Finite Elements: Fluid Dynamics*, Vol. IV (Prentice-Hall, Englewood Cliffs, NJ)

Carey, G.F., Barragy, E., Mclay, R., Sharma, M. (1988): Element-by-element vector and parallel computations. Commun. Appl. Num. Methods **4**, 299–307

Carey, G.F., Jiang, B.N. (1988): Least-squares finite elements for first-order hyperbolic systems. Int. J. Numer. Meth. Engrg. **26**, 81–93

Cartton, I. (1972): The effect of insulating vertical walls on the onset of motion in a fluid heated from below. Int. J. Heat Mass Trans. **15**, 665–572

Chan, D.C., Bailey, D.H., Bjorstad, P.E., Gilbert, J.R., Mascagni, M.V., Schreiber, R.S., Simon, H.D., Torczon, V.J., Watson, L.T. (1995): "A parallel least-squares finite element method for subsonic and supersonic flows", in Proc. the Seventh SIAM Conference on Parallel processing for Scientific Computing (SIAM, Philadelphia) 179–180

Chang, C.L., Gunzburger, M.D. (1987): A finite element method for first order elliptic systems in three dimensions. Appl. Math. Comput. **23**, 135–146

Chang, C.L. (1990): A mixed finite element method for Stokes problem: acceleration–pressure formulation. Appl. Math. Comput. **36**, 135–146

Chang, C.L., Jiang, B.N. (1990): An error analysis of least-squares finite element method of velocity–pressure–vorticity formulation for Stokes problem. Comput. Meth. Appl. Mech. Engrg. **84**, 247–255

Chang, C.L. (1992): Finite element approximation for grad-div type systems in the plane. SIAM J. Numer. Anal. **29**, 452–461

Chang, C.L., Yang, S.Y., Hsu, J.S. (1995): A least-squares finite element method for incompressible flow in stress–velocity–pressure version. Comput. Meth. Appl. Mech. Engrg. **128**, 1–9

Chang, Y.C., Hou, T.Y., Merriman, B., Osher, S. (1996): A level set formulation of Eulerian interface capturing methods for incompressible fluid flows. J. Comput. Phys. **124**, 449–462

Chattot, J.J., Guiu-roux, J., Lamine, J. (1982): Numerical solution of a first-order conservation equation by a least square method. Int. J. Num. Meth. Fluids **2**, 209–219

Chen, C.M. (1982): *Analysis of the Finite Element Method with Acceleration of Convergence* (Chinese) (Hunan Science and Technology, Changsha, China)

Chen, T.F. (1986): On least-squares approximations to compressible flow problems. Numerical Method for Partial Differential Equations **2**, 207–228

Chen, T.F., Fix, G.J. (1986): Least-squares finite element simulation of transonic flows. Appl. Numer. Math. **2**, 399–408

Chen, T.F. (1992): Semidiscrete least squares methods for linear convection–diffusion problem. Comput. Math. Applic. **24**, 29–44

Chenoweth, D.R., Paolucci, S. (1986): Natural convection in an enclosed vertical air layer with large horizontal temperature differences. J. Fluid Mech. **169**, 173–210

Chew, W.C. (1990): *Waves and Fields in Inhomogeneous Media* (Van Nostrand Reinhold, New York)

Choi, Y.-H., Merkle, C.L. (1993): The application of preconditioning in viscous flows. J. Comput. Phys. **105**, 207–223

Chorin, A.J. (1967): A numerical method for solving incompressible viscous flow problems. J. Comput. Phys. **2**, 12–26

Christie, I., Griffiths, D.F., Mitchell, A.R., Zienkiewicz, O.C. (1976): Finite element methods for second order differential equations with significant first derivatives. Int. J. Numer. Meth. Engrg. **10**, 1389–1376

Ciarlet, P.G. (1991): Basic error estimates for elliptic problems. in *Handbook of Numerical Analysis*, Vol. II, Finite Element Methods (Part 1), ed. by P.G. Ciarlet, J.L. Lions (North-Holland, Amsterdam) 23–351

Cloot, A., Lebon, G. (1984): A nonlinear stability analysis of the Bénard-Marangoni problem. J. Fluid Mech. **145**, 447–469

Corr, D.G., Davies, J.B. (1972): Computer analysis of fundamental and higher order modes in single and coupled microstrip. IEEE Trans. Microwave Theory and Tech. **20**, 669–678

Cox, C.L., Fix, G.J. (1984): On the accuracy of least squares methods in the presence of corner singularities. Comput. Math. Appl. **10**, 463–476

Crowley, C.W., Silvester, P.P., Hurwitz, H. (1988): Covariant projection elements for 3D vector field problems. IEEE Trans. Magn. **24**, 397–400

Daly, B.J. (1967): Numerical study of two fluid Rayleigh–Taylor instability. Phys. Fluids **10**, 297–307

Daly, B.J. (1969): A technique for including surface tension effects in hydrodynamic calculations. J. Comput. Phys. **4**, 97–117

Davis, S.H. (1967): Convection in a box: linear theory. J. Fluid Mech. **30**, 465–478

Davis, S.H. (1969): Buoyancy–surface tension instability by the method of energy. J. Fluid Mech. **39**, 347–359

Deconinck, H., Powell, K.G., Roe, P.L., Struijs, R. (1991): Multi-dimensional schemes for scalar advection. AIAA–91–1532–CP.

Dendy, J.E. (1974): Two methods of Galerkin type achieving optimal L^2 accuracy for first order hyperbolics. SIAM J. Numer. Anal. **11**, 637–653

Donea, J. (1984): A Taylor–Galerkin method for convective transport problems. Int. J. Numer. Meth. Engrg. **20**, 101–119

Donea, J., Quartapelle, L. (1992): An introduction to finite element methods for transient advection problems. Int. J. Num. Meth. Engrg. **95**, 169–203

Eason, E.D. (1976): A review of least-squares methods for solving partial differential equations. Int. J. Numer. Meth. Engrg. **10**, 1021–1046

Eidelman, S., Colella, P., Shreeve, R.P. (1984): Application of the Godunov method and its second-order extension to cascade flow modeling. AIAA J. **22**, 1609–1651

Einset, E.O., Jensen, K.F. (1992): A finite element solution of three-dimensional mixed convection gas flows in horizontal channels using preconditioned iterative matrix methods. Int. J. Num. Meth. Fluids **14**, 817–841

Engleman, M.S., Jamnia, M. (1990): Transient flow past a circular cylinder: a benchmark solution. Int. J. Num. Meth. Fluids **11**, 985–1000

Ern, A., Smooke, M.D. (1993): Vorticity–velocity formulation for three-dimensional steady compressible flows. J. Comput. Phys. **105**, 58–71

Farrar, A., Adams, A.T. (1976): Computation of propagation constants for the fundamental and higher modes in microstrip. IEEE Trans. Microwave Theory and Tech. **24**, 456–460

Feistauer, M. (1993): *Mathematical Methods in Fluid Dynamics* (Longman Scientific & Technical, Harlow, England)

Fix, G.J., Rose, M.E. (1985): A comparative study of finite element and finite difference methods for Cauchy–Reimann-type equations. SIAM J. Num. Analysis **22**, 250–260

Fix, G.J., Gunzburger, M.D. (1978): On least squares approximations to indefinite problems of the mixed type. Int. J. Numer. Meth. Engrg. **12**, 453–469

Fix, G.J., Gunzburger, M.D., Nicolaides, R.A. (1979): On finite element methods of the least squares type, Comp. & Maths. with Appls. **5**, 87-98

Fletcher, C.A.J. (1979): A primitive variable finite element formulation for inviscid compressible flow. J. Comput. Phys. **33**, 301–312

Foresti, S. Hassanzadeh, S., Murakami, H., Sonnad, V. (1990): Parallel implementation of iterative solution techniques with rapid operation on shared memory machines. IBM Kingston Research Report, KGN–217

Fox, R.L., Stanton, E.L. (1968): Developments in structural analysis by direct energy minimization. AIAA J. **6**, 1036–1042

Franca, L.P., Hughes, T.J.R., Loula, A.F.D., Miranda, I. (1989): "A new family of stable elements for the Stokes problem based on a mixed Galerkin/least squares finite element formulation", in Proc. Seventh International Conference on Finite Element Methods in Flow Problems, *Finite Element Analysis in Fluids*, ed. by T.J.Chung, G.R.Karrr (UAH press, Huntsville, Alabama) 1067–1074.

Fried, I. (1969a): More on generalized iterative methods in finite element analysis. AIAA J. **7**, 565–567

Fried, I. (1969b): Gradient methods for finite element eigenproblems. AIAA J., **7**, 739–741

Fried, I. (1970): A gradient computational procedure for the solution of large problems arising from the finite element discretization method. Int. J. Num. Meth. Engrg. **2**, 477–494

Friedrichs, K.O. (1958): Symmetric positive linear differential equations. Comm. Pure Appl. Math. **11**, 333–418

Fritts, M.J., Boris, J.P. (1979): The Lagrangian solution of transient problems in hydrodynamics using a triangular mesh. J. Comput. Phys. **31**, 173–215

Fyfe, D.E., Oran, E.S., Fritts, M.J. (1988): Surface tension and viscosity with Lagrangian hydrodynamics on a triangular mesh. J. Comput. Phys. **76**, 349–384

Ganguly, A.K., Spielman, B.E. (1977): Dispersion characteristics for arbitrarily configured transmission media, *IEEE Trans. Microwave Theory and Tech.* **25**, 1138–1141

Ghia, U., Ghia, K.N., Shin, C.T. (1982): High-Re solutions for incompressible flow using the Navier–Stokes equation and a multigrid method. J. Comput. Phys. **48**, 387–411

Girault, V., Raviart, P.-A. (1986): *Finite Element Methods for Navier–Stokes Equations* (Springer-Verlag, Berlin)

Gnoffo, P.A. (1983): A finite-volume, adaptive grid algorithm applied to planetary entry flow-fields. AIAA J. **21**, 1249–1254

Golub, G.H., Loan, C.F.V. (1989): *Matrix Computations*, 2nd Ed. (The Johns Hopkins University Press, Baltimore)

Gresho, P.M., Chan, S.T. (1988): Semi-consistent mass matrix technique for solving the incompressible Navier–Stokes equations. Lawrence Livermore National Laboratory, UCRL–99503.

Gresho, P.M. (1991a): Some current CFD issues relevant to the incompressible Navier–Stokes equations. Comput. Meth. Appl. Mech. Engrg. **87**, 201–252

Gresho, P.M. (1991b): Incompressible fluid dynamics: some fundamental formulation issues. Annu. Rev. Fluid Mech. **23**, 413–453

Gresho, P.M. (1992): Some interesting issues in incompressible fluid dynamics, both in the continuum and in numerical simulation. Advances in Applied Mechanics **28**, 45–137

Gresho, P.M., Chan, S.T. (1995): An update on projection methods for transient incompressible viscous flow. Lawrence Livermore National Laboratory, UCRL–JC–121525

Gunzburger, M.D. (1989): *Finite Element Methods for Viscous Incompressible Flows* (Academic Press, San Diego)

Hafez, M., Brucker, D. (1991): The effect of artificial vorticity on the discrete solution of Euler equations. AIAA paper 91–1553

Hageman, L.A., Young, D.M. (1981): *Applied Iterative Methods* (Academic Press, New York)

Haj-hariri, H., Shi, Q., Borhan, A. (1994): Effect of local property smearing on global variables: implication for numerical simulations of multiphase flows. Phys. Fluids **6**, 2555–2257

Hara, M., Wada, T., Fukasawa, T., Kikuchi, F. (1983): Three dimensional analysis of RF electromagnetic fields by finite element method. IEEE Trans. Magn. **19**, 2417–2420

Harbord, R., Gellert, M. (1991): A simple least-squares method for FE analysis of the Navier–Stokes problem. Comput. Mech. **8**, 19–24

Harlow, F.H., Welch, J.E. (1965): Numerical calculations of time-dependent viscous incompressible flow of fluid with free Surface. Phys. Fluids **8**, 2182–2189

Harlow, F.H., Shannon, J.P. (1967): The splash of a liquid drop. J. Comput. Phys. **38**, 3855–3866

Hayes, L.J., Devloo, P. (1985): "An element-by-element block iterative method for large non-linear problems", in Innovative Methods for Nonlinear Behavior, ed. by W.K. Liu et al. (Pineridge Press, Swansea)

Hayes, L.J., Devloo, P. (1986): A vectorized version of a sparse matrix–vector multiply. Int. J. Num. Meth. Engrg. **23**, 1043–1056

Hayes, L.J. (1989): "Advances and trends in element-by-element techniques", in State-of-the-Art Surveys on Computational Mechanics, ed. by A.K.Noor, J.T. Oden, 219–236

Hestenes, M.R., Stieffel, E.L. (1952): Methods of conjugate gradient for solving linear systems. Nat. Bur. Std. J. Res. **49**, 409–436

Hillion, P. (1997): Beware of Maxwell's divergence equations. J. Comput. Phys. **132**, 154–155

Hirsch, C. (1990): *Numerical Computation of Internal and External Flows*, Vol. 2 (John Wiley, Chichester)

Hirt, C.W., Nichols, B.D. (1981): Volume of fluid (VOF) method for the dynamics of free boundaries. J. Comput. Phys. **39**, 201–225

Hughes, T.J.R., Levit, I., Winget, J. (1983a): An element-by-element implicit algorithm for heat conduction. ASCE J. Engrg. Mech. Div. **109**, 576–583

Hughes, T.J.R., Levit, I., Winget, J. (1983b): An element-by-element solution algorithm for problems of structural and solid mechanics. Comp. Meth. Appl. Mech. Engrg. **36**, 241–254

Hughes, T.J.R. (1987): Recent progress in the development and understanding of SUPG methods with special reference to the compressible Euler and Navier–Stokes equations. Int. J. Numer. Meth. Fluids **7**, 1261–1275

Hughes, T.J.R., Ferencz, R.M., Hallquist, J.O. (1987): Large-scale vectorized implicit calculation in solid mechanics on a CRAY X-MP/48 utilizing EBE preconditioned conjugate gradients. Comp. Meth. Appl. Mech. Engrg. **61**, 215–248

Hughes, T.J.R., Franca, L.P., Hulbert, G.M. (1989): A new finite element formulation for computational fluid mechanics: VIII. The Galerkin/least-squares method for advective–diffusive equations. Comput. Methods Appl. Mech. Engrg. **73**, 173–189

Ikohagi, T., Shin, B.R., Daiguji, H. (1992): Application of an implicit time-marching scheme to a three-dimensional incompressible flow problem in curvilinear coordinate systems. Computer Fluids **21**, 163–175

Imai, I. (1941): On the flow of a compressible fluid past a circular cylinder, II. Proc. Phys-Math. Soc. Japan **23**, 180–193

Iwatsu, R., Ishii, K., Kawamura, T., Kuwahara, K., Hyun, J.M. (1989a): Simulation of transition to turbulence in a cubic cavity. AIAA papaer 89–0040

Iwatsu, R., Ishii, K., Kawamura, T., Kuwahara, K., Hyun, J.M. (1989b): Numerical simulation of three dimensional flow structure in a driven cavity. Fluid Dynamics Research **5**, 173–189

Jacqmin, D. (1995): Three-dimensional computations of droplet collisions, coalescence, and droplet/wall interactions using a continuum surface-tension method. AIAA-95-0883

Jiang, B.N., Chai, J.Z. (1980): Least squares finite element analysis of steady high subsonic plane potential flows. Acta Mechanica Sinica, No. 1, 90–93 (Chinese). [English transl.: Foreign Technology Div., Air Force Systems Command, Wright-Patterson AFB, OH., FTD-ID(RS)T-0708-81, 53–59]

Jiang, B.N., Carey, G.F. (1984): "Subcritical flow computation using an element-by-element conjugate-gradient method", in Proc. 5th Int. Symp. Finite Elements and Flow Problems (University of Texas at Austin) 103–106

Jiang, B.N. (1986): Least-squares finite element methods with element-by-element solution including adaptive refinement. Ph.D. dissertation (The University of Texas at Austin)

Jiang, B.N., Carey, G.F. (1987): Adaptive refinement for least-squares finite elements with element-by-element conjugate gradient solution. Int. J. Numer. Meth. Engrg. **24**, 569–580

Jiang, B.N., Carey, G.F. (1988): A stable least-squares finite element method for nonlinear hyperbolic problems. Int. J. Numer. Meth. Fluids **8**, 933–942

Jiang, B.N., Chang, C.L. (1988): Least-squares finite elements for Stokes problem. NASA TM 101308, ICOMP-88-1

Jiang, B.N., Chang, C.L. (1990): Least-squares finite elements for Stokes problem. Comput. Meth. Appl. Mech. Engrg. **78**, 297–311

Jiang, B.N., Carey, G.F. (1990): Least-squares finite element method for compressible Euler equations. Int. J. for Num. Fluids **10**, 557–568

Jiang, B.N., Povinelli, L.A. (1990): Least-squares finite element method for fluid dynamics. Comput. Meth. Appl. Mech. Engrg. **81**, 13–37. First appeared as NASA TM 102352, ICOMP-89-23

Jiang, B.N. (1991): The L_1 finite element method for pure convection problems. NASA TM 103773, ICOMP-91-03

Jiang, B.N. (1992): A least-squares finite element method for incompressible Navier–Stokes problems. Inter. J. Numer. Meth. Fluids **14**, 843–859. First appeared as NASA TM 102385, ICOMP-89-28

Jiang, B.N. (1993): Non-oscillatory and non-diffusive solution of convection problems by the iteratively reweighted least-squares finite element method. J. Comput. Phys. **105**, 108–121

Jiang, B.N., Povinelli, L.A. (1993): Optimal least-squares finite element method for elliptic problems. Comput. Meth. Appl. Mech. Engrg. **102**, 199–212. First appeared as NASA TM 105382, ICOMP-91-29

Jiang, B.N., Lin, T.L., Povinelli, L.A. (1994a): Large-scale computation of incompressible viscous flow by least-squares finite element method. Comput. Methods Appl. Mech. Engrg. **114**, 213–231. First appeared as NASA TM 105904, ICOMP-93-06

Jiang, B.N., Loh, C.Y., Povinelli, L.A. (1994b): Theoretical study of the incompressible Navier–Stokes equations by the least-squares method. NASA TM 106535, ICOMP-94-04

Jiang, B.N., Sonnad, V. (1994): Least-squares solution of incompressible Navier–Stokes equations with the p-version of finite elements. Comput. Mech. **15**, 129–136. First appeared as NASA TM 105203, ICOMP-91-14

Jiang, B.N., Hou, L.J., Lin, T.L., Povinelli, L.A. (1995): Least-squares finite element solutions for three-dimensional backward-facing step flow. Inter. J. Comput. Fluid Dynamics **4**, 1–19. First appeared as NASA TM 106353, ICOMP-93-31

Jiang, B.N., Wu, J., Povinelli, L.A. (1996): The origin of spurious solutions in computational electromagnetics. J. Comput. Phys. **125**, 104–123. First appeared as NASA TM 106921, ICOMP-95-8

Jiang, B.N. (1997): "The true origin of spurious solutions and their avoidance by the least-squares finite element method", in *Computational Electromagnetics and Its Applications*, ed. by Campbell, T.G., Nicolaides, R.A., Salas, M.D., ICASE/LaRC Interdisciplinary Series in Science and Engineering, Vol.5, (Kluwer Academic, Dordrecht, Netherlands), pp. 155–184

Jiang, B.N. (1998): On the least-squares method. Comput. Meth. Appl. Mech. Engrg. **152**, 239–257

Jin, J.M. (1993): *The Finite Element Method in Electro-Magnetics* (John Wiley and Sons, New York)

Johnson, C. (1987): *Numerical Solution of Partial Differential Equations by the Finite Element Method* (Cambridge University Press, Cambridge, England)

Johnson, C., Nävert, U., Pitkäranta, J. (1984): Finite element methods for linear hyperbolic problems. Comput. Meth. Appl. Mech. Engrg. **45**, 285–312

Jue, T.C., Ramaswamy, B. (1992): "Cavity natural convection with a deformable free surface", in *Advances in Finite Element Analysis in Fluid Dynamics,* ed. by M. N. Dhaubhadel, et al. (ASME, New York)

Kangro, U., Nicolaides, R. (1997): Spurious fields in time-domain computations of scattering problems. IEEE Trans. Antennas and Propagat. **45**, 228–234

Kececioglu, I., Rubinsky, B. (1989): A mixed-variable continuously deforming finite element method for parabolic evolution problems. Part II: The coupled problem of phase-change in porous media. Int. J. Num. Meth. Engrg. **28**, 2609–2634

Kong, J.A. (1990): *Electromagnetic Wave Theory* (John Wiley and Sons, New York)

Koschmieder, E.L. (1967): On convection under an air surface. J. Fluid Mech. **30**, 9–15

Koschmieder, E.L., Biggerstaff, M.I. (1986): Onset of surface-tension-driven Bénard convection. J. Fluid Mech. **167**, 49–64

Koschmieder, E.L., Prahl, S.A. (1990): Surface-tension-driven Bénard convection in small containers. J. Fluid Mech. **215**, 571–583

Koschmieder, E.L., Switzer, D.W. (1992): The wavenumbers of supercritical surface-tension-driven Bénard convection. J. Fluid Mech. **240**, 533–548

Koseff, J.R., Street, R.L. (1984a): Visualization studies of a shear driven three-dimensional recirculation flow. J. Fluids Eng. **106**, 21–29

Koseff, J.R., Street, R.L. (1984b): On end wall effects in a lid-driven cavity flow. J. Fluids Eng. **106**, 385–389

Koseff, J.R., Street, R.L. (1984c): The lid-driven cavity flow: a synthesis of qualitative and quantitative observations. J. Fluids Eng. **106**, 390–398

Křížeki, M., Neittaanmäki, P. (1984a): On the validity of Friedrich's inequalities. Math. Scand. **54**, 17–26

Křížeki, M., Neittaanmäki, P. (1984b): Finite element approximation for a div–rot system with mixed boundary conditions in non-smooth plane domains. Apl. Mat. **29**, 272–285

Křížeki, M., Neittaanmäki, P. (1990): *Finite Element Approximation of Variational Problems and Applications* (Pitman Scientific and Technical, Hawlow, England)

Ku, H.C., Hirsh, R.S., Taylor, T.D. (1987) A pseudospectral method for solution of the three dimensional incompressible Navier–Stokes equations. J. Comput. Phys. **70**, 439–462

Ku, H.C., Hirsh, R.S., Taylor, T.D., Rosenberg, A.P. (1989): A pseudospectral matrix element for solution of three-dimensional incompressible flows and its parallel implementation. J. Comput. Phys. **83**, 260–291

Kunz, K.S., Luebbers, R.J. (1993): *The Finite Difference Time Domain Method for Electro-Magnetics* (CRC Press, Boca Raton)

Lafaurie, B., Nardone, C., Scardovelli, R., Zaleski, S., Zanetti, G. (1994): Modeling merging and fragmentation in multiphase flows with SURFER. J. Comput. Phys. **113**, 134–147

Lager, I.E. (1996): Finite element modeling of static and stationary electric and magnetic fields. Thesis (Delft University Press, Delft, Netherlands)

Lager, I.E., Mur, G. (1996): The finite element modeling of static and stationary electric and magnetic field. IEEE Transactions on Magnetics **32**, 631–634

Laminie, J. (1994): Some aspect of computational fluid dynamics: finite element discretization, parallel implementation. Thesis (Université de Paris-Sud)

Landau, L. D. and Lifshitz, E. M. (1959): *Fluid Mechanics* (Pergammon, New York)

Lavery, J.E. (1988): Nonoscillatory solution of the steady-state inviscid Burgers' equation by mathematical programming. J. Comput. Phys. **79**, 436–448

Lavery, J.E. (1989): Solution of steady-state one-dimensional conservation laws by mathematical programming. SIAM J. Numer. Anal. **26**, 1081–1089

Lee, J.F., Lee, R., Cangellaris, A. (1997): Time-domain finite element methods. IEEE Transaction on Antennas and Propagat. **45**, 430–441

Lee, R.L., Madsen, N.K. (1990): A mixed finite element formulation for maxwell's equations in the time domain. J. Comput. Phys. **88**, 284–304

Lefebvre, D., Peraire, J., Morgan, K. (1992): Least squares finite element solution of compressible and incompressible flows. Int. J. Num. Meth. Heat Trans. Fluid Flow **2**, 99–113

Lefebvre, D., Peraire, J., Morgan, K. (1993): Finite element Least squares solution of the Euler equations using linear and quadratic approximations. Inter. J. Comput. Fluid Dynamics **1**, 1–23

Li, C.W. (1990): Least-squares characteristics and finite elements for advection dispersion simulation. Int. J. Num. Meth. Engrg. **29**, 1343–1364

Lighthill, J. (1986): *Informal Introduction to Theoretical Fluid Dynamics* (Clarendon Press, Oxford)

Lin, T.F., Huang, C.C., Chang, T.S. (1990): Transient binary mixture natural convection in square enclosures, Int. J. Heat Mass Trans. **33**, 287–299

Lions, J.L., Magenes, E. (1972): *Non-homogeneous Boundary-Value Problems and Applications*, Vol. I (Springer-Verlag, New York)

Lock, R.C. (1970): Test cases for numerical methods in two-dimensional transonic flows. AGARD-R-575.

Long, P.E., Peper, D.W. (1981): An examination of some simple numerical schemes for calculating scalar advection. J. Appl. Met. **20**, 1146–1156

Lowrie, R.B., Roe, P.L. (1994): On the numerical solution of conservation laws by minimizing residuals. J. Comput. Phys. **113**, 304–308

Lynn, P.P., Arya, S.K. (1973): Use of the least squares criterion in the finite element formulation. Int. J. Num. Meth. Engrg. **6**, 75–88

Madsen, N.K., Ziolkowski, R. (1988): Numerical solution of Maxwell's equations in the time domain using irregular nonorthogonal grids. Wave Motion **10**, 583–596

Martin, J.C., Moyce, W.J. (1952): An experimental study of the collapse of liquid columns on a horizontal plane. Philos. Trans. R. Soc. Lond. A **244**, 312–324

Mayergoyz, I.D., D'Angelo, J. (1993): A new point of view on the mathematical structure of Maxwell's equations. IEEE Trans. Magn. **29**, 1315–1320

Merkle, C.L., Choi, Y.H. (1987): Computation of low-speed flow with heat addition. AIAA J. **25**, 831–838

Merkle, C.L., Choi, Y.H. (1988): Computation of low-speed compressible flows with time-marching procedures. Int. J. Num. Meth. Engrg. **25**, 293–311

Monk, P. (1993): "Finite element time domain methods for Maxwell equations", in *Second International Conference on Mathematical and Numerical Aspects of Wave Propagation*, ed. by I. Stakgold *et al* (SIAM, Philadelphia) 380–389

Morgan, K., Hassan, O., Peraire, J. (1996): A time-domain unstructured grid approach to the simulation of electromagnetic scattering in piecewise homogeneous media. Int. J. Num. Meth. Engrg. **134**, 17–36

Morton, K.W., Parrot, A.K. (1980): Generalised Galerkin methods for first-order hyperbolic equations. J. Comput. Phys. **36**, 249–270

Mur, G. (1994): Edge elements, their advantages and their disadvantages. IEEE Trans. Magn. **30**, 3552–3557

Nakayama, T., Mori, M. (1996): An Eulerian finite element method for time-dependent free surface problems in hydrodynamics. Int. J. Num. Meth. Fluids **22**, 175–194

Nedelec, J. (1980): Mixed finite elements in $\mathbf{R}^3$. Numer. math. **35**, 315–341

Neittaanmäki, P., Sarannen, J. (1981a): Finite element approximation of vector fields given by curl and divergence. Math. Methods Appl. Sci. **3**, 328–335

Neittaanmäki, P., Sarannen, J. (1981b): Finite element approximation of electromagnetic fields in the three dimensional space. Numer. Funct. Anal. Optim. **2**, 487–506

Nelson, J.J., Chang, C.L. (1995): A mass conservative least-squares finite element method for the Stokes problem. Commun. Appl. Num. Methods **11**, 965–970

Nguyen, H., Reynen, J. (1984): A space-time least-squares finite element scheme for advection–diffusion equations. Comput. Meth. Appl. Mech. Engrg. **42**, 331–342

Nield, D. A. (1964): Surface tension and buoyancy effects in cellular convection. J. Fluid Mech., **19**, 341–352

Ni, R.-H. (1982): A multiple grid scheme for solving the Euler equations. AIAA J. **20**, 1565–1571

Noack, R.W., Anderson, D.A. (1992): Time-domain solution of Maxwell equations using a finite-volume formulation. AIAA Paper 92–0451

Nobari, M.R., Tryggvason, G. (1994a): Numerical simulation of drop collisions. NASA TM 106751, ICOMP–94–23

Nobari, M.R., Tryggvason, G. (1994b): The flow induced by the coalescence of two initially stationary drops. NASA TM 106752, ICOMP–94–24

Oden, J.T., Reddy, J.N. (1976): *An Introduction to the Mathematical Theory of Finite Elements* (John Wiley and Sons, New York)

Oden, J.T., Carey, G.F. (1983): *Finite Elements, Mathematical Aspects*, Vol. IV (Prentice-Hall, Englewood Cliffs, New Jersey)

Oden, J.T., Jacquotte, O.-P. (1984): Stability of some mixed finite element methods for Stokesian flows. Comput. Meth. Appl. Mech. Engrg. **43**, 231–248

Oden, J.T. (1991): "Finite elements: An introduction", in *Handbook of Numerical Analysis*, Vol II, Finite Element Methods (Part 1), ed. by P.G. Ciarlet, J.L. Lions (North-Holland, Amsterdam) 3–15

Oden, J.T., Demkowicz, L.F. (1996): *Applied Functional Analysis* (CRC, Boca Raton)

Odman, M.T., Russell, A.G. (1993): A nonlinear filtering algorithm for multidimensional finite element pollutant advection schemes. Atmospheric Environment **27A**, 793–799

Ortiz, M., Pinsky, P.M., Taloy, R.L. (1983): Unconditionally stable element-by-element algorithm for dynamic problems. Comp. Meth. Appl. Mech. Engrg. **36**, 223–239

Park, N.S., Liggett, J.A. (1990): Taylor-least squares finite element for two-dimensional advection dominated advection–diffusion problems. Int. J. Numer. Meth. Fluids **11**, 21–28

Park, N.S., Liggett, J.A. (1991): Application of Taylor-least squares finite element for three-dimensional advection–diffusion equation. Int. J. Num. Meth. Fluids **13**, 759–733

Paulsen, K.D., Lynch, D.R. (1991): Elimination of vector parasites in finite element Maxwell solutions. IEEE Trans. Microwave Theory and Tech. **39**, 395–404

Pearson, J.R. (1958): On convection cells induced by surface tension. J. Fluid Mech. **4**, 489–500

Pehlivanov, A.I., Carey, G.F., Lazarov, R.D., Shen, Y. (1993): Convergence of least-squares finite elements for first-order ODE systems. Computing **51**, 111–123

Pinchuk, A.R., Crowley, C.W., Silvester, P.P. (1988): Spurious solutions to vector diffusion and wave field problems. IEEE Trans. Magn. **24**, 158–161

Pironneau, O. (1989): *Finite Element Methods for Fluids* (John Wiley and Sons, Chichester)

Prenter, P.M. (1975): *Splines and Variational Methods* (John Wiley and Sons, Chichester)

Pulliam, T.H., Barton, J.T. (1985): Euler computation of AGARD working group 07, Airfoil test cases. AIAA Paper 85–0018

Pulliam, T.H. (1989): A computational challenge: Euler solution for ellipses. AIAA paper 89–0469

Quartapelle, L. (1993): *Numerical Solution of the Incompressible Navier–Stokes Equations* (Birkhäuser Verlag, Basel)

Rahman, B.M.A., Davies, J.B. (1984): Penalty function improvement of waveguide solution by finite elements. IEEE Trans. Microwave Theory and Tech. **32**, 922–928

Raviart, P.-A, Thoman, J.-M. (1977): Primal hybrid finite element methods for second order elliptic equations. Math. Comp. **31**, 391–413

Reddy, J.N., Gartling, D.K. (1994): *The Finite Element Method in Heat Transfer and Fluid Dynamics* (CRC Press, Boca Raton)

Reddy, M.P., Reddy, J.N., Akay, H.U. (1992): Penalty finite element analysis of incompressible flows using element by element solution algorithm. Comp. Meth. Appl. Mech. Engrg. **100**, 169–205

Richards, J. R., Lenhoff, A. M., Beris, A. N. (1994): Dynamics breakup of liquid-liquid jets. Phys. Fluids A **6**, 2640–2655

Roberts, J.E., Thomas, J.-M. (1991): "Mixed and hybrid methods", in: *Handbook of Numerical Analysis*, Vol. II. Finite Element Methods, ed. by P.G. Ciarlet, J.L. Lions (North-Holland, Amsterdam) 523–639

Roshko, A. (1953): On the development of turbulent wakes from vortex streets. NASA TN 2913

Saad, Y. (1986): GMRES: A generalized minimal residual algorithm for solving nonsymmetric linear systems. SIAM J. Sci. Stat. Comput. **7**, 856–869

Saranen, J. (1982): On an inequality of Friedrichs. Math. Scand. **51**, 310–322

Scanlon, J.W., Segal, L.A. (1967): Finite amplitude cellular convection induced by surface tension. J. Fluid Mech. **30**, 149–162

Schwieg, E., Bridges, W.B. (1984): Computer analysis of dielectric waveguides: a finite difference method. IEEE Trans. Microwave Theory and Tech. **32**, 531–541

Shakib, F., Hughes, T.J.R., Johan, Z. (1989): A multi-element group preconditioned GMRES algorithm for nonsymmetric systems arising in finite element analysis. Comp. Meth. Appl. Mech. Engrg. **75**, 415–456

Shang, J.S. (1995): A fractional-step method for solving 3-D time-domain Maxwell's equations. J. Comput. Phys. 109–119

Shang, J.S., Gaitonde, D. (1995): Scattered electromagnetic field of a re-entry vehicle. J. Spacecraft Rockets **32**, 294–301

Shankar, V., William, F.H., Mohammadian, A.H. (1989): A time-domain differential solver for electromagnetic scattering problems. Proceedings of IEEE **77**, 709–721

Shlager, K.L., Schneider, J.B. (1995) A selective survey of the finite-difference time-domain literature. IEEE Transactions on Antennas and Propagat. **37**, 39–56

Silvester, P.P., Pelosi, G. (1994) : *Finite Elements for Wave Electromagnetics* (IEEE Press, New York)

Sonnad, V., Foresti, S., Hassanzadeh, S., Murakami, H. (1990): Minimal storage solution of large 3-D problems using rapid operator application with the p-version of finite elements. IBM Kingston Research Report, KGN–209

Stork, K., Müller, U. (1972): Convection in a box: experiments. J. Fluid Mech. **54**, 599–611

Strang, G., Fix, G.J. (1973): *An Analysis of the Finite Element Method* (Prentice Hall, Englewood Cliffs, NJ)

Stratton, J. A. (1941): *Electromagnetic Theory* (McGraw-Hill, New York)

Su, C.-C. (1985): Origin of spurious modes in the analysis of optical fiber using the finite-element or finite-difference technique. Electron. Lett. **21**, 858–860

Sussman, M., Smereka, P., Osher, S.J. (1994): A level set method for computing solutions to incompressible two-phase flow. J. Comput. Phys. **114**, 146–159

Swaminathan, M., Sarkar, T.K., Adams, A.T. (1992): Computation of TM and TE modes in waveguides base on surface integral formulation. IEEE Trans. Microwave Theory and Tech. **40**, 285–297

Taflove, A., Umashankar, K.R. (1989): Review of FD-TD numerical modeling of electromagnetic wave scattering and radar cross section. Proceedings of IEEE **77**, 682–699

Taflove, A. (1995): *Computational Electrodynamics, The Finite-Difference Time-Domain Method* (Artech House, Boston)

Taghaddosi, F., Habashi, W.G., Guèvremont, G., Ait-Ali-Yahia, D. (1997): An adaptive least-squares method for the compressible Euler equations. AIAA–97–2097

Tang, L.Q., Tsang, T.T.H. (1993): A least-squares finite element method for time-dependent incompressible flows with thermal convection. Inter. J. Numer. Meth. Fluids **17**, 271–289

Tang, L.Q. (1994): A least-squares finite element method for time-dependent fluid flows and transport phenomena. Ph.D. Dissertation (The university of Kentucky, Lexington)

Tang, L.Q., Tsang, T.T.H. (1994): A least-squares finite element method for doubly-diffusive convection. Inter. J. Comput. Fluid Dynamics **3**, 1–17

Tang, L.Q., Cheng, T.W., Tsang,T.T.H. (1995): Transient solutions for three-dimensional lid-driven cavity flows by a least-squares finite element method. Int. J. Numer. Meth. Fluids **21**, 413–432

Tang, L.Q., Tsang, T.T.H. (1995): An efficient least-squares finite element method for incompressible flows and transport processes. Inter. J. Comput. Fluid Dynamics **4**, 21–39

Tang, L.Q., Tsang, T.T.H. (1997): Temporal, spatial and thermal features of 3-D Rayleigh–Bénard convection by a least-squares finite element method. Comput. Meth. Appl. Mech. Engrg. **140**, 201–219

Tezduyar, T.E., Liou, J. (1989): Grouped element-by-element iteration schemes for incompressible flow computations. Comput. Phys. Comm. **53**, 441–453

Tezduyar, T.E., Behr, M., Liou, J. (1992a): A new strategy for finite element computations involving moving boundaries and interfaces – the deforming-spatial-domain/space-time procedure: I. the concept and the preliminary numerical tests. Comp. Meth. Appl. Mech. Engrg. **94**, 339–351

Tezduyar, T.E., Behr, M., Mittal, S., Liou, J. (1992b): A new strategy for finite element computations involving moving boundaries and interfaces – the deforming-spatial-domain/space-time procedure: II. computation of free-surface flows, two-liquid flows, and flows with drifting cylinders. Comp. Meth. Appl. Mech. Engrg. **94**, 353–371

Tezduyar, T.E., Mittal, S., Ray S.E., Shih, R. (1992c): Incompressible flow computations with stablized bilinear and linear equal-order-interpolation velocity–pressure elements. Comput. Meth. Appl. Mech. Engrg. **95**, 221–242

Thess, A., Orszag, S.A. (1995): Surface-tension-driven Bénard convection at infinite Prandtl number. J. Fluid Mech. **283**, 201–230

Turkel, E. (1987): Preconditioning methods for solving the incompressible and low speed compressible equations. J. Comput. Phys. **72**, 277–298

Turkel, E. (1993): Review of preconditioning methods for fluid dynamics. Appl. Num. Math. **12**, 257–284

Umashankar, K.R., Taflove, A. (1993): *Computational Electromagnetics* (Artech House, Boston)

Unverdi, S.O., Tryggvason, G. (1992): A front-tracking method for viscous, incompressible, multi-fluid flows. J. Comput. Phys. **100**, 25–37

Van Leer, B., Lee, W.T., Roe, P. (1991): Characteristic time stepping or local preconditioning of the Euler equations. AIAA paper 91–1552

Vinh, H., Dwyer, H.A., van Dam, C.P. (1992): Finite-difference algorithms for the time-domain Maxwell equations – a numerical approach to RCS analysis. AIAA Paper 92–2989

Wahlbin, L.B. (1974): A dissipative Galerkin method applied to some quasilinear hyperbolic equations. RAIRO Anal. Numér. **8**, 109–117

Washizu, K. (1975): *Variational Methods in Elasticity and Plasticity*, 2nd ed. (Pergamon Press, Oxford)

Wathen, A.J. (1989): An analysis of some element-by-element techniques. Comput. Methods Appl. Mech. Engrg. **74**, 271–287

Wendland, W.L. (1979): *Elliptic Systems in the Plane* (Pitman, London)

Williams, P.T., Baker, A.J. (1997): Numerical simulations of laminar flow over a 3D backward-facing step. Int. J. Num. Meth. Fluids **24**, 1159–1183

Winget, J.M., Hughes, T.J.R. (1985): Solution algorithms for nonlinear transient heat conduction analysis employing element-by-element iterative strategies. Comp. Meth. Appl. Mech. Engrg. **52**, 711–815

Winterscheidt, D., Surana, K.S. (1993): P-version least squares finite element formulation for two dimensional incompressible fluid flow. Int. J. Numer. Meth. Engrg. **36**, 3629–3646

Winterstein, R., Hafez, M. (1993): Euler solution for blunt bodies using triangular meshes: artificial viscosity forms and numerical boundary conditions. AIAA 93–3333–cp

Wong, S.H., Cendes, Z.J. (1988): Combined finite element-model solution of three-dimensional eddy current problems. IEEE Trans. Magn., **24**, 2586–2687

Wong, S.H., Cendes, Z.J. (1989): Numerically stable finite element methods for Galerkin solution of eddy current problems. IEEE Trans. Magn. **25**, 3019–3021

Wu, J., Jiang, B.N., Yu, S.T., Liu, N.S. (1994): Time-accurate LSFEM for fluid flows and electromagnetic scattering problems. Least-Squares Finite Element Methods Workshop, Oct. 13–14, Cleveland.

Wu, J., Jiang, B.N. (1995a): "A least-squares finite element method for electromagnetic scattering problems", in Proceedings of the Third US National Congress on Computational Mechanics, ed. by J.N. Reddy (Texas A & M University, College Station) 295

Wu, J., Jiang, B.N. (1995b): A time-accurate least-squares finite-element method for incompressible inviscid and viscous flows. ICOMP report (unpublished)

Wu, J., Jiang, B.N. (1996): A least-squares finite element method for electromagnetic scattering problems. NASA Contract Report 202313, ICOMP–96–12

Wu, J., Yu, S.T., Jiang, B.N. (1996): Simulation of two-fluid flows by the least-squares finite-element method using a continuum surface tension model. NASA Contract Report 2023124, ICOMP–96–13

Yee, K.S. (1966): Numerical solution of initial boundary value problems involving Maxwell's equations in isotropic media. IEEE Trans. Antennas and Popagat. **14**, 302–307

Ying, L.A. (1988): *Lecture Notes on the Finite Element Method* (Chinese) (Beijing University Press, Beijing)

Yu, S.T., Jiang, B.N., Liu, N.S., Wu, J. (1995a): The least-squares finite-element method for low-Mach-number compressible viscous flows. Int. J. Num. Meth. Engrg. **38**, 3591–3610

Yu, S.T., Jiang, B.N., Wu, J., Jacqmin, D. (1995b): "A unified approach for simulating free-surface flows by the least-squares finite element method", in Proc. Sixth Int. Symp. on Computational Fluid Dynamics, A Collection of Technical Papers, Vol. III, 1467–1472

Yu, S.T., Jiang, B.N., Wu, J., Liu, N.S. (1996a): A div–curl–grad formulation for compressible buoyant flows solved by the least-squares finite-element method. Comp. Meth. Appl. Mech. Engrg. **137**, 59–88

Yu, S.T., Jiang, B.N., Wu, J. (1996b): A div–curl–grad formulation of three-dimensional low-Mach-number flows solved by the least-squares finite-element method. Comp. Meth. Appl. Mech. Engrg. (submitted)

Yu, S.T., Jiang, B.N., Wu, J., Duh, J.C. (1996c): Three-dimensional simulations of Marangoni–Bénard convection in small containers by the least-squares finite element method. AIAA paper 96–0863

Zhu, J., Sethian, J.A. (1992): Projection methods coupled to level set interface techniques. J. Comput. Phys. **102**, 128–138

Zienkiewicz, O.C., Owen, D.R.J., Lee, K.N. (1974): Least squares finite element for elasto-static problems – use of reduced integration. Int. J. Numer. Meth. Engrg. **8**, 341–358

Zienkiewicz, O.C. (1975): "Why finite elements", in *Finite Elements in Fluids*, Vol. 1, ed. by R. H. Gallagher et al. (John Wiley & Sons) 1–23

Zienkiewicz, O.C., Morgan, K (1983): *Finite Elements and Approximation* (John Wiley & Sons, Chichester)

Zienkiewicz, O.C., Zhu, J.Z. (1987): A simple error estimator and adaptive procedure for practical engineering analysis. Int. J. Num. Meth. Engrg. **24**, 337–357

Zienkiewicz, O.C., Taylor, R.L. (1988): *The Finite Element Method*, 4th ed., Vol. 1 (McGraw-Hill, London)

Zienkiewicz, O.C., Taylor, R.L. (1991): *The Finite Element Method*, 4th ed., Vol. 2 (McGraw-Hill, London)

Zienkiewicz, O.C., Zhu, J.Z. (1992): The superconvergent patch recovery and a posteriori error estimates. Part 1: the recovery technique. Int. J. Num. Meth. Engrg. **33**, 1331–1364

Index

Made in the USA
Monee, IL
13 September 2026

62975065R00240